An Introduction to Mathematical Proofs

Textbooks in Mathematics

Series editors:

Al Boggess and Ken Rosen

https://www.crcpress.com/Textbooks-in-Mathematics/book-series/CANDHTEXBOOMTH

An Introduction to Mathematical Proofs

Nicholas A. Loehr

CRC Press
Taylor & Francis Group
Boca Raton   London   New York

CRC Press is an imprint of the
Taylor & Francis Group, an **informa** business

CRC Press
Taylor & Francis Group
6000 Broken Sound Parkway NW, Suite 300
Boca Raton, FL 33487-2742

First issued in paperback 2022

ISBN 13: 978-1-03-247522-6 (pbk)
ISBN 13: 978-0-367-33823-7 (hbk)

DOI: 10.1201/9780429322587

Visit the Taylor & Francis Web site at
http://www.taylorandfrancis.com

and the CRC Press Web site at
http://www.crcpress.com

This book is dedicated to my father Frank.

Contents

Preface

This book contains an introduction to mathematical proofs, including fundamental material on logic, proof methods, set theory, number theory, relations, functions, cardinality, and the real number system. The book can serve as the main text for a proofs course taken by undergraduate mathematics majors. No specific prerequisites are needed beyond familiarity with high school algebra. Most readers are likely to be college sophomores or juniors who have taken calculus and perhaps some linear algebra, but we do not assume any knowledge of these subjects. Anyone interested in learning advanced mathematics could use this text for self-study.

Structure of the Book

This book evolved from classes given by the author over many years to students at the College of William & Mary, Virginia Tech, and the United States Naval Academy. I have divided the book into 8 chapters and 54 sections, including three review sections. Each section corresponds very closely to the material I cover in a single 50-minute lecture. Sections are further divided into many short subsections, so that my suggested pacing can readily be adapted for classes that meet for 75 minutes, 80 minutes, or other time intervals. If the instructor omits all sections and topics designated as "optional," it should be just possible to finish all of the core material in a semester class that meets for 2250 minutes (typically forty-five 50-minute meetings or thirty 75-minute meetings). More suggestions for possible course designs appear below.

I have tried to capture the best features of live mathematics lectures in the pages of this book. New material is presented to beginning students in small chunks that are easier to digest in a single reading or class meeting. The book maintains the friendly conversational style of a classroom presentation, without relinquishing the necessary level of precision and rigor. Throughout this text, you will find the personal pronouns "I" (the author), "you" (the reader), and "we" (the author and the reader, working together), reminding us that teaching and learning are fundamentally human activities. Teaching this material effectively can be as difficult as learning it, and new instructors are often unsure how much time to spend on the fundamentals of logic and proof techniques. The organization of this book shows at a glance how one experienced teacher of proofs allocates time among the various core topics. The text develops mathematical ideas through a continual cycle of examples, theorems, proofs, summaries, and reviews. A new concept may be introduced briefly via an example near the end of one section, then examined in detail in the next section, then recalled as needed in later sections. Every section ends with an immediate review of the key points just covered, and three review sections give detailed summaries of each major section of the book. The essential core material is supplemented by more advanced topics that appear in clearly labeled optional sections.

Contents of the Book

Here is a detailed list of the topics covered in each chapter of the book:

1. **Logic:** propositions, logical connectives (NOT, AND, OR, XOR, IF, IFF), truth tables, logical equivalence, tautologies, contradictions, universal and existential quantifiers, translating and denying complex logical statements, uniqueness.

2. **Proofs:** ingredients in mathematical theories (definitions, axioms, inference rules, theorems, proofs), proof by example, direct proof, contrapositive proof, contradiction proof, proof by cases, generic-element proofs, proofs involving multiple quantifiers.

3. **Set Theory:** set operations (union, intersection, set difference), subset proofs, set equality proofs, circle proofs, chain proofs, power sets, ordered pairs, product sets, unions and intersections of indexed collections.

4. **Integers:** recursive definitions, ordinary induction proofs, induction starting anywhere, backwards induction, strong induction, integer division with remainder, greatest common divisors, Euclid's GCD algorithm, primes, existence and uniqueness of prime factorizations.

5. **Relations and Functions:** relations, images, inverse of a relation, identity relation, composition of relations, formal definition of a function, function equality, operations on functions (pointwise operations, composition, restriction), direct images, preimages, injections, surjections, bijections, inverse functions.

6. **Equivalence Relations and Partial Orders:** reflexivity, symmetry, transitivity, equivalence relations, congruence modulo n, equivalence classes, set partitions, antisymmetry, partial orders, well-ordered sets.

7. **Cardinality:** finite sets, basic counting rules, countably infinite sets, countable sets, theorems on countability, uncountable sets, Cantor's Theorem.

8. **Real Numbers (Optional):** ordered field axioms for $\mathbb{R}$, algebraic properties, formal definition of $\mathbb{N}$ and $\mathbb{Z}$ and $\mathbb{Q}$, ordering properties, absolute value, distance, Least Upper Bound Axiom and its consequences (Archimedean ordering of $\mathbb{R}$, density of $\mathbb{Q}$ in $\mathbb{R}$, existence of real square roots, Nested Interval Theorem).

Possible Course Designs

A standard three-credit (2250 minute) proofs class could cover most of the topics in Chapters 1 through 7, which are essential for further study of advanced mathematics. When pressed for time, I have sometimes omitted or condensed the material on cardinality (Chapter 7) or prime factorizations (last half of Chapter 4). Many variations of the standard course are also feasible. Instructors wishing to preview ideas from abstract algebra could supplement the standard core with the following optional topics:

- the group axioms (end of Section 2.1);

- unique factorization properties for $\mathbb{Z}$ and $\mathbb{Q}$ (last four sections of Chapter 4);

- formal construction of the integers mod n and the rational numbers using equivalence relations (Section 6.6);

- algebraic properties of $\mathbb{R}$ developed from the ordered field axioms (Sections 8.1, 8.2, 8.3, and possibly 8.4).

A course introducing ideas from advanced calculus could include these topics:

- how to prove statements containing multiple quantifiers (Sections 2.6 and 2.7);

- general unions and intersections (Section 3.6);

- properties of preimages of sets under functions (Section 5.7);

- countable and uncountable sets (Sections 7.2, 7.3, and 7.4);

- rigorous development of the real numbers (and related number systems) from the ordered field axioms (Chapter 8).

A quarter-long (1500 minute) course focusing on basic proof methods might only cover Chapter 1, Chapter 2, and the early sections in Chapters 3 through 6. A quarter course on set theory, aimed at students with some prior familiarity with logic and proofs, might cover all of Chapters 3, 5, 6, and 7.

Topics can also be studied in several different orders. Chapter 1 on logic *must* come first, and Chapter 2 on proof methods *must* come second. Thereafter, some flexibility is possible. Chapter 4 (on induction and basic number theory) can be covered before Chapter 3 (on sets) or omitted entirely. Chapter 6 (on equivalence relations) can be covered before the last five sections of Chapter 5 (on functions). Chapter 7 (on cardinality) requires material from Chapter 5 on bijections, but it does not rely heavily on Chapter 6. Finally, the optional Chapter 8 (axiomatic development of the real numbers) could be covered anytime after Chapter 2, with minor adjustments to avoid explicit mention of functions and relations. However, Chapter 8 is more challenging than it may appear at first glance. We are all so familiar with basic arithmetic and algebraic facts about real numbers that it requires considerable intellectual discipline to deduce these facts from the axioms without accidentally using a property not yet proved. Nevertheless, it is rewarding and instructive (albeit somewhat tedious) to work through this logical development of $\mathbb{R}$ if time permits.

Book's Approach to Key Topics

This book adopts a methodical, detailed, and highly structured approach to teaching proof techniques and related mathematical topics. We start with basic logical building blocks and gradually assemble these ingredients to build more complex concepts. To give you a flavor of the teaching philosophy used here, the next few paragraphs describe my approach to explaining four key topics: proof-writing, functions, multiple quantifiers, and induction.

Skills for Writing Proofs

Like any other complex task, the process of writing a proof requires the synthesis of many small atomic skills. Every good proofs textbook develops the fundamental skill of breaking down a statement to be proved into its individual logical constituents, each of which contributes certain structure to the proof. For example, to begin a direct proof of a conditional statement "If P, then Q," we write: "Assume P is true; we must prove Q is true." I explain this particular skill in great detail in this text, introducing explicit *proof templates* for dealing with each of the logical operators.

But there are other equally crucial skills in proof-writing: memorizing and expanding definitions; forming useful denials of complex statements; identifying the logical status of each statement and variable in a proof via appropriate status words; using known universal and existential statements in the correct way; memorizing and using previously proved theorems; and so on. I cover each of these skills on its own, in meticulous detail, before assembling the skills to build increasingly complex proofs. Remarkably, this reduces the task of writing many basic proofs into an almost completely automatic process. It is very rewarding to see students gain confidence and ability as they master the basic skills one at a time and thereby develop proficiency in proof-writing.

Here is an example to make the preceding ideas concrete. Consider a typical practice problem for beginning proof-writers: *prove that for all integers x, if x is odd, then $x + 5$ is even.* In the proof below, I have annotated each line with the basic skill needed to produce that line.

Line in Proof	Skill Needed
1. Let x_0 be a fixed, arbitrary integer.	Prove $\forall$ statement using generic element.
2. Assume x_0 is odd; prove $x_0 + 5$ is even.	Prove an IF statement by direct proof.
3. We assumed there is $k \in \mathbb{Z}$ with $x_0 = 2k + 1$.	Expand a memorized definition.
4. We will prove there is $m \in \mathbb{Z}$ with $x_0 + 5 = 2m$.	Expand a memorized definition.
5. Doing algebra on the assumption gives:	Use logical status words.
6. $\quad x_0 + 5 = (2k + 1) + 5 = 2k + 6 = 2(k + 3)$.	Do basic algebraic manipulations.
7. Choose $m = k + 3$, so $x_0 + 5 = 2m$ holds.	Prove $\exists$ statement by giving an example.
8. Note m is in $\mathbb{Z}$, being the sum of two integers.	Verify a variable is in the required set.

Virtually every line in this proof is generated automatically using memorized skills; only the manipulation in line 6 requires a bit of creativity to produce the multiple of 2. Now, while many texts present a proof like this one, we seldom see a careful explanation of how the proof uses an assumed existential statement (line 3) to prove another existential statement (line 4) by *constructing an example* (lines 6 and 7) depending on the variable k in the assumption. This explanation may seem unnecessary in such a simple setting. But it is a crucial ingredient in understanding harder proofs in advanced calculus involving limits and continuity. There we frequently need to use an assumed multiply-quantified IF-statement to prove another multiply-quantified IF-statement. These proofs become much easier for students if they have already practiced the skill of using one quantified statement to prove another quantified statement in more elementary cases.

Similarly, there is not always enough prior coverage of the skill of *memorizing* and *expanding* definitions (needed to generate lines 3 and 4). This may seem to be a minor point, but it is in fact essential. Before writing this proof, students *must* have memorized the definition stating that "x is even" means "there exists $k \in \mathbb{Z}$ with $x = 2k$." But to generate line 4 from this definition, k must be replaced by a new variable m (since k was already given a different meaning in line 3), and x must be replaced by the expression $x_0 + 5$. I devote many pages to in-depth coverage of these separate issues, before integrating these skills into full proofs starting in Section 2.2.

Functions

A key topic in a proofs course is the rigorous definition of a function. A function is often defined to be a set of ordered pairs no two of which have the same first component. This definition is logically acceptable, but it causes difficulties later when studying concepts involving the codomain (set of possible outputs) for a function. Since the codomain cannot be deduced from the set of ordered pairs, great care is needed when talking about concepts that depend on the codomain (like surjectivity or the existence of a two-sided inverse). Furthermore, students accustomed to using the function notation $y = f(x)$ find the ordered pair notation $(x, y) \in f$ jarring and unpalatable. My approach includes the domain and codomain as part of the technical definition of a function; the set of ordered pairs by itself is called the *graph* of the function. This terminology better reflects the way most of us conceptualize functions and their graphs. The formal definition in Section 5.4 is preceded by carefully chosen examples (involving arrow diagrams, graphs in the Cartesian plane, and formulas) to motivate and explain the key elements of the technical definition. We introduce the standard function notations $y = f(x)$ and $f : X \to Y$ without delay, so students do not get bogged down with ordered triples and ordered pairs. Then we describe exactly what must be checked when a new function is introduced: single-valuedness and the fact that every x in the domain X has an associated output *in the claimed codomain Y*. We conclude with examples of formulas that do or do not give well-defined functions.

Multiple Quantifiers

A hallmark of this book is its extremely careful and explicit treatment of logical quantifiers: $\forall$ ("for all") and $\exists$ ("there exists"). The placement and relative ordering of these quantifiers has a big impact on the meaning of a logical statement. For example, the true statement $\forall x \in \mathbb{Z}, \exists y \in \mathbb{Z}, y > x$ ("for every integer x there is a larger integer y") asserts something very different from the false statement $\exists y \in \mathbb{Z}, \forall x \in \mathbb{Z}, y > x$ ("there exists an integer y larger than every integer x"). However, these doubly quantified examples do not reveal the full complexity of statements with three or more nested quantifiers. Such statements are quite common in advanced calculus, as mentioned earlier.

I give a very detailed explanation of multiple quantifiers in Sections 2.6 and 2.7. After examining many statements containing two quantifiers, I introduce more complicated statements with as many as six quantifiers, focusing on the structural outline of proofs of such statements. Using these big examples is the best way to explain the main point: an existentially quantified variable may only depend on quantified variables *preceding* it in the given statement. Other examples examine disproofs of multiply quantified statements, where the proof-writer must first form a useful denial of the given statement (which interchanges existential and universal quantifiers). Many exercises develop these themes using important definitions from advanced calculus such as continuity, uniform continuity, convergence of sequences, and least upper bounds.

Induction

Another vital topic in a proofs course is mathematical induction. Induction proofs are needed when working with recursively defined entities such as summations, factorials, powers, and sequences specified by a recursive formula. I discuss recursive definitions immediately before induction, and I carefully draw attention to the steps in an induction proof that rely on these definitions. Many expositions of induction do not make this connection explicit, causing some students to stumble at the point in the proof requiring the expansion of a recursive definition (for example, replacing a sum $\sum_{k=1}^{n+1} x_k$ by $[\sum_{k=1}^{n} x_k] + x_{n+1}$).

Induction proofs are often formulated in terms of *inductive sets*: sets containing 1 that are closed under adding 1. Students are told to prove a statement $\forall n \in \mathbb{Z}_{\geq 0}, P(n)$ by forming the set $S = \{n \in \mathbb{Z}_{\geq 0} : P(n) \text{ is true}\}$ and checking that S is inductive. This extra layer of translation confuses many students and is not necessary. Inductive sets do serve an important technical purpose: they provide a rigorous construction of the set of natural numbers as the intersection of all inductive subsets of $\mathbb{R}$. I discuss this advanced topic in the optional final chapter on real numbers (see Section 8.3), but I avoid mentioning inductive sets in the initial treatment of induction. Instead, induction proofs are based on the Induction Axiom, which says that the statements $P(1)$ and $\forall n \in \mathbb{Z}_{\geq 0}, P(n) \Rightarrow P(n+1)$ suffice to prove $\forall n \in \mathbb{Z}_{\geq 0}, P(n)$. This axiom is carefully motivated both with the visual metaphor of a chain of falling dominos and a more formal comparison to previously discussed logical inference rules.

Additional Pedagogical Features

(a) *Section Summaries and Global Reviews.* Every section ends with a concise recap of the key points just covered. Each major part of the text (logic and proofs; sets and integers; relations, functions, and cardinality) ends with a global review summarizing the material covered in that part. These reviews assemble many definitions, theorem statements, and proof techniques in one place, facilitating memorization and mastery of this vast amount of information.

(b) *Avoiding Logical Jargon.* This text avoids ponderous terminology from classical logic (such as conjunction, disjunction, modus ponens, modus tollens, modus tollendo ponens, hypothetical syllogism, constructive dilemma, universal instantiation, and existential instantiation). I use only those terms from logic that are essential for mathematical work (such as tautology, converse, contrapositive, and quantifier). My exposition replaces antiquated Latin phrases like "modus ponens" by more memorable English names such as "the Inference Rule for IF." Similarly, I refer to the *hypothesis* P and the *conclusion* Q of the IF-statement $P \Rightarrow Q$, rather than calling P the antecedent and Q the consequent of this statement.

(c) *Finding Useful Denials.* This is one of the most crucial skills students learn in a proofs course. Every good textbook states the basic denial rules, but students do not always realize (and texts do not always emphasize) that the rules must be applied *recursively* to find a denial of a complex statement. I describe this recursive process explicitly in Section 1.5 (see especially the table on page 35). Section 1.6 reinforces this key skill with many solved sample problems and exercises.

(d) *Annotated Proofs.* Advanced mathematics texts often consist of a series of definitions, theorems, and proofs with little explanation given for how the author found the proofs. This text is filled with explicit annotations showing the reader how we are generating the lines of a proof, why we are proceeding in a certain way, and what the common pitfalls are. These annotations are clearly delineated from the official proof by enclosing them in square brackets. Many sample proofs are followed by commentary discussing important logical points revealed by the proof.

(e) *Disproofs Contrasted with Proofs by Contradiction.* A very common student mistake is to confuse the disproof of a false statement P with a proof by contradiction of a true statement Q. This mistake occurs because of inattention to logical status words: the disproof of P begins with the goal of *proving* a denial of P, whereas a proof of Q by contradiction begins by *assuming* the denial of Q. We explicitly warn readers about this issue in Remark 2.60.

(f) *Set Definitions.* New sets and set operations are often defined using set-builder notation. For example, the *union* of sets S and T is defined by writing $S \cup T = \{x : x \in S \text{ or } x \in T\}$. This book presents these definitions in a format more closely matching how they arise in proofs, by explicitly stating what membership in the new set means. For instance, my definition of set union says that for all sets S and T and all objects x, the *defined term* $\boxed{x \in S \cup T}$ can be replaced by the *definition text* $\boxed{x \in S \text{ or } x \in T}$ at any point in a proof. This is exactly what the previous definition means, of course, but the extra layer of translation inherent in the set-builder notation causes trouble for many beginning students.

(g) *Careful Organization of Optional Material.* Advanced material and additional topics appear in clearly labeled optional sections. This organization provides maximum flexibility to instructors who want to supplement the material in the standard core, while signaling to readers what material may be safely skipped.

Exercises, Errata, and Feedback

The book contains more than 1000 exercises of varying scope and difficulty, which may be assigned as graded homework or used for self-study or review. I welcome your feedback about any aspect of this book, most particularly corrections of any errors that may be lurking in the following pages. Please send such communications to me by email at `nloehr@vt.edu`. I will post errata and other pertinent information on the book's website.

Words of Thanks

Some pedagogical elements of this book were suggested by the exposition in *A Transition to Advanced Mathematics* by Smith, Eggen, and St. Andre. My debt to this excellent text will be evident to anyone familiar with it. My development as a mathematician and a writer has also been deeply influenced by the superb works of James Munkres, Joseph Rotman, J. Donald Monk, and the other authors listed in the Suggestions for Further Reading (see page 383). I thank all the editorial staff at CRC Press, especially Bob Ross and Jose Soto, and the anonymous reviewers whose comments greatly improved the quality of my original manuscript.

I am grateful to many students, colleagues, friends, and family members who supported me during the preparation of this book. I especially thank my father Frank Loehr, my mother Linda Lopez, my stepfather Peter Lopez, Ken Zeger, Elizabeth Niese, Bill Floyd, Leslie Kay, and the students who took proofs classes from me over the years. Words cannot express how much I learned about teaching proofs from my students. Thank you all!

Very respectfully,
Nicholas A. Loehr

1

Logic

1.1 Propositions, Logical Connectives, and Truth Tables

Many people despise mathematics, believing it to be nothing more than a confusing jumble of arcane formulas and mind-numbing computations. This depressing view of the subject is understandable, when we consider how math is presented in grade school and many calculus classes. But in truth, mathematics is a beautiful, intricately structured *tower of knowledge* built up from a small collection of basic statements (called *axioms*) using the laws of *logic*. In this book, we shall study the foundation of this tower, as shown here:

$$
\vdots
$$

cardinality
functions
relations
integers
sets
proofs
logic

Propositions

We begin with *propositional logic*, which studies how the truth of a complex statement is determined by the truth or falsehood of its parts.

1.1. Definition: Propositions. A *proposition* is a statement that is either true or false, but not both.

Many things we say are not propositions, as seen in the next example.

1.2. Example. Which of these statements are propositions?
(a) 7 is positive.
(b) $1 + 1 = 7$.
(c) Memorize all definitions.
(d) Okra tastes great.
(e) Is it raining?
(f) This sentence is false.
(g) Paris is a city and $2 + 2$ is not 4, or Paris is not a city and $2 + 2$ is 4.
Solution. Statement (a) is a true proposition. Statement (b) is a false proposition. Commands, opinions, and questions do not have a truth value, so statements (c) through (e) are not propositions. Statement (f) is an example of a *paradox*: if you assume this statement is true, then the statement itself asserts that it is false. If you instead assume the statement is false, then the statement is also true. Since propositions are not allowed to be both true

1

and false, statement (f) is not a proposition. Finally, statement (g) is a false proposition, for reasons described below.

Propositional Forms

Statement (g) in the last example is a complex proposition built up from two shorter propositions using logical connective words such as AND, NOT, OR. The two shorter propositions are "Paris is a city" and "$2 + 2$ is 4." Let us abbreviate the proposition "Paris is a city" by the letter P, and let us abbreviate the proposition "$2 + 2$ is 4" by the letter Q. Then statement (g) has the form

$$(P \text{ AND } (\text{NOT } Q)) \text{ OR } ((\text{NOT } P) \text{ AND } Q).$$

We can create an even shorter expression to represent the logical form of statement (g) by introducing special symbols for the logical connectives. Propositional logic uses the six symbols shown in the following table, whose meaning is discussed in detail below.

Logical Symbol	English Translation
$\sim P$	P is not true.
$P \wedge Q$	P and Q.
$P \vee Q$	P or Q (or both).
$P \oplus Q$	P or Q, but not both.
$P \Rightarrow Q$	if P, then Q.
$P \Leftrightarrow Q$	P if and only if Q.

Using these symbols, we can describe the logical form of statement (g) by the expression

$$(P \wedge (\sim Q)) \vee ((\sim P) \wedge Q). \tag{1.1}$$

Any expression like this, which is built up by combining *propositional variables* (capital letters) using logical symbols, is called a *propositional form*.

Definitions of NOT and AND

In logic and mathematics, language is used in a very precise way that does not always coincide with how words are used in everyday conversation. In particular, before going further, we need to *define* the exact meaning of the logical connective words NOT, AND, OR, IF, etc. To do this, we give *truth tables* that show how to combine the truth values of propositions to obtain the truth value of a new proposition built from these using a logical connective. Here and below, italic capital letters such as P, Q, R are variables that represent propositions. The letters T and F stand for *true* and *false*, respectively.

1.3. Definition of NOT. For any proposition P, the truth value of $\sim P$ ("not P") is determined by the following table.

P	$\sim P$
T	F
F	T

Remember this table by noting that $\sim P$ always has the opposite truth value as P.

In our example above, where Q is the true proposition "$2 + 2$ is 4," $\sim Q$ is the false proposition "$2 + 2$ is not 4." If R is the false proposition "9 is negative," then $\sim R$ is the true proposition "9 is not negative."

The remaining logical connectives combine two propositions to produce a new proposition. As seen in the next definition, we need a four-row truth table to list all possible truth values of the two propositions that we start with.

1.4. Definition of AND. For any propositions P, Q, the truth value of $P \wedge Q$ ("P and Q") is determined by the following table.

P	Q	$P \wedge Q$
T	T	T
T	F	F
F	T	F
F	F	F

Remember this table by noting that $P \wedge Q$ is true only when both inputs P, Q are true; in all other cases, $P \wedge Q$ is false.

In our running example, where P is the true proposition "Paris is a city" and Q is the true proposition "$2 + 2$ is 4," note that $P \wedge (\sim Q)$ is false, because P is true and $\sim Q$ is false. Similarly, $(\sim P) \wedge Q$ is false, because $\sim P$ is false and Q is true. In this example, for any proposition R, we can deduce that $(\sim P) \wedge R$ *must* be false no matter what the truth value of R is, since $\sim P$ is already false. On the other hand, $P \wedge R$ is true for true propositions R, and false for false propositions R.

Definitions of OR and XOR

We now come to the mathematical definition of the word OR. In everyday English, the word OR can be used in two different ways. In the *inclusive* usage of this word, "P or Q" means that *at least one* of the propositions P, Q is true (possibly both). In the *exclusive* usage, "P or Q" means that *exactly one* of the propositions P, Q is true and the other one is false. Logic uses two different symbols for these two usages of the word OR: $\vee$ stands for *inclusive-OR*, and $\oplus$ stands for *exclusive-OR* (which is also abbreviated XOR). Remember this convention: *in mathematics, the English word OR always means inclusive-OR*, as defined by the truth table below. If you want to use XOR in mathematical English, you *must* use a longer phrase such as "P or Q, but not both."

1.5. Definition of OR. For any propositions P, Q, the truth value of $P \vee Q$ ("P or Q") is determined by the following table.

P	Q	$P \vee Q$
T	T	T
T	F	T
F	T	T
F	F	F

Remember this table by noting that $P \vee Q$ is false only when both inputs P, Q are false; in all other cases, $P \vee Q$ is true. To avoid confusing the symbols $\wedge$ and $\vee$, it may help to notice that $\wedge$ (the symbol for AND) resembles the diagonal strokes in the capital letter A.

To complete our analysis of statement (g) above, recall in this example that $(P \wedge (\sim Q))$ is false and $((\sim P) \wedge Q)$ is false. So the overall statement is false. On the other hand, $P \vee (\sim Q)$, $(\sim P) \vee Q$, and $P \vee Q$ all stand for true propositions in this example.

1.6. Definition of XOR. For any propositions P, Q, the truth value of $P \oplus Q$ ("P XOR Q") is determined by the following table.

P	Q	$P \oplus Q$
T	T	F
T	F	T
F	T	T
F	F	F

Remember this table by noting that $P \oplus Q$ is true when the input propositions P, Q have different truth values; $P \oplus Q$ is false when the input propositions P, Q have the same truth values.

1.7. Example. Let P be the false proposition "$1+1 = 3$," and let Q be the true proposition "2 is positive." Which propositions below are true?
(a) $P \oplus Q$;   (b) $P \oplus (\sim Q)$;   (c) $(\sim P) \oplus Q$;   (d) $P \vee Q$;   (e) $P \vee (\sim Q)$;   (f) $(\sim P) \vee Q$.
   Solution. Using the truth tables for $\oplus$, $\vee$, and $\sim$, we find that (a) is true, (b) is false, (c) is false, (d) is true, (e) is false, and (f) is true. The different answers for (c) and (f) illustrate the distinction between exclusive-OR and inclusive-OR.

Truth Tables for Propositional Forms

Given a complex propositional form built up from some variables using one or more logical connectives, we can determine the meaning of this form by making a *truth table* showing the truth value of the form for all possible combinations of truth values of the variables appearing in the form. We construct such a truth table step-by-step, making a column for each input variable and each smaller propositional form contained within the given form. We complete each column by using the defining truth tables (given above for $\sim$, $\wedge$, $\vee$, $\oplus$, and given in later sections for $\Rightarrow$ and $\Leftrightarrow$) to combine the truth values in one or two previously completed columns. We illustrate this process in the following examples.

1.8. Example: Truth Table for $(P \wedge Q) \vee (\sim P)$. This form has two input variables, P and Q, so we need a four-row truth table. We make columns for P, Q, $\sim P$, $P \wedge Q$, and $(P \wedge Q) \vee (\sim P)$, and fill them in as shown here:

P	Q	$\sim P$	$P \wedge Q$	$(P \wedge Q) \vee (\sim P)$
T	T	F	T	T
T	F	F	F	F
F	T	T	F	T
F	F	T	F	T

Note, in particular, that the final column is completed by comparing the truth values in each row of columns 3 and 4. In row 2 (and only row 2), the two input truth values are both false, so the output in this row is false. All other rows must have output that is true, by definition of OR.

1.9. Example: Truth Table for $P \wedge (Q \vee (\sim P))$. This form looks almost identical to the one in the last example, but the parentheses are in a different position. This changes the structure of the truth table, as shown here:

P	Q	$\sim P$	$Q \vee (\sim P)$	$P \wedge (Q \vee (\sim P))$
T	T	F	T	T
T	F	F	F	F
F	T	T	T	F
F	F	T	T	F

We fill in column 4 by looking for two Fs in columns 2 and 3 (which happens only in row 2), placing an F in that row, and filling the rest of column 4 with Ts. Then we fill in column 5 by looking for two Ts in columns 1 and 4 (which happens in row 1 only), placing a T in that row, and filling the rest of column 5 with Fs.

In these examples, observe that the truth tables for $(P \wedge Q) \vee (\sim P)$ and $P \wedge (Q \vee (\sim P))$ are not identical, since the outputs in rows 3 and 4 differ in the two tables. This shows that these two propositional forms are not logically interchangeable with one another in all situations. More briefly, we say that the two forms are not logically equivalent. On the other hand, if two propositional forms $\mathcal{A}$ and $\mathcal{B}$ have truth tables whose outputs agree in *every* row, we say that $\mathcal{A}$ and $\mathcal{B}$ are *logically equivalent* and write $\mathcal{A} \equiv \mathcal{B}$. (We study logical equivalence in detail in the next few sections.) Notice, for instance, that $P \wedge (Q \vee (\sim P))$ is logically equivalent to $P \wedge Q$, since column 5 in the previous example agrees with the truth table for $P \wedge Q$ in all four rows. So we could write $P \wedge (Q \vee (\sim P)) \equiv P \wedge Q$, enabling us to replace the complicated propositional form on the left by the shorter form $P \wedge Q$ on the right.

1.10. Example: Translation of XOR. We have noted that $P \oplus Q$ can be translated as "P or Q, but not both." This English phrase is built up using the logical connectives OR, NOT, BUT, and BOTH (the last two words have the same meaning as AND). So a more literal encoding of the phrase "P or Q, but not both" as a propositional form is $(P \vee Q) \wedge (\sim(P \wedge Q))$. The following truth table verifies that $(P \vee Q) \wedge (\sim(P \wedge Q)) \equiv P \oplus Q$. This logical equivalence justifies the use of this phrase as a translation of XOR.

P	Q	$P \oplus Q$	$P \vee Q$	$P \wedge Q$	$\sim(P \wedge Q)$	$(P \vee Q) \wedge (\sim(P \wedge Q))$
T	T	F	T	T	F	F
T	F	T	T	F	T	T
F	T	T	T	F	T	T
F	F	F	F	F	T	F

Note that we combine columns 4 and 6 to complete column 7 using the definition of AND. The claimed logical equivalence follows from the complete agreement of columns 3 and 7 in every row. In contrast, since columns 3 and 4 disagree in row 1, we see that $(P \oplus Q) \not\equiv (P \vee Q)$. In other words, inclusive-OR and exclusive-OR are not logically equivalent.

Formal Definition of Propositional Forms (Optional)

In the main text, we informally defined a propositional form to be any expression built up by combining propositional variables using logical symbols. To give a precise, rigorous definition of propositional forms, we need a *recursive definition* (definitions of this type are studied in detail later). Specifically, propositional forms are defined recursively via the following rules:

(a) A single capital italic letter is a propositional form.
(b) If $\mathcal{A}$ is any propositional form, then $(\sim\mathcal{A})$ is a propositional form.
(c) If $\mathcal{A}$ and $\mathcal{B}$ are any propositional forms, then $(\mathcal{A} \wedge \mathcal{B})$ is a propositional form.
(d) If $\mathcal{A}$ and $\mathcal{B}$ are any propositional forms, then $(\mathcal{A} \vee \mathcal{B})$ is a propositional form.
(e) If $\mathcal{A}$ and $\mathcal{B}$ are any propositional forms, then $(\mathcal{A} \oplus \mathcal{B})$ is a propositional form.
(f) If $\mathcal{A}$ and $\mathcal{B}$ are any propositional forms, then $(\mathcal{A} \Rightarrow \mathcal{B})$ is a propositional form.
(g) If $\mathcal{A}$ and $\mathcal{B}$ are any propositional forms, then $(\mathcal{A} \Leftrightarrow \mathcal{B})$ is a propositional form.
(h) An expression is a propositional form only if it can be formed by applying rules (a) through (g) finitely many times.

For example, rule (a) shows that P is a propositional form, as is Q. Then rule (b) shows that $(\sim P)$ and $(\sim Q)$ are propositional forms. By rule (c), $(P \wedge (\sim Q))$ and $((\sim P) \wedge Q)$ are propositional forms. By rule (d),

$$((P \wedge (\sim Q)) \vee ((\sim P) \wedge Q)) \tag{1.2}$$

is a propositional form, which is essentially the form studied in (1.1).

However, to be absolutely precise, the expression in (1.1) is *not* a propositional form. This is because (1.1) is missing the outermost pair of parentheses, which are required to be present when using rules (b) through (g). In practice, we often drop required parentheses to create abbreviated versions of propositional forms that are easier to read. For instance, we seldom write the outermost pair of parentheses. Logical equivalences called associativity rules (discussed in the next section) allow us to drop internal parentheses in an expression like $((P \vee Q) \vee R)$ or $(P \wedge (Q \wedge R))$. We can erase even more parentheses using the *precedence conventions* that $\sim$ has highest precedence, followed by $\wedge$, followed by $\vee$. For example, with these conventions, we could drop all parentheses from (1.2) and abbreviate this form as $P \wedge \sim Q \vee \sim P \wedge Q$. However, as shown in Examples 1.8 and 1.9, the placement of parentheses often affects the logical meaning of a propositional form (which is one reason why the official definition requires them). In this text, we often write parentheses that could be deleted by the precedence conventions, to reduce the chance of confusion.

1.11. Remark: Terminology for Propositional Forms. Let $\mathcal{A}$ and $\mathcal{B}$ be any propositional forms. In some logic texts, $(\sim \mathcal{A})$ is called the *negation* of $\mathcal{A}$; $(\mathcal{A} \wedge \mathcal{B})$ is called the *conjunction* of $\mathcal{A}$ and $\mathcal{B}$; $(\mathcal{A} \vee \mathcal{B})$ is called the *disjunction* of $\mathcal{A}$ and $\mathcal{B}$; $(\mathcal{A} \Rightarrow \mathcal{B})$ is called an *implication* or *conditional* with *hypothesis* (or *antecedent*) $\mathcal{A}$ and *conclusion* (or *consequent*) $\mathcal{B}$; $(\mathcal{A} \Leftrightarrow \mathcal{B})$ is called the *biconditional* of $\mathcal{A}$ and $\mathcal{B}$. We mostly avoid using these technical terms in this text, except for the term *negation*. Later, we study methods of finding propositional forms logically equivalent to $(\sim \mathcal{A})$, which are called *denials* of $\mathcal{A}$.

Section Summary

1. Memorize the definitions of the logical connectives, as summarized in the following truth table.

P	Q	NOT P $\sim P$	P AND Q $P \wedge Q$	P OR Q $P \vee Q$	P XOR Q $P \oplus Q$
T	T	F	T	T	F
T	F	F	F	T	T
F	T	T	F	T	T
F	F	T	F	F	F

   Note that NOT flips the truth value of the input; AND outputs true exactly when both inputs are true; OR outputs false exactly when both inputs are false; and XOR outputs true exactly when the two inputs differ.

2. Remember that in mathematics, "P or Q" always means $P \vee Q$ (inclusive-OR). To translate $P \oplus Q$ (exclusive-OR) into English, you *must* use an explicit phrase such as "P or Q, but not both."

3. Two propositional forms $\mathcal{A}$ and $\mathcal{B}$ are logically equivalent, denoted $\mathcal{A} \equiv \mathcal{B}$, when these forms have truth tables whose outputs agree in all rows of the truth table. If there is even a single row of disagreement, the forms are not logically equivalent. For instance, $P \vee Q \not\equiv P \oplus Q$.

Exercises

1. Is each statement true, false, or not a proposition? Explain.
   (a) $1 + 1 = 2$ or London is a city. (b) Math is fun. (c) France is a country, and Berlin is an ocean. (d) $0 < 1$ or $1 < 0$. (e) $0 < 1$ or $1 < 0$, but not both. (f) Chopin

was the greatest classical composer. (g) 2 is not even[1], or 2 is positive. (h) 15 is odd or positive, but not both.

2. Is each statement true, false, or not a proposition? Explain.
   (a) It is not the case that 2 is even or 2 is positive. (b) The following statement is a true proposition. (c) $\int_0^1 x^2\,dx = 1/3$ and $2^5 = 32$. (d) When do fish sleep? (e) The following statement is a true proposition. (f) The previous statement is a false proposition. (g) I am tired.

3. Construct truth tables for the following propositional forms.
   (a) $(\sim P) \wedge Q$ (b) $(P \wedge Q) \vee ((\sim P) \wedge (\sim Q))$ (c) $(P \vee (\sim Q)) \oplus ((\sim Q) \vee P)$
   (d) $(P \wedge Q) \vee R$ (e) $P \wedge (Q \vee R)$ (f) $P \vee (Q \oplus R)$

4. (a) Is $\sim(P \vee Q)$ logically equivalent to $(\sim P) \vee (\sim Q)$? Explain with a truth table.
   (b) Give English translations of the two propositional forms in (a).

5. Assume P is a true proposition, Q is a false proposition, and R is an arbitrary proposition. Say as much as you can about the truth or falsehood of each proposition below.
   (a) $P \wedge (\sim Q)$ (b) $(\sim P) \vee Q$ (c) $P \oplus Q$ (d) $P \oplus (\sim Q)$ (e) $R \vee P$ (f) $Q \vee R$
   (g) $Q \oplus R$ (h) $R \wedge (\sim R)$ (i) $R \vee (\sim R)$ (j) $R \oplus R$ (k) $R \oplus (\sim R)$.

6. (a) Make a truth table for the propositional form $(P \wedge (\sim Q)) \vee ((\sim P) \wedge Q)$. Be sure to show the columns for all intermediate forms. (b) Find a short propositional form that is logically equivalent to the form in (a). (c) Use (b) to write an English sentence with the same logical meaning as statement (g) in Example 1.2.

7. (a) Make a truth table for $P \wedge ((\sim Q) \vee P)$. (b) Make a truth table for $(P \wedge (\sim Q)) \vee P$.
   (c) Are the propositional forms in (a) and (b) logically equivalent? Why?
   (d) Find a short propositional form that is logically equivalent to the form in (a).

8. Which propositional forms (if any) are logically equivalent? Justify your answer with truth tables. (a) P (b) $\sim P$ (c) $P \vee P$ (d) $P \wedge P$ (e) $P \oplus P$ (f) $(P \wedge Q) \vee P$.

9. Which propositional forms (if any) are logically equivalent? Explain.
   (a) $P \wedge Q$ (b) $Q \wedge P$ (c) $(P \wedge Q) \wedge P$ (d) $\sim(P \oplus Q)$.

10. Which propositional forms (if any) are logically equivalent? Explain.
    (a) $\sim(P \wedge Q)$ (b) $(\sim P) \wedge (\sim Q)$ (c) $((\sim P) \wedge Q) \vee (P \wedge (\sim Q))$
    (d) $P \oplus (\sim Q)$ (e) $P \vee (\sim Q)$ (f) $(P \vee (\sim Q)) \wedge ((\sim P) \vee Q)$.

11. Which propositional forms (if any) are logically equivalent? Explain.
    (a) $P \vee (Q \vee R)$ (b) $P \oplus (Q \vee R)$ (c) $P \vee (Q \oplus R)$ (d) $P \oplus (Q \oplus R)$
    (e) $(P \vee Q) \vee R$ (f) $(P \oplus Q) \vee R$ (g) $(P \vee Q) \oplus R$ (h) $(P \oplus Q) \oplus R$.

12. Let x_0 be a fixed integer. Let P be the proposition "$x_0 < 8$," let Q be "$0 < x_0$," and let R be "$x_0 = 0$." Encode each statement as a propositional form involving P, Q, and R.
    (a) x_0 is strictly positive.   (b) $x_0 \geq 8$.   (c) x_0 is less than zero.
    (d) x_0 is nonnegative.   (e) $0 < x_0 < 8$.   (f) $0 \leq x_0 < 8$.   (g) $|x_0 - 4| \geq 4$.

13. Let P be the statement "$1+1 = 2$," let Q be "$0 = 1$," and let R be "71 is prime." Convert each propositional form into an English sentence *without parentheses*. Be sure that the sentence has the precise logical structure encoded by the form.
    (a) $(P \wedge Q) \vee R$ (b) $P \wedge (Q \vee R)$ (c) $\sim(P \wedge Q)$ (d) $(\sim P) \wedge Q$
    (e) $(P \wedge Q) \wedge R$ (f) $P \wedge (Q \wedge R)$ (g) $((\sim P) \wedge (\sim Q)) \oplus (\sim R)$

---

[1]Some exercises and examples in Chapter 1 assume familiarity with *even* integers $\ldots, -4, -2, 0, 2, 4, \ldots$ and *odd* integers $\ldots, -5, -3, -1, 1, 3, 5, \ldots$. Formal definitions of *even* and *odd* appear in §2.1.

14. Explain why each statement is not a proposition.
    (a) The last train leaves in five minutes. (b) We should raise the minimum wage.
    (c) Will you marry me? (d) My father's name is Frank. (e) She is smarter than
    he is. (f) Do your homework. (g) It is cold here. (h) This statement is true.

15. Find propositional forms using only the connectives $\vee$, $\wedge$, and $\sim$ that have the
    truth tables shown below. Try to use as few connectives as possible.

P	Q	R	(a)	(b)	(c)	(d)
T	T	T	F	F	F	F
T	T	F	T	T	F	T
T	F	T	T	F	F	T
T	F	F	T	T	T	T
F	T	T	F	F	T	F
F	T	F	T	F	F	T
F	F	T	T	F	F	F
F	F	F	T	F	F	T

16. (a) Find a simple verbal description characterizing when $((A \oplus B) \oplus (C \oplus D)) \oplus E$
    is true. (b) Does your answer to (a) change if we rearrange parentheses in the
    expression? What if we reorder A, B, C, D, and E?

17. Find a propositional form involving variables P, Q, R that is true precisely when
    exactly one of P, Q, R is true. Try to use as few logical connectives as possible.

18. Find a propositional form involving variables P, Q, R that is true when at least
    two of the inputs are true, and false otherwise. Try to use as few logical connectives
    as possible.

19. (a) How many truth tables are possible for propositional forms containing two
    variables P and Q? (b) For each possible truth table in (a), find a short proposi-
    tional form having that truth table.

1.2 Logical Equivalences and IF-Statements

You have probably been using the word IF for most of your life, but do you *really* know what this word means? The official answer, given in this section, may surprise you. We also look more closely at logical equivalence, developing transformation rules for propositional forms analogous to some basic laws of algebra. Everything is proved using our favorite proof technique, truth tables.

Logical Equivalence

We first recall the definition of logically equivalent propositional forms from the last section.

1.12. Definition: Logically Equivalent Propositional Forms. Two propositional forms $\mathcal{A}$ and $\mathcal{B}$ are *logically equivalent* when the truth tables for $\mathcal{A}$ and $\mathcal{B}$ have outputs that agree in every row. We write $\boxed{\mathcal{A} \equiv \mathcal{B}}$ when $\mathcal{A}$ and $\mathcal{B}$ are logically equivalent; we write $\boxed{\mathcal{A} \not\equiv \mathcal{B}}$ when $\mathcal{A}$ and $\mathcal{B}$ are not logically equivalent.

Our first theorem lists some logical equivalences that can often be used to replace complicated propositional forms by shorter ones.

1.13. Theorem on Logical Equivalence. For all propositional forms P, Q, and R, the following logical equivalences hold:

(a) *Commutative Laws:* $P \wedge Q \equiv Q \wedge P$, $P \vee Q \equiv Q \vee P$, and $P \oplus Q \equiv Q \oplus P$.

(b) *Associative Laws:* $P \wedge (Q \wedge R) \equiv (P \wedge Q) \wedge R$, $P \vee (Q \vee R) \equiv (P \vee Q) \vee R$, and $P \oplus (Q \oplus R) \equiv (P \oplus Q) \oplus R$.

(c) *Distributive Laws:* $P \wedge (Q \vee R) \equiv (P \wedge Q) \vee (P \wedge R)$, $P \vee (Q \wedge R) \equiv (P \vee Q) \wedge (P \vee R)$, and $P \wedge (Q \oplus R) \equiv (P \wedge Q) \oplus (P \wedge R)$.

(d) *Idempotent Laws:* $P \wedge P \equiv P$ and $P \vee P \equiv P$.

(e) *Absorption Laws:* $P \wedge (P \vee Q) \equiv P$ and $P \vee (P \wedge Q) \equiv P$.

(f) *Negation Laws:* $\sim(\sim P) \equiv P$, $\sim(P \wedge Q) \equiv (\sim P) \vee (\sim Q)$, and $\sim(P \vee Q) \equiv (\sim P) \wedge (\sim Q)$.

The names of the first three laws indicate the analogy between these logical equivalences and certain algebraic properties of real numbers. For example, the commutative laws for propositional forms resemble the commutative laws $x + y = y + x$ and $x \cdot y = y \cdot x$ for real numbers x, y; similarly, observe the resemblance between the distributive laws listed above and the algebraic distributive law $x \cdot (y + z) = (x \cdot y) + (x \cdot z)$.

All parts of this theorem are proved by constructing truth tables for the propositional forms on each side of each equivalence, and verifying that the truth tables agree in every row. We illustrate the technique for a few parts of the theorem, asking you to prove the other parts in the exercises.

1.14. Proof of the Idempotent Laws. The truth table is shown here:

P	P	$P \wedge P$	$P \vee P$
T	T	T	T
F	F	F	F

We list the column for P twice to make it easier to compute $P \wedge P$ and $P \vee P$. We compute the last two columns from the first two columns using the definition of AND and

OR, respectively. Since column 3 and column 2 agree in all rows, $P \wedge P \equiv P$. Since column 4 and column 2 agree in all rows, $P \vee P \equiv P$.

1.15. Proof of $\sim(P \vee Q) \equiv (\sim P) \wedge (\sim Q)$. For this equivalence, we need a four-row truth table:

P	Q	$P \vee Q$	$\sim(P \vee Q)$	$\sim P$	$\sim Q$	$(\sim P) \wedge (\sim Q)$
T	T	T	F	F	F	F
T	F	T	F	F	T	F
F	T	T	F	T	F	F
F	F	F	T	T	T	T

Columns 4 and 7 agree in every row, so the logical equivalence is proved. This equivalence and its companion $\sim(P \wedge Q) \equiv (\sim P) \vee (\sim Q)$ are called *de Morgan's laws*. In words, the law we just proved says that the negation of an OR-statement "P or Q" is equivalent to the statement "(not P) and (not Q)," whereas the dual law tells us that "not $(P$ and $Q)$" is logically equivalent to "(not P) or (not Q)." For a specific example, the statement "it is false that 10 is prime or 10 is odd" may be simplified to the statement "10 is not prime and 10 is not odd" without changing the truth value (both statements are true). Similarly, the statement that "John has brown hair AND blue eyes" is false precisely when John's hair is not brown OR John's eyes are not blue. In §1.5, we will say a lot more about simplifying statements that begin with NOT.

1.16. Proof of Associativity of XOR. This law has three input variables, so we need a truth table with eight rows, as shown below. One quick way to create the eight possible input combinations is to write 4 Ts followed by 4 Fs in the column for P; then write 2 Ts, 2 Fs, 2 Ts, 2 Fs in the column for Q; then write alternating Ts and Fs in the column for R. This pattern extends to truth tables with even more variables. In general, if there are n distinct input variables that stand for propositions, the truth table has 2^n rows.

P	Q	R	$Q \oplus R$	$P \oplus (Q \oplus R)$	$P \oplus Q$	$(P \oplus Q) \oplus R$
T	T	T	F	T	F	T
T	T	F	T	F	F	F
T	F	T	T	F	T	F
T	F	F	F	T	T	T
F	T	T	F	F	T	F
F	T	F	T	T	T	T
F	F	T	T	T	F	T
F	F	F	F	F	F	F

The quickest way to fill in this truth table is to look for disagreements between the inputs of each XOR (being sure to refer to the correct columns), writing Ts in those rows, and writing Fs in all other rows. Columns 5 and 7 agree in all eight rows, so the associative law for $\oplus$ is proved. The other associative laws and distributive laws are proved by similar eight-row truth tables.

What IF Means

We are now ready to define the precise logical meaning of the word IF. Be warned that the usage of IF in everyday conversational English often does *not* match the definition we are about to give. Nevertheless, this definition is universally used in logic, mathematics, and all technical communication. When discussing the statement $P \Rightarrow Q$, which is read "if P then Q," we call P the *hypothesis* and Q the *conclusion* of the IF-statement.

1.17. Definition of IF. For any propositions P and Q, the truth value of $P \Rightarrow Q$ ("if P then Q") is determined by the following table.

P	Q	$P \Rightarrow Q$
T	T	T
T	F	F
F	T	T
F	F	T

Remember this table by noting that $P \Rightarrow Q$ is false only when the hypothesis P is true and the conclusion Q is false; *in all other situations*, the IF-statement is true. In particular, when P is false, the IF-statement $P \Rightarrow Q$ is automatically true, whether Q is true or false; this fact often confuses beginners.

1.18. Example. True or false?
(a) If $1 + 1 = 3$, then $1 + 1 = 5$.
(b) If $0 \neq 0$, then $0 = 0$.
(c) If $3 > 2$, then London is the capital of England.
(d) If 9 is odd, then 9 is even.
Solution. Statement (a) has the form $P \Rightarrow Q$, where P is the false statement "$1 + 1 = 3$," and Q is the false statement "$1+1 = 5$." By definition of IF, statement (a) is *true*, no matter how unintuitive this statement may sound. Statement (b) has the form $(\sim Q) \Rightarrow Q$, where Q is the true proposition "$0 = 0$." Since the hypothesis $\sim Q$ is false, we can see that (b) is *true* without even reading the statement following THEN. Statement (c) is true (according to row 1 of the truth table), even though the two parts of the statement have no causal relation to each other. Finally, statement (d) is false because "9 is odd" is true, but "9 is even" is false.

These examples show that we must not rely upon an intuitive impression of the content of an IF-statement to determine its truth value; we must instead combine the truth values of the inputs according to the defining truth table. As shown in (c), there need not be any cause-and-effect relationship between the hypothesis and conclusion of an IF-statement.

It is fair to ask *why* IF is defined in such an apparently unnatural way. Most people agree that "if true then true" should be considered true, whereas "if true then false" should be considered false. Why, though, is "if false then anything" defined to be true? One possible answer draws an analogy between IF-statements, as they are used in logic, and IF-statements appearing in legal contracts or judicial proceedings. Suppose a painting company issues a contract to a homeowner containing the statement: "if you pay us $1000 by May 1, then we will paint your house by June 1." Under what conditions would we say that the company has violated this contract? For example, suppose the company does not paint the house because it was not paid on time. Both clauses of the IF-statement are false, yet no one would say that the company violated the contract in this situation. Similarly, supposing that the money is paid on May 3 and the house is painted on time, the company has not violated the contract even though the first part of the IF-statement is false.

We can give another, more mathematical motivation for the definition of IF by considering the following statement: "for all real numbers x, if $x > 2$ then $x^2 > 4$." (This statement uses variables and quantifiers, which are discussed later.) Most people would judge this statement to be true, on intuitive grounds, and this judgment is correct. But once you agree this statement should be true, you are forced to agree that it should be true after replacing each x in the statement by any specific real number. In particular, we see that each of these statements must be true:

- "if $3 > 2$ then $3^2 > 4$" (illustrating row 1 of the truth table);

- "if $-4 > 2$ then $(-4)^2 > 4$" (illustrating row 3 of the truth table);

- "if $0 > 2$ then $0^2 > 4$" (illustrating row 4 of the truth table).

Thus the whole truth table for IF is forced upon us, granting that row 2 of the truth table ought to be false.

In mathematics, a statement "if P then Q" is not making any claims about the truth of the conclusion Q in the event that the hypothesis P does not hold; so in this event, we should not declare the IF-statement to be false. But since propositions must have some truth value, we are forced to call the IF-statement true in such situations.

Converse and Contrapositive

Given an IF-statement $P \Rightarrow Q$, we can form two related IF-statements called the converse and contrapositive of the original statement.

1.19. Definition: Converse and Contrapositive. For any propositional forms P and Q: the $\boxed{\text{converse of } P \Rightarrow Q}$ is $\boxed{Q \Rightarrow P}$; the $\boxed{\text{contrapositive of } P \Rightarrow Q}$ is $\boxed{(\sim Q) \Rightarrow (\sim P)}$.

1.20. Example. The converse of "if 9 is negative, then 9 is odd" is "if 9 is odd, then 9 is negative." Observe that the first statement is true, whereas the converse is false. The contrapositive of "if 9 is negative, then 9 is odd" is "if 9 is not odd, then 9 is not negative," which is true.

The converse of "if $2 = 3$, then $4 = 9$" is "if $4 = 9$, then $2 = 3$." Both the original statement and its converse are true. The contrapositive of "if $2 = 3$, then $4 = 9$" is "if $4 \neq 9$, then $2 \neq 3$," which is also true.

In both examples, the contrapositive had the same truth value as the original statement, but this was not always the case for the converse. The next theorem explains what happens in general.

1.21. Theorem on IF. Let P and Q be distinct propositional variables.

(a) *Non-equivalence of Converse:* $P \Rightarrow Q \;\not\equiv\; Q \Rightarrow P$.

(b) *Equivalence of Contrapositive:* $P \Rightarrow Q \;\equiv\; (\sim Q) \Rightarrow (\sim P)$.

(c) *Elimination of IF:* $P \Rightarrow Q \;\equiv\; (\sim P) \vee Q$.

(d) *Denial of IF:* $\sim(P \Rightarrow Q) \;\equiv\; P \wedge (\sim Q)$.

Part (a) says that a general IF-statement is *not* logically equivalent to its converse. Part (b) says that a general IF-statement *is* logically equivalent to its contrapositive. Part (c) says that IF-statements are essentially special kinds of OR-statements in which the first input to the OR has been negated. We can use (c) to convert IF-statements to OR-statements and vice versa. Part (d) shows that the negation of an IF-statement is logically equivalent to an AND-statement. Parts (b), (c), and (d) hold for any propositional forms P and Q, not just individual letters. For instance, $(A \vee B) \Rightarrow (C \oplus D) \;\equiv\; (\sim(A \vee B)) \vee (C \oplus D)$ follows by replacing P by $A \vee B$ and replacing Q by $C \oplus D$ in part (c). *Memorize all parts of this theorem*, as they will be used constantly throughout our study of logic and proofs.

To prove the theorem, consider the following truth table:

P	Q	$P \Rightarrow Q$	$Q \Rightarrow P$	$\sim Q$	$\sim P$	$(\sim Q) \Rightarrow (\sim P)$	$(\sim P) \vee Q$
T	T	T	T	F	F	T	T
T	F	F	T	T	F	F	F
F	T	T	F	F	T	T	T
F	F	T	T	T	T	T	T

Column 3 does not match column 4 in rows 2 and 3, proving (a). Column 3 matches column 7 in all rows, proving (b). Column 3 also matches column 8 in all rows, proving (c). We can prove (d) by adding a few more columns to this truth table. Alternatively, (d) follows from (c) and the Negation Laws, since

$$\sim(P \Rightarrow Q) \;\equiv\; \sim((\sim P) \vee Q) \;\equiv\; (\sim(\sim P)) \wedge (\sim Q) \;\equiv\; P \wedge (\sim Q). \tag{1.3}$$

Five Remarks on Logical Equivalence (Optional)

(1) The Theorem on Logical Equivalence shows that many of the laws of logic have an algebraic character analogous to algebraic laws for arithmetical operations on real numbers. To make this analogy more explicit, we introduce the convention of representing true (T) by 1 and false (F) by 0. The symbols 0 and 1 are called *bits*. In this notation, the defining truth tables for the six logical operators (including IFF, discussed in the next section) look like this:

P	Q	$\sim P$	$P \wedge Q$	$P \vee Q$	$P \oplus Q$	$P \Rightarrow Q$	$P \Leftrightarrow Q$
0	0	1	0	0	0	1	1
0	1	1	0	1	1	1	0
1	0	0	0	1	1	0	0
1	1	0	1	1	0	1	1

If we think of P and Q as bits (the actual integers 0 and 1, rather than propositions), we see that $\sim P = 1 - P$; $P \wedge Q = P \cdot Q = \min(P, Q)$, the product or minimum of P and Q; $P \vee Q = \max(P, Q)$, the maximum of P and Q; and $P \oplus Q$ is the mod-2 sum of P and Q. (We discuss modular addition in detail later.) We can also think of $\oplus$, $\Rightarrow$, and $\Leftrightarrow$ as *relations* comparing their two inputs. In this interpretation, $P \oplus Q$ is true precisely when $P \neq Q$, so $\oplus$ models non-equality of bits; $P \Leftrightarrow Q$ is true precisely when $P = Q$, so $\Leftrightarrow$ models equality of bits; and $P \Rightarrow Q$ is true precisely when $P \leq Q$, so $\Rightarrow$ models ordering of bits. Now the various identities in the Theorem on Logical Equivalences translate into identities for these algebraic operations on bits. For instance, supposing we already know that $x(yz) = (xy)z$ for all integers x, y, z, we can deduce the associativity law $P \wedge (Q \wedge R) \equiv (P \wedge Q) \wedge R$ for propositional forms.

(2) Logical equivalence of propositional forms (denoted $\equiv$) is similar to, but not the same as, equality of forms (denoted $=$). Thinking of propositional forms $\mathcal{A}$ and $\mathcal{B}$ as strings of symbols, $\mathcal{A} = \mathcal{B}$ means that $\mathcal{A}$ has exactly the same symbols as $\mathcal{B}$, in the same order. For example, $P \vee Q$ is not *equal* to $Q \vee P$ since the first symbols do not match. Yet, as we know, $P \vee Q \equiv Q \vee P$ since the two truth tables agree. For most purposes of logic, two logically equivalent forms are interchangeable with each other (since their truth values must agree), even though such forms are often not equal as strings.

(3) Although logical equivalence of propositional forms (denoted $\equiv$) is not the same concept as logical equality (denoted $=$), these two concepts share many common properties. For example, given any objects x, y, z whatsoever, the following facts are true about equality:

$$x = x; \quad \text{if } x = y \text{ then } y = x; \quad \text{if } x = y \text{ and } y = z \text{ then } x = z.$$

These facts are called *reflexivity*, *symmetry*, and *transitivity* of equality. Three analogous facts also hold for logical equivalence: given any propositional forms $\mathcal{A}, \mathcal{B}, \mathcal{C}$,

$$\mathcal{A} \equiv \mathcal{A}; \quad \text{if } \mathcal{A} \equiv \mathcal{B} \text{ then } \mathcal{B} \equiv \mathcal{A}; \quad \text{if } \mathcal{A} \equiv \mathcal{B} \text{ and } \mathcal{B} \equiv \mathcal{C} \text{ then } \mathcal{A} \equiv \mathcal{C}. \tag{1.4}$$

These facts follow from the corresponding facts for equality, applied to the rows of the truth tables. For instance, if the output in every row of $\mathcal{A}$'s truth table equals the corresponding output of $\mathcal{B}$'s truth table, then the output in every row of $\mathcal{B}$'s truth table equals the corresponding output of $\mathcal{A}$'s truth table, which is why the second fact is true. The text implicitly uses the reflexivity, symmetry, and transitivity of logical equivalence in many places. For instance, the derivation (1.3) has the form $\mathcal{A} \equiv \mathcal{B} \equiv \mathcal{C} \equiv \mathcal{D}$, which is really an abbreviation for "$\mathcal{A} \equiv \mathcal{B}$ and $\mathcal{B} \equiv \mathcal{C}$ and $\mathcal{C} \equiv \mathcal{D}$." Using transitivity once, we see that $\mathcal{A} \equiv \mathcal{C}$. Using transitivity again, we see that $\mathcal{A} \equiv \mathcal{D}$, which is the required conclusion. Similarly, symmetry of $\equiv$ is needed whenever we want to use a known logical equivalence in reverse, e.g., to replace $(P \wedge Q) \vee (P \wedge R)$ by $P \wedge (Q \vee R)$ by invoking the Distributive Law. We will have much more to say about the properties of reflexivity, symmetry, and transitivity in Chapter 6, when we study equivalence relations.

(4) The logical equivalences in this section's theorems apply to *all* propositional forms P, Q, R, which means that we can replace each letter P, Q, R (wherever it appears in an identity) by a longer propositional form. For instance, replacing P by $A \Rightarrow B$ and Q by $B \oplus C$ in the second negation law, we see that $\sim((A \Rightarrow B) \wedge (B \oplus C)) \equiv (\sim(A \Rightarrow B)) \vee (\sim(B \oplus C))$. In contrast, consider the non-logical equivalence asserted in part (a) of the Theorem on IF. In that theorem, P and Q are distinct propositional variables (as opposed to general propositional forms), and we have $(P \Rightarrow Q) \not\equiv (Q \Rightarrow P)$ as shown in the proof. But for some choices of the form P and the form Q, the logical equivalence will hold; for instance, suppose P and Q are both R. It can be checked that when P and Q are logically equivalent forms, $(P \Rightarrow Q) \equiv (Q \Rightarrow P)$, because in this case we must always be in row 1 or row 4 of the truth table shown in the proof. (In fact, using terminology from the next section, $P \Rightarrow Q$ and $Q \Rightarrow P$ are both tautologies in this situation.) For example, $((A \vee B) \Rightarrow (B \vee A))$ is logically equivalent to $((B \vee A) \Rightarrow (A \vee B))$. On the other hand, when P and Q are not logically equivalent, there must exist some assignment of truth values to the propositional variables appearing in P and Q that cause P and Q themselves to have opposite truth values. In this row of the truth table, $P \Rightarrow Q$ and $Q \Rightarrow P$ disagree, so that these forms are not logically equivalent. For instance, $((A \vee B) \Rightarrow (A \oplus B))$ and $((A \oplus B) \Rightarrow (A \vee B))$ are not logically equivalent, as we see by considering the row of the truth table where A and B are both true.

(5) In algebra, if we know two variables represent the same number, we can replace one variable by the other in longer expressions. We can express this fact symbolically by *substitution rules* such as the following: for all numbers x, y, z, if $y = z$ then $x+y = x+z$ and $xy = xz$. There are analogous substitution rules for logically equivalent propositional forms. For instance, given any propositional forms $\mathcal{A}$, $\mathcal{B}$, and $\mathcal{C}$, if $\mathcal{B} \equiv \mathcal{C}$, then $(\mathcal{A} \vee \mathcal{B}) \equiv (\mathcal{A} \vee \mathcal{C})$, $(\mathcal{A} \wedge \mathcal{B}) \equiv (\mathcal{A} \wedge \mathcal{C})$, $(\mathcal{A} \Rightarrow \mathcal{B}) \equiv (\mathcal{A} \Rightarrow \mathcal{C})$, $(\mathcal{B} \Rightarrow \mathcal{A}) \equiv (\mathcal{C} \Rightarrow \mathcal{A})$, and so on. We freely use these substitution rules without justification in the main text, but it is possible to prove each rule by comparing truth tables. As an example, assume $\mathcal{B} \equiv \mathcal{C}$, and consider a fixed row in the truth tables for $\mathcal{A} \Rightarrow \mathcal{B}$ and $\mathcal{A} \Rightarrow \mathcal{C}$. Suppose that, for the values of the input propositional variables appearing in this row, $\mathcal{A}$ is true and $\mathcal{B}$ is false. On one hand, $\mathcal{A} \Rightarrow \mathcal{B}$ must be false in this row. On the other hand, since $\mathcal{B}$ and $\mathcal{C}$ were assumed to be logically equivalent, $\mathcal{C}$ is false in this row, so $\mathcal{A} \Rightarrow \mathcal{C}$ is also false in this row. Thus, $\mathcal{A} \Rightarrow \mathcal{B}$ and $\mathcal{A} \Rightarrow \mathcal{C}$ have the same truth value (namely, false) in this row. Considering the other possible cases, we see that $\mathcal{A} \Rightarrow \mathcal{B}$ and $\mathcal{A} \Rightarrow \mathcal{C}$ have the same truth value in every row of the truth table, so they are logically equivalent. We can summarize the cases in a *meta-truth table* that looks like this:

$\mathcal{A}$	$\mathcal{B}$	$\mathcal{C}$	$\mathcal{A} \Rightarrow \mathcal{B}$	$\mathcal{A} \Rightarrow \mathcal{C}$
T	T	T	T	T
T	F	F	F	F
F	T	T	T	T
F	F	F	T	T

We call this a meta-truth table because the underlying propositional variables appearing in $\mathcal{A}$, $\mathcal{B}$, and $\mathcal{C}$ are not explicitly listed. (For example, these forms might involve five propositional variables P, Q, R, S, and T, so that the actual truth table would have 32 rows.) However, to prove the logical equivalence of $\mathcal{A} \Rightarrow \mathcal{B}$ and $\mathcal{A} \Rightarrow \mathcal{C}$, we only need to know the truth values of the propositional forms $\mathcal{A}$, $\mathcal{B}$, and $\mathcal{C}$. Row 2 of the table is an abbreviation for the discussion preceding the table, and the other rows cover the remaining cases. The key point is that the table has four rows, not eight, because we have assumed that $\mathcal{B}$ and $\mathcal{C}$ are logically equivalent forms, hence always have the same truth value. You can construct similar meta-truth tables to prove the other substitution rules.

Section Summary

1. The logical operators satisfy the commutative, associative, distributive, idempotent, and absorption laws listed in the Theorem on Logical Equivalence. They also obey these negation rules:

$$\sim(P \wedge Q) \equiv (\sim P) \vee (\sim Q), \quad \sim(P \vee Q) \equiv (\sim P) \wedge (\sim Q),$$

$$\sim(\sim P) \equiv P, \quad \sim(P \Rightarrow Q) \equiv P \wedge (\sim Q).$$

We check all these logical equivalences by verifying that the truth tables for each side agree in every row.

2. $P \Rightarrow Q$ is false when P is true and Q is false; $P \Rightarrow Q$ is true in all other situations.

3. The *converse* of $P \Rightarrow Q$ is $Q \Rightarrow P$, which is *not* logically equivalent to $P \Rightarrow Q$. The *contrapositive* of $P \Rightarrow Q$ is $(\sim Q) \Rightarrow (\sim P)$, which *is* logically equivalent to $P \Rightarrow Q$. We have $P \Rightarrow Q \equiv (\sim P) \vee Q$, and hence $\sim(P \Rightarrow Q) \equiv P \wedge (\sim Q)$.

Exercises

1. Make a truth table for each propositional form.
   (a) $(P \vee (\sim Q)) \Rightarrow P$ (b) $P \Rightarrow (Q \wedge (\sim P))$ (c) $(P \oplus Q) \Rightarrow (P \vee Q)$.

2. Make a truth table for each propositional form.
   (a) $(P \Rightarrow Q) \wedge (Q \Rightarrow R)$ (b) $(P \Rightarrow Q) \Rightarrow (Q \vee R)$ (c) $(\sim(P \oplus Q)) \Rightarrow (\sim(R \wedge P))$.

3. Prove the negation laws $\sim(\sim P) \equiv P$ and $\sim(P \wedge Q) \equiv (\sim P) \vee (\sim Q)$ using truth tables.

4. Use truth tables to prove the commutative laws for $\wedge$, $\vee$, and $\oplus$ in the Theorem on Logical Equivalences.

5. Prove the associative laws for $\wedge$ and $\vee$.

6. Prove the absorption laws in part (e) of the Theorem on Logical Equivalences.

7. Prove the distributive laws in part (c) of the Theorem on Logical Equivalences.

8. Prove or disprove: $(\sim P) \oplus (Q \oplus P) \equiv (\sim Q)$.

9. True or false?
   (a) If Mars is a planet, then $2 + 2 = 4$.
   (b) If $2 \cdot 2 \neq 4$, then π is negative.
   (c) If Paris is a city, then Europe is an ocean.
   (d) If $0 = 0$, then 5 is even.
   (e) If some trees have roots, then all crows are white.
   (f) If $3 + 3 \neq 6$, then $\displaystyle\int_0^\infty \frac{1}{x^4 + 1} \, dx = \pi\sqrt{2}/4$.
   (g) If $1 + 1 = 3$, then $1 + 1 = 2$, but the converse is false.

10. Assume P is a true proposition, Q is a false proposition, and R is an arbitrary proposition. Say as much as you can about the truth or falsehood of each proposition below.
    (a) $P \Rightarrow R$ (b) $Q \Rightarrow R$ (c) $R \Rightarrow P$ (d) $R \Rightarrow Q$ (e) $R \Rightarrow R$ (f) $R \Rightarrow (\sim R)$.

11. Let x_0 be a fixed integer, and let P be the statement:
    "if x_0 is even and x_0 is prime, then $x_0 = 2$."
    (a) Write the converse of P.
    (b) Write the contrapositive of P.
    (c) Write the converse of the contrapositive of P.
    (d) Suppose x_0 is 8. Is P true or false? What about the converse of P?

12. The *inverse* of $P \Rightarrow Q$ is defined to be $(\sim P) \Rightarrow (\sim Q)$. Decide whether the inverse of $P \Rightarrow Q$ is logically equivalent to: (a) $P \Rightarrow Q$; (b) the converse of $P \Rightarrow Q$; (c) the contrapositive of $P \Rightarrow Q$.

13. Write the converse and the contrapositive of each statement.
    (a) If f is continuous or monotonic, then f is integrable.
    (b) If g is not continuous, then g is not differentiable.
    (c) If p is prime, then if p is even, then $p = 2$.

14. Use known logical equivalences to show that $(P \wedge R) \Rightarrow (\sim Q)$ is logically equivalent to $\sim(P \wedge (Q \wedge R))$ without making a truth table.

15. (a) Use known logical equivalences to show that $P \Rightarrow (Q \Rightarrow R)$ is logically equivalent to $(P \wedge Q) \Rightarrow R$ without making a truth table. (b) Is $(P \Rightarrow Q) \Rightarrow R$ logically equivalent to $P \Rightarrow (Q \Rightarrow R)$? Explain.

16. Write the converse of $(P \Rightarrow Q) \Rightarrow R$ and the contrapositive of $(P \Rightarrow Q) \Rightarrow R$.

17. Use the Theorem on IF to replace each propositional form with an equivalent form that does not use $\Rightarrow$. (a) $(P \wedge Q) \Rightarrow (R \vee S)$ (b) $P \Rightarrow ((\sim Q) \Rightarrow P)$
    (c) $[P \Rightarrow (Q \Rightarrow R)] \Rightarrow [(P \Rightarrow Q) \Rightarrow (P \Rightarrow R)]$

18. Which proposed distributive laws are correct? Prove your answers.
    (a) $P \vee (Q \oplus R) \equiv (P \vee Q) \oplus (P \vee R)$.
    (b) $P \Rightarrow (Q \wedge R) \equiv (P \Rightarrow Q) \wedge (P \Rightarrow R)$.
    (c) $P \Rightarrow (Q \vee R) \equiv (P \Rightarrow Q) \vee (P \Rightarrow R)$.
    (d) $P \Rightarrow (Q \oplus R) \equiv (P \Rightarrow Q) \oplus (P \Rightarrow R)$.
    (e) $P \wedge (Q \Rightarrow R) \equiv (P \wedge Q) \Rightarrow (P \wedge R)$.
    (f) $P \vee (Q \Rightarrow R) \equiv (P \vee Q) \Rightarrow (P \vee R)$.
    (g) $P \oplus (Q \Rightarrow R) \equiv (P \oplus Q) \Rightarrow (P \oplus R)$.

19. Prove $A \wedge (B \vee C \vee D \vee E) \equiv (A \wedge B) \vee (A \wedge C) \vee (A \wedge D) \vee (A \wedge E)$ without using truth tables.

20. Use logical equivalences from this section to find four different IF-statements that are logically equivalent to $(\sim P) \vee (\sim Q) \vee R$.

21. The FOIL Rule in algebra says that for all real numbers a, b, c, d, $(a+b) \cdot (c+d) = (ac+ad) + (bc+bd)$. (a) Prove this rule using laws of algebra. (b) Formulate three analogous FOIL rules for logical connectives, and prove these rules without truth tables using the Theorem on Logical Equivalences.

22. Use known logical equivalences (not truth tables) to prove:
    $(A \Rightarrow B) \wedge (C \Rightarrow D) \equiv (\sim(A \vee C)) \vee (A \Rightarrow D) \vee (C \Rightarrow B) \vee (B \wedge D)$.

23. True or false? Explain.
    (a) The contrapositive of $P \Rightarrow Q$ always has the same truth value as $P \Rightarrow Q$.
    (b) The converse of $P \Rightarrow Q$ always has the same truth value as the negation of

$P \Rightarrow Q$.

(c) If $P \Rightarrow Q$ is false, then the converse must be true.

(d) If $P \Rightarrow Q$ is true, then the converse must be true.

(e) If the hypothesis and conclusion of a conditional statement have the same truth value, then the conditional statement must be true.

24. Give specific examples of propositional forms $\mathcal{A}$ and $\mathcal{B}$ with each property below. Explain why your examples work. (a) $\mathcal{A} \vee \mathcal{B}$ is logically equivalent to $\mathcal{A} \oplus \mathcal{B}$. (b) $\mathcal{A} \Rightarrow \mathcal{B} \equiv \mathcal{B} \Rightarrow \mathcal{A}$, but $\mathcal{A} \Rightarrow \mathcal{B} \not\equiv \mathcal{A} \wedge \mathcal{B}$. (c) $\mathcal{A} \vee \mathcal{B}$ is logically equivalent to $\sim(\mathcal{A} \wedge \mathcal{B})$.

25. Prove that logical equivalence of propositional forms is reflexive and transitive, as stated in (1.4).

26. Prove the substitution rules for propositional forms given in Remark 5 on page 14.

1.3   IF, IFF, Tautologies, and Contradictions

Now that we know what IF means, it is time to talk about other English words and phrases that have the same logical meaning as an IF-statement. We also introduce the new logical connective IFF (pronounced "if and only if"). In everyday life, people often say IF when they mean IFF, so it is vital to understand the correct technical usage of these words. We close the section by introducing tautologies (propositional forms that always evaluate to true) and contradictions (propositional forms that always evaluate to false).

Translations of $P \Rightarrow Q$

The most straightforward English translation of the propositional form $P \Rightarrow Q$ is "if P, then Q." However, there are many other English phrases with the same logical meaning as $P \Rightarrow Q$. We begin by considering some translations of IF that often cause trouble. By definition:

- $\boxed{P \text{ if } Q}$ means $\boxed{Q \Rightarrow P}$.

- $\boxed{P \text{ only if } Q}$ means $\boxed{P \Rightarrow Q}$.

The phrase "P if Q" is obtained by changing the order of the clauses in "if Q then P." In both phrases, the proposition Q immediately follows the word IF, so Q is the hypothesis and P is the conclusion of the conditional statement. However, replacing IF by ONLY IF changes the meaning: in the sentence "P only if Q," P is the hypothesis and Q is the conclusion! We explain this counterintuitive phenomenon through an example. Suppose a professor says to a student, "You will pass my class ONLY IF you take the final exam." This statement means the same thing as, "If you do not take the final exam, then you will not pass the class." We recognize this as the *contrapositive* of the statement, "If you pass the class, then you took the final exam." The same reasoning holds in general: "P only if Q" really means "if Q is not true, then P is not true," which can be encoded as $(\sim Q) \Rightarrow (\sim P)$. This is the contrapositive of $P \Rightarrow Q$, so this statement is logically equivalent to $P \Rightarrow Q$.

Now consider the meaning of the words NECESSARY and SUFFICIENT. By definition:

- $\boxed{P \text{ is sufficient for } Q}$ means $\boxed{P \Rightarrow Q}$.

- $\boxed{P \text{ is necessary for } Q}$ means $\boxed{Q \Rightarrow P}$.

For example, "Differentiability of f is sufficient for continuity of f" means "If f is differentiable, then f is continuous." On the other hand, the word NECESSARY (like the word ONLY) has a hidden negating effect. The statement "Taking the final exam is necessary for passing the course" means, "If you do not take the final exam, then you will not pass the course," which in turn has the same meaning as, "If you pass the course, then you took the final exam." In general, saying "P is necessary for Q" means that if P does NOT hold, then Q does NOT hold, which can be encoded as $(\sim P) \Rightarrow (\sim Q)$. This is the contrapositive of $Q \Rightarrow P$, so this statement has the same logical meaning as $Q \Rightarrow P$.

One way to remember the correct meaning of NECESSARY is to note that "P is necessary for Q" is the *converse* of "P is sufficient for Q." Similarly, "P only if Q" is the *converse* of "P if Q." We must take care, however, because the passive voice (and similar constructions that change the order of phrases in a sentence) can reverse the positions of P and Q. For example, all of the following phrases can be used to translate $P \Rightarrow Q$:

- P is sufficient for Q.

- For Q to be true, it is sufficient that P be true.

- A sufficient condition for Q is P.

- Q is necessary for P.

- A necessary condition for P is Q.

- For P to be true, it is necessary that Q be true.

Notice the key preposition FOR, which can help you locate the correct hypothesis and conclusion in each phrase. Yet other possible translations of $P \Rightarrow Q$ include:

- P implies Q.

- Q is implied by P.

- When P is true, Q is true.

- Q whenever P.

- Q follows from P.

- Q is a consequence of P.

In particular, when reading a form such as $P \Rightarrow Q$ aloud, you may read the symbol $\Rightarrow$ as the word IMPLIES. But it is *wrong* to read $\Rightarrow$ as the word IF, since "P if Q" means $Q \Rightarrow P$, not $P \Rightarrow Q$. It is correct, but potentially confusing, to pronounce the symbol $\Rightarrow$ as ONLY IF. My personal recommendation is to read $P \Rightarrow Q$ as "if P, then Q." But you must still be familiar with all of the variants above, since they occur frequently in mathematical writing.

While we are discussing translation issues, we should mention the logical meaning of a few other common English phrases:

- "P but Q" means $P \wedge Q$.

- "Although P, Q" means $P \wedge Q$.

- "P, Q are both true" means $P \wedge Q$.

- "Neither P nor Q" means $(\sim P) \wedge (\sim Q)$, or equivalently $\sim(P \vee Q)$.

The words BUT and ALTHOUGH suggest some kind of contrast between the two clauses joined by these words, but this contrast is ignored in logic. For logical purposes, BUT and ALTHOUGH (and BOTH) have exactly the same meaning as AND.

There are some other English phrases whose mathematical meaning does not seem to be completely standardized. I assert that "either P or Q" means $P \vee Q$ (inclusive-OR), "P provided that Q" means $Q \Rightarrow P$, and "P unless Q" means $P \vee Q$, but not all mathematical writers agree with these conventions.

What IFF Means

Now that we understand IF and its many synonyms, it is time to define the related logical connective IFF (pronounced "if and only if" and symbolized by $\Leftrightarrow$). The formal definition of this word appears in the truth table below.

1.22. Definition of IFF. For any propositions P and Q, the truth value of $P \Leftrightarrow Q$ ("P iff Q") is determined by the following table.

P	Q	$P \Leftrightarrow Q$
T	T	T
T	F	F
F	T	F
F	F	T

Remember this table by noting that $P \Leftrightarrow Q$ is true exactly when the two inputs P, Q have the same truth values; $P \Leftrightarrow Q$ is false when the inputs P, Q have different truth values. Comparing the defining truth tables for IF and IFF, we see that $P \Rightarrow Q \not\equiv P \Leftrightarrow Q$.

1.23. Example. True or false?
(a) $1 + 1 = 3$ only if $1 + 1 = 2$.
(b) $0 < 1$ if $1 < 0$.
(c) $0 < 1$ iff $1 < 0$.
(d) A necessary condition for 7 to be even is that 5 be odd.
(e) For 7 to be even, it is sufficient that 5 be odd.
(f) 5 is negative iff -5 is positive.
Solution. The suggested approach is to encode each English sentence symbolically, then refer to the definitions of IF and IFF. Thus, (a) becomes "$1 + 1 = 3 \Rightarrow 1 + 1 = 2$," which is true since F $\Rightarrow$ T is true. Similarly, (b) becomes "$1 < 0 \Rightarrow 0 < 1$," which is true. But (c) is "$0 < 1 \Leftrightarrow 1 < 0$," which is false since T $\Leftrightarrow$ F is false. (d) says that "7 is even $\Rightarrow$ 5 is odd," which is true. (e) says that "5 is odd $\Rightarrow$ 7 is even," which is false. Finally, (f) is true since both inputs to the $\Leftrightarrow$ operator are false.

1.24. Theorem on IFF. For all propositional forms P and Q:

(a) *Reduction to IF:* $P \Leftrightarrow Q \equiv (P \Rightarrow Q) \wedge (Q \Rightarrow P)$.

(b) *Reduction to AND and OR:* $P \Leftrightarrow Q \equiv (P \wedge Q) \vee ((\sim P) \wedge (\sim Q))$.

(c) *Symmetry of IFF:* $P \Leftrightarrow Q \equiv Q \Leftrightarrow P$.

(d) *Denial of IFF:* $\sim(P \Leftrightarrow Q) \equiv P \oplus Q$.

(e) *Negating Both Sides:* $P \Leftrightarrow Q \equiv (\sim P) \Leftrightarrow (\sim Q)$.

Part (a) relates IFF to IF and explains why the symbol $\Leftrightarrow$ is used for IFF. Part (b) lets us replace IFF by a form built from other logical connectives.

As with previous theorems, we prove these logical equivalences with truth tables. As a sample, we prove parts (a) and (e) as shown:

P	Q	$P \Leftrightarrow Q$	$P \Rightarrow Q$	$Q \Rightarrow P$	$(P \Rightarrow Q) \wedge (Q \Rightarrow P)$	$\sim P$	$\sim Q$	$(\sim P) \Leftrightarrow (\sim Q)$
T	T	T	T	T	T	F	F	T
T	F	F	F	T	F	F	T	F
F	T	F	T	F	F	T	F	F
F	F	T	T	T	T	T	T	T

Columns 3 and 6 agree, proving (a); columns 3 and 9 agree, proving (e).

1.25. Translations of IFF. We can translate $P \Leftrightarrow Q$ by several English phrases, including:
(a) P iff Q.
(b) P if and only if Q.
(c) P is necessary and sufficient for Q.
(d) P when and only when Q.
(e) P precisely when Q.
(f) P is equivalent to Q.
(g) P has the same truth value as Q.

Regarding (b), note that "P if and only if Q" is short for "P if Q, and P only if Q," which translates to $(Q \Rightarrow P) \wedge (P \Rightarrow Q)$. We know this is equivalent to $P \Leftrightarrow Q$ by part (a) of the theorem and commutativity of $\wedge$. A similar analysis applies to the phrases "necessary and sufficient" and "when and only when."

From part (e) of the Theorem on IFF, negating both sides of an IFF-statement does not deny the entire statement, but actually produces a *logically equivalent* statement. To deny an IFF-statement, we must replace the IFF by XOR, as stated in part (d) of the theorem. By negating each side of the equivalences in the Theorem on IFF, or by constructing more truth tables, we obtain the following equivalences for XOR.

1.26. Theorem on XOR (Optional). For all propositional forms P, Q, R:

(a) *Reduction to AND and OR:* $P \oplus Q \equiv (P \wedge (\sim Q)) \vee (Q \wedge (\sim P))$.

(b) *Negating Both Sides:* $P \oplus Q \equiv (\sim P) \oplus (\sim Q)$.

(c) *Second Reduction to AND and OR:* $P \oplus Q \equiv (P \vee Q) \wedge (\sim(P \wedge Q))$.

(d) *Commutativity of XOR:* $P \oplus Q \equiv Q \oplus P$.

(e) *Denial of XOR:* $\sim(P \oplus Q) \equiv P \Leftrightarrow Q$.

(f) *Reduction to IFF:* $P \oplus Q \equiv P \Leftrightarrow (\sim Q) \equiv (\sim P) \Leftrightarrow Q$.

(g) *Associativity of XOR:* $P \oplus (Q \oplus R) \equiv (P \oplus Q) \oplus R$.

Part (a) shows that we can translate $P \oplus Q$ as "exactly one of P and Q is true." Part (c), which we proved in an earlier section, justifies the translation of $P \oplus Q$ as "P or Q, but not both." Part (f) indicates how to transform XOR-statements into IFF-statements by negating one of the inputs.

Tautologies and Contradictions

Certain propositional forms, called tautologies and contradictions, play a special role in logic and proofs. These are defined as follows.

1.27. Definition: Tautologies and Contradictions. A propositional form $\mathcal{A}$ is called a *tautology* iff every row of the truth table for $\mathcal{A}$ has output true. A propositional form $\mathcal{B}$ is called a *contradiction* iff every row of the truth table for $\mathcal{B}$ has output false.

Note that most propositional forms have a mixture of true and false outputs, so *most propositional forms are neither tautologies nor contradictions.*

1.28. Example. The propositional form $R \vee (\sim R)$ is a tautology, whereas the form $R \wedge (\sim R)$ is a contradiction, as shown in the following truth table:

R	$\sim R$	$R \vee (\sim R)$	$R \wedge (\sim R)$
T	F	T	F
F	T	T	F

Intuitively, "R or not R" must be true, regardless of the truth value of R, whereas "R and not R" must be false. More generally, for any propositional form $\mathcal{A}$ (not just the individual variable R), $\mathcal{A} \vee (\sim\!\mathcal{A})$ is a tautology, whereas $\mathcal{A} \wedge (\sim\!\mathcal{A})$ is a contradiction.

1.29. Example. Classify each propositional form as a tautology, a contradiction, or neither.
(a) $P \Rightarrow (Q \Rightarrow P)$ (b) $(P \Rightarrow Q) \Rightarrow P$ (c) $(P \oplus Q) \wedge (P \Leftrightarrow Q)$

Solution. In each case, we make a truth table for the given form and see if every output is true (for a tautology) or false (for a contradiction). The truth table for (a) and (b) is shown here:

P	Q	$Q \Rightarrow P$	$P \Rightarrow (Q \Rightarrow P)$	$P \Rightarrow Q$	$(P \Rightarrow Q) \Rightarrow P$
T	T	T	T	T	T
T	F	T	T	F	T
F	T	F	T	T	F
F	F	T	T	T	F

Column 4 is true in every row, so (a) is a tautology. Column 6 has a mixture of Ts and Fs, so (b) is neither a tautology nor a contradiction. For (c), we make the following truth table:

P	Q	$P \oplus Q$	$P \Leftrightarrow Q$	$(P \oplus Q) \wedge (P \Leftrightarrow Q)$
T	T	F	T	F
T	F	T	F	F
F	T	T	F	F
F	F	F	T	F

The final output is always false, so (c) is a contradiction.

Section Summary

1. All of the following phrases mean $P \Rightarrow Q$: if P then Q; P implies Q; Q if P; P only if Q; when P, Q; Q whenever P; P is sufficient for Q; Q is necessary for P; and other variations involving passive voice or changes in word order.

2. $P \Leftrightarrow Q$ ("P iff Q") is true when P and Q are both true or both false; $P \Leftrightarrow Q$ is false when exactly one of P and Q is true. Some useful equivalences involving IFF include:

$$P \Leftrightarrow Q \equiv (P \Rightarrow Q) \wedge (Q \Rightarrow P), \qquad P \Leftrightarrow Q \equiv (P \wedge Q) \vee ((\sim\!P) \wedge (\sim\!Q)),$$

$$P \Leftrightarrow Q \equiv (\sim\!P) \Leftrightarrow (\sim\!Q) \equiv Q \Leftrightarrow P, \qquad \sim\!(P \Leftrightarrow Q) \equiv P \oplus Q.$$

3. "P only if Q" is the converse of "P if Q." "P is necessary for Q" is the converse of "P is sufficient for Q." So, $P \Leftrightarrow Q$ can be translated "P if and only if Q" or "P is necessary and sufficient for Q." Other possible translations include: P iff Q; P precisely when Q; P is equivalent to Q.

4. In logic, the words BUT, ALTHOUGH, and BOTH have the same meaning as AND. "Neither P nor Q" means $(\sim\!P) \wedge (\sim\!Q)$.

5. A *tautology* is a propositional form that has output true in all rows of its truth table. A *contradiction* is a propositional form that has output false in all rows of its truth table. Most propositional forms are neither tautologies nor contradictions.

Exercises

1. Use truth tables to prove parts (b), (c), and (d) of the Theorem on IFF.

2. Which expressions below are logically equivalent to the statement "if P then Q"?

$P \Rightarrow Q$	$Q \Rightarrow P$	$P \Leftrightarrow Q$
P implies Q.	$(\sim P) \Rightarrow (\sim Q)$	$(\sim Q) \Rightarrow (\sim P)$
$(\sim P) \Leftrightarrow (\sim Q)$	$(P \wedge Q) \vee ((\sim P) \wedge (\sim Q))$	$(\sim P) \vee Q$
$\sim(P \oplus Q)$.	P if Q.	Q if P.
P iff Q.	P is sufficient for Q.	P is necessary for Q.
P is nec. and suff. for Q.	$(P \Rightarrow Q) \wedge (Q \Rightarrow P)$	$Q \Leftrightarrow P$

3. Which expressions listed in the previous exercise are logically equivalent to the statement "P if and only if Q"?

4. Use truth tables to classify each propositional form below as a tautology, a contradiction, or neither.
   (a) $P \Leftrightarrow (\sim P)$ (b) $P \Rightarrow (\sim P)$ (c) $(P \oplus Q) \Leftrightarrow ((\sim P) \Leftrightarrow Q)$
   (d) $Q \wedge (\sim(P \Rightarrow Q))$.

5. Is each propositional form a tautology, a contradiction, or neither? Explain.
   (a) $(P \wedge Q) \Rightarrow Q$ (b) $(P \vee Q) \Rightarrow Q$
   (c) $\sim(Q \Rightarrow (P \vee Q))$ (d) $(P \Leftrightarrow Q) \Leftrightarrow (Q \Leftrightarrow P)$
   (e) $(P \Leftrightarrow Q) \wedge (P \Leftrightarrow (\sim Q))$ (f) $(P \Rightarrow Q) \wedge (P \Rightarrow (\sim Q))$.

6. Rewrite each statement, replacing all logical connective words (such as IF, BUT, SUFFICIENT, and so on) with appropriate logical symbols (such as $\Rightarrow$).
   (a) If x_0 is even but prime, then $x_0 = 2$.
      [*Answer.* "(x_0 is even $\wedge$ x_0 is prime)$\Rightarrow x_0 = 2$."]
   (b) f is continuous only if f is bounded.
   (c) x is real whenever x is rational.
   (d) A necessary condition for m to be perfect is that m be even.
   (e) X is compact iff X is both closed and bounded.
   (f) Although $x > 2$ implies $x^2 > 4$, the converse is not true. [Don't use the word "converse" in your answer.]
   (g) When f has compact domain, continuity of f is equivalent to uniform continuity of f.
   (h) Continuity of f is sufficient but not necessary for integrability of f.

7. Write seven different English sentences with the same meaning as "If $\overline{AC}$ and $\overline{BC}$ are equal in length, then $\angle A = \angle B$."

8. Write five different English sentences equivalent to "x_0 is odd iff $x_0 + 3$ is even."

9. Rewrite each statement in the standard form "If P, then Q."
   (a) G cyclic implies G is commutative.
   (b) Normality of X is sufficient for regularity of X.
   (c) Measurability of f is necessary for continuity of f.
   (d) f is integrable only if f is bounded.
   (e) For K to be path-connected, it is necessary that K be connected.
   (f) A sufficient condition for T to be linear is that T be orthogonal.

10. True or false? (It may help to first rewrite complex statements in symbolic form.)
    (a) Columbus discovered America in 1942 iff World War I ended in 1981.
    (b) Mongolia is a vegetable if Paris is a city.
    (c) Mongolia is a vegetable iff Paris is a city.

(d) Snakes are not reptiles whenever $2 + 2 = 4$.

(e) 3 is neither even nor positive.

(f) Grand pianos are heavier than oboes only if cannonballs are lighter than feathers.

(g) For Venus to be a planet, it is necessary that $1 + 1 = 3$.

(h) $1 + 1 = 2$ is a sufficient condition for anvils to float.

(i) Although whales are mammals, they live in the ocean.

(j) January has 32 days only if March has 31 days.

(k) For Hillary Clinton to be 45th President of the U.S., it is sufficient that George Washington was the 1st President of the U.S.

11. For each sentence below, write a logically equivalent sentence using the word NECESSARY.

    (a) If x is rational, then x is real.

    (b) For f to be continuous, it is sufficient that f be differentiable.

    (c) $\det(B) \neq 0$ implies B is invertible.

    (d) G is cyclic only if G is commutative.

    (e) A sufficient condition for T to be isosceles is that T be equilateral.

12. For each sentence below, write a logically equivalent sentence using the phrase ONLY IF.

    (a) Whenever n ends in the digit 8, n is not prime.

    (b) If F is a field, then F is an integral domain.

    (c) A necessary condition for a sequence (a_n) to converge is that (a_n) be bounded.

    (d) Having a zero column is sufficient for a matrix B to be singular.

    (e) K is compact or K is connected.

13. (a) Prove that IFF is transitive by showing that $[(P \Leftrightarrow Q) \wedge (Q \Leftrightarrow R)] \Rightarrow (P \Leftrightarrow R)$ is a tautology. (b) Is the converse of (a) a tautology? Explain.

14. Is IFF associative? In other words, is $(P \Leftrightarrow Q) \Leftrightarrow R$ logically equivalent to $P \Leftrightarrow (Q \Leftrightarrow R)$? Explain.

15. True or false?

    (a) For 57 to be negative, it is necessary and sufficient that George Bush was the 1st President of the United States.

    (b) Peking is the capital of Venezuela only if Boise is the capital of Idaho.

    (c) 3 is odd if and only if the positivity of 7 implies that 9 is even.

    (d) Whenever pigs can fly or fish have gills, it is false that money grows on trees iff Los Angeles is located in California.

    (e) If $\mathbb{R}$ is paracompact, then red is a color.

16. True or false? Explain.

    (a) Every propositional form is either a tautology or a contradiction.

    (b) The negation of $P \Leftrightarrow Q$ is logically equivalent to "P or Q."

    (c) $\sim(P \Rightarrow Q)$ is logically equivalent to $(\sim P) \oplus (\sim Q)$.

    (d) "P is necessary for Q" is logically equivalent to the converse of "P is sufficient for Q."

    (e) "P only if Q" is logically equivalent to the contrapositive of "Q if P."

17. (a) Prove parts (a) and (b) of the Theorem on XOR using the Theorem on IFF and known logical equivalences (not truth tables). (b) Prove the rest of the Theorem on XOR by any convenient technique.

18. Suppose $\mathcal{A}$, $\mathcal{B}$, and $\mathcal{C}$ are propositional forms such that $\mathcal{B} \equiv \mathcal{C}$. (a) Prove the substitution rule $(\mathcal{A} \Leftrightarrow \mathcal{B}) \equiv (\mathcal{A} \Leftrightarrow \mathcal{C})$ using a meta-truth table (see Remark 5 on

page 1.2). (b) Deduce from (a) that $(\mathcal{B} \Leftrightarrow \mathcal{A}) \equiv (\mathcal{C} \Leftrightarrow \mathcal{A})$ without using a truth table.

19. Let $\mathcal{A}$ and $\mathcal{B}$ be any propositional forms. (a) Suppose $\mathcal{A} \equiv \mathcal{B}$. Show that $\mathcal{A} \Leftrightarrow \mathcal{B}$ is a tautology. (b) Now suppose $\mathcal{A} \Leftrightarrow \mathcal{B}$ is a tautology. Show that $\mathcal{A} \equiv \mathcal{B}$.

20. For each propositional form, find a logically equivalent form that only uses the logical symbols $\sim$ and $\vee$. (a) $P \Rightarrow Q$ (b) $\sim(P \wedge Q)$ (c) $P \wedge Q$ (d) $P \oplus Q$.

21. For each propositional form, find a logically equivalent form that only uses the connectives $\sim$ and $\wedge$. (a) $\sim(P \Rightarrow Q)$ (b) $P \Rightarrow Q$ (c) $P \vee Q$ (d) $P \Leftrightarrow Q$ (e) $P \oplus Q$.

22. For each propositional form, find a logically equivalent form that only uses the connectives $\sim$ and $\Rightarrow$. (a) $P \vee Q$ (b) $P \wedge Q$ (c) $P \Leftrightarrow Q$ (d) $P \oplus Q$.

23. Write the converse and the contrapositive of each statement. Give the truth value of the statement, its converse, and its contrapositive.
(a) If milk is a beverage, then chowder is a soup.
(b) If 9 is negative, then 10 is odd.
(c) If $1 + 1 = 5$, then the cosine function is continuous.
(d) Sugar is sweet if lemons are sour.
(e) A necessary condition for pigs to fly is that eagles oink.
(f) For π to be negative, it is sufficient that -1000 be positive.
(g) An argon atom is lighter than a helium atom whenever a sodium atom is heavier than a hydrogen atom.
(h) If 7 is positive iff 8 is odd, then 9 is a square iff 10 is even.
(i) The oddness of 13 implies that $5 > 11$ only if the liquidity of water at room temperature is a sufficient condition for fish to have lungs.

24. Show that any propositional form using only the logical connectives $\sim$ and $\oplus$ cannot be logically equivalent to $P \wedge Q$.

1.4   Tautologies, Quantifiers, and Universes

Elementary logic has two main parts: the logic of propositions and the logic of quantifiers. After a few more examples to round out our treatment of propositional logic, we begin our discussion of quantifiers in the second part of this section. Intuitively, quantifiers allow us to extend the propositional connectives AND and OR to apply to variables that can range over potentially infinite sets of values.

More on Tautologies and Contradictions

Recall that a tautology is a propositional form that is always true, regardless of the truth values of the variables appearing in the form; similarly, a contradiction is a propositional form that is always false. Recall also that two propositional forms are logically equivalent iff they have the same truth value for all possible values of the input variables.

1.30. Example. Show that the following propositional form is a tautology:

$$(P \Leftrightarrow Q) \Rightarrow ((P \wedge R) \Leftrightarrow (Q \wedge R)).$$

Solution. Call the given propositional form $\mathcal{A}$. We make an eight-row truth table, shown here:

P	Q	R	$P \Leftrightarrow Q$	$P \wedge R$	$Q \wedge R$	$(P \wedge R) \Leftrightarrow (Q \wedge R)$	$\mathcal{A}$
T	T	T	T	T	T	T	T
T	T	F	T	F	F	T	T
T	F	T	F	T	F	F	T
T	F	F	F	F	F	T	T
F	T	T	F	F	T	F	T
F	T	F	F	F	F	T	T
F	F	T	T	F	F	T	T
F	F	F	T	F	F	T	T

We fill in the Ts in columns 5 and 6 by looking for two Ts in the appropriate input columns; the remaining entries in these columns are Fs. We fill in columns 4 and 7 by looking for agreement between the two input columns, putting Ts in these rows, and putting Fs elsewhere. Finally, we fill column 8 by looking for places where column 4 is true and column 7 is false. No such places exist, so column 8 is all Ts. This shows that $\mathcal{A}$ is a tautology.

1.31. Example. Let $\mathcal{A}$ be any propositional form, and let $\mathcal{C}$ be any contradiction.
(a) Show that $\mathcal{A} \wedge \mathcal{C}$ is a contradiction.
(b) Show that $\mathcal{A} \vee \mathcal{C} \equiv \mathcal{A}$.
(c) Show that $((\sim\!\mathcal{A}) \Rightarrow \mathcal{C}) \Rightarrow \mathcal{A}$ is a tautology.
Solution. We cannot make the actual, complete truth tables for the propositional forms appearing here, since we do not know which specific propositional variables appear within the forms $\mathcal{A}$ and $\mathcal{C}$. However, we can still make a table showing the possible truth values of various expressions built from $\mathcal{A}$ and $\mathcal{C}$. The key point is to recall that $\mathcal{C}$ *must* be false, whereas (without further information about $\mathcal{A}$), $\mathcal{A}$ might be true or false. So we prepare the following two-row table:

$\mathcal{A}$	$\mathcal{C}$	$\mathcal{A} \wedge \mathcal{C}$	$\mathcal{A} \vee \mathcal{C}$	$\sim\!\mathcal{A}$	$((\sim\!\mathcal{A}) \Rightarrow \mathcal{C})$	$((\sim\!\mathcal{A}) \Rightarrow \mathcal{C}) \Rightarrow \mathcal{A}$
T	F	F	T	F	T	T
F	F	F	F	T	F	T

For any assignment of truth values to the propositional variables appearing within $\mathcal{A}$ and $\mathcal{C}$, $\mathcal{A}$ must be true or false, whereas $\mathcal{C}$ must be false. Column 3 therefore proves that $\mathcal{A} \wedge \mathcal{C}$ must also be false in all situations, so this form is a contradiction. Comparison of columns 1 and 4 shows that $\mathcal{A} \vee \mathcal{C}$ always has the same truth value as $\mathcal{A}$, hence these two forms are logically equivalent. Finally, the last three columns show that the form in (c) is always true, so this form is a tautology. We return to this tautology later, using it to justify a method of proof called proof by contradiction.

1.32. Remark. By arguments similar to those in the last example, you can establish the following facts. Let $\mathcal{A}$ and $\mathcal{B}$ be any propositional forms, and let $\mathcal{T}$ be any tautology.
(a) $\mathcal{A} \wedge \mathcal{T} \equiv \mathcal{A}$.
(b) $\mathcal{A} \vee \mathcal{T}$ is a tautology.
(c) $\mathcal{A}$ is a tautology iff $\sim\mathcal{A}$ is a contradiction.
(d) $\mathcal{A} \equiv \mathcal{B}$ iff $\mathcal{A} \Leftrightarrow \mathcal{B}$ is a tautology.
Part (d) lets us harvest many tautologies from the logical equivalences proved earlier. For instance, the logical equivalence $P \wedge Q \equiv Q \wedge P$ leads to the tautology $(P \wedge Q) \Leftrightarrow (Q \wedge P)$.

Variables and Quantifiers

Quantified logic studies the truth or falsehood of statements containing variables such as x and y. To see how this differs from propositional logic, consider the following example.

1.33. Example. True or false?
(a) For all real numbers x, $x^2 = x$.
(b) There is a real number x satisfying $x^2 = x$.
(c) $x^2 = x$ (where x is a *variable* varying over all real numbers).
(d) $x_0^2 = x_0$ (where x_0 is a *constant*, designating one particular real number).
Intuitively, (a) is false, since the real number $x = 2$ does not satisfy $x^2 = x$ (as $2^2 = 4 \neq 2$). Similarly, (b) is true, since $x = 1$ satisfies $1^2 = 1$ (note $x = 0$ works also). Statement (c) is a bit problematic: if the variable x has not been assigned a particular value, but is varying through all real numbers, then (c) has no truth value (it is not a proposition). Put another way, (c) is true for some values of the variable x and false for other values of x, so statement (c) has no definite truth value of its own. In contrast, the constant x_0 in (d) represents one, particular, fixed real number. So $x_0^2 = x_0$ *is* a proposition, although we cannot tell without more information about x_0 whether this proposition is true or false.

In general, a statement containing a variable x, such as statement (c) above, is called an *open sentence* and is denoted by a symbol such as $P(x)$, $Q(x)$, etc. Open sentences can involve more than one variable; for instance, we could let $R(x, y)$ be the formula $x^2 + y^2 = 25$. An open sentence does not have a truth value; it is not a proposition. One way to turn an open sentence into a proposition is to assign specific values to every variable appearing in it. In our example, $R(5, 0)$ is the true proposition $5^2 + 0^2 = 25$; $R(2, 4)$ is the false proposition $2^2 + 4^2 = 25$; $R(3, -4)$ is true; $R(7, 10)$ is false; and so on. $R(3, z)$ is the open sentence $3^2 + z^2 = 25$ involving the new variable z. We can also substitute letters that represent constants into an open sentence. In this text, the letters a, b, c and letters with subscripts (like x_0) denote constants. Thus $R(x_0, b)$ is the proposition $x_0^2 + b^2 = 25$, but the truth value of this proposition cannot be determined without more information about x_0 and b.

There are two other ways to turn an open sentence $P(x)$ into a proposition, which are illustrated in (a) and (b) above: we may *quantify* the variable x with a phrase such as "for all x" or "there exists an x." We use the *universal quantifier symbol* $\forall$ to abbreviate "for all," and the *existential quantifier symbol* $\exists$ to abbreviate "there exists." These symbols are explained further in the next two definitions.

1.34. Definition of $\forall$ (Universal Quantifier). Let $P(x)$ be an open sentence involving the variable x. The statement "$\forall x, P(x)$" is a proposition that is true iff for all objects x_0, $P(x_0)$ is true. The quantifier $\forall x$ may be translated "for all x," "for every x," "for each x," "for any x," among other possibilities. The symbol $\forall$ looks like an upside-down capital A, reminding us of the word ALL.

1.35. Definition of $\exists$ (Existential Quantifier). Let $P(x)$ be an open sentence involving the variable x. The statement "$\exists x, P(x)$" is a proposition that is true iff there is at least one object x_0 making $P(x_0)$ true. The quantifier $\exists x$ may be translated "there is an x," "there exists an x," "there is at least one x," "for some x," among other possibilities. The symbol $\exists$ looks like a backwards capital E, reminding us of the word EXISTS.

Universes; Restricted Quantifiers

In most discussions, we do not want to be making statements about literally *all* objects x. Instead, we want to restrict attention to a particular set of objects, such as the set of real numbers, or the set of integers, or the set of continuous real-valued functions. A set U of objects that contains all objects we want to discuss in a particular situation is called a *universe* or *universal set*. We write $x \in U$ to mean that x is a member of the universal set U, and we write $x \notin U$ to mean that x is not in U. (This notation comes from set theory, which we study in Chapter 3.) By combining this notation with quantifiers, we obtain the restricted quantifiers defined next.

1.36. Definition: Restricted Quantifiers. Let U be any set, and let $P(x)$ be an open sentence involving x. The statement "$\forall x \in U, P(x)$" means that for every object x_0 in U, $P(x_0)$ is true. The statement "$\exists x \in U, P(x)$" means there exists at least one object x_0 in U for which $P(x_0)$ is true. The symbols "$\forall x \in U$" and "$\exists x \in U$" are *restricted* quantifiers; in contrast, the previous symbols "$\forall x$" and "$\exists x$" are *unrestricted* quantifiers.

Next we introduce some notation for certain universes of numbers that arise frequently in mathematics.

1.37. Definition: Notation for Number Systems.

(a) $\mathbb{Z}$ is the set of all *integers*, namely $\mathbb{Z} = \{\ldots, -3, -2, -1, 0, 1, 2, 3, \ldots\}$.

(b) $\mathbb{Q}$ is the set of all *rational numbers*, which are ratios a/b where a is an integer and b is a nonzero integer.

(c) $\mathbb{R}$ is the set of all *real numbers*.

(d) $\mathbb{C}$ is the set of all *complex numbers*, which are expressions $a + bi$ with a, b real, and where i satisfies $i^2 = -1$.

(e) We write $\mathbb{Z}_{\geq 0} = \{0, 1, 2, 3, \ldots\}$ for the set of nonnegative integers, and $\mathbb{Z}_{>0} = \{1, 2, 3, \ldots\}$ for the set of positive integers. Some authors use $\mathbb{N}$ (the set of *natural numbers*) to denote the set $\mathbb{Z}_{\geq 0}$, whereas other authors use $\mathbb{N}$ to denote the set $\mathbb{Z}_{>0}$. To avoid confusion, we do not use the symbol $\mathbb{N}$ except in §8.3.

(f) By analogy with (e), let $\mathbb{R}_{>0}$ be the set of strictly positive real numbers (not including zero). For a fixed integer b, let $\mathbb{Z}_{\geq b}$ be the set of integers that are at least b, and let $\mathbb{Z}_{>b}$ be the set of integers strictly greater than b. We similarly define $\mathbb{Z}_{\neq b}$, $\mathbb{R}_{>x_0}$, $\mathbb{R}_{\leq x_0}$, and so on.

(g) Given n distinct objects $x_1, x_2, \ldots, x_n$, let $U = \{x_1, x_2, \ldots, x_n\}$ denote the finite universal set consisting of these n objects.

This definition is merely setting up notation. In Chapter 8, we give more precise definitions as part of a formal development of the number systems $\mathbb{Z}$, $\mathbb{Q}$, $\mathbb{R}$, and $\mathbb{C}$.

1.38. Example. True or false?

(a) $\exists x \in \mathbb{Z}, x + 3 = 2$.

(b) $\exists x \in \mathbb{Z}_{\geq 0}, x + 3 = 2$.

(c) $\exists x \in \mathbb{R}, x^2 = -1$.

(d) $\exists x \in \mathbb{C}, x^2 = -1$.

(e) $\forall x \in \mathbb{Z}_{>0}, x + 5 > 3$.

(f) $\forall x \in \mathbb{Q}, x + 5 > 3$.

(g) $\forall x \in \mathbb{R}, x + 5 > 3$.

(h) $\forall x \in \mathbb{Z}, x^3 = x$.

(i) $\exists x \in \mathbb{Z}, x^3 = x$.

(j) $\forall x \in \{-1, 0, 1\}, x^3 = x$.

Solution. First, (a) is true, since $x = -1$ is in $\mathbb{Z}$ and satisfies $x + 3 = 2$. In fact, $x = -1$ is the only real solution of this equation. So (b) is false, since -1 is not in the universe $\mathbb{Z}_{\geq 0}$. Similarly, (c) is false, since the square of any real number is at least 0, so cannot equal -1. But (d) is true, since $x = i$ is a complex number making the equation true ($x = -i$ also works). Adding 5 to any positive integer produces an integer 6 or greater, which is also greater than 3, so (e) is true. But (f) is false, since $x + 5 > 3$ fails to hold for the rational number $x = -11/2$ (among many other examples). Similarly, (g) is false. Next, (h) is false since $2 \in \mathbb{Z}$ and $2^3 = 2$ is not true. But (i) is true, since 1 is an integer such that $1^3 = 1$. Also (j) is true, since each of the three objects b in the finite universe $\{-1, 0, 1\}$ does satisfy $b^3 = b$. These examples show that *changing to a different universe can affect the truth of a statement with a restricted quantifier.* The type of quantifier ($\exists$ or $\forall$) certainly makes a big difference as well. We also remark that it would not make sense to use unrestricted quantifiers in statements (a) through (j). For, if x is an arbitrary object (not necessarily a number), it is not clear what statements such as $x + 3 = 2$ and $x^3 = x$ should mean.

Quantifiers vs. Logical Operators

The quantifiers $\forall$ and $\exists$ can be viewed as generalizations of the logical operators $\wedge$ (AND) and $\vee$ (inclusive-OR), respectively. To see why, consider a finite universe $U = \{z_1, z_2, \ldots, z_n\}$. The universal statement "$\forall x \in U, P(x)$" is true iff $P(a)$ is true for each fixed object $a \in \{z_1, \ldots, z_n\}$. In other words, "$\forall x \in U, P(x)$" is true iff $P(z_1)$ is true and $P(z_2)$ is true and ... and $P(z_n)$ is true. Similarly, "$\exists x \in U, P(x)$" is true iff there exists an object $b \in \{z_1, \ldots, z_n\}$ making $P(b)$ true, which holds iff $P(z_1)$ is true or $P(z_2)$ is true or ... or $P(z_n)$ is true. In summary,

$$\boxed{\forall x \in \{z_1, z_2, \ldots, z_n\}, P(x)} \quad \Leftrightarrow \quad \boxed{P(z_1) \wedge P(z_2) \wedge \cdots \wedge P(z_n)};$$

$$\boxed{\exists x \in \{z_1, z_2, \ldots, z_n\}, P(x)} \quad \Leftrightarrow \quad \boxed{P(z_1) \vee P(z_2) \vee \cdots \vee P(z_n)}.$$

In the general case of an infinite universe U, "$\forall x \in U, P(x)$" is a statement that (informally speaking) AND's together the infinitely many statements $P(a)$ as a ranges through all objects in U. Similarly, "$\exists x \in U, P(x)$" is a statement that OR's together all the statements $P(b)$ for b ranging through U. This analogy linking $\forall$ with $\wedge$ and $\exists$ with $\vee$ can help us understand and remember some of the quantifier properties presented later (such as the negation rules in the next section).

Section Summary

1. Given any tautology $\mathcal{T}$, any contradiction $\mathcal{C}$, and any propositional form $\mathcal{A}$, $\mathcal{A} \wedge \mathcal{T} \equiv \mathcal{A} \equiv \mathcal{A} \vee \mathcal{C}$; $\mathcal{A} \vee \mathcal{T}$ is a tautology; $\mathcal{A} \wedge \mathcal{C}$ is a contradiction. For any forms $\mathcal{A}$ and $\mathcal{B}$, $\mathcal{A} \equiv \mathcal{B}$ iff $\mathcal{A} \Leftrightarrow \mathcal{B}$ is a tautology.

2. The universal quantifier $\forall$ and the existential quantifier $\exists$ have the following meaning.

Symbol	Meaning
$\forall x, P(x)$	For all objects x_0, $P(x_0)$ is true.
$\forall x \in U, P(x)$	For all objects x_0 in U, $P(x_0)$ is true.
$\exists x, P(x)$	There exists an object x_0 such that $P(x_0)$ is true.
$\exists x \in U, P(x)$	There exists an object x_0 in U such that $P(x_0)$ is true.

Words such as "all," "any," "each," and "every" signal the use of $\forall$; phrases such as "there is," "there exists," "for some," and "at least one" signal the use of $\exists$. Common choices of the universal set U include $\mathbb{Z}$ (integers), $\mathbb{Q}$ (rational numbers), $\mathbb{R}$ (real numbers), $\mathbb{Z}_{\geq 0}$ (nonnegative integers), $\mathbb{Z}_{> 0}$ (positive integers), and finite sets.

3. For finite universes, quantifiers can be replaced by propositional logic operators as follows:

$$\forall x \in \{z_1, z_2, \ldots, z_n\}, P(x) \quad \Leftrightarrow \quad P(z_1) \wedge P(z_2) \wedge \cdots \wedge P(z_n);$$
$$\exists x \in \{z_1, z_2, \ldots, z_n\}, P(x) \quad \Leftrightarrow \quad P(z_1) \vee P(z_2) \vee \cdots \vee P(z_n).$$

Exercises

1. Is each propositional form a tautology, a contradiction, or neither? Explain.
   (a) $(P \Rightarrow (Q \Rightarrow R)) \wedge (P \Rightarrow Q) \wedge (P \wedge (\sim R))$.
   (b) $((P \vee R) \Leftrightarrow (Q \vee R)) \Rightarrow (P \Leftrightarrow Q)$.
   (c) $(P \Leftrightarrow Q) \Rightarrow ((R \Rightarrow P) \Leftrightarrow (R \Rightarrow Q))$.

2. Is each propositional form a tautology, a contradiction, or neither? Explain.
   (a) $[(P \vee Q) \wedge (P \Rightarrow R) \wedge (Q \Rightarrow R)] \Rightarrow R$.
   (b) $[(P \Leftrightarrow Q) \Leftrightarrow R] \Leftrightarrow [P \Leftrightarrow (Q \Leftrightarrow R)]$.
   (c) $[P \Rightarrow (Q \Rightarrow R)] \Rightarrow [(P \Rightarrow Q) \Rightarrow (P \Rightarrow R)]$.

3. Prove Remark 1.32.

4. Prove: for any propositional forms $\mathcal{A}$ and $\mathcal{B}$, if $\mathcal{A} \Leftrightarrow \mathcal{B}$ is a tautology, then $\mathcal{A} \Rightarrow \mathcal{B}$ is a tautology. Is the converse always true?

5. Let $\mathcal{T}$ be a tautology, $\mathcal{C}$ be a contradiction, and $\mathcal{A}$ be any propositional form. Prove: (a) $\mathcal{A} \Rightarrow \mathcal{T}$ is a tautology. (b) $\mathcal{A} \oplus \mathcal{C} \equiv \mathcal{A}$. (c) $\mathcal{A} \oplus \mathcal{T} \equiv (\sim \mathcal{A})$. (d) $\mathcal{T} \Leftrightarrow \mathcal{C}$ is a contradiction.

6. Rewrite each sentence using symbols, not words. (a) The square of any real number is nonnegative. (b) There is an integer whose cube is two. (c) Every positive rational number has the form m/n for some positive integers m and n.

7. Rewrite each statement using words, not symbols.
   (a) $\forall x \in \mathbb{Q}, x^2 \neq 7$. (b) $\exists y \in \mathbb{R}, \forall v \in \mathbb{R}, yv = v$. (c) $\exists a \in \mathbb{Z}, \exists b \in \mathbb{Z}, 25 = a^2 + b^2$.

8. True or false? (a) $\exists x \in \mathbb{Z}, x^2 = 2$. (b) $\exists x \in \mathbb{R}, x^2 = 2$. (c) $\forall x \in \mathbb{Q}, x^2 > 0$.
   (d) $\forall x \in \mathbb{Z}, x^2 \geq x$. (e) $\forall x \in \mathbb{R}, x^2 \geq x$. (f) $\exists x \in \mathbb{Z}, x^2 - 3x - 10 = 0$.

9. True or false? Explain.
   (a) $\forall x \in \{3, 5, 7\}, x$ is an odd integer.
   (b) $\forall x, x \in \{3, 5, 7\} \Rightarrow x$ is an odd integer.
   (c) $\forall x, x \in \{3, 5, 7\} \wedge x$ is an odd integer.
   (d) $\exists x \in \{3, 5, 7\}, x$ is an even integer.

(e) $\exists x, x \in \{3, 5, 7\} \wedge x$ is an even integer.

(f) $\exists x, x \in \{3, 5, 7\} \Rightarrow x$ is an even integer.

10. True or false? Explain.

(a) $\forall x \in \mathbb{Z}, 4x = 7$. (b) $\forall x, x \in \mathbb{Z} \Rightarrow 4x = 7$. (c) $\forall x, x \in \mathbb{Z} \wedge 4x = 7$.

(d) $\exists x \in \mathbb{Z}, 4x = 7$. (e) $\exists x, x \in \mathbb{Z} \wedge 4x = 7$. (f) $\exists x, x \in \mathbb{Z} \Rightarrow 4x = 7$.

11. For which universes U is the statement $\forall x \in U, \exists y \in U, x + y = 0$ true?

(a) $\mathbb{Z}_{>0}$ (b) $\mathbb{Z}$ (c) $\mathbb{Q}$ (d) $\mathbb{R}$ (e) $\{0\}$ (f) $\{-1, 0, 1\}$

12. For which universes U is the statement $\forall x \in U, \exists y \in U, xy = 1$ true?

(a) $\mathbb{Z}_{>0}$ (b) $\mathbb{Q}_{>0}$ (c) $\mathbb{R}_{>0}$ (d) $\mathbb{Z}$ (e) $\mathbb{R}$ (f) $\{-1\}$

13. (a) Rewrite the statement $\forall x \in \{-1, 0, 1\}, x^3 = x$ without using quantifiers. (b) Rewrite the statement $\exists x \in \{1, 2, 3\}, x^2 = 2x$ without using quantifiers. (c) Rewrite the statement "11 is prime and 13 is prime and 17 is prime and 19 is prime" using a quantifier instead of the word AND.

14. Let x_0 be a fixed real number, and let x be a variable ranging through real numbers. (a) Explain the logical difference between these two sentences: I. "if $x_0 > 0$, then $x_0^2 \geq x_0$." II. "if $x > 0$, then $x^2 \geq x$." (b) One of the sentences in (a) is not a proposition. Use quantifiers to turn this sentence into a proposition in two different ways, and state whether these propositions are true or false. (c) Do the truth values of the propositions in (b) change if we restrict x to vary through $\mathbb{Z}$ instead of $\mathbb{R}$?

15. (a) Express the statement $\forall x, P(x)$ using only $\exists$ and propositional logic operators. (b) Express the statement $\exists x, Q(x)$ using only $\forall$ and propositional logic operators.

16. Suppose U and V are sets such that every member of U is also a member of V. Which statements below must be true for every open sentence $P(x)$? For those that are true, explain why. For those that are false, illustrate with a specific choice of U, V, and $P(x)$.

(a) $\forall x \in U, P(x)$ implies $\forall x \in V, P(x)$.

(b) $\forall x \in V, P(x)$ implies $\forall x \in U, P(x)$.

(c) $\exists x \in U, P(x)$ implies $\exists x \in V, P(x)$.

(d) $\exists x \in V, P(x)$ implies $\exists x \in U, P(x)$.

17. We know $Q \vee (\sim Q)$ is a tautology. Which statements below must be true for all choices of U and $P(x)$? Explain.

(a) $\forall x \in U, [P(x) \vee (\sim P(x))]$.

(b) $(\forall x \in U, P(x)) \vee (\forall x \in U, \sim P(x))$.

(c) $(\forall x \in U, P(x)) \vee \sim (\forall x \in U, P(x))$.

18. We know $Q \wedge (\sim Q)$ is a contradiction. Which statements below must be false for all choices of U and $P(x)$? Explain.

(a) $\exists x \in U, [P(x) \wedge (\sim P(x))]$.

(b) $(\exists x \in U, P(x)) \wedge (\exists x \in U, \sim P(x))$.

(c) $(\exists x \in U, P(x)) \wedge \sim (\exists x \in U, P(x))$.

19. True or false? Explain each answer.

(a) For all propositional forms $\mathcal{A}$ and all contradictions $\mathcal{C}$, $\mathcal{C} \Rightarrow \mathcal{A}$ is a tautology.

(b) For all propositional forms $\mathcal{A}$ and $\mathcal{B}$, $\mathcal{A} \Rightarrow \mathcal{B}$ is not logically equivalent to $\mathcal{B} \Rightarrow \mathcal{A}$.

(c) Any two contradictions are logically equivalent.

(d) For all logically equivalent forms $\mathcal{A}$ and $\mathcal{B}$, $(\mathcal{A} \Rightarrow \mathcal{B}) \equiv (\mathcal{A} \Leftrightarrow \mathcal{B})$.

20. Show that any propositional form using only the logical connectives $\wedge$ and $\vee$ is neither a tautology nor a contradiction.

1.5   Quantifier Properties and Useful Denials

This section develops more properties of the existential quantifier $\exists$ and the universal quantifier $\forall$. We look at rules for converting restricted quantifiers to unrestricted quantifiers, translation examples illustrating how English phrases encode quantifiers, and rules for transforming negated quantified statements. We then discuss how to find useful denials of complex statements, which is a crucial skill needed to analyze and create proofs.

Conversion Rules for Restricted Quantifiers

We start with some rules for eliminating restricted quantifiers. Let U be any set, and consider the statement $\exists x \in U, P(x)$. This statement is true iff there is an object x_0 *in the set U* for which $P(x_0)$ is true. Intuitively, this condition holds iff there is an object x_0 (not initially required to be in U) for which "$x_0 \in U$ and $P(x_0)$" is true. Therefore, we obtain the rule

$$\boxed{\exists x \in U, P(x)} \qquad \text{iff} \qquad \boxed{\exists x, (x \in U \wedge P(x))}$$

for converting a restricted existential quantifier to an unrestricted existential quantifier. The next example shows that this rule *does not work* if we blindly replace both $\exists$ symbols by $\forall$ symbols.

1.39. Example. Let U be the set of polar bears, and let $P(x)$ be the open sentence "x is white." The two statements $\exists x \in U, P(x)$ and $\exists x, (x \in U \wedge P(x))$ both assert that a white polar bear exists. Similarly, $\forall x \in U, P(x)$ says that all polar bears are white, which (ignoring evolutionary anomalies) is a true proposition. But $\forall x, (x \in U \wedge P(x))$ means something entirely different: this statement says that *every* object x is a white polar bear!

   We can fix the example by replacing $\wedge$ (AND) by $\Rightarrow$ (IMPLIES). Consider the statement $\forall x, (x \in U \Rightarrow P(x))$. This says that for all objects x_0, IF x_0 is a polar bear, THEN x_0 is white. Imagine testing the truth of this statement by considering each object x_0 in the whole universe, one at a time. Some objects x_0 are polar bears, and for these objects the IF-statement is true (having a true hypothesis and true conclusion). All remaining objects x_0 are not polar bears, and for such objects the IF-statement is automatically true (since its hypothesis is false). Thus these objects are essentially irrelevant when determining the truth of the quantified statement.

   The rule suggested by the example holds in general:

$$\boxed{\forall x \in U, P(x)} \qquad \text{iff} \qquad \boxed{\forall x, (x \in U \Rightarrow P(x))}.$$

Intuitively, if the left side is true, then the implication on the right side must be true for each fixed object x_0 satisfying the hypothesis $x_0 \in U$. But the implication is always true for all other objects x_0 (those making the hypothesis $x_0 \in U$ false), so the right side is true. Similarly, if the right side is true, then applying the right side to each object $x_0 \in U$ shows that $P(x_0)$ is true for all such objects x_0. So the left side is also true. Later, after discussing proof methods, we will see that this paragraph is essentially a proof of the conversion rule stated above. We also remark that a major reason for defining the IF truth table as we did was to make sure that this quantifier rule would work (compare to our discussion of the statement $\forall x \in \mathbb{R}, x > 2 \Rightarrow x^2 > 4$ on page 11; this statement is equivalent to $\forall x \in \mathbb{R}_{>2}, x^2 > 4$).

Translation Examples

Here are some examples illustrating the process of translating back and forth between English statements and logical symbolism. We use these universes and open sentences: U is the set of roses; V is the set of violets; C is the set of carrots; $R(x)$ means x is red; $P(x)$ means x is purple; $O(x)$ means x is orange; and $B(x)$ means x is beautiful. In the first few examples, we encode English statements, first using restricted quantifiers, and then converting to unrestricted quantifiers.

(a) "All violets are purple." The word "all" signifies a universal statement, so we write $\forall x \in V, P(x)$. Eliminating the restricted quantifier gives $\forall x, (x \in V \Rightarrow P(x))$.

(b) "Some roses are orange." Here, "some" indicates an existential statement, so we get $\exists x \in U, O(x)$, which in turn becomes $\exists x, (x \in U \wedge O(x))$. Note that the logical statement would be true even if there were only one orange rose, despite the use of the plural "roses" in the English statement.

(c) "Every carrot is orange or purple." We encode this as $\forall x \in C, (O(x) \vee P(x))$ or equivalently $\forall x, (x \in C \Rightarrow (O(x) \vee P(x)))$.

(d) "There is a purple and orange carrot." We encode this as $\exists x \in C, (P(x) \wedge O(x))$ or equivalently $\exists x, (x \in C \wedge P(x) \wedge O(x))$. On the other hand, "Purple carrots exist, and orange carrots exist" is a different statement, which could be encoded as

$$(\exists x, (x \in C \wedge P(x))) \wedge (\exists y, (y \in C \wedge O(y))).$$

(e) "Any red rose is beautiful." We have not introduced a universe of red roses, but we can use unrestricted quantifiers to write $\forall x, (R(x) \wedge x \in U) \Rightarrow B(x)$. We could also say $\forall x \in U, (R(x) \Rightarrow B(x))$.

(f) "If purple carrots exist, then each white rose is beautiful." In this sentence, two quantifiers appear within an IF-THEN construction. One possible encoding is

$$[\exists x, (x \in C \wedge P(x))] \Rightarrow [\forall y, ((W(y) \wedge y \in U) \Rightarrow B(y))].$$

Note that adjectives modifying a noun produce $\wedge$ symbols in the encoding, whether the noun is existentially or universally quantified.

(g) "$x + y = y + x$ for all real numbers x and y." Although English grammar allows a quantifier to appear after the variable it quantifies, formal logic requires the quantifier to come first. Thus we write $\forall x \in \mathbb{R}, \forall y \in \mathbb{R}, x + y = y + x$. Note that the word "and" in this sentence does not translate into the logical operator $\wedge$, but allows the single phrase "for all" to stand for two universal quantifiers.

(h) "$n = 2k$ for some integer k." Here too, the quantifier at the end of the English sentence must be moved to the front in the logical version, giving $\exists k \in \mathbb{Z}, n = 2k$.

(i) "Roses are red, carrots are orange." This poetic sentiment reveals a new and unpleasant feature of English: *hidden* quantifiers and logical operators. The use of the plural "roses" and "carrots" implicitly indicates that these statements are intended to apply to *all* roses and *all* carrots. Moreover, the comma joining the two phrases has the same meaning as AND. Thus we write $(\forall x, (x \in U \Rightarrow R(x)) \wedge (\forall y, (y \in C \Rightarrow O(y))$. In this case, we could get away with a single quantifier for both clauses, writing

$$\forall z, ((z \in U \Rightarrow R(z)) \wedge (z \in C \Rightarrow O(z))).$$

But if $\wedge$ had been $\vee$ in both statements, the two encodings above would *not* be equivalent. The first encoding now says "All roses are red OR all carrots are orange," which is false (as white roses and purple carrots do exist!). But the second encoding (upon eliminating IF) states that every object is not a rose or is red or is not a carrot or is orange, which is true.

(j) "Each even integer greater than 2 is the sum of two primes." This sentence (called *Goldbach's Conjecture*) contains two hidden existential quantifiers signaled by the verb "is." We can encode it as follows:

$$\forall x \in \mathbb{Z}, (x \text{ is even} \wedge x > 2) \Rightarrow [\exists y \in \mathbb{Z}, \exists z \in \mathbb{Z}, (y \text{ is prime} \wedge z \text{ is prime} \wedge x = y + z)].$$

As this example illustrates, we often allow ourselves to use restricted quantifiers involving a "standard" universe (like $\mathbb{Z}$), but we apply the conversion rules to move the non-standard restrictions on the variables (like being even, or being prime) into the propositional part of the statement, rather than inventing new universes for the set of even integers larger than 2 or the set of prime integers.

Negating Quantified Statements

We continue our translation examples, letting U be the universe of roses and letting $R(x)$ mean "x is red." The next family of examples reveals the subtleties that arise when words with a negating effect are mixed with quantifiers.

1.40. Example. Encode each sentence as literally as possible, using only unrestricted quantifiers.
(a) All roses are not red.
(b) Not all roses are red.
(c) No roses are red.
(d) There does not exist a red rose.
(e) Red roses do not exist.
(f) Some roses are not red.
(g) There is a non-red rose.

Solution. (a) We initially say $\forall x \in U, \sim R(x)$, and convert this to $\forall x, (x \in U \Rightarrow \sim R(x))$. (b) In this example, the word "not" applies to everything following it, so we get $\sim \forall x \in U, R(x)$, which converts to $\sim \forall x, (x \in U \Rightarrow R(x))$. When we say "no roses" in part (c), we are making a negative statement about all roses, leading first to $\forall x \in U, \sim R(x)$ and then to $\forall x, (x \in U \Rightarrow \sim R(x))$. (d) becomes $\sim \exists x, (x \in U \wedge R(x))$; (e) is encoded in exactly the same way, even though the word "exist" appears at the end. (f) says $\exists x, (x \in U \wedge \sim R(x))$, and (g) is encoded in the same way. Intuitively, each of the four statements (a), (c), (d), and (e) is saying the same thing. Similarly, statements (b), (f), and (g) say the same thing but are not the same as the other four statements. Using restricted quantifiers now, these observations mean that $\forall x \in U, \sim R(x)$ is equivalent to $\sim \exists x \in U, R(x)$, whereas $\exists x \in U, \sim R(x)$ is equivalent to $\sim \forall x \in U, R(x)$.

The equivalences at the end of the last example hold in general. In other words, for any universe U and any open sentence $P(x)$, we have the *quantifier negation rules*:

$$\boxed{\sim \exists x \in U, P(x)} \qquad \text{iff} \qquad \boxed{\forall x \in U, \sim P(x)};$$

$$\boxed{\sim \forall x \in U, P(x)} \qquad \text{iff} \qquad \boxed{\exists x \in U, \sim P(x)}.$$

We justify the first rule intuitively, as follows. The left side of this rule is the negation of the statement "There is an x_0 in U making $P(x_0)$ true." How can this existential statement fail to hold? The statement fails precisely when every single object x_0 in U fails to make $P(x_0)$ true. This is exactly the universal statement on the right side of the rule.

Similarly, the left side of the second rule asserts the falsehood of the statement, "For every x_0 in U, $P(x_0)$ is true." When does this universal statement fail? It fails when, and only when, at least one object x_0 in U fails to make $P(x_0)$ true. This is exactly the existential statement on the right side of the second rule.

Negations and Denials

Given any proposition P, recall that the *negation* of P is the proposition $\sim P$, which has the opposite truth value as P. We say that a proposition Q is a *denial* of P iff Q is logically equivalent to $\sim P$. In general, if P is a complicated statement built from logical operators and quantifiers, then the negation $\sim P$ can be difficult to work with directly. Thus we need to develop techniques for passing from the particular denial $\sim P$ of P to another denial that does not begin with the negation symbol. It turns out that the most useful denials of P are those in which the negation symbol is not applied to any substatement involving a logical operator or quantifier symbol.

How can we find a *useful denial* of P? In this section and previous ones, we have already derived the rules we need to simplify expressions that begin with NOT. These rules are summarized in the following table. The tricky aspect of this table is that we usually need to apply several rules recursively to convert the negation $\sim P$ into the most useful denial in which no subexpression begins with NOT.

Statement	Denial of Statement	Symbolic Version of Rule
A and B	(denial of A) or (denial of B)	$\sim(A \wedge B) \equiv (\sim A) \vee (\sim B)$
A or B	(denial of A) and (denial of B)	$\sim(A \vee B) \equiv (\sim A) \wedge (\sim B)$
if A, then B	A and (denial of B)	$\sim(A \Rightarrow B) \equiv A \wedge (\sim B)$
not A	A	$\sim(\sim A) \equiv A$
For all x, $P(x)$	There is x, (denial of $P(x)$)	$\sim \forall x \in U, P(x)$ iff $\exists x \in U, \sim P(x)$
There is x, $P(x)$	For all x, (denial of $P(x)$)	$\sim \exists x \in U, P(x)$ iff $\forall x \in U, \sim P(x)$
A iff B	A or B, not both	$\sim(A \Leftrightarrow B) \equiv (A \oplus B)$
A or B, not both	A iff B	$\sim(A \oplus B) \equiv (A \Leftrightarrow B)$

Informally, applying a negation operator converts AND to OR, OR to AND, FOR ALL to EXISTS, EXISTS to FOR ALL, and in these cases we must continue recursively, denying the inputs to these operators. On the other hand, negation converts IFF to XOR and XOR to IFF without needing to negate the inputs (see also Exercises 17 and 18). Applying a negation to something that already begins with NOT removes the NOT. Finally, the *most troublesome* entry in the table is the denial rule for IF. For emphasis, we repeat the key fact here:

**The denial of an IF-statement is an AND-statement,
not another IF-statement!**

One way to understand this is to recall that "if A then B" is equivalent to "(not A) or B." Negating the latter expression produces "A and (not B)," as asserted in the table.

1.41. Example. Find useful denials of each statement.
(a) 3 is odd or 5 is not prime.
(b) If $a < b < c$, then $a < c$.
(c) $a < c$ only if $a < b < c$.
(d) Not everyone likes math.
Solution. A denial of (a) is "3 is not odd and 5 is prime." Denying the IF-statement in (b) produces "$a < b < c$ and $a \geq c$" (assuming a, b, c are real constants). Statement (c) contains two traps. First, the "only if" construction in (c) is equivalent (before negating) to the

statement "If $a < c$, then $a < b < c$." Second, the phrase $a < b < c$ is really an abbreviation for "$a < b$ and $b < c$." Thus, a denial of (c) is "$a < c$ and ($a \geq b$ or $b \geq c$)." It would be incorrect to replace the parenthesized expression by $a \geq b \geq c$, which still contains an implicit AND. Note also that (for real numbers) "not ($a < b$)" becomes $a \geq b$, as opposed to $a > b$.

Finally, the *negation* of (d) is "It is false that not everyone likes math." A *useful denial* of (d) is "Everyone likes math." (Astute readers may have noticed that (d) and its denials are not really propositions.)

Rant on Hidden Quantifiers (Optional)

We have now seen several examples of *hidden* or *implicit* quantifiers in English sentences, in which universal or existential quantifiers are suggested by grammatical constructions (like plural subjects) rather than explicitly designated by phrases like "for all" or "there exists." My personal position is that *all mathematical writers should avoid the use of all hidden quantifiers in all circumstances.* The reason is that one of the most common sources of error and confusion among beginners (and experts!) in this subject is forgetting to quantify a variable, confusing universal quantifiers with existential quantifiers, changing the meaning of a statement by moving a quantifier to a new location, or misinterpreting the meaning or scope of a quantifier. The correct use of quantifiers is one of the keys to success in reading and writing proofs, and it is one of the hardest skills for newcomers to master. Why, then, should we make it even harder by gratuitously omitting or disguising the quantifiers used in our own writing? I have tried very hard in this text (and elsewhere) to quantify every variable appearing in all proofs and other formal discussions, and I implore you to exert similar efforts in your own writing.

Nevertheless, you must be aware that the majority of mathematical authors do not share my view and use hidden quantifiers all over the place. Here is a particularly common convention that you must know: *in all theorem statements, any unquantified variables are understood to be universally quantified.* The universe of objects often must be inferred from context as well. For example, an algebra text might announce as a theorem the commutative law for addition: $x + y = y + x$. Filling in quantifiers, the actual theorem being presented is $\forall x \in \mathbb{R}, \forall y \in \mathbb{R}, x + y = y + x$. However, this convention must be used with care, keeping in mind all the various grammatical constructions that can serve the purpose of quantification. Consider, for instance, the statement: "An even integer $n > 2$ can be written as the sum of two primes p and q." The article "an" indicates universal quantification of n, whereas the verb "can be" indicates existential quantification of p and q. To reduce the chance of confusion, and to draw attention to the fact that quantifiers are being used, in this text I prefer the symbols $\forall$ and $\exists$ (or the standard English phrases "for all" and "there exists") to introduce quantified variables. So I would have written the statement under consideration like this: "For all even integers $n > 2$, there exist primes p and q such that $n = p + q$." One final grammatical point: note the plural verb form EXIST (not EXISTS) agrees with the plural subject PRIMES occurring later in the sentence.

Exercise: Find all the places in this text where, despite my comments here, I inadvertently used a hidden quantifier.

Section Summary

1. *Rules for Converting and Negating Quantifiers.*

$$\boxed{\exists x \in U, P(x)} \quad \Leftrightarrow \quad \boxed{\exists x, (x \in U \land P(x))}$$

$$\boxed{\forall x \in U, P(x)} \quad \Leftrightarrow \quad \boxed{\forall x, (x \in U \Rightarrow P(x))}$$

$$\boxed{\sim\exists x \in U, P(x)} \quad \Leftrightarrow \quad \boxed{\forall x \in U, \sim P(x)}$$

$$\boxed{\sim\forall x \in U, P(x)} \quad \Leftrightarrow \quad \boxed{\exists x \in U, \sim P(x)}$$

2. *Translation Points.* (a) In English sentences, changing the relative positions of quantifiers and logical keywords (especially negative words like "not") can affect the meaning.
   (b) In symbolic logic (unlike English), quantifiers must precede the variables they modify.
   (c) Quantifiers can occur within other logical constructions (*example:* "if all roses are red, then some violets are blue"). Moving these quantifiers to the front of the symbolic version of the statement can change the meaning.
   (d) Quantifiers often occur implicitly in English; for instance, a plural subject may signal a universal quantifier, whereas the verb "is" can disguise one or more existential quantifiers.

3. *Finding Useful Denials.* Applying a negation operator to a complex statement has the following effect on the outermost logical operator: $\land$ becomes $\lor$, $\lor$ becomes $\land$, $\forall$ becomes $\exists$, $\exists$ becomes $\forall$, $\sim$ cancels out, $\Leftrightarrow$ becomes $\oplus$, and $\oplus$ becomes $\Leftrightarrow$. In the case of $\land$, $\lor$, $\forall$, and $\exists$, we must continue by recursively denying the inputs to these operators. Regarding IF, chant this statement aloud five times: $\boxed{\text{THE DENIAL OF } A \Rightarrow B \text{ IS } A \land (\sim B).}$ The denial rules are summarized in the table on page 35.

Exercises

1. Convert each statement to an equivalent statement using unrestricted quantifiers.
   (a) $\forall n \in \mathbb{Z}, n$ is even $\oplus n$ is odd. (b) $\exists n \in \mathbb{Z}, n$ is odd $\land n$ is perfect.
   (c) $\forall x \in \mathbb{R}, x > 3 \Rightarrow x + 2 > 5$. (d) $\exists q \in \mathbb{Q}_{>0}, \exists r \in \mathbb{Q}_{>0}, q^3 + r^3 = 1$.
   (e) $\forall x \in \mathbb{R}, \exists n \in \mathbb{Z}, x < n$.

2. Eliminate all propositional operators from the following expressions by converting to restricted quantifiers. (a) $\exists x, x \in \mathbb{Q} \land x^2 = 3$. (b) $\forall x, x \in \mathbb{Z} \Rightarrow x^2 \in \mathbb{Z}_{\geq 0}$.
   (c) $\forall x, \forall y, (x \in \mathbb{R} \land y \in \mathbb{R}) \Rightarrow x + y \in \mathbb{R}$. (d) $\forall x, [x \in \mathbb{Q} \Rightarrow (\exists n, n \in \mathbb{Z}_{>0} \land xn \in \mathbb{Z})]$.
   (e) $[\exists k, k \in \mathbb{Z} \land n = 2k] \Rightarrow [\exists m, m \in \mathbb{Z} \land n + 3 = 2m + 1]$.

3. Let U be the set of people, let $T(x)$ mean x is 20 feet tall, and let P be the statement "$\sim\exists x \in U, T(x)$." Write statements equivalent to P using: (a) an unrestricted existential quantifier; (b) a restricted universal quantifier; (c) an unrestricted universal quantifier. Give literal English translations of P and your three answers.

4. Let L be the set of lobsters, let $R(x)$ mean x is red, and let Q be the statement "$\sim\forall x \in L, R(x)$." Write statements equivalent to Q using: (a) an unrestricted universal quantifier; (b) a restricted existential quantifier; (c) an unrestricted existential quantifier. Give literal English translations of Q and your three answers.

5. Consider the true sentence "Every human is male or female." Someone incorrectly encodes this sentence as: $(\forall x \in H, M(x)) \vee (\forall x \in H, F(x))$, where H is the set of humans, $M(x)$ means x is male, and $F(x)$ means x is female. (a) Translate this encoding back into English (do not use variables or symbols), and explain how the meaning differs from the original sentence. (b) Give a correct encoding of the original sentence.

6. Give a specific example of a universe U and open sentences $P(x)$ and $Q(x)$ such that the two statements $\exists x \in U, (P(x) \wedge Q(x))$ and $[\exists x \in U, P(x)] \wedge [\exists x \in U, Q(x)]$ have different truth values. Explain the difference in meaning between these two statements.

7. Let H be the set of humans, and let $M(x)$ mean x is mortal. Translate each symbolic formula into an English sentence that does not use propositional logic keywords such as AND, IF, etc. Which sentences are true? (a) $\forall x, x \in H \Rightarrow M(x)$. (b) $\forall x, x \in H \wedge M(x)$. (c) $\exists x, x \in H \wedge (\sim M(x))$. (d) $\exists x, x \in H \Rightarrow (\sim M(x))$.

8. Find useful denials of each statement. All answers should be English sentences. (a) John has brown hair and blue eyes. (b) If x_0 is even then $3x_0$ is odd. (c) All prime numbers are odd. (d) Purple cows do not exist. (e) There are deserts with red sand. (f) The sine function is continuous or the secant function is not bounded.

9. Find useful denials of each statement. All answers should be English sentences. (a) Jamestown was founded in 1607 or Williamsburg is not the capital of North Dakota. (b) $2 + 2 = 4$ if $3 + 3 = 7$. (c) All rubies are red. (d) Some diamonds are blue. (e) Not all elephants are pink. (f) All carrots are orange iff all rabbits are white. (g) Every integer is either positive or negative, but not both. (h) For all natural numbers x, if x is prime and even, then $x = 2$.

10. Convert each proposition to the standard form $P \Rightarrow Q$, and then give a useful denial of the statement. (a) G has a proper normal subgroup only if G is not simple. (b) A sufficient condition for normality of X is metrizability of X. (c) For f to be Lebesgue integrable, it is necessary that f be measurable. (d) Invertibility of A implies $\det(A) \neq 0$. (e) Whenever n_0 is a power of an odd prime, n_0 has a primitive root. (f) Injectivity of g is sufficient for g to have an inverse.

11. True or false? Explain.
    (a) $\exists x \in \mathbb{Z}, x^2 = -4$. (b) $\exists x, x \in \mathbb{Z} \wedge x^2 = -4$. (c) $\exists x, x \in \mathbb{Z} \Rightarrow x^2 = -4$.
    (d) $\forall x \in \mathbb{Z}, x+1 = 1+x$. (e) $\forall x, x \in \mathbb{Z} \Rightarrow x+1 = 1+x$. (f) $\forall x, x \in \mathbb{Z} \wedge x+1 = 1+x$.

12. Encode each English sentence in symbolic form. Only use unrestricted quantifiers in your answers. Let $B(x)$ mean x is black, $R(x)$ mean x is a raven, and $E(x)$ mean x is evil.
    Sample Question: All black ravens are evil.   *Answer:* $\forall x, [(B(x) \wedge R(x)) \Rightarrow E(x)]$.
    (a) Some ravens are evil. (b) Non-black ravens exist. (c) Although every raven is black, not every raven is evil. (d) Each raven is either black or evil, but not both. (e) If some ravens are black, then all ravens are evil. (f) Not all ravens are evil if some ravens are not black.

13. Write useful denials of each statement in the previous exercise; your answers should be English sentences.

14. Let M be the set of melons, let V be the set of vegetables, let $G(x)$ mean x is green, let $O(x)$ mean x is orange, let $T(x)$ mean x is tasty, and let $J(x)$ mean x is juicy. Encode each English sentence in symbolic form; your final answers should not use restricted quantifiers. (a) There exists an orange vegetable. (b) Every melon is juicy. (c) No green vegetable is tasty. (d) Some melons are green, some

are orange, but all are tasty. (e) If some orange vegetable is juicy, then all green melons are tasty. (f) All tasty melons are orange. (g) Juiciness is a necessary condition for a melon to be tasty. (h) Vegetables are green only if they are not juicy.

15. Write useful denials of each statement in the previous exercise; your answers should be English sentences.

16. Find all hidden quantifiers in the following sentences, and rewrite the sentences to make all quantifiers explicit. (a) Eagles have wings, but pigs do not. (b) The square of a real number is nonnegative. (c) Positive integers are expressible as sums of four squares. (d) A cyclic group G can be generated by an element x in G.

17. Use truth tables to show that all of the following propositional forms are possible denials of $P \Leftrightarrow Q$.   (a) $P \Leftrightarrow (\sim Q)$ (b) $(\sim P) \Leftrightarrow Q$ (c) $(\sim P) \oplus (\sim Q)$

18. Which of the following propositional forms are correct denials of $P \oplus Q$? Explain.
    (a) $(\sim P) \wedge (\sim Q)$ (b) $(\sim P) \oplus Q$ (c) $(\sim P) \Leftrightarrow (\sim Q)$
    (d) $(\sim P) \oplus (\sim Q)$ (e) $P \oplus (\sim Q)$ (f) $\sim (P \oplus Q)$

19. Let $F(x)$ mean x is a fish, let $A(x)$ mean x is an animal, let $S(x)$ mean x can swim, let $L(x)$ mean x has lungs, and let $W(x)$ mean x is white. Use these open sentences to encode each English sentence in symbolic form. (a) Some fish are white. (b) Not all animals have lungs. (c) Every white fish can swim. (d) Some animals that can swim are not fish, but all fish can swim. (e) Being a fish is a sufficient condition to be able to swim. (f) If no fish have lungs, then some white animals cannot swim.

20. Write useful denials of each statement in the previous exercise; your answers should be English sentences.

21. Encode each sentence in symbolic form, using only unrestricted quantifiers. Let $R(x)$ mean x is a rabbit, $U(x)$ mean x is a unicorn, $W(x)$ mean x is white, and $C(x)$ mean x is cute. (a) All white rabbits are cute. (b) Some rabbits are not cute. (c) All unicorns are white but not cute. (d) There are non-white unicorns. (e) Although not every rabbit is white, every unicorn is cute. (f) All rabbits are cute only if some unicorns are white. (g) Some white objects are unicorns or rabbits, but not both. (h) For all objects, a necessary condition for being a unicorn is being white. (i) The non-cuteness of some rabbits is sufficient for the whiteness of all unicorns.

22. Write useful denials of each statement in the previous exercise; your answers should be English sentences.

23. Find all the hidden quantifiers in the true-false exercises of §1.3. Consider whether this changes any of your answers to those exercises (e.g., if some whales live in aquariums rather than in the ocean).

1.6   Denial Practice and Uniqueness Statements

Finding useful denials of complex logical statements is an absolutely fundamental skill
needed to understand and create mathematical proofs. The only way to master this skill,
like so many others, is *practice*! You will practice translating and denying various statements
in the first half of this section. Afterward, we consider one of the more subtle ideas in
quantified logic — the concept of *uniqueness*.

Practice with Useful Denials

In the last section, we gave a table summarizing the rules needed to find useful denials of
logical statements. Review that table now, memorizing all the rules, and noting particularly
that *the denial of an IF-statement is an AND-statement*. Next, spend some time working
out useful denials of the statements in the next example. Solutions and discussion appear in
the next subsection. For complicated sentences, you might begin by converting the sentence
(partially or completely) into symbolic form, to see how logical operators nest within the
sentence. It can also help to rewrite potentially confusing phrases such as "only if" or "it
is necessary" into standard IF-THEN form.

1.42. Example: Denial Practice. Write a useful denial of each statement.
(a) If $a > b$, then $a^2 > b^2$.
(b) $a > 3$ iff $-2a < -4$.
(c) All carrots are orange.
(d) There is an integer x such that x is odd and x is perfect.
(e) All roses are not red.
(f) Roses are red, yet some violets are not blue.
(g) $1 + 1 = 2$ and $2 \times 2 = 4$; or $3^2 = 8$.
(h) Some cherries are not red, or all apples are red, but not both.
(i) Some pigs can fly, if water is wet or the sky is blue.
(j) For a to be prime, it is necessary that a is odd or $a > 2$.
(k) If a is prime, then: a is even implies $a = 2$.
(l) A sufficient condition for the existence of white ravens is the non-existence of black doves.
(m) $\exists x \in \mathbb{Z}, \exists y \in \mathbb{Z}, \exists z \in \mathbb{Z}, (x > 0 \text{ and } y > 0 \text{ and } x^3 + y^3 = z^3)$.
(n) $a > b$ is necessary and sufficient for $b^{-1} > a^{-1}$ only if $b > 0$ and $a > 0$.
(o) $\forall x, \exists y, \forall z, ((x < y + z) \Rightarrow (z < xy < y))$.

Solutions to Denial Exercise

Do not read any further until you have tried all parts of the preceding example yourself!

(a) A denial of "if P then Q" is "P and not Q." So one answer is "$a > b$ and $a^2 \le b^2$."

(b) Denying an IF-statement produces an XOR-statement. So one answer is "$a > 3$ or
$-2a < -4$, but not both."

(c) Denying a universal statement produces an existential statement. So one answer is "Some
carrot is not orange."

(d) Denials convert $\exists$ to $\forall$ and $\wedge$ to $\vee$. So one answer is "For all integers x, x is not odd or
x is not perfect."

(e) *Answer:* "Some rose is red." In this example, avoid the temptation to declare that a
denial of "All roses are not red" is "All roses are red." Observe that both of the latter state-

ments are false, so they cannot be denials of each other. The double negation rule can only be used to cancel a NOT operator applying to the entire statement, including quantifiers.

(f) This sentence features an implicit universal quantifier (signaled by the plural subject "roses"), as well as the word YET, which has the same logical meaning as AND. Denying the sentence produces "Some rose is not red, or all violets are blue."

(g) Notice the semicolon in the original statement, which serves to group the clauses of the sentence as shown by the parentheses here: "$(1 + 1 = 2$ and $2 \times 2 = 4)$ or $3^2 = 8$." Using the denial rules for OR and then AND, we get "$(1 + 1 \neq 2$ or $2 \times 2 \neq 4)$, and $3^2 \neq 8$." To avoid parentheses, we could say: $1 + 1 \neq 2$ or $2 \times 2 \neq 4$; and $3^2 \neq 8$.

(h) This sentence combines "Some cherries are not red" and "All apples are red" with the exclusive-OR operation. Denying gives "Some cherries are not red iff all apples are red."

(i) This sentence is an IF-statement where the hypothesis comes at the end. Denying gives "Water is wet or the sky is blue, and all pigs cannot fly."

(j) Before denying, the sentence can be rewritten: "If a is prime, then (a is odd or $a > 2$)." After denying, we get "a is prime and a is not odd and $a \leq 2$."

(k) The conclusion of this IF-statement is another IF-statement. Denying each IF in turn produces "a is prime and a is even and $a \neq 2$."

(l) Before denying, we rewrite the sentence as: "If black doves do not exist, then white ravens exist." Denying gives "Black doves do not exist and white ravens do not exist." Since both new clauses begin with NOT, we can further simplify the denial to: "All doves are not black, and all ravens are not white."

(m) In statements starting with multiple quantifiers, imagine forming the negation by placing the negation symbol $\sim$ at the far left. Now move this symbol to the right past each quantifier, flipping quantifiers from $\exists$ to $\forall$ and vice versa. Finally, the negation symbol acts on the ANDs in the middle of the statement, flipping them into ORs. The final denial is

$$\forall x \in \mathbb{Z}, \forall y \in \mathbb{Z}, \forall z \in \mathbb{Z}, (x \leq 0 \vee y \leq 0 \vee x^3 + y^3 \neq z^3).$$

(n) The statement can be recast as "If ($a > b$ iff $b^{-1} > a^{-1}$), then ($b > 0$ and $a > 0$)." Denying the statement produces "($a > b$ iff $b^{-1} > a^{-1}$) and ($b \leq 0$ or $a \leq 0$)."

(o) Flipping the initial quantifiers and denying the IF, we get "$\exists x, \forall y, \exists z, (x < y + z) \wedge (z \geq xy \vee xy \geq y)$." Note that the original formula $z < xy < y$ contains an implicit AND, which becomes OR in the denial.

Uniqueness

An object is *unique* iff it is the only one of its kind. We now introduce a modified version of the existential quantifier that can be used to assert the existence of a unique object satisfying some property.

1.43. Definition: Uniqueness Symbol. Let U be a set, and let $P(x)$ be an open sentence. The statement $\boxed{\exists ! x \in U, P(x)}$ means that there exists exactly one object x_0 in U for which $P(x_0)$ is true. Equivalently, there exists a unique object x_0 in U making $P(x_0)$ true. The unrestricted quantifier $\exists ! x, P(x)$ is defined similarly; this statement means that there exists one and only one object x_0 making $P(x_0)$ true. The exclamation mark following the existential quantifier is called the *uniqueness symbol*.

1.44. Example. True or false?
(a) $\exists ! x \in \mathbb{R}, x^2 = x$.

(b) $\exists\,!\,x \in \mathbb{Z}_{>0}, x^2 = x$.
(c) $\exists\,!\,x \in \mathbb{R}, x^3 = 8$.
(d) $\exists\,!\,x \in \mathbb{Z}, 3x = 5$.
(e) $\exists\,!\,x \in \mathbb{Q}, 3x = 5$.

Solution. (a) is false, because there are *two* real numbers x that satisfy $x^2 = x$, namely $x = 0$ and $x = 1$. On the other hand, (b) is true, since $x^2 = x$ has one and only one solution (namely $x = 1$) belonging to the given universe $\mathbb{Z}_{>0}$. Part (c) is true, since $x = 2$ is the unique real solution to $x^3 = 8$. (Note that $(-2)^3 = -8$.) Part (c) would be false if we enlarged the universe from $\mathbb{R}$ to $\mathbb{C}$. Part (d) is false because there is no *integer* x for which $3x = 5$. On the other hand, $x = 5/3$ is in $\mathbb{Q}$ and satisfies this equation, so "$\exists x \in \mathbb{Q}, 3x = 5$" is true. Moreover, $x = 5/3$ is the *only* solution to $3x = 5$ in $\mathbb{Q}$, so that the stronger existence-and-uniqueness statement in (e) is also true. (e) would remain true in the larger universes $\mathbb{R}$ or $\mathbb{C}$.

It is possible to eliminate the uniqueness symbol, replacing $\exists\,!\,x, P(x)$ by an equivalent statement using previously introduced logical symbols. This elimination rule gives us insight into the precise meaning of uniqueness, and we often need the rule when giving proofs of uniqueness. To derive the rule, let us consider the logical encoding of several related statements first.

<u>Step 1.</u> How can we encode the following statement? "There exist *at least two* objects x_0 making $P(x_0)$ true." A first attempt might be to write $\exists x, \exists y, P(x) \wedge P(y)$. However, this does not quite work, because we are allowed to pick the same object for x and for y. For instance, we see that $\exists x, \exists y, x + 1 = 3 \wedge y + 1 = 3$ is true by taking $x = y = 2$. If we intend the variables x and y to represent different (distinct) objects, we must explicitly say so by saying $x \neq y$. So the given statement can be encoded as follows:

$$\exists x, \exists y, [(P(x) \wedge P(y)) \wedge x \neq y].$$

<u>Step 2.</u> Now consider how to encode "There exists *at most one* object x_0 making $P(x_0)$ true." The key is to recognize that the situation described here (having at most one object that works) is the exact logical opposite of the situation in Step 1 (having at least two objects that work). Thus we can obtain the answer by denying the statement from Step 1. One possible denial looks like this:

$$\forall x, \forall y, [\sim(P(x) \wedge P(y)) \vee x = y].$$

We could continue to simplify the denial by replacing $\sim(P(x) \wedge P(y))$ by $(\sim P(x)) \vee (\sim P(y))$. However, another way to proceed is to remember the equivalence $A \Rightarrow B \equiv (\sim A) \vee B$ from the Theorem on IF. Using this equivalence in reverse, we obtain the following IF-statement as a possible denial of the statement in Step 1:

$$\forall x, \forall y, [(P(x) \wedge P(y)) \Rightarrow x = y]. \tag{1.5}$$

Reading this in English, with some words added for emphasis, (1.5) says that "for all objects x_0 and y_0, if $P(x_0)$ and $P(y_0)$ both happen to be true, then it must be the case that x_0 actually equals y_0." This statement does have the intended effect of preventing more than one object from satisfying the open sentence $P(x)$.

Although it already follows from the denial rules, let us confirm directly that (1.5) is true when $P(x)$ is satisfied by zero objects, or by exactly one object. First suppose no objects make $P(x)$ true. Then for any objects x_0 and y_0, $P(x_0) \wedge P(y_0)$ is false, so the IF-statement is true. Thus, (1.5) is true in this situation. Now suppose exactly one object z_0 makes $P(x)$ true. In this case, when $x = z_0$ and $y = z_0$, the IF-statement says $(P(z_0) \wedge P(z_0)) \Rightarrow z_0 = z_0$, which is true. On the other hand, for all other choices of x and y, the IF-statement has a false hypothesis, hence is true.

Step 3. We are now ready to encode the target statement, "There exists *exactly one* object x_0 making $P(x_0)$ true." The key is to rewrite this assertion using AND, as follows: "There exists *at least one* object x_0 making $P(x_0)$ true, AND there exists *at most one* object x_0 making $P(x_0)$ true." The first clause can be handled by an ordinary existential quantifier, and we encoded the second clause in Step 2. To summarize, our *elimination rule for the uniqueness symbol* is:

$$\boxed{\exists\,!\,x, P(x)} \quad \Leftrightarrow \quad \boxed{(\exists x, P(x)) \wedge (\forall x, \forall y, [(P(x) \wedge P(y)) \Rightarrow x = y])}.$$

A similar rule holds for restricted quantifiers. For instance, $\exists\,!\,x \in \mathbb{R}, x^3 = 8$ can be rewritten as

$$(\exists x \in \mathbb{R}, x^3 = 8) \wedge (\forall x \in \mathbb{R}, \forall y \in \mathbb{R}, (x^3 = 8 \wedge y^3 = 8) \Rightarrow x = y).$$

We can also adapt the reasoning in Steps 1 through 3 to make statements such as "there are at least three objects x_0 making $P(x_0)$ true," "there are at most two objects x_0 making $P(x_0)$ true," "there are exactly two objects x_0 making $P(x_0)$ true," and so on. We also mention that a statement such as, "the solution is unique, if it exists" is asserting the existence of *at most one* solution, hence could be encoded by the formula (1.5).

1.45. Remark. The English word UNIQUE is a special kind of adjective called an *absolute* (or *non-gradable*) adjective. Absolute adjectives *cannot* be modified by adverbs indicating the degree to which that adjective applies. For example, although it might be *very* cold today, or *somewhat* cold, or *really* cold, or *colder* than yesterday, we cannot say that an object is *very* unique, *somewhat* unique, *really* unique, *more* unique, or *almost* unique. An object either is unique (the one and only one object in a given universe having a specified property), or it is not.

Section Summary

1. *Denials.* Have you fully memorized the table of denial rules on page 35? Keep practicing until the rules become second nature.

2. *Uniqueness.* $\exists\,!\,x \in U, P(x)$ means there exists exactly one object x_0 in U making $P(x_0)$ true. We could say "unique" or "one and only one" instead of "exactly one" here. We can eliminate the uniqueness symbol (the exclamation mark "!") with this rule:

$$\boxed{\exists\,!\,x \in U, P(x)} \Leftrightarrow \boxed{(\exists x \in U, P(x)) \wedge (\forall x \in U, \forall y \in U, [(P(x) \wedge P(y)) \Rightarrow x = y])}.$$

To say there is *at most one* x_0 in U making $P(x_0)$ true, we write:

$$\forall x \in U, \forall y \in U, [(P(x) \wedge P(y)) \Rightarrow x = y].$$

Exercises

1. True or false? Explain.
   (a) $\exists\,!\,x \in \mathbb{R}, x^2 = 2.$ (b) $\exists\,!\,x \in \mathbb{R}_{<0}, x^2 = 2.$
   (c) $\exists\,!\,x \in \mathbb{Z}, x^2 - 3x - 10 = 0.$ (d) $\exists\,!\,x \in \mathbb{Z}_{>0}, x^2 - 3x - 10 = 0.$
   (e) $\exists\,!\,x \in \mathbb{Z}_{\geq 0}, x^2 + 4x + 4 = 0.$ (f) $\exists\,!\,x \in \mathbb{R}, x^2 + 4x + 4 = 0.$

2. True or false? Explain.
   (a) $\exists\,!\,x \in \mathbb{Z}, 2 < x < 4.$ (b) $\exists\,!\,x \in \mathbb{Q}, 2 < x < 4.$ (c) $\exists\,!\,x \in \mathbb{Z}, 2 \leq x \leq 4.$
   (d) $\exists\,!\,y \in \mathbb{Z}, 5y - 7 = 2.$ (e) $\exists\,!\,y \in \mathbb{Q}, 5y - 7 = 2.$ (f) $\exists\,!\,y \in \mathbb{R}, 5y - 7 = 2.$

3. Write a useful denial of each statement. (Assume all unquantified letters are constants.)
   (a) If $ab \neq 0$, then $a \neq 0$ and $b \neq 0$.
   (b) All perfect numbers have last digit 6 or 8.
   (c) (X is compact) $\Leftrightarrow$ (X is closed and bounded).
   (d) When X is compact, X is closed and bounded.
   (e) $\forall x \in \mathbb{Z}, \exists y \in \mathbb{Q}, x < y < x + 1$.
   (f) If every odd number is prime, then some raven is not black.
   (g) Not all elephants are pink.
   (h) $2 + 2 = 4$ only if $2 \times 3 = 9$.

4. Write a useful denial of each statement. (Assume all unquantified letters are constants.)
   (a) If $ab = 0$, then $a = 0$ or $b = 0$.
   (b) Some prime numbers p satisfy $10 < p < 20$.
   (c) X is compact if X is both complete and totally bounded.
   (d) X is connected and compact whenever X is a closed interval.
   (f) If some carrot is not orange, then every even number is prime.
   (g) Mauve pigs do not exist.
   (h) A necessary condition for $1 + 1 = 2$ is that $2 \times 5 = 11$.
   (i) Some violets are blue only if all oranges are green or yellow.
   (j) $((P \oplus Q) \Leftrightarrow R) \wedge (Q \Rightarrow (R \vee (\sim P)))$.

5. Write a useful denial of each statement.
   (a) For every positive real number y, there is a real number x with $e^x = y$.
   (b) For all real x and z, if $\tan x = \tan z$, then $x = z$.
   (c) For all real x and y, $y = 3x + 1$ iff $x = (y - 1)/3$.

6. Write a useful denial of each statement.
   (a) A necessary condition for all roses to be red is that some carrots are white or blue.
   (b) $\forall \epsilon \in \mathbb{R}_{>0}, \exists \delta \in \mathbb{R}_{>0}, \forall x \in \mathbb{R}, \forall y \in \mathbb{R}, (|x - y| < \delta \Rightarrow |\sin(x) - \sin(y)| < \epsilon)$.
   (c) For all functions f, if the continuity of f is sufficient for the differentiability of f, then the integrability of f is necessary for the continuity of f.
   (d) $((P \Rightarrow Q) \oplus R) \vee (Q \wedge (R \Rightarrow (\sim P)))$.
   (e) For some real x, $x > 2$ implies $x^2 > 4$ but not conversely.

7. Write a useful denial of this statement: $0 \in H$ and $\forall a, b \in H, a + b \in H$ and $\forall c \in H, -c \in H$.

8. For each set H, decide whether the statement in the previous exercise or its denial is true for H. (a) $H = \mathbb{Q}$. (b) $H = \mathbb{R}_{\geq 0}$. (c) $H = \mathbb{Z}_{\neq 0}$. (d) $H = \{-1, 0, 1\}$. (e) H is the set of even integers.

9. Write a useful denial of each statement.
   (a) $\exists e \in G, \forall x \in G, e \star x = x = x \star e$.
   (b) $\forall x \in P, \forall y \in P, \forall z \in P, x \leq y \leq z \Rightarrow x \leq z$.
   (c) $\exists L \in \mathbb{R}, \forall \epsilon \in \mathbb{R}_{>0}, \exists N \in \mathbb{Z}_{>0}, \forall n \in \mathbb{Z}_{>0}, n \geq N \Rightarrow |a_n - L| < \epsilon$.
   (d) $\exists y \in \mathbb{R}, [(\forall x \in S, y \leq x) \wedge (\forall z \in \mathbb{R}, y < z \Rightarrow \exists u \in S, u < z)]$.

10. Write a useful denial of each statement.
    (a) $\exists M \in \mathbb{R}, \forall x \in S, \forall y \in S, d(x, y) \leq M$.
    (b) $\forall y \in G, \exists z \in G, y \star z = e = z \star y$.
    (c) $\forall \epsilon \in \mathbb{R}_{>0}, \forall N \in \mathbb{Z}_{>0}, \exists n \in \mathbb{Z}_{>0}, n \geq N \wedge d(x_n, x) < \epsilon$.

11. Let x_0 be a fixed object. Encode the following sentence using logical symbols: "x_0 is the unique object in U making $P(x)$ true."

12. Encode the following sentence using logical symbols: "for *all but one* object x_0 in U, $Q(x_0)$ is true."

13. Find a useful denial of $\exists! x \in U, P(x)$ by eliminating the uniqueness symbol and applying the denial rules. Give your answer in symbols and then in English.

14. Consider this statement: "If $\exists! x, P(x)$ is true, then $\exists x, P(x)$ must be true."
    (a) Is this statement true for all open sentences $P(x)$? Explain.
    (b) Repeat (a) for the converse of the given statement.
    (c) Repeat (a) for the contrapositive of the given statement.

15. Consider this statement: "If $\exists x \in U, P(x)$ is true, then $\exists x, P(x)$ must be true."
    (a) Is this statement true for all choices of U and $P(x)$? Explain.
    (b) Repeat (a) for the converse of the given statement.

16. Consider this statement: "If $\exists! x \in U, P(x)$ is true, then $\exists! x, P(x)$ must be true."
    (a) Is this statement true for all choices of U and $P(x)$? Explain.
    (b) Repeat (a) for the converse of the given statement.

17. Let $P(x)$ be a fixed open sentence. Find symbolic versions of each statement below that do not use the uniqueness symbol.
    (a) There exist at least three objects x_0 making $P(x_0)$ true.
    (b) There exist at most two objects x_0 making $P(x_0)$ true.
    (c) There exist exactly two objects x_0 making $P(x_0)$ true.

18. Encode the following sentence using logical symbols: "There exist exactly four objects x_0 making $P(x_0)$ true."

2

Proofs

2.1 Definitions, Axioms, Theorems, and Proofs

In most branches of science, such as physics or chemistry, we develop and test theories based on experiments. Mathematics, on the other hand, is a deductive science, in which we use logical rules to derive a large body of truths from a small set of initial statements, called axioms. This section introduces the main ingredients in mathematical theories: undefined terms, definitions, axioms, theorems, inference rules, and proofs.

Undefined Terms and Definitions

In mathematics, we would like to give a precise and rigorous meaning to every word and symbol that we use. This is typically done by giving formal definitions for various mathematical terms. However, all words and symbols are defined by means of other words and symbols, so it is not possible to define everything! The solution is to begin our theory with a collection of *undefined terms*, such as "true," "false," "for all," "set," and "membership in a set." We can often convey an intuitive idea of what an undefined term is supposed to mean by giving synonyms for the term or specific examples. For instance, we might say that a set is a collection of objects, or we might give examples of sets such as $\{1, 2, 3\}$ or the set of all cows. But, these informal explanations are not mathematical definitions. Ultimately, the meaning of the undefined terms is captured by the axioms we use to describe their properties.

Most of the words and symbols we use are, in fact, *defined terms*. For instance, we have already defined the precise meaning of words such as AND, OR, and IF by means of truth tables that show how these words are related to the undefined concepts of "true" and "false." Typically, a mathematical definition describes a new object, relationship, or symbol via an expression consisting of previously defined symbols. Almost every definition contains quantifiers and logical keywords. For example, the following IFF-statement can be viewed as a definition of the *uniqueness symbol*:

$$\boxed{\exists\,!\,x \in U, P(x)} \quad \Leftrightarrow \quad \boxed{(\exists x \in U, P(x)) \wedge (\forall x \in U, \forall y \in U, ([P(x) \wedge P(y)] \Rightarrow x = y)}.$$

In the next few sections, we will be practicing proofs using four basic concepts that you may have already seen informally: even numbers, odd numbers, divisibility, and rational numbers. We now give formal definitions of these concepts.

2.1. Definition of EVEN, ODD, DIVIDES, and RATIONAL.

(a) For all n, $\boxed{n \text{ is } even}$ means: $\boxed{\exists k \in \mathbb{Z}, n = 2k.}$

(b) For all n, $\boxed{n \text{ is } odd}$ means: $\boxed{\exists k \in \mathbb{Z}, n = 2k + 1.}$

(c) For all integers a and b, $\boxed{a \ divides \ b}$ means: $\boxed{\exists c \in \mathbb{Z}, b = ac.}$

(d) For all x, $\boxed{x \ is \ rational}$ means: $\boxed{\exists m \in \mathbb{Z}, \exists n \in \mathbb{Z}, (n \neq 0 \wedge x = m/n).}$

Before continuing, we offer some advice and remarks about definitions. When studying any mathematical subject, you must **memorize all definitions perfectly** as soon as they occur. Do so now for the four definitions listed above. Note that each definition consists of the *defined term* (appearing in the box on the left) followed by the *definition text* (appearing in the box on the right). The overall definition asserts the logical equivalence of the defined term and the definition text, for all choices of the variables in the defined term. The quantifiers and universes appearing on the right side of each definition are a *crucial* part of the definition; do not forget them! For example, when asked what "n is odd" means, the answer "$\exists k, n = 2k + 1$" is incorrect (the universe is omitted), and the answer "$n = 2k + 1$" is even worse (the quantifier on k is omitted).

The only *unimportant* feature of the definitions are the letters used on each side. Different letters may be used instead, as long as there is no conflict with previously introduced letters. Sometimes we must use a different letter on the right side, if the letters in the original definition already have a different meaning. Furthermore, expressions can be substituted for the letters on the left side of a definition. These observations are basic, but they are so crucial that we give an example.

2.2. Example. The phrase "m is odd" means $\exists k \in \mathbb{Z}, m = 2k + 1$. We could also have said $\exists z \in \mathbb{Z}, m = 2z + 1$ or $\exists q \in \mathbb{Z}, m = 2q + 1$ as the definition of this phrase. The phrase "k is even" means $\exists u \in \mathbb{Z}, k = 2u$; here, we needed to change the quantified variable from k to some other letter (in this case, u) because k already had a different meaning. The phrase "$5x + 7$ is odd" means $\exists k \in \mathbb{Z}, 5x + 7 = 2k + 1$. The phrase "$c$ divides c^3" means $\exists d \in \mathbb{Z}, c^3 = cd$ (we had to change the quantified variable from c to d here). The phrase "$m^2 + n^2$ is rational" means $\exists a \in \mathbb{Z}, \exists b \in \mathbb{Z}, (b \neq 0 \wedge m^2 + n^2 = a/b)$.

We reiterate that *formal mathematical definitions must be meticulously memorized, including all quantifiers and universes.* It is fine, and indeed helpful, to have intuitive descriptions of what a formal definition is saying. For example, the intuition for the definition of "rational" is that a rational number is the ratio of two integers. However, this phrase does not capture the full content of the definition (the existential quantifiers should be explicit, and the denominator must be nonzero).

Axioms

In developing a mathematical theory, we would ideally like to derive every true statement as a logical consequence of previously known statements. However, as in the case of definitions and undefined terms, it is impossible to prove literally every statement in this way. For, at the very beginning, we have not derived any true statements, so we have nothing to work with. The solution is to begin our theory with a small collection of initial statements, called *axioms* or *postulates*, that are assumed to be true without proof. Each axiom is a proposition that may use quantifiers, logical symbols, undefined terms, and defined terms.

What are some examples of axioms? We obtain a large supply of axioms by agreeing that *any instance of a tautology is an axiom.* Recall that a tautology is a propositional form, such as $P \vee (\sim P)$, that is true for all values of the propositional variables appearing in the form. An *instance* of a tautology is any statement obtained from the propositional form by replacing all propositional variables by specific propositions. For example, here are some instances of the tautology $P \vee (\sim P)$: "$1 + 1 = 2$ or $1 + 1 \neq 2$;" "All roses are red or not all roses are red;" "The sine function is continuous or the sine function is not continuous;" "$(0 = 1$ and some pigs can fly) or it is false that $(0 = 1$ and some pigs can fly)." All of these statements are axioms. Instances of tautologies are particularly trustworthy axioms, since the truth of these statements is virtually forced upon us by the way we have agreed to use the logical connectives such as NOT, AND, and OR.

Every new definition can be viewed as a special kind of axiom, called a *definitional axiom*. For example, we can rewrite the definition of EVEN (given above) as the following axiom:

$$\forall n, [n \text{ is even} \Leftrightarrow \exists k \in \mathbb{Z}, n = 2k].$$

On one hand, the forward implication of this IFF-statement lets us eliminate the defined term "n is even" and replace it with the definition text "$\exists k \in \mathbb{Z}, n = 2k$" (where k is a new variable). On the other hand, the converse implication lets us go back from the definition text to the defined term, when convenient. The definitions of ODD, DIVIDES, and RATIONAL can similarly be recast as axioms involving universally quantified IFF-statements. For instance:

$$\forall a \in \mathbb{Z}, \forall b \in \mathbb{Z}, [a \text{ divides } b \Leftrightarrow \exists c \in \mathbb{Z}, b = ac].$$

In what follows, we state all new definitions as definitional axioms in this way.

2.3. Remark. Some mathematical texts present new definitions as IF-statements rather than IFF-statements. For instance, a text might say, "We define an integer n to be *even* IF $n = 2k$ for some integer k." This practice, while very common, is logically wrong! Each definition needs to be an "if and only if" statement telling us precisely when the newly defined term is to be considered true.

All mathematical theories share a certain common core of logical axioms (e.g., instances of tautologies, and axioms concerning quantifiers and the equality symbol). Each particular theory, such as the theory of sets or the theory of integers or the theory of Euclidean geometry, contains additional axioms giving specific properties of the objects under investigation. Some of the axioms in the theory of sets will be introduced later when we study set theory. In the theory of integers, here is a sample of some of the possible axioms we might start with. (We assume $\mathbb{Z}$, $+$, $\cdot$, 0, and 1 are undefined terms.)

2.4. Some Axioms for $\mathbb{Z}$.

(a) *Closure under Addition:* $\forall x \in \mathbb{Z}, \forall y \in \mathbb{Z}, x + y \in \mathbb{Z}$.

(b) *Commutativity of Addition:* $\forall x \in \mathbb{Z}, \forall y \in \mathbb{Z}, x + y = y + x$.

(c) *Associativity of Addition:* $\forall x \in \mathbb{Z}, \forall y \in \mathbb{Z}, \forall z \in \mathbb{Z}, (x + y) + z = x + (y + z)$.

(d) *Additive Identity:* $\forall x \in \mathbb{Z}, x + 0 = x$.

(e) *Additive Inverses:* $\forall x \in \mathbb{Z}, \exists y \in \mathbb{Z}, x + y = 0$.

(f) *Closure under Multiplication:* $\forall x \in \mathbb{Z}, \forall y \in \mathbb{Z}, x \cdot y \in \mathbb{Z}$.

(g) *Commutativity of Multiplication:* $\forall x \in \mathbb{Z}, \forall y \in \mathbb{Z}, x \cdot y = y \cdot x$.

(h) *Associativity of Multiplication:* $\forall x \in \mathbb{Z}, \forall y \in \mathbb{Z}, \forall z \in \mathbb{Z}, (x \cdot y) \cdot z = x \cdot (y \cdot z)$.

(i) *Multiplicative Identity:* $\forall x \in \mathbb{Z}, x \cdot 1 = x$.

(j) *Distributive Law:* $\forall x \in \mathbb{Z}, \forall y \in \mathbb{Z}, \forall z \in \mathbb{Z}, x \cdot (y + z) = (x \cdot y) + (x \cdot z)$.

(k) *No Zero Divisors:* $\forall x \in \mathbb{Z}, \forall y \in \mathbb{Z}, (x \neq 0 \wedge y \neq 0) \Rightarrow x \cdot y \neq 0$.

These axioms describe some of the basic properties of integer addition and integer multiplication. We could list more axioms giving properties of equalities and inequalities among integers. In fact, all the axioms we have listed can be proved from even more basic axioms that we do not state here. However, for our initial practice with proofs, we assume that all of the above statements and similar basic algebraic facts are already known. This includes arithmetical facts such as $2 + 2 = 4$, $\sim(1/2 \in \mathbb{Z})$, etc., as well as properties of inequalities such as:

$$\forall x \in \mathbb{Z}, \forall y \in \mathbb{Z}, \forall z \in \mathbb{Z}, \text{if } x < y \text{ and } z > 0 \text{ then } xz < yz.$$

We also assume that analogous facts about the real number system are already known. However, unless otherwise noted, we do not allow ourselves to use any fact that specifically concerns the concepts of even numbers, odd numbers, divisibility, and rational numbers. In Chapter 8, we give a detailed development of the real number system starting from a list of 19 axioms. This development makes extensive use of the proof techniques to be presented in the coming sections.

Theorems, Inference Rules, and Proofs

Starting from the axioms, we use rules of logic to deduce new true statements called *theorems*. Each axiom is also considered a theorem. The sequence of steps leading from the axioms to a given theorem is called a *proof* of that theorem. An individual step in a proof uses an *inference rule* to combine one or more previous theorems to obtain another theorem. Our first inference rule involves the logical keyword IF.

2.5. Inference Rule for IF. Suppose $P \Rightarrow Q$ and P are already known to be theorems. Then we may deduce Q as a new theorem.

We can informally justify this rule using the truth table for IF, shown here:

P	Q	$P \Rightarrow Q$
T	T	T
T	F	F
F	T	T
F	F	T

In the situation described in the rule, P is a known theorem, and hence is a true proposition. Thus we must be in one of the first two rows of the truth table. But $P \Rightarrow Q$ is also known to be true, forcing us to be in the first row of the truth table. In this row, Q is a true proposition, so we may safely conclude that Q is a theorem. Similar reasoning with the truth table for IF establishes the following inference rule.

2.6. Contrapositive Inference Rule for IF. Suppose $P \Rightarrow Q$ and $\sim Q$ are already known to be theorems. Then we may deduce $\sim P$ as a new theorem.

We can also justify the Contrapositive Inference Rule as follows. Given that $P \Rightarrow Q$ is a known theorem, the logically equivalent statement $(\sim Q) \Rightarrow (\sim P)$ must also be true. Given that $\sim Q$ is also known to be true, we can deduce the truth of $\sim P$ by applying the Inference Rule for IF to the theorems $(\sim Q) \Rightarrow (\sim P)$ and $\sim Q$.

On the other hand, suppose $P \Rightarrow Q$ and Q are known theorems. Is it always safe to deduce P as a new theorem? Consulting the truth table for IF, we see that the answer is no! More specifically, in row 3 of the truth table, $P \Rightarrow Q$ and Q are both true, yet P is false. The incorrect deduction of P from known theorems $P \Rightarrow Q$ and Q is sometimes called the *converse error*.

Some inference rules involve quantifiers, as we see in the next rule. This rule formally recasts our earlier intuitive description of what a universal quantifier means.

2.7. Inference Rule for ALL. Suppose $\forall x \in U, P(x)$ is a known theorem and $c \in U$ is a known theorem, where c is a variable or expression denoting a particular object. Then we may deduce the new theorem $P(c)$.

Intuitively, if we already know that property P is true for *all* objects x in U, and if we also know that the expression c represents an object in U, then we can conclude that property P is true for the particular object c.

2.8. Example. Suppose we have already proved the theorem $\forall x \in \mathbb{R}, x^2 \geq 0$, and suppose a and b are fixed positive real numbers. Then $\sqrt{a}$ and $\sqrt{b}$ are also positive real numbers, and hence $c = \sqrt{a} - \sqrt{b}$ is in $\mathbb{R}$ because $\mathbb{R}$ is closed under subtraction. Now we can apply the Inference Rule for ALL, replacing the quantified variable x in "$x^2 \geq 0$" by the expression c. We deduce the new theorem $(\sqrt{a} - \sqrt{b})^2 \geq 0$. Now we manipulate this inequality using known algebraic facts. Expanding the square, we get $a - 2\sqrt{a}\sqrt{b} + b \geq 0$. Rearranging terms, we get $a + b \geq 2\sqrt{ab}$ and then $(a+b)/2 \geq \sqrt{ab}$. This inequality says that the arithmetic mean of two positive numbers is greater than or equal to the geometric mean of those numbers.

We can obtain more inference rules for AND, OR, IFF, etc., by analyzing truth tables. Some examples of these rules are considered in the exercises. But these new inference rules turn out to be redundant, because they already follow by combining tautology-based axioms with the Inference Rule for IF. For instance, consider this inference rule for AND: given that $P \wedge Q$ is a known theorem, we can deduce Q as a new theorem. To see why this rule is redundant, note that $(P \wedge Q) \Rightarrow Q$ is a tautology and hence an axiom. Suppose that we have proved $P \wedge Q$ already. Using the Inference Rule for IF, we see that Q is a new theorem. Similarly, the Contrapositive Inference Rule for IF follows from the Inference Rule for IF and an appropriate tautology.

2.9. Remark. In many logic texts, the Inference Rule for IF is called the *modus ponens* rule, abbreviated MP. "Modus ponens" is a Latin phrase meaning "method of affirming." The Contrapositive Inference Rule for IF is called the *modus tollens* rule (MT), which means "method of raising." The Inference Rule for ALL is called *universal instantiation* (UI). We shall not use these terms.

More on Mathematical Theories (Optional)

This optional subsection gives a bit more detail on how logicians use axiom systems and inference rules to develop formal mathematical theories. The problem of not being able to define the initial words and symbols is dealt with in the following way. First, we invent a *formal language* using a highly restricted set of symbols (such as $\sim$, $\Rightarrow$, $\forall$, $=$, $\in$, and letters for variables and constants). There are concrete rules determining which strings of symbols in this language constitute *terms* and *formulas*. Certain strings of symbols in the language are designated as *axioms*, and certain *inference rules* are given. The key is that each inference rule can be executed by performing very basic mechanical manipulations on strings of symbols, without needing to know what any of the symbols mean. For instance, one inference rule (corresponding to the Inference Rule for IF discussed earlier) says: if the string of symbols consisting of a formula P followed by the symbol $\Rightarrow$ followed by a formula Q is a known theorem, and if the string of symbols P is a known theorem, then the string of symbols Q is a new theorem. The inference rule for universal quantifiers says (roughly): if one of our known theorems consists of the symbol $\forall$, then a variable symbol x, then a formula P, and if t is any term, then we get a new theorem by writing the string P with all occurrences of the symbol x replaced by the string of symbols t. We see that axioms, proofs, and theorems are all *syntactic* constructions — properties of the formal language that make no specific reference to the underlying meaning of the symbols.

Of course, the only reason we care about proving theorems is because of what they *mean*! Thus, every formal language has an associated *semantic* component that attempts to define the *meaning* of the terms, formulas, and theorems in the language. The idea is to create an *interpretation* for the formal theory by introducing a specific universe (set) of objects for the variables to range through, and stipulating the meaning (within this specific model) of each symbol in the language. For instance, the meaning of the logical connectives is defined using truth tables, as before. A given formal theory can have many different interpretations, so the meaning of each formula in the theory depends on which interpretation is used. In some interpretations, all of the axioms of the theory become true statements; such interpretations are called *models* of the theory. (Actually, this is not quite true: some theories are *inconsistent*, and have no models at all.)

It can be shown that all theorems of a formal language (not just the axioms) become true statements in any model of that theory. A celebrated result of mathematical logic (*Gödel's Completeness Theorem*) asserts that, conversely, any formula of a theory that is true in *all* models of that theory is provable from the axioms of that theory. This result is absolutely amazing, because it means that the apparently mindless symbol-manipulation rules of a formal theory are powerful enough to discover *all* true statements expressible within that theory! Put another way, the concrete *syntactic* machinery of proofs and formal theorems delivers abstract *semantic* results applicable to a wide variety of situations. Many theorems of modern mathematical logic have a similar flavor, establishing the equivalence of a syntactic concept and a semantic concept. For instance, another version of the Completeness Theorem states that a theory has no models iff a contradictory formula of the form $P \wedge (\sim P)$ is a formal theorem within that theory.

One practical benefit of organizing mathematical theories using abstract axiom systems is economy of thought. By selecting appropriate general axioms at the outset, we can simultaneously prove facts about many different situations all at once. As an example of this, we now describe the axioms used to define an abstract algebraic structure called a *group*. The undefined symbols are G (representing the set of objects in the group), $\star$ (representing an *algebraic operation* that combines two elements of G to produce a new element of G), and e (representing an *identity element* for the operation). The axioms are listed next.

2.10. Group Axioms.

(a) *Closure:* $\forall x \in G, \forall y \in G, x \star y \in G$.

(b) *Associativity:* $\forall x \in G, \forall y \in G, \forall z \in G, x \star (y \star z) = (x \star y) \star z$.

(c) *Identity:* $\forall x \in G, x \star e = x \wedge e \star x = x$.

(d) *Inverses:* $\forall x \in G, \exists y \in G, x \star y = e \wedge y \star x = e$.

Starting from these axioms, we can prove many theorems about abstract groups. One such theorem is the Left Cancellation Law:

$$\forall a \in G, \forall x \in G, \forall y \in G, (a \star x = a \star y) \Rightarrow x = y.$$

The advantage of the axiomatic setup is that all theorems proved for abstract groups from the group axioms are automatically available for any particular concrete group we may happen to be working with. For example, it can be shown that the integers under addition, the nonzero real numbers under multiplication, and invertible 3×3 matrices under matrix multiplication all satisfy the group axioms. Each of these mathematical structures therefore has a left cancellation law. Rather than reproving this law (and all the other theorems of group theory) each time a new system comes along, we need only verify that the new system satisfies the four initial axioms of a group, and then we know that all the theorems are true for this system.

Section Summary

1. *Elements of Mathematical Theories.* A mathematical theory starts with a small collection of undefined terms and introduces definitions for new concepts that combine logical operators, undefined terms, and previous definitions. Similarly, the theory starts with axioms (statements assumed to be true without proof) and uses chains of inference rules called proofs to deduce new theorems. When learning a new theory, it is critical to memorize definitions, axioms, and theorem statements as soon as they arise.

2. *Definitions of Even, Odd, Divides, Rational.* **Memorize** these definitional axioms, which will be the subject of practice proofs in the next few sections.

   (a) For all n, $\boxed{n \text{ is } even}$ iff $\boxed{\exists k \in \mathbb{Z}, n = 2k}$.

   (b) For all n, $\boxed{n \text{ is } odd}$ iff $\boxed{\exists k \in \mathbb{Z}, n = 2k + 1}$.

   (c) For all $a, b \in \mathbb{Z}$, $\boxed{a \text{ } divides \text{ } b}$ iff $\boxed{\exists c \in \mathbb{Z}, b = ac}$.

   (d) For all x, $\boxed{x \text{ is } rational}$ iff $\boxed{\exists m \in \mathbb{Z}, \exists n \in \mathbb{Z}, (n \neq 0 \wedge x = m/n)}$.

   Note that quantifiers and universes are critical features of the definition, whereas the letters used are not important. Expressions can be substituted for the letters in the defined term, and letters in the definition text must be changed if they already have a different meaning.

3. *Axioms.* Every instance of a tautology is an axiom. Every new definition can be viewed as an axiom. Some axioms for $\mathbb{Z}$ are listed in item 2.4. The closure axioms state that the sum or product of any two integers is also an integer. There are axioms asserting the commutative, associative, distributive, and identity laws for addition and multiplication. Another axiom says that the product of any two nonzero integers is always nonzero.

4. *Inference Rules.* The Inference Rule for IF says that if $P \Rightarrow Q$ and P are already known theorems, we may deduce Q as a new theorem. The Inference Rule for ALL says that if $\forall x \in U, P(x)$ and $c \in U$ are already known theorems, we may deduce $P(c)$ as a new theorem. Many more inference rules follow from the rule for IF using appropriate tautologies.

Exercises

1. Rewrite each statement by expanding definitions. For statements with negative logic, give a useful denial of the expanded definition. (Do not prove anything here.) [*Sample Question:* u^2 is not odd. *Answer:* $\forall k \in \mathbb{Z}, u^2 \neq 2k + 1$.]
   (a) 37 is odd. (b) $k^2 + k$ is even. (c) $m^3 - 5$ is rational. (d) 7 divides c. (e) 3 does not divide 5. (f) $\sqrt{2}$ is not rational.

2. Rewrite each statement by expanding definitions. For statements with negative logic, give a useful denial of the expanded definition.
   (a) -18 is even. (b) $\ell^2 + 3k + 1$ is odd. (c) $\sqrt{m^2 + n^2}$ is rational. (d) $c + 1$ divides $c^2 - 1$. (e) ab does not divide $8b + c$. (f) $e + \pi$ is not rational.

3. Informally justify each inference rule by analyzing a truth table, imitating our discussion of the Inference Rule for IF.
   (a) Given known theorems $P \Rightarrow Q$ and $\sim Q$, deduce the new theorem $\sim P$.
   (b) Given known theorems A and B, deduce the new theorem $A \wedge B$.

    (c) Given the known theorem $A \wedge B$, deduce the new theorem B.

    (d) Given the known theorem $(\sim Q) \Rightarrow (R \wedge (\sim R))$, deduce the new theorem Q.

4. Show that each proposed inference rule is not correct by analyzing truth tables.
   (a) Given the known theorem $A \vee B$, deduce the new theorem A.
   (b) Given known theorems $P \Rightarrow Q$ and $\sim P$, deduce the new theorem $\sim Q$.
   (c) Given the known theorem $P \Rightarrow Q$, deduce the new theorem $P \Leftrightarrow Q$.

5. Decide, with explanation, whether each proposed inference rule is always valid.
   (a) Given the known theorem Q, deduce the new theorem $P \vee Q$.
   (b) Given the known theorem Q, deduce the new theorem $P \oplus Q$.
   (c) Given the known theorem $P \Leftrightarrow Q$, deduce the new theorem $P \Rightarrow Q$.
   (d) Given known theorems $P \Leftrightarrow Q$ and $\sim P$, deduce the new theorem $\sim Q$.
   (e) Given the known theorem $P \Rightarrow (Q \Rightarrow R)$, deduce the new theorem $(P \Rightarrow Q) \Rightarrow R$.
   (f) Given the known theorem $(P \Rightarrow Q) \Rightarrow R$, deduce the new theorem $P \Rightarrow (Q \Rightarrow R)$.

6. Show that the following inference rule is valid by a truth table analysis: given known theorems $P \vee Q \vee R$, $P \Rightarrow S$, $Q \Rightarrow S$, and $R \Rightarrow S$, deduce the new theorem S.

7. (a) Write the contrapositive of the IF-statement in Axiom 2.4(k). (b) Write the converse of your answer to part (a). Is this converse true for all $x, y \in \mathbb{Z}$?

8. Suppose "$\forall x \in \mathbb{Z}, 4x$ is even" is a known theorem, and a and b are fixed positive integers. Which of the following conclusions can be drawn from this theorem using the Inference Rule for ALL? Explain. (a) 8 is even. (b) 6 is even. (c) -100 is even. (d) 0.4 is even (taking $x = 1/10$). (e) $4b$ is even. (f) $4(a + 3b)$ is even. (g) $4(a/b)$ is even.

9. Suppose "$\forall k \in \mathbb{Z}, k^2 - k$ is even" is a known theorem, and a and b are fixed positive integers. Which conclusions can be drawn from this theorem using the Inference Rule for ALL? Explain. (a) 20 is even. (b) 6 is even. (c) 8 is even. (d) -30 is even. (e) $-1/4$ is even (taking $k = 1/2$). (f) $(a+b)^2 - (a+b)$ is even. (g) $(a/b)^2 - (a/b)$ is even. [*Sample Answer:* (a) CAN be deduced by taking $k = 5$ in the known theorem, since $k^2 - k = 25 - 5 = 20$.]

10. Suppose $\leq$, $=$, and the logical operators are undefined concepts in a mathematical theory of real numbers. Use these symbols to give formal definitions of: (a) $x < y$; (b) $x > y$; (c) $x \geq y$; (d) z is strictly between x and y.

11. Using only arithmetic operators $(+, \cdot, <)$ and logical symbols, give definitions of the following concepts, assuming $m, n \in \mathbb{Z}$ and $x \in \mathbb{R}$: (a) m is a perfect square (examples: 0, 1, 4, 9, 100); (b) n is a multiple of m; (c) x is a half-integer (examples: $5/2$, -22.5, $100/2$); (d) n is prime (examples: 2, 3, 5, 7, 11, 101).

12. Consider this proposed inference rule involving propositional forms P, Q, R, and S: "given known theorems P, Q, and R, deduce the new theorem S." (a) Assume $(P \wedge Q \wedge R) \Rightarrow S$ is a tautology. Argue informally that the proposed inference rule is valid. (b) Now assume that the proposed inference rule always works. Argue informally that $(P \wedge Q \wedge R) \Rightarrow S$ is a tautology.

13. Suppose the words IF and NOT are undefined concepts in a certain logical theory. Using only these words, complete the following definitions of other logical keywords. (a) $\boxed{P \text{ OR } Q}$ means... (b) $\boxed{P \text{ AND } Q}$ means... (c) $\boxed{P \text{ IFF } Q}$ means... (d) $\boxed{P \text{ XOR } Q}$ means...

14. Suppose the symbols $\wedge$, $\sim$, and $\exists$ are undefined symbols in a certain logical theory. Using only these symbols and parentheses, complete the following definitions of other logical symbols. (a) $\boxed{P \vee Q}$ means... (b) $\boxed{P \Rightarrow Q}$ means... (c) $\boxed{P \Leftrightarrow Q}$ means... (d) $\boxed{\forall x, P(x)}$ means...

15. Let x, y, a, b be fixed real numbers. Show how to deduce the new theorem $(ax + by)^2 \leq (a^2 + b^2)(x^2 + y^2)$ from the known theorem $\forall z \in \mathbb{R}, 0 \leq z^2$ using the Inference Rule for ALL and algebraic manipulations.

16. Which structures $(G, \star, e)$ satisfy all the group axioms? Explain.
    (a) $G = \{-1, 0, 1\}$, $a \star b$ means $a + b$, e is 0.
    (b) $G = \{-1, 1\}$, $a \star b$ means $a \cdot b$, e is 1.
    (c) $G = \mathbb{Z}$, $a \star b$ is always 4, e is 4.
    (d) $G = \{0, 1, 2\}$, $a \star 0 = a = 0 \star a$ for all a in G, $a \star b = 0$ for all nonzero $a, b \in G$, e is 0.
    (e) $G = \{0, 1, 2, \ldots\}$, $a \star b$ is the maximum of a and b, e is 0.
    (f) $G = \mathbb{Z}$, $a \star b = a + b - 2$, e is 2.

17. Assume $(G, \star, e)$ is a group and a, b, c are fixed elements of G such that $a \star b = e = b \star c$. Use the group axioms and the Inference Rule for ALL to prove $a = c$.

18. In a certain logical theory, suppose we adopt the following axioms for equality:
    (i) $\forall x, x = x$;   (ii) $\forall a, \forall b, \forall c, (a = c \wedge b = c) \Rightarrow a = b$.
    (a) Let r and s be fixed objects. Prove $r = s \Rightarrow s = r$, justifying every step using axiom (i), axiom (ii), an instance of a tautology, the Inference Rule for IF, or the Inference Rule for ALL. [*Hint:* $P \Rightarrow [((P \wedge Q) \Rightarrow R) \Rightarrow (Q \Rightarrow R)]$ is a tautology.]
    (b) Let r, s, t be fixed objects. Prove $(r = s \wedge s = t) \Rightarrow r = t$.

19. Let n be a fixed integer. Use the axioms for $\mathbb{Z}$ given in the section to prove $0 \cdot n = 0$. Which axioms are used in the proof?

20. Suppose, instead of using all tautologies, we only allow instances of the following three tautologies as axioms for our theory:
    (i) $P \Rightarrow (Q \Rightarrow P)$;
    (ii) $(P \Rightarrow (Q \Rightarrow R)) \Rightarrow ((P \Rightarrow Q) \Rightarrow (P \Rightarrow R))$;
    (iii) $((\sim Q) \Rightarrow (\sim P)) \Rightarrow (P \Rightarrow Q)$.
    Let A and B be specific, fixed propositions. Using only instances of (i), (ii), and (iii) and the Inference Rule for $\Rightarrow$, prove: (a) $A \Rightarrow A$; (b) $(\sim B) \Rightarrow (B \Rightarrow A)$. [Remarkably, it can be shown that all tautologies are derivable in this way from (i), (ii), and (iii).]

2.2   Proving Existence Statements and IF Statements

Our goal in the next several sections is to learn how to write precise, formal, mathematical proofs. The cornerstone of our exposition is the idea that the *logical structure of a statement dictates the structure of the proof of that statement.* We will give a series of rules, called *proof templates,* for automatically constructing the structural outline of a proof of a statement based on the main logical operator appearing in that statement. For instance, one of the proof templates says that to prove an IF-statement $P \Rightarrow Q$, we may write a proof that starts: "Assume P is true. Prove Q is true." We can then continue to use proof templates recursively to generate a proof of the new goal, Q, aided by the extra information that P is true. We introduce proof templates for existential statements and IF-statements in this section; other logical operators are covered later.

Initial Advice on Proofs

To gain facility at writing proofs, it is crucial to *memorize the proof templates,* and *practice using the proof templates.* Equally crucial is to *memorize all definitions* (did I say that already?), since a very common step in proofs is to replace a defined concept with its definition. You cannot carry out this step if you have not memorized the relevant definition.

   An essential feature of proofs is that *every formula, statement, and letter in a proof has a **logical status** that needs to be explicitly indicated with appropriate words and phrases.* For instance, the logical status of a given statement might be a known result, something to be proved, or a temporary assumption. The logical status of a letter in a proof might be a fixed (constant) object not chosen by us, a specific (constant) object that we choose, or a quantified variable. Here are some examples of sentences in which the logical status words have been written in bold: "**Assume** n is even." "**We must prove** $n + 3$ is odd." "**We know** $n \geq 0$ or $n < 0$." "**Let** x_0 be a **fixed, but arbitrary** object." The proof templates often direct us to write certain sentences that contain logical status words — do not omit these words! Note, for instance, that the IF-template mentioned above contains the logical status words "**assume**" and "**prove.**" In our initial examples of proofs below, we stress the use of logical status words by placing them in **bold type**.

   *At any point in a proof, we can replace any statement by an equivalent statement, which has the same **logical status** as the original statement.* For instance, suppose $A \Leftrightarrow B$ is already known (in the most common case, this might be a definitional axiom in which a new term appearing in A is defined via the statement B). If we have **assumed** A in a proof, then we can continue the proof by saying that we have **assumed** B. If we already **know** A, then we can continue by saying that we **know** B. Finally, if we are trying to **prove** A, then we can instead try to **prove** B. These replacement rules are used constantly in proofs, e.g., every time we expand a definition.

   In many places in a proof, we may need to introduce new variables and constants. Whenever this occurs, we must take care *not to reuse* a letter that has already been given a meaning. For instance, suppose we have just written a line in a proof that says: "**Assume** n is even and m is even." To continue the proof, we might try expanding the definition of "even" in both places. An *incorrect* way to do this would be to say: "**We have assumed** $\exists k \in \mathbb{Z}, n = 2k$ and $\exists k \in \mathbb{Z}, m = 2k$." The error arises because there is no guarantee that the *same* integer k will work for both n and m (indeed, this would only be true if $n = m$). One possible correct way to proceed would be to say: "**We have assumed** $\exists k \in \mathbb{Z}, n = 2k$ and $\exists j \in \mathbb{Z}, m = 2j$," provided that k and j have not already been given a meaning.

Proving Existence Statements: Proof By Example

Here is our first official proof template, which tells us that we can prove existential statements by constructing a specific example of an object that works.

2.11. Proof Template for Proving $\exists x \in U, P(x)$.
Choose a specific object x_0 that you think satisfies the statement. (This x_0 may depend on constants that have already appeared in the proof.)
Prove that x_0 is in the universe U.
Prove that $P(x_0)$ is true.

The proof template does not tell you *how* to choose the object x_0 — that is up to you! Often, the hardest part of applying this template is deciding on an appropriate x_0 to use. This proof method is called *proof by construction* or *proof by example*. Let us turn to some illustrations of this proof template, which use the definitions given in the previous section. Note that the statements we prove in our initial examples are very basic and may seem obvious. We have deliberately done this to allow the reader to focus on each new proof template and the structural features of proofs, without getting sidetracked by complex mathematical content. That will come soon enough!

2.12. Example. Prove that 52 is even.
Proof. By definition of "even," **we must prove** that $\exists k \in \mathbb{Z}, 52 = 2k$. **Choose** $k = 26$. **We know** that 26 is an integer, and $52 = 2 \times 26$ by arithmetic.
Comments: The first step in this proof is replacing a defined term by the text of its definition. We do this step *all the time* in proofs. (Did you memorize the definitions?) This produced an existence statement that needed to be proved. Here we could immediately see which specific k would work, but it was still necessary to point out that k was in the universe $\mathbb{Z}$. If you think this is not really necessary, try imitating this proof to prove that 52 is odd — what goes wrong?

2.13. Example. Prove that 0.36 is rational.
Proof. By definition of "rational," **we must prove** $\exists m \in \mathbb{Z}, \exists n \in \mathbb{Z}, n \neq 0 \wedge 0.36 = m/n$. **Choose** $m = 36$ and $n = 100$. **We know** 36 is an integer, 100 is an integer, $100 \neq 0$, and $0.36 = 36/100$.
Comments: Other choices of m and n also work (for instance, $m = 9$ and $n = 25$), but the proof template only requires us to find a single valid example.

2.14. Example. Prove that 7 divides 0.
Proof. By definition of "divides," **we must prove** $\exists w \in \mathbb{Z}, 0 = 7w$. **Choose** $w = 0$. **Note that** $0 \in \mathbb{Z}$ and $0 = 7 \cdot 0$.
Comments: Was it obvious to you before reading the proof that 7 would divide 0? Also observe that logical status can be indicated in many ways; here we said "note that" to introduce some known arithmetical facts.

In the next example, and throughout this text, we shall sometimes include commentary within a proof indicating what we are thinking and how we are generating the next step of the proof. This commentary appears in square brackets, and is not part of the proof itself.

2.15. Example. Suppose a is a fixed (constant) integer. Prove that $4a^2 + 11$ is odd.
Proof. By definition of "odd," **we must prove** $\exists k \in \mathbb{Z}, 4a^2 + 11 = 2k + 1$. [It is probably not evident at this stage what k to choose. So we instead continue by manipulating $4a^2 + 11$ as follows.] By algebra, **we know** that $4a^2 + 11 = 4a^2 + 10 + 1 = 2(2a^2 + 5) + 1$. **Choose** $k = 2a^2 + 5$; the previous equation **shows** that $4a^2 + 11 = 2k + 1$ is true. Moreover, **since**

a and 2 and 5 are all integers, a^2 and $2a^2$ and $k = 2a^2 + 5$ are also integers because $\mathbb{Z}$ is closed under addition and multiplication.

Comments: We frequently need the closure properties of $\mathbb{Z}$ to see that the objects we choose in proofs do belong to the universe $\mathbb{Z}$ as needed. Also note that the integer k that we chose in the proof was allowed to be an expression involving the previously introduced constant a. This is what we mean in the proof template when we say that the chosen object x_0 is allowed to "depend on" previous constants.

2.16. Example. Let c be a fixed integer. Prove that c divides $5c^3$.
Proof. **We must prove** $\exists m \in \mathbb{Z}, 5c^3 = cm$. **Choose** $m = 5c^2$, which is an integer since it is the product of the integers 5, c, and c. **By algebra,** $cm = c(5c^2) = 5c^3$, as needed.
Comments: When we expanded the definition of "divides," we used m (not c) as the quantified variable since c already had a meaning. As in the last proof, our choice of m in this proof is allowed to depend on the previously introduced constant c.

Direct Proof of an IF-Statement

Theorems involving IF-statements are ubiquitous in mathematics, and we will develop several different methods for proving IF-statements. The most basic method, called *direct proof*, is presented in the following fundamental proof template.

2.17. Proof Template for a Direct Proof of $P \Rightarrow Q$.
Assume P is true. **Prove** Q is true. [To prove Q, use the assumed statement P and other known facts.]

This proof template features a new idea — the notion of *temporarily assuming* a statement that may or may not actually be true. We can explain informally why the proof template is valid by considering the truth table for IF one more time:

P	Q	$P \Rightarrow Q$
T	T	T
T	F	F
F	T	T
F	F	T

When using the direct proof template, our goal is to show that $P \Rightarrow Q$ is true, but we do not yet know that this statement is true. We do know that P, being a proposition, must be either true or false. Now, in the event that P is false, we see from the truth table that $P \Rightarrow Q$ is automatically true; we do not need to prove anything in this case. So we may as well **assume** we are in the other case, where the hypothesis P is true. If we can then deduce that Q must be true, we must be in the first row of the truth table, and in that row, $P \Rightarrow Q$ is true. Note carefully that, at the end of the proof, we have *not* proved that Q itself must be true. Our proof of Q relies on the assumption that P is true, which may not actually be correct. The final conclusion of the proof, i.e., that $P \Rightarrow Q$ is true, does not rely on the temporary assumption that P is true.

Let us start with an example using defined terms you probably have not seen, to show how a proof template can guide you in an unfamiliar setting.

2.18. Example. Outline the proof of this statement: "If f is continuous, then f is measurable." Here, f is a fixed function. Even without knowing the meaning of "continuous" or "measurable," we can still begin the proof by saying: **Assume** that f is continuous. **We must prove** that f is measurable. To proceed, we would have to know the definitions of continuity and measurability.

2.19. Example. Let x_0 be a fixed integer. Prove: if x_0 is odd, then $x_0 + 7$ is even.

Proof. **Assume** x_0 is odd. **We will prove** $x_0 + 7$ is even. [How can we continue? Our only real option is to start expanding definitions; note how we carry along logical status words.] **We have assumed** there is an integer k with $x_0 = 2k+1$. **We must prove** $\exists m \in \mathbb{Z}, x_0 + 7 = 2m$. **From our assumption**, we see that $x_0 + 7 = 2k + 1 + 7 = 2k + 8 = 2(k+4)$. **Choose** $m = k+4$, which is an integer since k is an integer. Our calculation **shows** that $x_0 + 7 = 2m$, as needed.

Comments: We did not get to *choose* k, which came from an *assumed* existential statement. But, we did get to *use* k when we chose m in the existential statement *to be proved*. In general, statements that are already known (or assumed) are treated entirely differently from statements that have yet to be proved. This is one reason it is crucial to indicate the logical status of every statement in a proof.

2.20. Example. Let a, b, c be fixed (constant) integers. Prove: if a divides b and b divides c, then a divides c.

Proof. **Assume** a divides b and b divides c. **We must prove** a divides c. Writing out the definition three times, **we have assumed** that for some integer x, $b = ax$; and for some integer y, $c = by$; **we must prove** $\exists w \in \mathbb{Z}, c = aw$. **Combining our assumptions**, we see that $c = by = (ax)y = a(xy)$. So, **choosing** $w = xy$, $c = aw$ holds. **Note that** $w \in \mathbb{Z}$, since w is the product of the two integers x and y.

Comments: When we wrote out the definition of "divides" in three places, note that we used three different letters (x, y, and w) for the quantified variable. The original definition used c for this variable; but that letter already has a meaning in this setting, so we must avoid it.

The Inference Rule for EXISTS (Optional)

Recall that the Inference Rule for ALL (sometimes called universal instantiation or UI) allows us to pass from a known *universal* statement $\forall x \in U, P(x)$ to particular *instances* of that statement. Specifically, we are allowed to choose any object c known to be in U and deduce that $P(c)$ is a true proposition. We now discuss a related rule, the Inference Rule for EXISTS, which is sometimes called *existential instantiation* or EI. This rule lets us extract information from a known *existential* statement. Suppose we know (or have assumed) the statement $\exists x \in U, P(x)$. The EI rule allows us to invent a new constant symbol, say x_0, and assert that $x_0 \in U$ and $P(x_0)$ are true. It is crucial to note that we do *not* have the freedom to make a specific choice of x_0 (in contrast to the UI rule). In other words, we know that x_0 exists, but we do not have any control over what it is (other than knowing it is in U and makes $P(x)$ true). The symbol x_0 created by the EI rule must be a *new* constant, which does not already have some other meaning assigned to it.

In the main text, we tacitly use the EI rule without explicit mention, since the rule often confuses beginners. To see how this rule works, let us return to the first few lines of the proof in Example 2.19. We initially assumed x_0 is odd, and stated the new goal of proving $x_0 + 7$ is even. Invoking the definition of ODD, we see that we have assumed an *existential* statement, namely $\exists k \in \mathbb{Z}, x_0 = 2k + 1$. To proceed with the proof, we must process the existential quantifier using the Inference Rule for EXISTS (EI). That rule provides us with a new *constant* k_0 that we cannot choose, but which we know satisfies $k_0 \in \mathbb{Z}$ and $x_0 = 2k_0 + 1$. Here k (with no subscript) is a quantified *variable*, whereas k_0 is a fixed, but unknown *constant*. Technically, $x_0 = 2k + 1$ is not a proposition (it is an open sentence), but $x_0 = 2k_0 + 1$ is a proposition. The proof given earlier in this section did not draw attention to this logical nuance, instead passing directly from the assumption "x_0 is odd" to the existence of a constant integer (called k) satisfying $x_0 = 2k + 1$. Here is a version of the proof where EI is invoked explicitly, and where subscripts are used throughout to distinguish quantified variables from constants.

Proof. Assume x_0 is odd; we must prove $x_0 + 7$ is even. We assumed $\exists k \in \mathbb{Z}, x_0 = 2k + 1$. By EI, let k_0 be a fixed integer satisfying $x_0 = 2k_0 + 1$. We must prove $\exists m \in \mathbb{Z}, x_0 + 7 = 2m$. Choose $m_0 = k_0 + 4$, which is in $\mathbb{Z}$ since it is the sum of two integers. Now calculate $x_0 + 7 = 2k_0 + 1 + 7 = 2k_0 + 8 = 2(k_0 + 4) = 2m_0$. We have now proved $\exists m \in \mathbb{Z}, x_0 + 7 = 2m$, so $x_0 + 7$ is indeed even.

Formal Justification of Proof by Example (Optional)

We hope the proof template for proving an existential statement $\exists x, P(x)$ is very plausible at the intuitive level: if you are able to choose a specific object x_0 and prove $P(x_0)$ is true, then surely $\exists x, P(x)$ is true! However, we can also give a more formal justification of this proof technique based on an unexpected connection to the Inference Rule for ALL. One way to formulate this rule is to introduce axioms of the form $(\forall x, P(x)) \Rightarrow P(x_0)$. There is one such axiom for each open sentence $P(x)$, variable x, and constant x_0. We can replace the Inference Rule for ALL by these new axioms, since the Inference Rule for IF allows us to combine the axiom $(\forall x, P(x)) \Rightarrow P(x_0)$ and a known statement $\forall x, P(x)$ to deduce the new theorem $P(x_0)$. Now recall that every IF-statement is logically equivalent to its contrapositive. Apply this comment to the new axiom $(\forall x, \sim P(x)) \Rightarrow \sim P(x_0)$ to see that $(\sim\sim P(x_0)) \Rightarrow \sim\forall x, \sim P(x)$ has the logical status of an axiom. Finally, use denial rules to simplify this statement into the form $P(x_0) \Rightarrow \exists x, P(x)$. We can now justify "proof by example" as follows: once we have proved $P(x_0)$ is true, we use the previous formula and the Inference Rule for IF to obtain the new theorem $\exists x, P(x)$. For simplicity, we gave this discussion for unrestricted quantifiers, but the same argument works for restricted quantifiers.

Section Summary

1. *Proof Templates.* The logical structure of a statement dictates the structure of the proof of that statement. For each logical operator discussed earlier (like $\Rightarrow$, $\exists$, etc.), we will give a proof template for proving statements built from that operator. Statements containing several operators are proved by applying several proof templates recursively.

2. *General Advice for Proofs.* Memorize all definitions! Memorize and practice the proof templates! Do not forget to introduce every statement, formula, and letter in a proof with words indicating their logical status! Do not reuse letters that already have a meaning! Do not overuse exclamation marks! You can always replace a statement by an equivalent statement, which has the same logical status as the original statement. We use this step all the time when expanding definitions.

3. *Proof by Example.* To prove $\exists x \in U, P(x)$: **choose** a specific object x_0 (which can depend on previously introduced constants); **prove** $x_0 \in U$; and **prove** $P(x_0)$. We often need closure properties of U to see that the chosen object does belong to U.

4. *Direct Proof.* To prove $P \Rightarrow Q$: **assume** P; **prove** Q. After writing these two lines, we often continue by expanding definitions appearing in P and Q, or using another proof template to begin proving Q.

Exercises

1. Let a and b be fixed integers. Give careful proofs of each statement, making sure to use appropriate logical status words. (a) -53 is odd. (b) $\sqrt{2.25}$ is rational. (c) 1 divides a. (d) $8a - 6b$ is even. (e) b is rational.

2. Let r and s be fixed integers. Prove the following statements. (a) 4298 is even. (b) $1.3/5.2$ is rational. (c) $-s$ divides s. (d) $6r + 4s - 3$ is odd. (e) $r - s$ divides $r^3 - s^3$.

3. Let x, y be fixed integers. Prove the following statements. (a) $\sqrt{169}$ is rational. (b) $10^7 - 1$ is odd. (c) x divides x^3. (d) If x is even, then $3x^2$ is even. (e) If x is odd and y is even, then $y - x$ is odd.

4. Let m and n be fixed integers, and let x be a fixed real number. Prove: (a) If m is odd, then $5m - 3$ is even. (b) If 5 divides n, then 5 divides mn. (c) If x is rational, then $mx + n$ is rational. (d) $m/(m^2 + n^2 + 1)$ is rational.

5. Let a, b, c be fixed integers. Prove: (a) If a is even, then $3a + 5$ is odd. (b) If a divides b and b divides c, then ab divides c^2. (c) If $a \neq b$, then $(a + b)/(a - b)$ is rational.

6. Let a, b, c, r, s be fixed integers. Prove: if c divides r and c divides s, then c divides $ar + bs$.

7. Let y, z be fixed real numbers. Prove: if y and z are both rational, then $y + z$ is rational.

8. Prove the following existential statements. (a) $\exists m \in \mathbb{Z}, \exists n \in \mathbb{Z}, 30m + 7n = 1$. (b) $\exists x \in \mathbb{Z}, \exists y \in \mathbb{Z}, 13x + 5y = 1$. (c) $\exists a \in \mathbb{Z}, \exists b \in \mathbb{Z}, 100a + 39b = 1$.

9. Prove the following statements. (a) $\exists a \in \mathbb{Z}_{>0}, \exists b \in \mathbb{Z}_{>0}, a^2 + b^2 = 100$. (b) $\exists x \in \mathbb{Z}, \exists y \in \mathbb{Z}, \exists z \in \mathbb{Z}, x^5 + y^5 = z^5$. (c) $\exists a \in \mathbb{Z}, \exists b \in \mathbb{Z}, \exists c \in \mathbb{Z}, \exists d \in \mathbb{Z}, a^2 + b^2 + c^2 + d^2 = 1001$.

10. Let x be a fixed real number. Use facts from algebra to prove: (a) If $3x + 7 = 13$, then $x = 2$. (b) If $x = 2$, then $x^3 - 6x^2 + 11x = 6$. (c) If $x^2 + 2x - 15 = 0$, then $x = 3$ or $x = -5$. (d) If $x > 3$, then $x^2 + 5x - 4 > 20$.

11. Write the converse of each statement in the previous exercise. Explain informally whether each converse is true or false.

12. Write just the outline of a proof of each statement (imitate Example 2.18). Do not expand unfamiliar definitions.
(a) If a group G has 49 elements, then G is commutative.
(b) If a metric space X is complete and totally bounded, then X is compact.
(c) If the series $\sum_{n=1}^{\infty} a_n$ converges, then $\lim_{n \to \infty} a_n = 0$.
(d) If a function f has a right inverse, then f is surjective.

13. Outline the proof of each statement using the templates for IF and $\exists$.
(a) If a group G has 700 elements, then $\exists x \in G$, x has order 7.
(b) If p is prime and $\exists k \in \mathbb{Z}, p = 4k + 1$, then $\exists y \in \mathbb{Z}, \exists z \in \mathbb{Z}, p = y^2 + z^2$.
(c) If g is a polynomial of odd degree, then for some real number r, $g(r) = 0$.

14. Let m and n be fixed positive integers. Carefully explain the error in the following incorrect proof of the statement: "if m is even, then m/n is even." *Proof.* Assume m is even. We must prove m/n is even. We have assumed there exists an integer k with $m = 2k$. We must prove $\exists p \in \mathbb{Z}, m/n = 2p$. Dividing the assumed equation $m = 2k$ by n (which is not zero) produces $m/n = (2k)/n = 2(k/n)$. Hence, choosing $p = k/n$, we do indeed have $m/n = 2p$. So m/n is even.

15. Let a, b, and c be fixed positive integers. We are trying to prove: "if a divides b and a divides c, then a divides $b + c$." Find and correct the error in the proof below. *Proof.* Assume a divides b and a divides c; we must prove a divides $b + c$. We have assumed there is an integer k with $b = ak$. We have also assumed there is an integer k with $c = ak$. We must prove $\exists m \in \mathbb{Z}, b + c = am$. Choose $m = 2k$, which is an integer. Use the assumptions to compute $b + c = ak + ak = a(2k) = am$, as needed.

16. Let n be a fixed integer. Find the error in the following proof that $n^2 + n$ is even. *Proof.* We must prove there exists k such that $n^2 + n = 2k$. Choose $k = (n^2 + n)/2$. Multiplying both sides by 2, we see that $n^2 + n = 2k$ does hold.

17. Let x be a fixed integer. Criticize the following proof of the statement "if x is even then $x^3 - 1$ is odd." *Proof.* $x = 2k$, $k \in \mathbb{Z}$, $x^3 - 1 = (2k)^3 - 1 = 8k^3 - 1 = 2(4k^3 - 1) + 1$, $4k^3 - 1 \in \mathbb{Z}$.

18. Let c be a fixed integer. Criticize the following proof of the statement "if 10 divides $c + 4$, then 10 divides $c - 6$." Assume 10 divides $c - 6$, which means $\exists a \in \mathbb{Z}, c - 6 = 10a$. Prove 10 divides $c + 4$, which means $\exists b \in \mathbb{Z}, c + 4 = 10b$. Choose $b = a + 1$, which is in $\mathbb{Z}$ by closure. Use algebra to compute $10b = 10(a + 1) = 10a + 10 = c - 6 + 10 = c + 4$.

19. Let r and s be fixed real numbers. Find the error in the following proof that if r and s are rational and $s \neq 0$, then r^2/s^2 is rational. *Proof.* Assume r and s are rational and $s \neq 0$. Prove r^2/s^2 is rational. We must prove $\exists m \in \mathbb{Z}, \exists n \in \mathbb{Z}, n \neq 0$ and $r^2/s^2 = m/n$. Choose $m = r^2$ and $n = s^2$; note that $n \neq 0$ because $s \neq 0$.

20. Give a correct proof of the statement in the previous problem.

21. Let a and b be fixed nonzero integers. Find the error in the following proof of the statement "if a divides b and b divides a then $b = \pm a$." *Proof.* Assume a divides b and b divides a. Prove $b = \pm a$. By assumption, there is an integer c with $b = ca$ and $a = cb$. By substitution, $a = cb = c(ca) = c^2 a$. Since $a \neq 0$, we can divide by a to get $c^2 = 1$. Then $c = 1$ or $c = -1$, hence $b = a$ or $b = -a$.

22. Define a rational number r to be *dyadic* iff $\exists m \in \mathbb{Z}, \exists k \in \mathbb{Z}_{\geq 0}, r = m/2^k$. Let x and y be fixed rational numbers. Prove: (a) If x and y are dyadic, then xy is dyadic. (b) If x and y are dyadic, then $x + y$ is dyadic.

23. Use the group axioms in §2.1 to prove this theorem about fixed elements a, b, c in a group $(G, \star, e)$: if $a \star b = a \star c$, then $b = c$.

24. Define an integer m to be *bisquare* iff $\exists a \in \mathbb{Z}, \exists b \in \mathbb{Z}, m = a^2 + b^2$. Let x, y be fixed integers. Prove: (a) 13 is bisquare. (b) 200 is bisquare. (c) 610 is bisquare. (d) x^2 is bisquare. (e) If x is bisquare, then xy^2 is bisquare. (f) Not every integer is bisquare.

25. Let r and s be fixed integers. Prove: if r and s are bisquare, then rs is bisquare.

2.3   Contrapositive Proofs and IFF Proofs

Recall our direct proof template for proving $P \Rightarrow Q$: "**Assume** P. **Prove** Q." Let us see what happens when we use this template in the following example.

2.21. Example. Let x be a fixed integer. Prove: if x^2 is even, then x is even.

Proof. **Assume** x^2 is even; **prove** x is even. **We have assumed** that for some integer k, $x^2 = 2k$. **We must prove** $\exists m \in \mathbb{Z}, x = 2m$. Taking the square root of the **known** equation $x^2 = 2k$, **we deduce**

$$x = \sqrt{2k} = \sqrt{4(k/2)} = 2\sqrt{k/2}.$$

So **choosing** $m = \sqrt{k/2}$, we have $x = 2m$, as needed.

   This proof has a minor flaw and a major flaw — can you spot them? The minor flaw is that x might be negative, but $2\sqrt{k/2}$ is always nonnegative. This can be corrected by using the negative of the square root when $x < 0$. The major flaw is that the proof did not check that m was in $\mathbb{Z}$, as required by the definition. In general, dividing an integer k by 2 and then taking a square root produces a real number that may *not* be an integer. There is no straightforward way to correct this error. We need a new approach, called the contrapositive proof method.

Contrapositive Proofs

Recall that to prove any statement A, we can instead prove a new statement B that is known to be equivalent to A. In particular, taking A to be the IF-statement $P \Rightarrow Q$, we know that the contrapositive IF-statement $(\sim Q) \Rightarrow (\sim P)$ is equivalent to A. Using the direct proof template to prove this new statement, we arrive at the following new proof template for proving the original IF-statement, called *proof by contrapositive*.

2.22. Contrapositive Proof Template to Prove $P \Rightarrow Q$.
Assume $\sim Q$. **Prove** $\sim P$.
[We often replace $\sim Q$ and $\sim P$ by useful denials of Q and P, respectively. These are found using the denial rules.]

   We can repair the direct proof in Example 2.21 by doing a contrapositive proof instead. For this new proof, and throughout our initial discussion of proofs, we assume that the following fact about odd and even numbers is already known:

$$\forall n \in \mathbb{Z}, n \text{ is not even} \Leftrightarrow n \text{ is odd.} \tag{2.1}$$

(Though you are probably aware of this fact, it is not so easy to prove! We provide a proof of this fact later.) Since $(\sim P) \Leftrightarrow Q$ is equivalent to $P \Leftrightarrow (\sim Q)$, the fact also says that for all integers n, n is even iff n is not odd. Since $(\sim P) \Leftrightarrow Q$ is also equivalent to $P \oplus Q$, yet another way to rephrase the fact is to say that every integer is even or odd, but not both. With this fact in hand, let us return to our initial example.

2.23. Example. Let x be a fixed integer. Prove: if x^2 is even, then x is even.

Proof by Contrapositive. **Assume** x is not even. **Prove** x^2 is not even. Using the known fact, **we have assumed** x is odd, **which means** there is an integer k with $x = 2k + 1$. Similarly, **we must prove** x^2 is odd, **which means** $\exists m \in \mathbb{Z}, x^2 = 2m + 1$. Squaring the **known** equation $x = 2k + 1$ gives $x^2 = 4k^2 + 4k + 1 = 2(2k^2 + 2k) + 1$. So **choosing** $m = 2k^2 + 2k$, which is an integer by closure properties of $\mathbb{Z}$, we have $x^2 = 2m + 1$ as needed.

Comments: We can see retroactively why the contrapositive proof works and the direct proof fails. Namely, proving the contrapositive allows us to pass from assumed information about x to needed information about x^2 by squaring an equation. This step succeeds because $\mathbb{Z}$ is closed under addition and multiplication. On the other hand, in the direct proof we tried to go from assumptions about x^2 to needed information about x by taking square roots. Here we did not have closure properties to guarantee that the integer m (chosen at the end of the proof) must be an integer.

2.24. Example. Let x and y be fixed real numbers. Prove: if $x^2y + xy^2 < 30$, then $x < 2$ or $y < 3$. *Proof by Contrapositive.* **Assume** $x \geq 2$ and $y \geq 3$. **Prove** $x^2y + xy^2 \geq 30$. To proceed, we need to use two **known theorems** from algebra regarding inequalities:

$$\forall a, b, c, d \in \mathbb{R}, (a \geq b \wedge c \geq d) \Rightarrow a + c \geq b + d; \tag{2.2}$$

$$\forall a, b, c, d \in \mathbb{R}, (a \geq b \geq 0 \wedge c \geq d \geq 0) \Rightarrow ac \geq bd \geq 0. \tag{2.3}$$

Taking $a = x$, $b = 2$, $c = y$, and $d = 3$ in the first theorem, we **deduce** that $x + y \geq 2 + 3$, i.e., $x + y \geq 5$. [We have just used the Inference Rules for ALL and IF.] Using the same values in the second theorem, we **deduce** that $xy \geq 2 \cdot 3 = 6$. Now, since $x + y \geq 5 \geq 0$ and $xy \geq 6 \geq 0$, we can use the second theorem again to **deduce** that $(x + y)xy \geq 30$. By algebra, **we know** $(x + y)xy = x^2y + xy^2$. **So** $x^2y + xy^2 \geq 30$, completing the proof.

Comments: How did we know to try a contrapositive proof here, rather than a direct proof? Perhaps we started a direct proof by saying: "Assume $x^2y + xy^2 < 30$. Prove $x < 2$ or $y < 3$." At this point, we probably got stuck because it was unclear how to move forward from the rather complicated assumption. Doing a contrapositive proof instead provides the much more tractable assumption "$x \geq 2$ and $y \geq 3$."

2.25. Example. Let x and y be fixed integers. Prove: if xy is irrational, then x is irrational or y is irrational.

Proof. [The given statement has a lot of negative logic in it (note "irrational" means "not rational"), so we try simplifying it with a contrapositive proof. Denying the conclusion, we begin with the line:] **Assume** x is rational AND y is rational. [Denying the hypothesis, we continue by saying:] **Prove** xy is rational. [We continue by expanding definitions:] **We have assumed** $x = m/n$ and $y = p/q$ for some integers m, n, p, q with n and q nonzero. **We must prove** $\exists a \in \mathbb{Z}, \exists b \in \mathbb{Z}, b \neq 0 \wedge (xy = a/b)$. Multiplying together our **assumptions**, we **deduce** that $xy = (mp)/(nq)$. **Choose** $a = mp$ and $b = nq$; **note** a and b are integers by closure of $\mathbb{Z}$, and $b \neq 0$ since b is the product of two nonzero integers. Also $xy = a/b$ by the above calculation. [We have proved our goal, so the proof is now done.]

Proving an AND Statement

Let us take a break from the logical subtleties of IF-statements to introduce one of the more straightforward proof templates. This template tells us how to prove an AND-statement.

2.26. Proof Template to Prove $P \wedge Q$.
Part 1. **Prove** P.
Part 2. **Prove** Q.

I hope this proof template is fully believable to you at the intuitive level. But we can still justify it by appealing to the truth table defining $\wedge$, as follows. Consulting the truth table, we see that the composite statement $P \wedge Q$ is true if and only if the two separate statements P (on the one hand) and Q (on the other hand) are both true. Thus to prove the composite statement, it is enough to prove P by itself and also to prove Q by itself. These proofs often proceed by recursively using more proof templates. More generally, to prove an AND-statement of the form $P_1 \wedge P_2 \wedge \cdots \wedge P_n$, we must carry out n independent subproofs: **prove** P_1; next, **prove** P_2; ...; finally, **prove** P_n.

2.27. Example. Let b be a fixed integer. Prove 1 divides b and b divides b.
Proof. First, we **prove** 1 divides b. By definition, **we must prove** $\exists c \in \mathbb{Z}, b = 1c$. **Choose** $c = b$; **note** c is an integer and $1c = 1b = b$. Second, we **prove** b divides b. **We must prove** $\exists d \in \mathbb{Z}, b = bd$. **Choose** $d = 1$; this is an integer, and $bd = b1 = b$.

2.28. Example. Let x and y be fixed real numbers. Prove: if $x < y$ then $x < (x+y)/2 < y$.
Proof. **Assume** $x < y$. **Prove** $x < (x+y)/2 < y$. The goal contains a hidden AND; so **we must prove** $x < (x+y)/2$ AND $(x+y)/2 < y$.
<u>Part 1.</u> **Prove** $x < (x+y)/2$. Adding x to both sides of the **assumption** $x < y$, **we deduce** $2x < x + y$. Multiplying this by the positive number $1/2$, **we get** $x < (x+y)/2$.
<u>Part 2.</u> **Prove** $(x+y)/2 < y$. Adding y to both sides of the **assumption** $x < y$, **we deduce** $x + y < 2y$. Multiplying this by the positive number $1/2$, **we get** $(x+y)/2 < y$.

Proving IFF Statements

Our next proof template tells us how to prove an "if and only if" statement $P \Leftrightarrow Q$. The template arises by recalling that $P \Leftrightarrow Q$ is equivalent to $(P \Rightarrow Q) \wedge (Q \Rightarrow P)$. We already know how to prove this new statement, using the templates for $\wedge$ and $\Rightarrow$. Writing this out, we get the following.

2.29. Proof Template for Proving $P \Leftrightarrow Q$.
<u>Part 1.</u> **Prove** $P \Rightarrow Q$ (by any method).
<u>Part 2.</u> **Prove** $Q \Rightarrow P$ (by any method).

We should make a few comments about this template. The proof method is called a *two-part proof* of the IFF-statement, since the proof requires us to prove two separate IF-statements. Part 1 is often called the *forward direction* or the *direct implication*. Part 2 is often called the *backward direction* or the *converse implication*. If we are proving "P is necessary and sufficient for Q," part 1 can be called the *proof of sufficiency*, and part 2 can be called the *proof of necessity*.

Beginners sometimes forget to write part 2 by the time they finish writing part 1 of the proof. Thus, they have omitted half of the proof! To avoid this, I recommend setting up the entire proof structure shown above, by writing "<u>Part 1.</u> Prove $P \Rightarrow Q$," leaving lots of space, writing "<u>Part 2.</u> Prove $Q \Rightarrow P$," and then filling in both parts. Another remark is that there are multiple ways of completing part 1 — we could use direct proof, contrapositive proof, or proof by contradiction (to be discussed later). Independently of part 1, we have similar choices for how to prove part 2. The natural choice is to try a direct proof for each part. Often, however, one direction of the proof may be easier using a contrapositive proof (or perhaps a contradiction proof). Remember that these options are available. One final point: any temporary assumptions made in part 1 (e.g., "assume P") are no longer available when we reach part 2. Similarly, things proved in part 1 using the assumptions in part 1 can no longer be used in part 2 (unless we can reprove them using the assumptions in part 2). On the other hand, variables and constants introduced in part 1 of the proof are allowed to be reused (with fresh meanings) in part 2 — although it may be preferable to use new letters instead to avoid confusion.

2.30. Example. Let x be a fixed integer. Prove x is even iff x^2 is even.
Proof. <u>Part 1.</u> **Assume** x is even; **prove** x^2 is even. **We have assumed** there is an integer

k with $x = 2k$. **We must prove** $\exists m \in \mathbb{Z}, x^2 = 2m$. Squaring the **assumption** gives $x^2 = 4k^2 = 2(2k^2)$. **Choose** $m = 2k^2$; then $m \in \mathbb{Z}$ and $x^2 = 2m$, as needed.

<u>Part 2.</u> **We must prove:** "if x^2 is even, then x is even." We did this earlier in the section (using a contrapositive proof), so we do not repeat the proof here.

2.31. Example. Let r be a fixed real number. Prove r is rational iff $-r$ is rational.

Proof. <u>Part 1.</u> **Assume** r is rational. **Prove** $-r$ is rational. **We have assumed** there are integers p, q with $q \neq 0$ and $r = p/q$. **We must prove** $\exists m, n \in \mathbb{Z}, n \neq 0 \wedge -r = m/n$. Since $-r = (-p)/q$, we can **choose** $m = -p$ and $n = q \neq 0$; these are integers with $-r = m/n$.

<u>Part 2.</u> **Assume** $-r$ is rational. **Prove** r is rational. **We have assumed** there are integers p, q with $q \neq 0$ and $-r = p/q$. **We must prove** $\exists m, n \in \mathbb{Z}, n \neq 0 \wedge r = m/n$. Since $r = -(-r) = (-p)/q$, we can **choose** $m = -p$ and $n = q \neq 0$; these are integers with $r = m/n$.

Comments: Part 2 is totally separate from part 1, so we are allowed to reuse the letters p, q, m, n in part 2 (which no longer have the meanings they did in part 1). Despite the extreme similarity of the two parts of this proof, both parts are needed to have a complete proof of the IFF-statement. On the other hand, we will see later that part 1 of the proof really proved a *universal* statement: $\forall r \in \mathbb{R}, (r \in \mathbb{Q} \Rightarrow -r \in \mathbb{Q})$. This leads to a shorter proof of part 2 that relies on the result proved in part 1. Using the Inference Rule for ALL, we can replace r by $-r$ in the universal statement written above, to see that $(-r) \in \mathbb{Q} \Rightarrow -(-r) \in \mathbb{Q}$. Since $-(-r) = r$ by algebra, we get $(-r) \in \mathbb{Q} \Rightarrow r \in \mathbb{Q}$, which proves part 2 without expanding any definitions.

2.32. Example. Outline the proof of this statement:
"if $X \subseteq \mathbb{R}^k$, then: X is compact iff X is closed and bounded."
Solution. We can use the proof templates to generate the structural skeleton of the proof without knowing the definitions of "compact," "closed," and "bounded." Here is the proof outline:

> **Assume** $X \subseteq \mathbb{R}^k$. **Prove** X is compact iff X is closed and bounded.
>> <u>Part 1.</u> **Assume** X is compact.
>>> *Part 1a.* **Prove** X is closed. (...)
>>> *Part 1b.* **Prove** X is bounded. (...)
>> <u>Part 2.</u> **Assume** X is closed and bounded. **Prove** X is compact. (...)

Comments: The main logical operator in the given statement is IF, so we generate the first line of the solution using the direct proof template for IF. This reduces us to proving an IFF-statement, which requires a two-part proof. We use the direct proof template two more times (along with the template for AND in part 1) to finish the proof outline. We write (...) for omitted portions of the proof, which would require us to know the definitions of the terms involved. Note that the assumption $X \subseteq \mathbb{R}^k$ is available throughout the entire proof, whereas the assumption that X is compact can be used only in part 1. Other outlines are possible. For instance, we could use a contrapositive proof in part 2 and write "**Assume** X is not compact; **prove** X is not closed or not bounded." This new goal could be expanded further using the OR-proof template, given later.

Formally Justifying the Proof Templates (Optional)

We have given intuitive justifications for the new proof templates in this section based on truth tables. We can give a completely rigorous derivation of these templates based on certain tautologies and the Inference Rule for IF. For example, consider the template for proving the AND-statement $P \wedge Q \wedge R$. Suppose we can prove P by itself, Q by itself, and R by itself. Note that $(P \Rightarrow (Q \Rightarrow (R \Rightarrow (P \wedge Q \wedge R))))$ is an axiom, because it is a tautology (as one may check with a truth table). Applying the Inference Rule for IF to this tautology and the known statement P, we get $(Q \Rightarrow (R \Rightarrow (P \wedge Q \wedge R)))$. Applying the Inference Rule for IF to this new statement and the known statement Q, we get $(R \Rightarrow (P \wedge Q \wedge R))$. Finally, applying the Inference Rule for IF to this statement and the known statement R, we get the sought-for theorem $P \wedge Q \wedge R$. This chain of deductions, using the Inference Rule for IF three times in a row, is the same for any AND-statement $P \wedge Q \wedge R$. The proof template for AND lets us condense this argument by just proving P, Q, and R separately.

Next, consider the contrapositive proof template for proving $P \Rightarrow Q$. Assume the direct proof template for IF-statements is already available. Then assuming $\sim Q$ and proving $\sim P$ proves the new theorem $(\sim Q) \Rightarrow (\sim P)$. Using this theorem and the tautology

$$((\sim Q) \Rightarrow (\sim P)) \Rightarrow (P \Rightarrow Q),$$

the Inference Rule for IF gives the new theorem $P \Rightarrow Q$. The proof template for IFF can be justified similarly (see Exercise 24).

Section Summary

1. *Contrapositive Proof Template.* To prove $P \Rightarrow Q$ via a contrapositive proof:
   Assume $\sim Q$. **Prove** $\sim P$.

2. *AND Proof Template.* To prove $P_1 \wedge P_2 \wedge \cdots \wedge P_n$:
   Part 1. **Prove** P_1.
   Part 2. **Prove** P_2.

   . . .

   Part n. **Prove** P_n.

3. *IFF Proof Template.* To prove $P \Leftrightarrow Q$:
   Part 1. **Prove** $P \Rightarrow Q$ (by any method).
   Part 2. **Prove** $Q \Rightarrow P$ (by any method).

4. *Points of Advice.* (a) Memorize the new proof templates! (b) Given a complex statement built up using multiple logical operators, you need to combine multiple proof templates to generate the structure of the proof. (c) When do we use contrapositive proofs? Try this method when a direct proof fails, or when the original statement contains negative logic, or when the hypothesis of the given IF-statement seems unwieldy. (d) When proving an IFF-statement, do not forget that *both* implications must be proved! Also, the two parts of the proof are completely independent, so that assumptions made in part 1 cannot be used in part 2, and vice versa. But letters used in part 1 can be reused in part 2, with new meanings.

Exercises

1. Use the proof templates studied so far to write just the structural outline of a proof of each statement. Do not expand definitions of unfamiliar words.
   (a) If G is a cyclic group, then G is commutative. [use contrapositive proof]
   (b) f is a bijection iff f is injective and surjective. [use direct proof for both parts]
   (c) If f is continuous or monotonic, then f is integrable. [use contrapositive proof]

2. Let a, b, c be fixed integers. Write a contrapositive proof of each statement.
   (a) If ab is even, then a is even or b is even.
   (b) If a does not divide bc, then a does not divide c.

3. Let x and y be fixed real numbers. Prove: if $x - y$ is irrational, then x is irrational or y is irrational.

4. Let m, n be fixed integers, and let x be a fixed real number. Prove each IFF-statement.
   (a) $2x + 3 > 11$ iff $x > 4$.
   (b) n is odd if and only if $-3n$ is odd.
   (c) n divides m iff n divides $-m$.

5. Let x be a fixed real number. Prove:
   (a) $3x + 7 = 5x - 4$ iff $x = 11/2$;
   (b) $(x + 3)^2 = (x - 1)^2$ iff $x = -1$.

6. Let x be a fixed real number. Prove:
   (a) $9x + 4 < 2x - 10$ iff $x < -2$;
   (b) $2 < x < 5$ iff $-6 < 4 - 2x < 0$;
   (c) $(x + 2)^2 > (x + 5)^2$ iff $x < -3.5$.

7. Let x be a fixed real number. Prove: if $x^5 + 2x^3 + x < 50$, then $x < 2$.

8. Let x and y be fixed real numbers. Prove: if $x < y$ then $x < (2x + y)/3 < (x + 2y)/3 < y$.

9. Let x be a fixed integer. Prove that x is odd iff x^2 is odd: (a) using (2.1) and the definitions; (b) using (2.1) and the theorem proved in Example 2.30.

10. Let c be a fixed integer. Prove $\exists a \in \mathbb{Z}, c = 10a + 2$ iff $\exists b \in \mathbb{Z}, c = 5b - 3$ and b is odd.

11. Let a, j, r, s be fixed integers. Prove: if j divides $r + s$, then: $a = bj + r$ for some integer b iff $a = cj - s$ for some integer c.

12. Let a, b, d, q, r be fixed integers. Using only the definitions, prove: if $b = aq + r$, then: (d divides a and d divides r) iff (d divides a and d divides b).

13. (a) Informally justify the contrapositive proof template for proving $P \Rightarrow Q$ by analyzing the truth table for $P \Rightarrow Q$. (b) Informally justify the proof template for proving $P \Leftrightarrow Q$ by analyzing the truth table for $P \Leftrightarrow Q$.

14. Write a proof outline for each statement based on proof templates studied so far.
    (a) $(P \Leftrightarrow Q) \Rightarrow (S \Leftrightarrow T)$.
    (b) $(\exists x, P(x)) \Leftrightarrow (Q \wedge \exists y, R(y))$.
    (c) $\exists x \in U, [P(x) \Rightarrow (\sim Q(x) \vee R(x))]$.

15. Find the errors (if any) in each proof outline below.
    (a) *Statement:* If $ab \in P$, then $a \in P$ or $b \in P$. *Proof Outline.* We use a contrapositive proof. Assume $a \notin P$ or $b \notin P$. Prove $ab \notin P$. (...)
    (b) *Statement:* F is closed iff $X - F$ is open. *Proof Outline.* <u>Part 1.</u> Assume F is closed. Prove $X - F$ is open. (...) <u>Part 2.</u> Assume F is not closed. Prove $X - F$ is not open. (...)
    (c) *Statement:* A is invertible iff $\det(A) \neq 0$. *Proof Outline.* <u>Part 1.</u> Assume A is invertible. Prove $\det(A) \neq 0$. (...) <u>Part 2.</u> Assume $\det(A) = 0$. Prove A is not invertible. (...)

16. Identify and correct all errors in the following proof that x is odd iff $3x + 7$ is even (where x is a fixed integer). *Proof.* Assume x is odd, which means $x = 2k + 1$. We must prove $3x + 7$ is even, which means $3x + 7 = 2k$ for some integer k. By algebra, we know $3x + 7 = 3(2k + 1) + 7 = 6k + 8 = 2(3k + 4)$. So we can choose $k = 3k + 4$.

17. (a) For fixed positive real numbers x and y, prove: if $x \neq y$ then $x^2 \neq y^2$.
    (b) State the converse of part (a). Is the converse always true?
    (c) Is $\forall x, y \in \mathbb{R}, x = y \Leftrightarrow x^2 = y^2$ true or false? Explain.

18. Prove (2.2) and (2.3) using only the ordering axioms for the set of real numbers (see axioms O1 through O6 in §8.1).

19. Use (2.3) (and the analogous fact for $>$) to prove: for any fixed $x, y \in \mathbb{R}_{\geq 0}$, $x \leq y$ iff $x^2 \leq y^2$.

20. Let r and s be fixed positive real numbers. Prove: if $r \leq s$, then $r \leq \sqrt{rs} \leq s$.

21. Suppose $U = \{z_1, z_2, \ldots, z_n\}$ is a finite set. Create a proof template for proving $\forall x \in U, P(x)$ by converting this quantified statement to an AND-statement.

22. Create and justify two proof templates for proving $P \vee Q$ by replacing this OR-statement by an equivalent IF-statement.

23. Create and justify a proof template for proving $P \oplus Q$ by replacing this XOR-statement by an equivalent IFF-statement.

24. Carefully justify the proof template for IFF using the Inference Rule for IF and an appropriate tautology.

2.4   Proofs by Contradiction and Proofs of OR-Statements

This section introduces a powerful general technique for proving statements called *proof by contradiction*. After studying some proofs using this technique, we introduce some proof templates for proving XOR-statements and OR-statements. We will see that previous proof templates for proving IF-statements (the direct proof and contrapositive proof methods) can be viewed as special cases of the OR-proof templates. This should not be too surprising, since we know an IF-statement can always be converted to an equivalent OR-statement.

Template for Proof by Contradiction

The method of proof by contradiction is a versatile, but sometimes confusing, technique for proving many different kinds of statements. The intuition for this proof method is as follows. Suppose we are trying to prove that some statement P is true. One way to proceed is to assume that P is false, and see what goes wrong. More specifically, if the assumption that P is false leads to a contradiction, then P must be true after all. These ideas suggest the following proof template.

2.33. Proof Template to Prove Any Statement P by Contradiction.
Assume, to get a contradiction, $\sim P$.
Use this assumption and other known facts to **deduce** a contradiction $\mathcal{C}$.
Conclude that P must be true after all.

2.34. Remarks on the Contradiction Proof Template.

(a) As always, the logical status word "**assume**" in Step 1 is *mandatory*! The next phrase in Step 1, "to get a contradiction," while not absolutely mandatory, is highly recommended. The reader of your proof is likely to get confused if you suddenly assume a false statement $\sim P$, which is the very opposite of the statement to be proved. You can allay this confusion by stating at the outset that the sole purpose of this assumption is to reach a contradiction. Similarly, upon reaching a contradiction at the end of the proof, it is good to finish with a line such as: "This contradiction shows that the original statement P is true."

(b) In Step 1, we almost always replace the assumption $\sim P$ with a *useful* denial of P, found using the denial rules.

(c) Step 2 of the proof template tells you to "deduce a contradiction," but it does not tell you exactly what contradiction you are aiming for. This makes proofs by contradiction harder to finish compared to other types of proofs, since you do not have a specific goal to reach. You must explore the consequences of the false assumption $\sim P$, combining it with other known facts to see what happens.

(d) Remember that a contradiction is a propositional form that evaluates to false in every row of its truth table; a typical example is $P \wedge (\sim P)$. To be perfectly precise, Step 2 is really deducing an *instance* of a contradiction, which is a proposition obtained by replacing every propositional variable in the contradiction by specific statements.

Before looking at examples of proof by contradiction, we give a technical justification for why the proof method works. Suppose we succeed in completing Steps 1 and 2 of the contradiction proof template. Then, according to the direct proof template discussed earlier, we have really proved the statement $(\sim P) \Rightarrow \mathcal{C}$. Now, recall from Example 1.31 that $((\sim P) \Rightarrow \mathcal{C}) \Rightarrow P$ is a tautology and hence an axiom. Since we have also just proved $(\sim P) \Rightarrow \mathcal{C}$, it follows from the Inference Rule for IF that P is a theorem, as needed.

Examples of Proof by Contradiction

We now give examples of the method of proof by contradiction.

2.35. Example. Prove that no integer is both even and odd.

Proof. **We must prove:** $\sim\exists x \in \mathbb{Z}, x$ is even $\wedge\, x$ is odd. **Assume, to get a contradiction,** that there does exist $x \in \mathbb{Z}$ such that x is even and x is odd. This **assumption** means that for some integer k, $x = 2k$, and for some integer m, $x = 2m + 1$. **Combining** these assumptions, we see that $2k = 2m + 1$, **hence** $2(k - m) = 1$, so $k - m = 1/2$. On one hand, **we know** $1/2$ is not an integer. On the other hand, since $k, m \in \mathbb{Z}$, $1/2 = k - m$ is an integer by closure properties of $\mathbb{Z}$. We have reached the contradiction "$1/2$ is not an integer and $1/2$ is an integer." Thus, our assumption is wrong, meaning that no integer is both even and odd.

Comments: The given statement P begins with NOT, so we began the proof by using the double negation rule to deny P. We then expanded definitions until discovering a contradiction.

2.36. Example. Prove: for all positive integers a and b, if a divides b then $a \leq b$.

Proof. **Assume, to get a contradiction,** that there exist positive integers a, b such that a divides b and $a > b$. **We know** there is an integer c with $b = ac$. **Since** a and b are both positive, we see that $c = b/a$ must also be positive. **As** c is an integer, $c \geq 1$ **follows.** But now, by a **known** theorem (see (2.3) in §2.3), we deduce

$$b = ac \geq a \cdot 1 = a > b.$$

We have reached the contradiction $b > b$, so the original statement is true.

Comments: To see the contradiction more explicitly, note that $b > b$ is the negation of $b \leq b$. Since $b \leq b$ is a known fact, the proof has derived the contradiction "$(b \leq b)\wedge \sim(b \leq b)$."

2.37. Example. Prove: for all $x, y \in \mathbb{R}$, if x is rational and y is irrational, then $x + y$ is irrational.

Proof. **Assume, to get a contradiction,** that there exist real numbers x, y such that x is rational and y is not rational and $x + y$ is rational. Expanding definitions, **our assumption means** that $x = m/n$ and $x + y = p/q$ for some integers m, n, p, q with $n \neq 0 \neq q$. By algebra, we **deduce** that

$$y = (x + y) - x = (p/q) - (m/n) = \frac{pn - qm}{qn},$$

where $pn - qm \in \mathbb{Z}$, $qn \in \mathbb{Z}$, and $qn \neq 0$. Comparing to the definition of "y is rational," this calculation **proves** that y is rational. We now have the contradiction "y is rational and y is not rational." So the original statement must be true.

Next we prove a famous theorem using proof by contradiction. We need one fact in the proof that we have not yet justified from basic principles: any fraction m/n with $m, n \in \mathbb{Z}$ and $n \neq 0$ can be written in *lowest terms*, meaning that no integer larger than 1 divides both m and n. In particular, this means that m and n can be chosen so that m and n are not both even (divisible by 2). For one proof of this fact, see Exercise 9 in §4.6.

2.38. Theorem: $\sqrt{2}$ is irrational.

Proof. **Assume, to get a contradiction,** that $\sqrt{2}$ is rational. **This assumption means** that for some integers m, n with $n \neq 0$, $\sqrt{2} = m/n$. By the fact mentioned above, we are allowed to **assume** that m and n are not both even. We reach a contradiction by **proving** that, in fact, m and n must both be even. Squaring $\sqrt{2} = m/n$ and rearranging, we **deduce** that $m^2 = 2n^2$, where n^2 is an integer. This equation **means** that m^2 must be even. Recall a theorem **already proved** in the last section: "if x^2 is even, then x is even." Using this result and the Inference Rule for IF, we **see** that m is even. **So**, there is an integer k with $m = 2k$. Now, $m^2 = 2n^2$ **becomes** $(2k)^2 = 2n^2$, so $4k^2 = 2n^2$, so $2k^2 = n^2$. Here, k^2 is an integer, **so** n^2 is even. By the same theorem used earlier, we **deduce** that n is even. We now have the contradiction "m and n are not both even, and m is even, and n is even." The only way out of the contradiction is to admit that $\sqrt{2}$ is not rational.

So far, we have checked that propositional forms are tautologies by writing out truth tables. The next example shows that proof by contradiction can also be used to demonstrate that a propositional form is a tautology without making a giant truth table.

2.39. Example. Use proof by contradiction to show that

$$[(P \vee Q \vee R) \wedge (P \Rightarrow S) \wedge (Q \Rightarrow S) \wedge (R \Rightarrow S)] \Rightarrow S \tag{2.4}$$

is a tautology.

Proof. [Recall, by definition, that $\mathcal{A}$ is a tautology iff *every* row of the truth table for $\mathcal{A}$ evaluates to true. Note the universal quantifier in the definition text. So, to begin the proof by contradiction, we deny this and say:] **Assume, to get a contradiction,** that *some* row of the truth table for the propositional form (2.4) evaluates to false. In such a row, let us see what we can deduce about the truth values of various subexpressions of (2.4). The overall propositional form is a false IF-statement in this row, which (by the truth table for IF) forces

$$(P \vee Q \vee R) \wedge (P \Rightarrow S) \wedge (Q \Rightarrow S) \wedge (R \Rightarrow S)$$

to be true and S to be false. Next, the truth of the AND-statement just written forces each input to the AND to be true, so we see that

$$P \vee Q \vee R, \quad P \Rightarrow S, \quad Q \Rightarrow S, \quad R \Rightarrow S$$

are all true in this row. Now, since $P \Rightarrow S$ is true and S is false, it must be that P is false in this row (by the truth table for IF or the Contrapositive Inference Rule for IF). Similarly, Q must be false and R must be false in this row. But now $P \vee Q \vee R$ must be false in this row, and yet also true. This contradiction shows that the original propositional form must be a tautology.

Proving XOR Statements and OR Statements

We close this section by introducing proof templates for proving XOR-statements and OR-statements. The template for XOR arises from the logical equivalence $P \oplus Q \equiv P \Leftrightarrow (\sim Q)$, together with the IFF template discussed earlier.

2.40. Proof Template to Prove $P \oplus Q$.
<u>Part 1.</u> Prove $P \Rightarrow (\sim Q)$ (by any method).
<u>Part 2.</u> Prove $(\sim Q) \Rightarrow P$ (by any method).

Similarly, we can create templates for proving $P \vee Q$ by converting the OR-statement to an IF-statement and invoking previous proof templates. Specifically, note that

$$P \vee Q \quad \equiv \quad (\sim P) \Rightarrow Q \quad \equiv \quad (\sim Q) \Rightarrow P.$$

This leads to the following templates.

2.41. First Proof Template to Prove $P \vee Q$.
Assume $\sim P$. **Prove** Q.

2.42. Second Proof Template to Prove $P \vee Q$.
Assume $\sim Q$. **Prove** P.

2.43. Example. Let x, y be fixed real numbers. Prove: if $x \neq 0$ and $y \neq 0$, then $xy \neq 0$.
Proof. [The inequalities in the given statement seem hard to work with; we can convert these to equalities by doing a contrapositive proof. So we begin:] **Assume** $xy = 0$. **Prove** $x = 0$ or $y = 0$. Our new goal is an OR-statement, so we use the first template and **assume** $x \neq 0$. **We must prove** $y = 0$. [We must now combine our two assumptions ($xy = 0$ and $x \neq 0$) to reach the new goal $y = 0$. We do so as follows:] **It is known** that the nonzero real number x has a multiplicative inverse $x^{-1} \in \mathbb{R}$ satisfying $x^{-1}x = 1$. Multiplying both sides of $xy = 0$ by x^{-1} gives $x^{-1}(xy) = x^{-1}0$. By algebra, this becomes $1y = 0$, giving $y = 0$.

We can prove an OR-statement with more than two alternatives as follows.

2.44. Proof Template to Prove $P \vee Q \vee R \vee S$.
Assume $\sim P$ and $\sim Q$ and $\sim R$. **Prove** S.

This proof template can be justified via truth tables, or by the Inference Rule for IF and the tautology

$$([(\sim P) \wedge (\sim Q) \wedge (\sim R)] \Rightarrow S) \Rightarrow (P \vee Q \vee R \vee S).$$

When using the template, we can assume the negations of any three of the four alternatives and then prove the remaining alternative. The pattern seen in this template extends to a template for proving $A_1 \vee A_2 \vee \cdots \vee A_k$ where $k \geq 2$. To prove such an OR-statement, assume all but one of the A_j is false, and then prove the remaining A_j.

Section Summary

1. *Template for Proof by Contradiction.* To prove a statement P by contradiction:
   Assume, to get a contradiction, $\sim P$.
   Combine the assumption with other known facts to **deduce** a contradiction.
   Say: "This contradiction shows that P is true, after all."

2. *Advice on Contradiction Proofs.* (a) Signal the start and end of a contradiction proof with phrases similar to those in the proof template, to lessen the risk of confusing the reader (or writer) of the proof. (b) Use the denial rules to find a useful form of $\sim P$. (c) The proof template does not give you a specific contradiction to aim for, which makes this template hard to use compared to others. You need creativity and patience to explore the consequences of the given assumptions until a contradiction is found.

3. *Proof Template for XOR.* To prove $P \oplus Q$:
   <u>Part 1.</u> Prove $P \Rightarrow (\sim Q)$ by any method.
   <u>Part 2.</u> Prove $(\sim Q) \Rightarrow P$ by any method.

4. *Proof Template for OR.* To prove $P \vee Q \vee R \vee S$:
   Assume $\sim P$ and $\sim Q$ and $\sim R$. **Prove** S.
   (Similarly for OR-statements with any number of alternatives.)

Exercises

1. Use proof templates for IF, OR, etc., to write a proof outline for each statement.
    (a) If f is continuous or monotonic, then f is integrable. [use contradiction proof]
    (b) If P is a prime ideal and $ab \in P$, then $a \in P$ or $b \in P$. [use direct proof]
    (c) If P is a prime ideal and $ab \in P$, then $a \in P$ or $b \in P$. [use contrapositive proof]
    (d) $\forall x \in \mathbb{Z}, \forall y \in \mathbb{Z}, \forall z \in \mathbb{Z}, \forall n \in \mathbb{Z}, (x > 0$ and $y > 0$ and $n > 2) \Rightarrow x^n + y^n \neq z^n$. [use proof by contradiction]
    (e) $\forall n \in \mathbb{Z}, \exists p \in \mathbb{Z}, p > n \land p$ is prime $\land p + 2$ is prime [use proof by contradiction]

2. Let x be a fixed real number. Prove each statement by contradiction.
    (a) If $x \neq 0$, then $-x \neq 0$. (b) If $x \neq 0$, then $1/x \neq 0$.

3. Prove by contradiction: $\forall x \in \mathbb{R}, \forall y \in \mathbb{R}$, if $x \in \mathbb{Q}$ and $y \notin \mathbb{Q}$ and $x \neq 0$, then $xy \notin \mathbb{Q}$.

4. Prove by contradiction: $\forall m \in \mathbb{Z}, \forall n \in \mathbb{Z}, 6m + 8n \neq 5$.

5. Prove by contradiction: $P \Rightarrow (Q \Rightarrow P)$ is a tautology.

6. Suppose, in a certain proof, we assume $\sim P$, go through a chain of reasoning, and eventually deduce P. What, if anything, can we conclude about the truth or falsehood of P?

7. Let x be a fixed real number. (a) Prove: if $x^2 + 4x - 60 = 0$, then $x = -10$ or $x = 6$. (b) Is the converse of part (a) true?

8. For fixed $x \in \mathbb{R}$, prove: if $x^2 \geq x$, then $x \leq 0$ or $x \geq 1$.

9. For fixed $m \in \mathbb{Z}$, use (2.1) to prove: 2 divides m XOR 2 divides $m + 7$.

10. For fixed $x, y \in \mathbb{Z}_{>0}$, prove: if $x + y = 5$, then $x = 1$ or $x = 2$ or $x = 3$ or $x = 4$.

11. (a) Find a statement equivalent to $(P \land Q) \Rightarrow R$ that does not use $\land$ or $\Rightarrow$. (b) Give four possible proof outlines of $(P \land Q) \Rightarrow R$, each with different assumptions and a different goal.

12. Give four possible proof outlines of $A \Rightarrow (B \lor (\sim C))$.

13. Suppose this fact is a known theorem: for all $a, b \in \mathbb{Z}$, if 3 divides ab, then 3 divides a or 3 divides b. Use this to prove that $\sqrt{3}$ is irrational, by imitating a proof in this section.

14. Suppose we imitate the proof of Theorem 2.38 in an attempt to prove that $\sqrt{4}$ is irrational. Find the exact point in the proof that no longer works.

15. (a) Use the truth table for XOR to justify the XOR proof template. (b) Use the truth table for OR to justify the two proof templates for proving $P \lor Q$. (c) Use any method to justify the proof template for proving $P \lor Q \lor R \lor S$.

16. Make a new proof template for proving $P \oplus Q$ based on the equivalence $(P \oplus Q) \equiv (P \lor Q) \land \sim(P \land Q)$; prove both parts of the AND-statement by contradiction.

17. Assume p, a, b, c are fixed integers, and p is known to satisfy: $\forall x, y \in \mathbb{Z}$, if p divides xy, then p divides x or p divides y. Prove: if p divides abc, then p divides a or p divides b or p divides c.

18. Fix $m \in \mathbb{Z}$. Find the error in the following proof that $6m$ is odd. *Proof.* Assume, to get a contradiction, that $6m$ is not odd. So $6m$ is even, which means $\exists k \in \mathbb{Z}, 6m = 2k$. Choose $k = 3m$, which is in $\mathbb{Z}$ by closure. By algebra, $2k = 2(3m) = 6m$, proving that $6m$ is even.

19. Find the error in the following proof outline of the statement $(P \Rightarrow Q) \Rightarrow (\sim S)$.
    Proof. To get a contradiction, prove $(P \Rightarrow Q) \wedge S$. <u>Part 1.</u> Assume P; prove Q. <u>Part 2.</u> Prove S.

20. Prove by contradiction: $[P \Rightarrow (Q \Rightarrow R)] \Rightarrow [(P \Rightarrow Q) \Rightarrow (P \Rightarrow R)]$ is a tautology.

21. Let m be a fixed integer. Recall the theorem that every integer is even or odd, but not both. Using this, prove:
    if m is even, then $(\exists k \in \mathbb{Z}, m = 4k) \oplus (\exists j \in \mathbb{Z}, m = 4j + 2)$;
    but if m is odd, then $(\exists a \in \mathbb{Z}, m = 4a + 1) \oplus (\exists b \in \mathbb{Z}, m = 4a + 3)$.

22. Let b be a fixed odd positive integer. Prove $\sqrt{2b}$ is not rational.

2.5   Proofs by Cases and Disproofs

We ended the last section by describing ways to *prove* OR-statements that are not yet known
to be true. Now, we consider the related question of how to *use* an OR-statement that is
already known to be true. We introduce a new proof method called *proof by cases* that is
often needed in situations where we already know (or have assumed) an OR-statement and
need to prove some other statement. The second half of the section considers the question
of how we may disprove statements we believe to be false.

Proof by Cases

We introduce proof by cases in the situation where we know (or have assumed) an OR-
statement with three alternatives.

2.45. Proof Template for Proof by Cases. Suppose $P \vee Q \vee R$ is a known OR-statement.
To prove any statement S using proof by cases:
Say "Since $P \vee Q \vee R$ is **known**, we consider 3 cases."
Case 1. **Assume** P. **Prove** S.
Case 2. **Assume** Q. **Prove** S.
Case 3. **Assume** R. **Prove** S.
Conclude by saying "So, S is true in all cases."

2.46. Remarks on Proof by Cases.

(a) The proof outlined above has three completely separate subproofs, called the three *cases*.
In each subproof, we must prove the target statement S. Each of these three subproofs begins
with a different temporary assumption (P, Q, or R), so the details of these subproofs are not
the same. Assumptions and conclusions appearing in the proof of one case cannot be reused
in another case unless they are reproved under the assumptions for that case. Temporary
variables used in one case can be reused (with new meanings) in other cases.

(b) The proof template extends to the situation where we start with a known OR-statement
$P_1 \vee P_2 \vee \cdots \vee P_k$ with k alternatives, where $k \geq 2$, and we are trying to prove S. In this
setting, there are k cases. The proof outline looks like this:

Since $P_1 \vee \cdots \vee P_k$ is **known**, consider k cases.
Case 1. **Assume** P_1. **Prove** S.
. . .
Case i. **Assume** P_i. **Prove** S.
. . .
Case k. **Assume** P_k. **Prove** S.
So, S **is true** in all cases.

(c) Here is a technical justification for why the proof-by-cases template is legitimate. On
one hand, in order to use this template, we must already know (or have proved) $P \vee Q \vee R$.
On the other hand, Case 1 of the template proves the IF-statement $P \Rightarrow S$; Case 2 proves
$Q \Rightarrow S$; and Case 3 proves $R \Rightarrow S$. By the AND-template, we have proved

$$(P \vee Q \vee R) \wedge (P \Rightarrow S) \wedge (Q \Rightarrow S) \wedge (R \Rightarrow S).$$

We also have the following axiom (a tautology proved in (2.4) of the last section):

$$[(P \vee Q \vee R) \wedge (P \Rightarrow S) \wedge (Q \Rightarrow S) \wedge (R \Rightarrow S)] \Rightarrow S.$$

Applying the Inference Rule for IF to the two displayed theorems, we see that S is a new theorem, as needed.

Examples of Proof by Cases

Let us look at some sample proofs where proof by cases can be used. We remark that you will generally not be told: "use proof by cases here." Writers of proofs must recognize situations where proof by cases can be applied. As a general rule, if you encounter a situation where you *know* or *have assumed* an OR-statement (say because a proof template directed you to write one), then proof by cases gives you one potential way to utilize that OR-statement.

2.47. Example. Let a and b be fixed integers. Prove: if a is even or b is even, then ab is even. *Proof.* Assume a is even or b is even. Prove ab is even. [Noting that we have just assumed an OR-statement, we continue by using proof by cases.]
<u>Case 1.</u> Assume a is even. Prove ab is even. We have assumed there is an integer k with $a = 2k$. We must prove $\exists m \in \mathbb{Z}, ab = 2m$. Multiplying our assumption by b gives $ab = 2kb$. Choosing m to be kb, which is an integer, we get $ab = 2m$ as needed.
<u>Case 2.</u> Assume b is even. Prove ab is even. We have assumed there is an integer k with $b = 2k$. We must prove $\exists m \in \mathbb{Z}, ab = 2m$. Multiplying our assumption by a gives $ab = 2ka$. Choosing m to be ka, which is an integer, we get $ab = 2m$ as needed.
Comment: Although the two cases had nearly identical proofs, both were logically necessary.

2.48. Example. For a fixed integer n, prove: if 30 divides n or 70 divides n, then 10 divides n. *Proof.* Assume 30 divides n or 70 divides n. Prove 10 divides n. [We have assumed an OR-statement, so we decide to use proof by cases.]
<u>Case 1.</u> Assume 30 divides n. Prove 10 divides n. We have assumed $n = 30k$ for some integer k. We must prove $n = 10m$ for some integer m. Note $n = 30k = 10(3k)$. So we choose $m = 3k$, which is an integer such that $n = 10m$.
<u>Case 2.</u> Assume 70 divides n. Prove 10 divides n. We have assumed $n = 70k$ for some integer k. We must prove $n = 10m$ for some integer m. Note $n = 70k = 10(7k)$. So we choose $m = 7k$, which is an integer such that $n = 10m$.

2.49. Example. Let n be a fixed integer. Prove: $n^2 - n$ is even.
Proof. We must prove $\exists k \in \mathbb{Z}, n^2 - n = 2k$. [Where do we go from here? Thinking of the next sentence in this proof requires a leap of creativity, even though the sentence itself seems to be stating a rather mundane fact.] **We know** n is even or n is not even (this is an axiom). Since we have stated a known OR-statement, we can try proof by cases.
<u>Case 1.</u> Assume n is even. Prove $\exists k \in \mathbb{Z}, n^2 - n = 2k$. We have assumed $n = 2m$ for some integer m. We deduce from this that $n^2 - n = (2m)^2 - 2m = 4m^2 - 2m = 2(2m^2 - m)$. We now see that we can choose $k = 2m^2 - m$ (which is an integer since m is) in order to have $n^2 - n = 2k$.
<u>Case 2.</u> Assume n is not even, so n is odd by a known theorem. Prove $\exists k \in \mathbb{Z}, n^2 - n = 2k$. We have assumed $n = 2m + 1$ for some integer m. Now, algebra gives

$$n^2 - n = (2m + 1)^2 - (2m + 1) = 4m^2 + 4m + 1 - 2m - 1 = 4m^2 + 2m = 2(2m^2 + m).$$

Choose $k = 2m^2 + m$; this is an integer such that $n^2 - n = 2k$. So, $n^2 - n$ is even in all cases.

2.50. Example. Let x be a fixed real number. Prove: $x^3 = 4x$ iff $x = -2$ or $x = 0$ or $x = 2$. *Proof.* This IFF-statement requires a two-part proof.
<u>Part 1.</u> Assume $x^3 = 4x$. Prove $x = -2$ or $x = 0$ or $x = 2$. We are **proving** an OR-statement now, so we follow the OR-template. Assume $x \neq -2$ and $x \neq 0$. Prove $x = 2$. Since $x \neq 0$,

we can divide both sides of the assumption $x^3 = 4x$ by x to deduce $x^2 = 4$, so $x^2 - 4 = 0$, so $(x-2)(x+2) = 0$. Since $x \neq -2$, we can divide both sides of this equation by the nonzero quantity $x + 2$ to get $x - 2 = 0$, hence $x = 2$, as needed.

Part 2. Assume $x = -2$ or $x = 0$ or $x = 2$. Prove $x^3 = 4x$. Here we have **assumed** an OR-statement, so we use proof by cases.

   Case 1. Assume $x = -2$. Prove $x^3 = 4x$. By arithmetic, $x^3 = -8 = 4x$ in this case.
   Case 2. Assume $x = 0$. Prove $x^3 = 4x$. By arithmetic, $x^3 = 0 = 4x$ in this case.
   Case 3. Assume $x = 2$. Prove $x^3 = 4x$. By arithmetic, $x^3 = 8 = 4x$ in this case.
Since $x^3 = 4x$ in all cases, we are done with part 2.

Definition by Cases

Sometimes cases are used to define a new mathematical concept. Proofs involving such concepts often rely on proof by cases. We give a few examples here involving the sign and absolute value of real numbers.

2.51. Definition: sgn(x). For each real number x, define $\mathrm{sgn}(x) = +1$ if $x > 0$, $\mathrm{sgn}(x) = 0$ if $x = 0$, and $\mathrm{sgn}(x) = -1$ if $x < 0$.

   For this definition by cases to make sense, we need to know that for each real x, *exactly one* of the statements "$x > 0$," "$x = 0$," or "$x < 0$" is true. We take this to be a known fact from algebra. When we prove theorems about $\mathrm{sgn}(x)$, we often start the proof with the known fact "$x > 0$ or $x = 0$ or $x < 0$," which leads into a proof by cases.

2.52. Example. For fixed $z \in \mathbb{R}$, prove $\mathrm{sgn}(-z) = -\mathrm{sgn}(z)$.
Proof. We know $z > 0$ or $z = 0$ or $z < 0$, so use cases.
Case 1. Assume $z > 0$; prove $\mathrm{sgn}(-z) = -\mathrm{sgn}(z)$. Because $z > 0$, we know $-z < 0$ by algebra. By definition of sgn, we know $\mathrm{sgn}(-z) = -1$ and $\mathrm{sgn}(z) = +1$, so $\mathrm{sgn}(-z) = -1 = -\mathrm{sgn}(z)$ holds in this case.
Case 2. Assume $z = 0$; prove $\mathrm{sgn}(-z) = -\mathrm{sgn}(z)$. Here $-z = -0 = 0$, so $\mathrm{sgn}(-z) = \mathrm{sgn}(0) = 0 = -0 = -\mathrm{sgn}(0) = -\mathrm{sgn}(z)$ follows.
Case 3. Assume $z < 0$; prove $\mathrm{sgn}(-z) = -\mathrm{sgn}(z)$. Because $z < 0$, we know $-z > 0$ by algebra. Therefore, $\mathrm{sgn}(-z) = +1 = -(-1) = -\mathrm{sgn}(z)$ in this case.
We see that $\mathrm{sgn}(-z) = -\mathrm{sgn}(z)$ holds in all cases.

2.53. Definition: Absolute Value. For each real number r, define $|r| = r$ if $r \geq 0$, and define $|r| = -r$ if $r < 0$.

   For example, $|0.23| = 0.23$, whereas $|-3| = -(-3) = 3$. Note carefully that when r is negative, the expression $-r$ denotes a positive real number.

2.54. Example: For fixed $x, y \in \mathbb{R}$, prove $|xy| = |x| \cdot |y|$.
Proof. [Here we need cases for both x and y. It turns out to be easier to create special cases when one variable is zero.] We know $x = 0$ or $y = 0$ or $x, y > 0$ or $x > 0 > y$ or $y > 0 > x$ or $x, y < 0$, so use cases.
Case 1. Assume $x = 0$; prove $|xy| = |x| \cdot |y|$. In this case, we know $xy = 0y = 0$, so $|xy| = |0| = 0$ and $|x| \cdot |y| = 0|y| = 0$.
Case 2. Assume $y = 0$; prove $|xy| = |x| \cdot |y|$. As in Case 1, $xy = 0$, so $|xy| = 0 = |x|0 = |x| \cdot |y|$.
Case 3. Assume $x > 0$ and $y > 0$; prove $|xy| = |x| \cdot |y|$. By algebra, $xy > 0$. Using the definition of absolute value three times, we see that $|xy| = xy = |x| \cdot |y|$.
Case 4. Assume $x > 0$ and $y < 0$, so $xy < 0$. We compute $|xy| = -(xy) = x \cdot (-y) = |x| \cdot |y|$.
Case 5. Assume $x < 0$ and $y > 0$, so $xy < 0$. We compute $|xy| = -(xy) = (-x) \cdot y = |x| \cdot |y|$.
Case 6. Assume $x < 0$ and $y < 0$, so $xy > 0$. We compute $|xy| = xy = (-x) \cdot (-y) = |x| \cdot |y|$.
So $|xy| = |x| \cdot |y|$ holds in all cases, as needed.

Disproving Statements

So far, we have given many proof templates that enable us to prove true statements P. Of course, many statements P are false, and we cannot prove such statements. Instead, we would like a method for *disproving* a statement we believe to be false. Since P is false iff $\sim P$ is true, we are led to the following template.

2.55. Proof Template to Disprove a False Statement P.
Prove $\sim P$.

The next question is, how do we prove a true statement of the form $\sim P$? We offer the following template as an answer.

2.56. Proof Template to Prove $\sim P$.
Use the denial rules to replace $\sim P$ with an equivalent statement Q that does not begin with NOT. **Prove** Q.

To complete Step 2, we recursively apply other templates, expand definitions appearing in Q, and so on.

2.57. Example. Disprove: $\forall x \in \mathbb{R}, x^2 \geq x$.
Solution. By the disproof template, we must **prove** $\sim \forall x \in \mathbb{R}, x^2 \geq x$. Using the denial rules, we find that we must **prove** the statement: $\exists x \in \mathbb{R}, x^2 < x$. [If you cannot initially find an example of this existential assertion, try drawing the graphs of $y = x^2$ and $y = x$.] We now see that we may **choose** $x = 1/2$, which is a real number such that $x^2 = 1/4 < 1/2 = x$.

2.58. Example. Disprove: for all $a, b \in \mathbb{Z}$, if a divides b, then $a \leq b$.
Solution. Using the denial rules, we must prove there exist integers a, b such that a divides b and $a > b$. [Before reading further, can you think of an example of this situation? The key is to realize that a and b come from the universe $\mathbb{Z}$, which includes negative integers.] We can now choose $a = 4$ and $b = -4$ (say). Note that $-4 = 4(-1)$ where -1 is an integer, verifying that 4 divides -4. We also know $4 > -4$, finishing the proof.

In the next example, the notation "$x \notin \mathbb{Q}$" means "$\sim(x \in \mathbb{Q})$," i.e., x is not rational.

2.59. Example. Disprove: for all real x, y, if $x \notin \mathbb{Q}$ and $y \notin \mathbb{Q}$, then $x + y \notin \mathbb{Q}$.
Solution. We must prove there exist $x, y \in \mathbb{R}$ with $x \notin \mathbb{Q}$ and $y \notin \mathbb{Q}$ and $x + y \in \mathbb{Q}$. Recall from Theorem 2.38 that $\sqrt{2}$ is not rational. So, let us choose $x = \sqrt{2}$ and $y = -\sqrt{2}$. The fact that y is also irrational follows from Example 2.31, taking $r = \sqrt{2}$. We know $x + y = \sqrt{2} + -\sqrt{2} = 0 = 0/1$, where 0 and 1 are integers and $1 \neq 0$. So $x + y \in \mathbb{Q}$, as needed.

The last example shows that the set of irrational real numbers is *not* closed under addition. Similarly, we ask the reader to check below that the set of irrational numbers is not closed under multiplication.

2.60. Remark. Beginners sometimes confuse disproofs and proofs by contradiction. Let us contrast these two situations. To **prove** a **true** statement P by the contradiction method, we begin with the line "**Assume** $\sim P$ to get a contradiction." Our goal in the proof is to deduce a contradiction (not specified in advance). In contrast, to **disprove** a **false** statement P by the disproof method, we begin with the line "**Prove** $\sim P$." The next step in both proofs is to find a useful denial of $\sim P$ by the denial rules. But the two proofs use this denial in totally different ways, since $\sim P$ is the *assumption* in the contradiction proof and the *goal* in the disproof. To avoid confusion, always write (and pay attention to!) these logical status words.

Section Summary

1. *Proof by Cases.* When you **know** or **have assumed** an OR-statement $P \vee Q \vee R$ and need to **prove** another statement S, say this:
   "Since $P \vee Q \vee R$ is known/assumed, we consider 3 cases.
   Case 1. **Assume** P. Prove S.
   Case 2. **Assume** Q. Prove S.
   Case 3. **Assume** R. Prove S.
   So, S is true in all cases."
   (Similarly, a known OR-statement $P_1 \vee P_2 \vee \cdots \vee P_n$ leads to a proof with n cases.)

2. *When to Use Proof by Cases.* If some proof template directs you to **assume** an OR-statement, you often need proof by cases to use this assumption. To prove properties of concepts defined via cases, you typically must use proof by cases based on the same cases appearing in the definition. Axioms that are instances of the tautology $P \vee (\sim P)$ are **known** OR-statements that often lead to proof by cases. For example, given a fixed integer n_0, we know "n_0 is even or n_0 is not even." Given a fixed real number r_0, we know "$r_0 \geq 0$ or $r_0 < 0$." Given a fixed object x_0 and universe U, we know "$x_0 \in U$ or $x_0 \notin U$."

3. *Disproving False Statements.* To **disprove** a false statement P, **prove** the negation $\sim P$. To **prove** a statement $\sim P$ that begins with NOT, use the denial rules to replace this statement with an equivalent statement not starting with NOT. Remember to flip any quantifiers that occur at the beginning of the statement being denied.

Exercises

1. Give proof outlines of the following statements.
   (a) If f is continuous or monotonic, then f is integrable. [use direct proof]
   (b) It is false that every injective function has a left inverse.
   (c) $(P \vee Q) \oplus (R \wedge (\sim S))$.

2. Consider the statement $(A \vee B \vee (\sim C)) \Rightarrow (D \wedge E)$. Give proof outlines of this statement based on: (a) a direct proof and proof by cases; (b) a contrapositive proof and proof by cases; (c) a contradiction proof.

3. Let n be a fixed integer. Prove: if 6 divides n or $n + 5$ is odd, then n is even.

4. Let k be a fixed integer. Prove: $k^2 + 3k$ is even.

5. (a) Let x and y be fixed integers. Prove: if x is even or y is even, then $x^2 y^3$ is even. (b) Disprove: $\forall x \in \mathbb{Z}, \forall y \in \mathbb{Z}$, if x is odd or y is odd, then $x^2 y^3$ is odd.

6. Let x be a fixed real number.
   (a) Prove: if $x < 1$ or $x > 3$, then $x^2 - 4x + 3 > 0$. Must the converse hold?
   (b) Prove: $x^3 - x > 0$ iff $-1 < x < 0$ or $x > 1$.

7. Let $z \in \mathbb{R}$ be fixed. Prove: for $z^2 - 7z + 10$ to be positive, it is necessary and sufficient that $z < 2$ or $z > 5$.

8. Let x be a fixed real number.
   (a) Prove: if x^2 is rational or x^3 is rational, then x^6 is rational.
   (b) Prove: if $x^5 \notin \mathbb{Q}$, then $x^2 \notin \mathbb{Q}$ or $x^3 \notin \mathbb{Q}$.

9. (a) Disprove: for all $m, n \in \mathbb{Z}$, if 10 divides mn, then 10 divides m or 10 divides n. (b) Disprove: for all $a, b \in \mathbb{Z}_{\geq 0}$, if a divides b, then $a \leq b$.

10. (a) Disprove: for all $x, y \in \mathbb{R}$, if $x \notin \mathbb{Q}$ and $y \notin \mathbb{Q}$, then $xy \notin \mathbb{Q}$.
    (b) Disprove: for all $x, y \in \mathbb{R}_{>0}$, if $x \notin \mathbb{Q}$ and $y \notin \mathbb{Q}$, then $x + y \notin \mathbb{Q}$.

11. For fixed $x, y \in \mathbb{R}$, prove: $\mathrm{sgn}(xy) = \mathrm{sgn}(x)\,\mathrm{sgn}(y)$.

12. For fixed $x \in \mathbb{R}$, is $\mathrm{sgn}(x^3) = \mathrm{sgn}(x)$ true or false? Explain.

13. For fixed $x \in \mathbb{R}$, prove: (a) $|x| \geq 0$; (b) $|-x| = |x|$; (c) $x = \mathrm{sgn}(x)|x|$.

14. (a) Disprove: for all $x, y \in \mathbb{R}$, $|x + y| = |x| + |y|$.
    (b) Disprove: $\forall x \in \mathbb{R}, \mathrm{sgn}(x^2) = +1$.
    (c) Disprove: for all $x, y \in \mathbb{R}$, $\mathrm{sgn}(x + y) = \mathrm{sgn}(\mathrm{sgn}(x) + \mathrm{sgn}(y))$.

15. For fixed $x, y \in \mathbb{R}$, prove: $|x + y| \leq |x| + |y|$.

16. Use the previous exercise (which holds for *all* $x, y \in \mathbb{R}$) to deduce the following facts for fixed real numbers r, s, t.
    (a) $|r + s + t| \leq |r| + |s| + |t|$.
    (b) $|r - t| \leq |r - s| + |s - t|$.
    (c) $|r - s| \leq |r| + |s|$.

17. For fixed $x, y \in \mathbb{R}$, prove: $||x| - |y|| \leq |x - y|$.

18. Let x, a, r be fixed real numbers. (a) Prove: $|x - a| < r$ iff $a - r < x < a + r$.
    (b) Use (a) and logic to deduce a condition equivalent to $|x - a| \geq r$.

19. Find the error in this attempted disproof of "for some positive integers x, y, z, $x^3 + y^3 = z^3$." *Disproof:* We must prove that for some positive integers x, y, z, $x^3 + y^3 \neq z^3$. Choose $x = y = 1$ and $z = 2$. We compute $x^3 + y^3 = 1 + 1 = 2$, while $z^3 = 2^3 = 8$.

20. Find and correct the error in this outline of a disproof of "$P \Rightarrow (Q \Rightarrow R)$." *Disproof:* Assume P and Q are true and R is false. Deduce a contradiction.

21. Define $f(x) = x^2 + 1$ for $0 \leq x \leq 2$, and $f(x) = 3 - x$ for $2 < x \leq 3$. For a fixed $x \in \mathbb{R}$, prove: if $0 \leq x \leq 3$, then $0 \leq f(x) \leq 5$.

22. For a fixed integer m, prove $m^2 + 5m - 3$ is odd.

23. Suppose we know this theorem: $\forall m \in \mathbb{Z}, \exists j \in \mathbb{Z}, m = 4j \vee m = 4j + 1 \vee m = 4j + 2 \vee m = 4j + 3$. Let b be a fixed integer. Prove: 4 divides b^2 or 8 divides $b^2 - 1$.

24. For fixed $n \in \mathbb{Z}$, prove: if n is odd, then 8 divides $n^3 - n$.

25. Suppose "$P \vee Q \vee R$" is a known theorem. Justify the following proof template, which proves a statement S by combining the contradiction method with proof by cases.
    Proof of S. We know $P \vee Q \vee R$, so use cases.
    <u>Case 1.</u> Assume P and $\sim S$. Deduce a contradiction $\mathcal{C}_1$.
    <u>Case 2.</u> Assume Q and $\sim S$. Deduce a contradiction $\mathcal{C}_2$.
    <u>Case 3.</u> Assume R and $\sim S$. Deduce a contradiction $\mathcal{C}_3$.
    Since all three cases led to a contradiction, S must be true.

2.6   Proving Quantified Statements

So far, we have discussed proof templates involving the following logical operators: $\exists$, $\Rightarrow$, $\wedge$, $\Leftrightarrow$, $\oplus$, $\vee$, and $\sim$. It is finally time to present templates for proving a universally quantified statement. This leads us to a tricky but essential topic: how to manipulate statements containing multiple quantifiers.

Proof by Exhaustion

Our first proof template for proving a universal statement "$\forall x \in U, P(x)$" is called *proof by exhaustion*. It can only be used when the universe U is a *finite* set.

2.61. Proof Template for Proof by Exhaustion. Suppose $U = \{z_1, z_2, \ldots, z_n\}$ is a finite set. To prove $\forall x \in U, P(x)$ in this situation:
Step 1. Prove $P(z_1)$.
Step 2. Prove $P(z_2)$.
. . .
Step n. Prove $P(z_n)$.

Intuitively, we can verify the universal statement by explicitly testing the truth of $P(c)$ for every possible object c in the finite universe U. We can justify this proof method more formally by recalling the following rule from §1.4:

$$\boxed{\forall x \in \{z_1, z_2, \ldots, z_n\}, P(x)} \quad \Leftrightarrow \quad \boxed{P(z_1) \wedge P(z_2) \wedge \cdots \wedge P(z_n)}.$$

To prove the universal statement, we may instead prove the equivalent AND-statement. Following the proof template for an AND-statement, we obtain the proof outline written above. The name "exhaustion" is used because we are exhaustively checking all possible cases, one at a time.

2.62. Example. Every time we use a truth table to verify that a propositional form is a tautology, we are really doing a proof by exhaustion. To see why, recall that a propositional form $\mathcal{T}$ is a tautology iff *every row* in the truth table for $\mathcal{T}$ evaluates to true. This definition has the form $\forall x \in U, P(x)$, where U is the set of rows in the truth table, and $P(x)$ is the statement "the propositional form $\mathcal{T}$ has output TRUE in row x of the truth table." Since there are only finitely many rows in the truth table, U is a finite set. By filling in the truth table, we are checking the truth of $P(x)$ for each row x, one at a time.

Generic Element Proofs of Universal Statements

When the universe U is not finite, proof by exhaustion does not work. We need a new strategy for proving a universal statement "$\forall x \in U, P(x)$," which says that $P(x_0)$ is true for all objects x_0 in the potentially infinite set U. This strategy is called the *generic-element method*.

2.63. Generic-Element Proof Template to Prove $\forall x \in U, P(x)$.
Let x_0 be a **fixed**, but **arbitrary**, object in U.
Prove $P(x_0)$.

Before using this template, let us discuss its logical nuances by contrasting it with the template for proving $\exists x \in U, P(x)$. When we prove the *existence* statement $\exists x \in U, P(x)$, we do so by *choosing* one **specific**, constant object x_0 in U that makes $P(x_0)$ true. For instance, we might say: "choose $x_0 = 7$." More generally, x_0 could be an expression involving other constants introduced earlier in the proof. We complete the existence proof by showing that x_0 really is in U and proving that $P(x_0)$ really is true.

In the new template to prove $\forall x \in U, P(x)$, we also introduce a constant object x_0 in U, but we **do not choose** a specific value for this object. The letter x_0 used here represents a "generic" object in U that we do not control. We know nothing about x_0 initially, except that it comes from the universe U. If we can prove the proposition $P(x_0)$ using only this information, we may safely conclude that $P(x)$ does hold for every object x in U. Here is a basic example.

2.64. Example. Prove: $\forall x \in \mathbb{Z}$, x divides $3x^2$.
Proof. Let x_0 be a **fixed, but arbitrary** integer. We must **prove** x_0 divides $3x_0^2$. Expanding the definition, we must **prove** $\exists k \in \mathbb{Z}, 3x_0^2 = kx_0$. **Choose** $k_0 = 3x_0$, which is an integer **since** $x_0 \in \mathbb{Z}$ and $3 \in \mathbb{Z}$ and $\mathbb{Z}$ is closed under multiplication. We **know** $k_0 x_0 = (3x_0)x_0 = 3x_0^2$, so the proof is done.

In this proof, we used subscripts (x_0 and k_0) to distinguish the *constants* x_0 and k_0 from the *quantified variables* x and k. Observe that x_0 (coming from the universally quantified variable x) is a *generic* constant, whose value is not chosen or controlled by us. But k_0 (coming from the existentially quantified variable k in the definition of "divides") is a *specific* constant that we choose in a way to make the proof work. Observe that k_0 can depend on the previously introduced constant x_0; the expression $3x_0$ used for k_0 is a specific choice depending on the previous generic constant x_0.

We remark that many of the examples in previous sections were really proving universally quantified statements. Since we had not yet introduced the generic-element template, we introduced those examples with statements like "Let n be a fixed integer." So, for instance, Example 2.20 on page 59 actually proves the theorem

$$\forall a \in \mathbb{Z}, \forall b \in \mathbb{Z}, \forall c \in \mathbb{Z}, \text{if } a \text{ divides } b \text{ and } b \text{ divides } c \text{ then } a \text{ divides } c.$$

Statements with Multiple Quantifiers

We now discuss statements with multiple quantifiers, which reveal further distinctions between the proof templates for universal and existential quantifiers. The first point is that we may safely reorder two universal quantifiers or two existential quantifiers, as stated below.

2.65. Quantifier Reordering Rules. For any open sentence $P(x, y)$:

$$\boxed{\forall x, \forall y, P(x, y)} \quad \text{is equivalent to} \quad \boxed{\forall y, \forall x, P(x, y)};$$

$$\boxed{\exists x, \exists y, P(x, y)} \quad \text{is equivalent to} \quad \boxed{\exists y, \exists x, P(x, y)}.$$

Intuitively, the statements $\forall x, \forall y, P(x, y)$ and $\forall y, \forall x, P(x, y)$ are equivalent since both statements assert that $P(x_0, y_0)$ is true for all possible choices of the objects x_0 and y_0. Similarly, $\exists x, \exists y, P(x, y)$ and $\exists y, \exists x, P(x, y)$ are equivalent since both statements assert the existence of two objects x_0 and y_0 (possibly equal) such that $P(x_0, y_0)$ is true. We can give more formal proofs of these rules using the proof templates we have developed; see §2.7 for details. We can also use restricted quantifiers here; for instance,

$$\boxed{\exists x \in U, \exists y \in V, P(x, y)} \quad \text{is equivalent to} \quad \boxed{\exists y \in V, \exists x \in U, P(x, y)}.$$

Recall from our discussion of definitions that *we can change the letter used for a quantified variable, as long as the new letter does not already have a different meaning.* For instance, the three quantified statements $\forall z, P(z)$ and $\forall w, P(w)$ and $\forall s, P(s)$ are equivalent, since each statement says that $P(c)$ is true for every specific object c. We can do the same thing for statements with multiple quantifiers. For example,

$$\boxed{\forall x, \exists y, \forall z, P(x, y, z)} \quad \text{is equivalent to} \quad \boxed{\forall r, \exists s, \forall t, P(r, s, t)}.$$

Mixed Quantifiers

A crucial fact of quantified logic is that *the relative order of universal and existential quantifiers in a multiply-quantified statement has a dramatic effect on the meaning of the statement.* The next examples illustrate this fact.

2.66. Example. Prove or disprove: $\forall x \in \mathbb{Z}, \exists y \in \mathbb{Z}, y > x$.
Solution. This statement is true; it can be translated into English by saying: "for every integer, there is a larger integer." To prove it with the templates, we say: let x_0 be a **fixed, but arbitrary** integer. We must **prove** $\exists y \in \mathbb{Z}, y > x_0$. **Choose** $y_0 = x_0 + 1$. **We know** y_0 is an integer (by closure), and $x_0 + 1 > x_0$, completing the proof. In this proof, note carefully that the specific constant y_0 chosen in the existence proof was allowed to depend on the previously introduced generic constant x_0. Contrast this situation to the next example.

2.67. Example. Prove or disprove: $\exists y \in \mathbb{Z}, \forall x \in \mathbb{Z}, y > x$. (The only difference from the previous example is the order of the initial quantifiers.)
Solution. If we tentatively try to prove this statement by using the templates, we must begin by choosing a specific, constant integer y_0, and continue by proving $\forall x \in \mathbb{Z}, y_0 > x$. What y_0 shall we pick? If we try $y_0 = 7$, we must prove the statement $\forall x \in \mathbb{Z}, 7 > x$, which is evidently false. If we try $y_0 = 100$, we must prove the statement $\forall x \in \mathbb{Z}, 100 > x$, which is also false. We might try $y_0 = \infty$, but ∞ is not an integer. Why can't we choose $y_0 = x_0 + 1$, as we did in the previous proof? The reason is that in this proof, the fixed constant integer x_0 has not been introduced yet, so we cannot use it when choosing y_0. That is the key difference between the two statements. In fact, the current statement is false; in English, it says that "there is an integer larger than every integer."

Now, since we think the statement is false, we ought to be able to disprove it. We do so by proving a useful denial of the statement. Applying the denial rules, we must prove $\forall y \in \mathbb{Z}, \exists x \in \mathbb{Z}, y \leq x$. This denial resembles the original example, since it starts with $\forall$ followed by $\exists$. To prove the denial, we fix an arbitrary integer y_0, and prove $\exists x \in \mathbb{Z}, y_0 \leq x$. We now choose x_0 to be y_0, which works since $y_0 \in \mathbb{Z}$ and $y_0 \leq y_0$. We could have also chosen x_0 to be $y_0 + 1$ or $y_0 + 2$, for instance. Note that x_0 is allowed to depend on the previously introduced constant y_0.

In general, the statement $\exists y, \forall x, P(x, y)$ is harder to prove than the statement $\forall x, \exists y, P(x, y)$, because the y we choose is allowed to depend on x in the latter statement, but y cannot depend on x in the former. Nevertheless, statements of the first type are sometimes true, as we see next.

2.68. Example. Prove or disprove: $\exists y \in \mathbb{Z}, \forall x \in \mathbb{Z}, y < x^2$.

Proof. The existential quantifier comes first, so we must begin by choosing y. We choose $y_0 = -1$, which is a specific integer. We must prove $\forall x \in \mathbb{Z}, -1 < x^2$. Let x_0 be a fixed, but arbitrary integer. We must prove $-1 < x_0^2$. By algebra, $x_0^2 \geq 0 > -1$. [We could give a more detailed proof that $x_0^2 \geq 0$ by looking at the two cases where $x_0 \geq 0$ or $x_0 < 0$, and using properties of inequalities to deduce $x_0^2 \geq 0$ in each case.] □

2.69. Example. Prove or disprove: $\forall x \in \mathbb{Z}, \exists y \in \mathbb{Z}, y < x^2$.

Proof. [This statement is true; we proved an even stronger statement in the previous example. To see why, let us prove the current statement.] Let x_0 be a fixed, but arbitrary integer. We must prove $\exists y \in \mathbb{Z}, y < x_0^2$. Choose y_0 to be -1 [note that y_0 is *allowed* to depend on x_0, but our choice is not *required* to use x_0]. We know $-1 \in \mathbb{Z}$ and $-1 < 0 \le x_0^2$, as before. □

The last two examples illustrate the following general fact.

2.70. Theorem. For any open sentence $P(x, y)$ and any universes U and V,

$$\exists y \in U, \forall x \in V, P(x, y) \quad \Rightarrow \quad \forall x \in V, \exists y \in U, P(x, y).$$

We prove this theorem in the next section.

Proving Statements with Multiple Quantifiers

It is very common in mathematics to have statements in which three or more quantifiers (some existential, some universal) appear at the beginning of a complex statement. The next example illustrates how the rules above generalize to such a situation. The main fact to remember is that when *proving an existential statement by choosing a value for the variable, that value is only allowed to depend on all previously introduced constants, which correspond to quantified variables appearing to the left of the given variable.*

2.71. Example. Outline the proof of this statement:

$$\forall u, \forall \delta, \exists w, \forall x, \exists y, \forall z, P(u, \delta, w, x, y, z).$$

The second quantified variable is the Greek letter δ (delta).
Solution. We follow the proof templates, processing the quantifiers from left to right, obtaining the following lines in this order:

Let u_0 be a fixed, but arbitrary object.
Let δ_0 be a fixed, but arbitrary object.
Choose $w_0 = (\ldots)$ [some expression that may involve u_0 and δ_0].
Let x_0 be a fixed, but arbitrary object.
Choose $y_0 = (\ldots)$ [some expression that may involve u_0, δ_0, w_0 and x_0].
Let z_0 be a fixed, but arbitrary object.
Prove $P(u_0, \delta_0, w_0, x_0, y_0, z_0)$.

Suppose someone tries to shorten the preceding text by writing the following variation of the proof outline:

Let u_0, δ_0, x_0, and z_0 be fixed, but arbitrary objects.
Choose $w_0 = (\ldots)$ and $y_0 = (\ldots)$.
Prove $P(u_0, \delta_0, w_0, x_0, y_0, z_0)$.

This variation is not correct, since the order in which variables are introduced indicates that w_0 and y_0 might potentially depend on all the previously mentioned constants u_0, δ_0, x_0, and z_0. In fact, the alternate proof outline would prove the statement

$$\forall u, \forall \delta, \forall x, \forall z, \exists w, \exists y, P(u, \delta, w, x, y, z),$$

which is weaker than the original statement in general.

2.72. Example. Prove: $\forall x \in \mathbb{R}, \forall y \in \mathbb{R}$, if $x < y$, then $\exists z \in \mathbb{R}$, $x < z < y$.

Proof. Let x_0 and y_0 be fixed, but arbitrary real numbers. Assume $x_0 < y_0$. Prove $\exists z \in \mathbb{R}, x_0 < z < y_0$. Intuitively, the average of x_0 and y_0 should lie between these two numbers. So, we choose $z_0 = (x_0 + y_0)/2$, which is a real number since $\mathbb{R}$ is closed under addition and division by 2. We must prove $x_0 < z_0 < y_0$, which is really the AND-statement $x_0 < z_0$ and $z_0 < y_0$. To do this, divide the assumed inequality $x_0 < y_0$ by 2 to get $x_0/2 < y_0/2$. On one hand, adding $x_0/2$ to both sides produces $x_0/2 + x_0/2 < x_0/2 + y_0/2$, so $x_0 < z_0$. On the other hand, adding $y_0/2$ to both sides produces $x_0/2 + y_0/2 < y_0/2 + y_0/2$, so $z_0 < y_0$. This completes the proof.

Section Summary

1. *Proof by Exhaustion.* To prove $\forall x \in U, P(x)$, where U is the *finite* set $\{z_1, \ldots, z_n\}$: **Prove** $P(z_1)$. **Prove** $P(z_2)$. ... **Prove** $P(z_n)$.

2. *Generic-Element Template for Universal Statements.* To prove $\forall x \in U, P(x)$:
   Step 1. **Let** x_0 be a **fixed**, but **arbitrary**, object in U.
   Step 2. **Prove** $P(x_0)$.
   Here x_0 is a new letter denoting a *generic* constant. We *cannot* choose a specific value for x_0. We have no specific initial knowledge about x_0, except we know $x_0 \in U$. Step 1 is often abbreviated by saying: "fix $x_0 \in U$."

3. *Constructive Proof of $\exists x \in U, P(x)$.*
   Choose a specific object x_0; **prove** x_0 is in U; **prove** $P(x_0)$.
   In contrast to the generic-element method, we *must* choose a *specific* object x_0 here. This object is allowed to depend on constants previously introduced in the proof, but it cannot involve constants introduced later.

4. *Quantifier Reordering Rules.* The following rules hold for all open sentences $R(x, y)$:

   1. $\forall x, \forall y, R(x, y)$ $\Leftrightarrow$ $\forall y, \forall x, R(x, y)$.
   2. $\exists x, \exists y, R(x, y)$ $\Leftrightarrow$ $\exists y, \exists x, R(x, y)$.
   3. $\exists x, \forall y, R(x, y)$ $\Rightarrow$ $\forall y, \exists x, R(x, y)$.

   The converse of rule 3 is *not* universally valid: in general, changing the order of the quantifiers $\exists$ and $\forall$ affects the meaning of the statement.

5. *Sample Proof Outlines of Statements with Mixed Quantifiers.* Observe how differences in the types and ordering of quantifiers leads to contrasting proof structures in the following examples.

   (a) To prove $\forall x, \forall y, \exists z, P(x, y, z)$: Fix x_0. Fix y_0. Choose $z_0 = (\ldots)$ (an expression that can involve x_0 and y_0). Prove $P(x_0, y_0, z_0)$.

   (b) To prove $\forall x, \exists z, \forall y, P(x, y, z)$: Fix x_0. Choose $z_0 = (\ldots)$ (an expression that can involve x_0 only). Fix y_0. Prove $P(x_0, y_0, z_0)$.

   (c) To prove $\exists z, \forall x, \forall y, P(x, y, z)$: Choose $z_0 = (\ldots)$ (a specific object not depending on x or y). Fix x_0. Fix y_0. Prove $P(x_0, y_0, z_0)$.

   (d) To disprove $\forall x, \exists z, \forall y, P(x, y, z)$: Prove $\exists x, \forall z, \exists y, \sim P(x, y, z)$. Choose $x_0 = (\ldots)$ (a specific object not depending on anything else). Fix z_0. Choose $y_0 = (\ldots)$ (an expression that can involve x_0 and z_0). Prove $\sim P(x_0, y_0, z_0)$.

Exercises

1. Prove each universal statement.
   (a) For all $x, y \in \mathbb{Z}$, if x divides y, then x^2 divides y^2.
   (b) For all $x \in \{1, 3, 5, 7\}$, 8 divides $x^2 - 1$.
   (c) For all $x \in \mathbb{R}$, if $x^3 \notin \mathbb{Q}$, then $x \notin \mathbb{Q}$.

2. Give proof outlines for each statement similar to those in summary item 5 on page 86.
   (a) $\forall r, \forall s, \exists t, \exists u, Q(r, s, t, u)$.
   (b) $\forall s, \exists t, \forall r, \exists u, Q(r, s, t, u)$.
   (c) $\forall s, \exists t, \exists u, \forall r, Q(r, s, t, u)$.
   (d) $\exists u, \exists t, \forall r, \forall s, Q(r, s, t, u)$.
   (e) Find all pairs of statements in (a) through (d) such that the first statement always implies the second statement.

3. Outline *disproofs* of statements (a) through (d) in the previous exercise.

4. True or false? Prove each true statement, and disprove each false statement.
   (a) $\exists y \in \mathbb{Z}, \forall x \in \mathbb{Z}, xy = 0$.
   (b) $\forall x \in \mathbb{R}, \exists y \in \mathbb{R}, xy = 1$.
   (c) $\exists y \in \mathbb{R}, \forall x \in \mathbb{R}, xy = 1$.
   (d) $\forall x \in \mathbb{R}_{>0}, \exists y \in \mathbb{R}_{>0}, y < x$.
   (e) $\exists y \in \mathbb{R}_{>0}, \forall x \in \mathbb{R}_{>0}, y \leq x$.
   (f) $\forall x \in \mathbb{Z}, \exists y \in \mathbb{Z}_{\geq 0}, x^2 = y^2$.

5. Prove or disprove each statement.
   (a) $\exists m \in \mathbb{Z}, \forall n \in \mathbb{Z}, mn$ is even.
   (b) $\exists m \in \mathbb{Z}, \forall n \in \mathbb{Z}, mn$ is odd.
   (c) $\forall n \in \mathbb{Z}, \exists m \in \mathbb{Z}, mn$ is even.
   (d) $\forall n \in \mathbb{Z}, \exists m \in \mathbb{Z}, mn$ is odd.
   (e) $\forall n \in \mathbb{Z}, \exists m \in \mathbb{Z}, m + n$ is odd.

6. Prove or disprove each statement.
   (a) $\exists a \in \mathbb{Z}, \forall b \in \mathbb{Z}, a$ divides b.
   (b) $\exists a \in \mathbb{Z}, \forall b \in \mathbb{Z}, b$ divides a.
   (c) $\forall a \in \mathbb{Z}, \forall b \in \mathbb{Z}, \exists c \in \mathbb{Z}, a$ divides $b + c$.
   (d) $\forall a \in \mathbb{Z}, \exists c \in \mathbb{Z}, \forall b \in \mathbb{Z}, a$ divides $b + c$.
   (e) $\forall a \in \mathbb{Z}, \forall b \in \mathbb{Z}, \exists c \in \mathbb{Z}, b + c$ divides a.
   (f) $\forall a \in \mathbb{Z}, \exists c \in \mathbb{Z}, \forall b \in \mathbb{Z}, b + c$ divides a.

7. Prove or disprove each statement.
   (a) $\exists x \in \mathbb{R}, \forall y \in \mathbb{R}, xy$ is rational.
   (b) $\forall x \in \mathbb{R}, \exists y \in \mathbb{R}, x + y$ is rational.
   (c) $\forall q \in \mathbb{Q}, \exists n \in \mathbb{Z}_{>0}, nq \in \mathbb{Z}$.
   (d) $\exists n \in \mathbb{Z}_{>0}, \forall q \in \mathbb{Q}, nq \in \mathbb{Z}$.
   (e) $\exists q \in \mathbb{Q}, \forall n \in \mathbb{Z}, nq \in \mathbb{Z}$.

8. Prove or disprove each statement.
   (a) $\forall x \in \mathbb{R}, \exists y \in \mathbb{R}, y \neq x \wedge |x| = |y|$.
   (b) $\exists z \in \mathbb{R}, \forall w \in \mathbb{R}, |z| \leq |w|$.
   (c) $\forall x \in \mathbb{R}_{>0}, \exists y \in \mathbb{R}, |x + y| \neq |x| + |y|$.
   (d) $\exists y \in \mathbb{R}, \forall x \in \mathbb{R}_{>0}, |x + y| \neq |x| + |y|$.

9. (a) Prove: $\forall y \in \mathbb{R}, \exists x \in \mathbb{R}, y = 2x - 7$.
   (b) Disprove: $\exists x \in \mathbb{R}, \forall y \in \mathbb{R}, y = 2x - 7$.
   (c) Disprove: $\forall y \in \mathbb{Z}, \exists x \in \mathbb{Z}, y = 2x - 7$.

10. (a) Prove: $\forall y \in \mathbb{R}_{\neq 3}, \exists x \in \mathbb{R}_{\neq 5}, xy - 5y - 3x + 13 = 0$.
    (b) Prove: $\forall x \in \mathbb{R}_{\neq 5}, \exists y \in \mathbb{R}_{\neq 3}, xy - 5y - 3x + 13 = 0$.
    (c) Prove: $\exists x \in \mathbb{R}, \exists b \in \mathbb{R}, \forall y \in \mathbb{R}, xy - 5y - 3x + 13 = b$.

11. Prove or disprove the following statements, which are variations of Example 2.72.
    (a) $\forall x \in \mathbb{R}, \exists z \in \mathbb{R}, \forall y \in \mathbb{R}, x < y \Rightarrow x < z < y$.
    (b) $\forall x \in \mathbb{R}, \forall z \in \mathbb{R}, \exists y \in \mathbb{R}, x < y \Rightarrow x < z < y$.
    (c) $\forall x \in \mathbb{R}, \forall z \in \mathbb{R}, \exists y \in \mathbb{R}, x < z \Rightarrow x < z < y$.
    (d) $\exists z \in \mathbb{R}, \forall x \in \mathbb{R}_{<0}, \forall y \in \mathbb{R}_{>0}, x < z < y$.
    (e) $\forall x \in \mathbb{Q}, \forall y \in \mathbb{Q}, \exists z \in \mathbb{Q}, x \leq z \leq y \vee y \leq z \leq x$.

12. Prove or disprove each statement.
    (a) $\forall x \in \mathbb{R}, \forall y \in \mathbb{R}, \exists z \in \mathbb{R}, x + y + z = 0$.
    (b) $\forall x \in \mathbb{R}, \exists y \in \mathbb{R}, \forall z \in \mathbb{R}, x + y + z = 0$.
    (c) $\exists x \in \mathbb{R}, \forall y \in \mathbb{R}, \forall z \in \mathbb{R}, x + y + z = 0$.
    (d) $\exists y \in \mathbb{R}, \forall z \in \mathbb{R}, \exists x \in \mathbb{R}, x + yz = 3$.
    (e) $\exists x \in \mathbb{R}, \exists y \in \mathbb{R}, \forall z \in \mathbb{R}, x + yz = 3$.

13. When choosing objects in this problem, indicate explicitly which other objects the chosen object may depend upon.
    (a) Outline a proof of this statement:
    $\forall \epsilon \in \mathbb{R}_{>0}, \forall x \in \mathbb{R}, \exists \delta \in \mathbb{R}_{>0}, \forall y \in \mathbb{R}, |x - y| < \delta \Rightarrow |f(x) - f(y)| < \epsilon$.
    (b) Outline a proof of this statement:
    $\forall \epsilon \in \mathbb{R}_{>0}, \exists \delta \in \mathbb{R}_{>0}, \forall x \in \mathbb{R}, \forall y \in \mathbb{R}, |x - y| < \delta \Rightarrow |f(x) - f(y)| < \epsilon$.
    (c) Outline a disproof of the statement in (b).
    [*Remark:* (a) is the definition of *continuity* of the function f, whereas (b) is the definition of *uniform continuity* of the function f. These concepts are studied in advanced calculus.]

14. Let $(a_1, a_2, \ldots, a_m, \ldots)$ be a fixed sequence of real numbers. Let $L \in \mathbb{R}$ be fixed.
    (a) Outline a proof of: $\forall \epsilon \in \mathbb{R}_{>0}, \exists N \in \mathbb{Z}_{>0}, \forall m \in \mathbb{Z}_{>0}, (m \geq N \Rightarrow |a_m - L| < \epsilon)$.
    (b) Outline a disproof of the statement in (a).

15. (a) Outline a proof of: $\exists e \in G, \forall x \in G, (e \star x = x \wedge \exists y \in G, y \star x = e)$.
    (b) Outline a disproof of the statement in (a).

16. (a) Prove or disprove the statement in part (a) of the previous problem, taking $G = \mathbb{Z}$, and replacing $\star$ by $+$ (integer addition).
    (b) Repeat (a) taking $G = \mathbb{R}$ and replacing $\star$ by $\cdot$ (real multiplication).
    (c) Repeat (a) taking $G = \mathbb{Q}_{>0}$ and replacing $\star$ by $\cdot$ (multiplication of rational numbers).

17. (a) Is $\exists x, \exists y, P(x, y)$ equivalent to $\exists x, \exists y, P(y, x)$ for all $P(x, y)$? Explain.
    (b) Is $\forall x, \forall y, P(x, y)$ equivalent to $\forall x, \forall y, P(y, x)$ for all $P(x, y)$? Explain.
    (c) Is $\forall x, \exists y, P(x, y)$ equivalent to $\forall x, \exists y, P(y, x)$ for all $P(x, y)$? Explain.

18. Create your own proof template to prove a uniqueness statement of the form $\exists! x \in U, P(x)$. [*Hint:* Use the elimination rule for the uniqueness symbol, and then combine several previously discussed proof templates to outline a proof of the new statement.]

19. Use the template in the previous problem to prove each uniqueness statement.
    (a) $\exists! x \in \mathbb{R}, 3x + 5 = 0$.
    (b) $\forall y \in \mathbb{R}, \forall a \in \mathbb{R}_{\neq 0}, \forall b \in \mathbb{R}, \exists! x \in \mathbb{R}, ax + b = y$.
    (c) $\exists! x \in \mathbb{Z}_{>0}, x^2 - 5x - 14 = 0$.

20. Let $P(x, y)$ be a fixed open sentence.

(a) Outline a proof of: $\forall x \in \mathbb{R}, \exists ! y \in \mathbb{R}, P(x, y)$.

(b) Outline a disproof of the statement in (a).

21. For each choice of $P(x, y)$, prove or disprove the statement in part (a) of the previous problem. (a) $P(x, y)$ is "$y = x^2$." (b) $P(x, y)$ is "$x = y^2$." (c) $P(x, y)$ is "$x = y^3$." (d) $P(x, y)$ is "$xy = 1$." (e) $P(x, y)$ is "$x^2 + y^2 = 4 \wedge y \geq 0$."

22. Let S be a fixed set of real numbers.

(a) Outline a proof of: $\exists y \in \mathbb{R}, [(\forall x \in S, x \leq y) \wedge (\forall z \in \mathbb{R}, z < y \Rightarrow \exists u \in S, z < u)]$.

(b) Outline a disproof of the statement in (a).

23. For each choice of the set S, prove or disprove the statement in part (a) of the previous problem. (a) $S = \mathbb{R}$. (b) S is the set of $x \in \mathbb{R}$ with $0 \leq x \leq 1$. (c) S is the set of $x \in \mathbb{R}$ with $x < 3$. (d) S is the set of real numbers of the form $2 - 1/n$ for some $n \in \mathbb{Z}_{>0}$. (e) S is the empty set. (f) $S = \mathbb{Z}$.

24. Prove $\forall \epsilon \in \mathbb{R}_{>0}, \exists \delta \in \mathbb{R}_{>0}, \forall x \in \mathbb{R}, \forall y \in \mathbb{R}, |x - y| < \delta \Rightarrow |(3x + 5) - (3y + 5)| < \epsilon$ by completing the outline in part (b) of Exercise 13.

25. Prove $\forall \epsilon \in \mathbb{R}_{>0}, \forall x \in \mathbb{R}, \exists \delta \in \mathbb{R}_{>0}, \forall y \in \mathbb{R}, |x - y| < \delta \Rightarrow |x^2 - y^2| < \epsilon$ by completing the outline in part (a) of Exercise 13.

26. Disprove $\forall \epsilon \in \mathbb{R}_{>0}, \exists \delta \in \mathbb{R}_{>0}, \forall x \in \mathbb{R}, \forall y \in \mathbb{R}, |x - y| < \delta \Rightarrow |x^2 - y^2| < \epsilon$ by completing the outline in part (c) of Exercise 13.

27. Prove: $\forall \epsilon \in \mathbb{R}_{>0}, \forall x \in \mathbb{R}_{>0}, \exists \delta \in \mathbb{R}_{>0}, \forall y \in \mathbb{R}_{>0}, |x - y| < \delta \Rightarrow |(1/x) - (1/y)| < \epsilon$.

28. Disprove:

$$\forall \epsilon \in \mathbb{R}_{>0}, \exists \delta \in \mathbb{R}_{>0}, \forall x \in \mathbb{R}_{>0}, \forall y \in \mathbb{R}_{>0}, |x - y| < \delta \Rightarrow |(1/x) - (1/y)| < \epsilon.$$

29. (a) Prove: $\exists ! y \in \mathbb{R}, \exists ! x \in \mathbb{R}, x^2 = y$.

(b) Prove or disprove: $\exists ! x \in \mathbb{R}, \exists ! y \in \mathbb{R}, x^2 = y$.

2.7 More Quantifier Properties and Proofs (Optional)

We begin this section with some logical rules showing how the quantifiers $\forall$ and $\exists$ interact with propositional operators such as $\wedge$ and $\vee$. After some informal discussion, we use the proof templates to prove these rules and some other properties stated in the last section. These proofs highlight the difference between how multiple quantifiers are treated in *known* statements vs. statements *to be proved*. The material presented here is a bit more abstract and challenging compared to earlier sections, so I have designated this section as optional. But it is a great opportunity to understand, at a deeper level, the nuances of how quantifiers and proof templates work.

How Quantifiers Interact with Propositional Operators

We have remarked that the universal quantifier $\forall$ can be viewed as a generalization of the logical AND operator $\wedge$, whereas the existential quantifier $\exists$ generalizes the OR operator $\vee$. This observation may give some algebraic intuition for the following result.

2.73. Theorem. For all open sentences $P(x)$ and $Q(x)$ and all universes U:

(a) $\forall x \in U, [P(x) \wedge Q(x)] \quad \Leftrightarrow \quad (\forall y \in U, P(y)) \wedge (\forall z \in U, Q(z))$.

(b) $\exists x \in U, [P(x) \vee Q(x)] \quad \Leftrightarrow \quad (\exists y \in U, P(y)) \vee (\exists z \in U, Q(z))$.

We give an intuitive explanation for this theorem here, followed by a formal proof later in this section. The universal statement on the left side of (a) asserts that every object c in U makes both $P(c)$ and $Q(c)$ true. The right side of statement (a) is an AND-statement asserting that, first, every object c in U makes $P(c)$ true; and second, every object c in U makes $Q(c)$ true. Intuitively, the two sides of statement (a) mean the same thing, so they should be logically equivalent. Statement (b) has a similar intuitive justification: the left side says that at least one object c in U makes $P(c)$ true or $Q(c)$ true; the right side says that either some object c in U makes $P(c)$ true, or some object c in U makes $Q(c)$ true.

Next we present two more subtle rules, involving the interaction of $\forall$ and $\vee$ on one hand, and the interaction of $\exists$ and $\wedge$ on the other hand.

2.74. Theorem. For all open sentences $P(x)$ and $Q(x)$ and all universes U:

(a) $[(\forall y \in U, P(y)) \vee (\forall z \in U, Q(z))] \quad \Rightarrow \quad [\forall x \in U, (P(x) \vee Q(x))]$.

(b) $[\exists x \in U, (P(x) \wedge Q(x))] \quad \Rightarrow \quad [(\exists y \in U, P(y)) \wedge (\exists z \in U, Q(z))]$.

Warning: The converse of (a) is NOT always true; the converse of (b) is NOT always true.

Let us begin with specific examples illustrating the failure of the converse statements. For statement (a), let U be the set of all humans, let $P(y)$ mean "y is female," and let $Q(z)$ mean "z is male." The statement $\forall y \in U, P(y)$ says that every human is female, which is false. The statement $\forall z \in U, Q(z)$ says that every human is male, which is also false. So the OR-statement appearing as the hypothesis of statement (a) is false in this example. On the other hand, for any fixed, particular human x_0, the statement that "x_0 is female or x_0 is male" is true, so the universal statement appearing as the conclusion of statement (a) is true in this example. So, statement (a) is indeed true (as $F \Rightarrow T$ is true), but the converse of statement (a) is false in this example (as $T \Rightarrow F$ is false).

For statement (b), we give a more mathematical example. Let $U = \mathbb{Z}$, let $P(x)$ mean "$x > 0$," and let $Q(x)$ mean "$x < 0$." Does there exist at least one fixed integer x_0 making the statement "$x_0 > 0$ and $x_0 < 0$" true? The answer is no, so the hypothesis of statement (b) is false in this example. On the other hand, consider the AND-statement in the conclusion of statement (b). Does there exist an integer y_0 such that $y_0 > 0$? Yes, for example, $y_0 = 1$. So the first half of the AND-statement is true. Does there exist an integer z_0 such that $z_0 < 0$? Yes, for example, $z_0 = -1$. So the second half of the AND-statement is true. Thus, statement (b) has a true conclusion. We see now that statement (b) is true but its converse is false, in this example.

We now give an informal justification of Theorem 2.74; see below for a formal proof. Intuitively, the hypothesis of (a) demands that one of two very strong conditions hold: either every object in the universe satisfies P, or every object in the universe satisfies Q. The conclusion of statement (a) is the much weaker assertion that for each individual object, considered one at a time, that object satisfies either P or Q. Which one of the two properties is true can vary from object to object.

Here is intuition for statement (b). The hypothesis of statement (b) is the rather strong statement that we can find one particular object in U that simultaneously satisfies both P and Q. The conclusion of statement (b) only asserts that some object in U satisfies P, and some (possibly different) object in U satisfies Q.

Proofs of the Quantifier Interaction Properties

We now use the proof templates to generate formal proofs of Theorem 2.73 and 2.74. Recall that letters with subscripts (like x_0) denote constants, whereas the corresponding letter without a subscript is a quantified variable. Throughout these proofs, we let $P(x)$ and $Q(x)$ be fixed, but arbitrary, open sentences, and let U be a fixed universe. [Note that we have just used the generic-element proof template three times to deal with the universal quantifiers that begin the theorem statements.]

2.75. Proof of $\forall x \in U, [P(x) \wedge Q(x)] \quad \Leftrightarrow \quad (\forall y \in U, P(y)) \wedge (\forall z \in U, Q(z))$.
[This is an IFF-statement, so we need a two-part proof.]

Part 1. Assume $\forall x \in U, [P(x) \wedge Q(x)]$. Prove $(\forall y \in U, P(y)) \wedge (\forall z \in U, Q(z))$.
[The new goal is an AND-statement, so we prove each part separately.]
First, prove $\forall y \in U, P(y)$. Fix an arbitrary object $y_0 \in U$. Prove $P(y_0)$. To do so, take $x = y_0$ in the assumed universal statement (we are using the Inference Rule for ALL here). The assumption tells us that $P(y_0) \wedge Q(y_0)$ is true. By definition of AND, $P(y_0)$ is true, as needed.
Second, prove $\forall z \in U, Q(z)$. Fix an arbitrary object $z_0 \in U$. Prove $Q(z_0)$. To do so, take $x = z_0$ in the assumed universal statement (using the Inference Rule for ALL). The assumption tells us that $P(z_0) \wedge Q(z_0)$ is true. By definition of AND, $Q(z_0)$ is true, as needed. Part 1 is now complete.

Part 2. Assume $(\forall y \in U, P(y)) \wedge (\forall z \in U, Q(z))$. Prove $\forall x \in U, [P(x) \wedge Q(x)]$.
[The goal of part 2 is a universal statement, so we use the generic-element proof template.]
Let x_0 be a fixed, but arbitrary element of U. We must prove $P(x_0) \wedge Q(x_0)$. To prove this AND-statement, we first prove $P(x_0)$, and then prove $Q(x_0)$. Note that the AND-statement we have assumed tells us that $\forall y \in U, P(y)$ is true, and $\forall z \in U, Q(z)$ is true. Taking $y = x_0$ in the first of these statements, we see that $P(x_0)$ is true by the Inference Rule for ALL. Taking $z = x_0$ in the second of these statements, we see that $Q(x_0)$ is true by the Inference Rule for ALL. Part 2 is now complete.

2.76. Proof of $\exists x \in U, [P(x) \vee Q(x)] \quad \Leftrightarrow \quad (\exists y \in U, P(y)) \vee (\exists z \in U, Q(z))$.
We could prove this statement from first principles, but it is quicker to *use* the statement we have just proved, namely Theorem 2.73(a). Since this statement was proved for *all* open

sentences $P(x)$ and $Q(x)$, we can replace $P(x)$ by $\sim P(x)$ and $Q(x)$ by $\sim Q(x)$ in part (a). We see that we already know this fact:

$$\forall x \in U, [(\sim P(x)) \wedge (\sim Q(x))] \quad \Leftrightarrow \quad [\forall y \in U, \sim P(y)) \wedge (\forall z \in U, \sim Q(z)].$$

Recall the logical equivalence $A \Leftrightarrow B \equiv (\sim A) \Leftrightarrow (\sim B)$. We can use this equivalence to transform the fact written above by negating the statements on the left and right sides of the $\Leftrightarrow$ symbol. Using the denial rules to further simplify each side, we eventually get

$$\exists x \in U, [P(x) \vee Q(x)] \quad \Leftrightarrow \quad (\exists y \in U, P(y)) \vee (\exists z \in U, Q(z)),$$

which is precisely the result we needed to prove.

2.77. Proof of $[(\forall y \in U, P(y)) \vee (\forall z \in U, Q(z))] \quad \Rightarrow \quad [\forall x \in U, (P(x) \vee Q(x))]$.
[We give a direct proof of the IF-statement.]
Assume $(\forall y \in U, P(y)) \vee (\forall z \in U, Q(z))$. Prove $\forall x \in U, (P(x) \vee Q(x))$.
[We have just assumed an OR-statement, so use proof by cases.]
Case 1. Assume $\forall y \in U, P(y)$. Prove $\forall x \in U, (P(x) \vee Q(x))$. Let x_0 be a fixed, but arbitrary element of U. Prove $P(x_0) \vee Q(x_0)$. Taking $y = x_0$ in the assumption, the Inference Rule for ALL tells us that $P(x_0)$ is true. Now, by definition of OR, "$P(x_0) \vee Q(x_0)$" must also be true.
Case 2. Assume $\forall z \in U, Q(z)$. Prove $\forall x \in U, (P(x) \vee Q(x))$. Let x_0 be a fixed, but arbitrary element of U. Prove $P(x_0) \vee Q(x_0)$. Taking $z = x_0$ in the assumption, the Inference Rule for ALL tells us that $Q(x_0)$ is true. Now, by definition of OR, "$P(x_0) \vee Q(x_0)$" must also be true.
So the needed conclusion holds in all cases.

We could deduce Theorem 2.74(b) by taking the contrapositive of what we just proved and simplifying (see the exercises), but proving part (b) from scratch is also instructive.

2.78. Proof of $[\exists x \in U, (P(x) \wedge Q(x))] \quad \Rightarrow \quad [(\exists y \in U, P(y)) \wedge (\exists z \in U, Q(z))]$.
Assume $\exists x \in U, P(x) \wedge Q(x)$. Our assumption means there is at least one constant object x_0 in U such that $P(x_0) \wedge Q(x_0)$ is true [we are using existential instantiation (EI) to give a name x_0 to the object whose existence is assumed; see page 59]. Prove $(\exists y \in U, P(y)) \wedge (\exists z \in U, Q(z))$. [To prove an AND-statement, we prove each separate part.] We must first prove $\exists y \in U, P(y)$. Choose $y_0 = x_0$, which is in U. By definition of AND, $P(x_0)$ is true. We must next prove $\exists z \in U, Q(z)$. Choose $z_0 = x_0$, which is in U. By definition of AND, $Q(x_0)$ is true.

The reader is invited to attempt to prove the converse of each statement in Theorem 2.74 for general open sentences $P(x)$ and $Q(x)$, to see exactly where the proof templates get stuck.

Proofs of Quantifier Reordering Rules

This section provides proofs of the rules for reordering quantifiers, which were stated in the previous section. Throughout this section, we let x and y be arbitrary distinct variables, and we let $P(x, y)$ be a fixed, but arbitrary, open sentence.

2.79. Proof of $\forall x, \forall y, P(x, y) \quad \Leftrightarrow \quad \forall y, \forall x, P(x, y)$.
[This is an IFF-statement, so we do a two-part proof.]
Part 1. Assume $\forall x, \forall y, P(x, y)$. Prove $\forall y, \forall x, P(x, y)$.
[The new goal is a universal statement, so use the generic-element template.]
Let y_0 be a fixed, but arbitrary object. Let x_0 be a fixed, but arbitrary object. Prove $P(x_0, y_0)$ is true. [To continue, note that the assumption is a known universal statement, so

we can use the Inference Rule for ALL, abbreviated IRA.] In the assumption, take x to be the object x_0 to deduce (by IRA) $\forall y, P(x_0, y)$. Then take y to be the object y_0 to deduce (by IRA) that $P(x_0, y_0)$ is true. This completes part 1. We have now proved

$$\forall x, \forall y, P(x, y) \quad \Rightarrow \quad \forall y, \forall x, P(x, y).$$

<u>Part 2.</u> [This part is almost identical to part 1, so we let you fill in the details. Alternatively, the IF-statement to be proved in part 2 follows from the IF-statement already proved in part 1 by interchanging x and y, after replacing $P(x, y)$ by $P(y, x)$.]

In the exercises below, we ask you to prove

$$\exists x, \exists y, P(x, y) \quad \Leftrightarrow \quad \exists y, \exists x, P(x, y) \tag{2.5}$$

in two ways.

2.80. Proof of $[\exists x, \forall y, P(x, y)] \Rightarrow [\forall y, \exists x, P(x, y)]$.
Assume $\exists x, \forall y, P(x, y)$. Prove $\forall y, \exists x, P(x, y)$.
By Existential Instantiation (EI), the assumption tells us there is some fixed object x_0 that makes $\forall y, P(x_0, y)$ true. To begin proving the new goal, let y_0 be a fixed, but arbitrary object. We must show $\exists x, P(x, y_0)$. We choose x here to be x_0; so now we must show $P(x_0, y_0)$ is true. This follows by applying the Inference Rule for ALL to $\forall y, P(x_0, y)$, letting y be the object y_0.

To help understand what happened in that last proof, let's try to prove the converse and see what goes wrong.

2.81. Attempted Proof of $[\forall y, \exists x, P(x, y)] \Rightarrow [\exists x, \forall y, P(x, y)]$.
Assume $\forall y, \exists x, P(x, y)$. Prove $\exists x, \forall y, P(x, y)$. [Here, at the very beginning of the proof, we are required to choose a specific object x_0 making $\forall y, P(x_0, y)$ true. What are we going to choose? The assumption cannot help us unless we can figure out a helpful constant to substitute for y using the Inference Rule for ALL (IRA). Suppose we blindly try $y = y_0$, where y_0 is some particular object. By IRA and then EI, the assumption does give us an object x_0 making $P(x_0, y_0)$ true for this particular y_0. But now, to prove $\forall y, P(x_0, y)$, we need to fix a new *arbitrary* object y_1, which need not equal the fixed object y_0 selected earlier in the proof. So there is no obvious way to complete the proof. We have already seen examples where the original statement fails, so we will not be able to find a proof.]

The moral of this section is that the relative ordering of quantifiers and other logical symbols often has a big impact on the meaning of the statement. So great care is needed when working with a statement containing multiple quantifiers.

Section Summary

1. *Quantifier Rules.* The following rules hold for all open sentences $P(x)$ and $Q(x)$:

$$\forall x \in U, [P(x) \wedge Q(x)] \quad \Leftrightarrow \quad (\forall y \in U, P(y)) \wedge (\forall z \in U, Q(z)).$$
$$\exists x \in U, [P(x) \vee Q(x)] \quad \Leftrightarrow \quad (\exists y \in U, P(y)) \vee (\exists z \in U, Q(z)).$$
$$[(\forall y \in U, P(y)) \vee (\forall z \in U, Q(z))] \quad \Rightarrow \quad [\forall x \in U, (P(x) \vee Q(x))].$$
$$[\exists x \in U, (P(x) \wedge Q(x))] \quad \Rightarrow \quad [(\exists y \in U, P(y)) \wedge (\exists z \in U, Q(z))].$$

In the last two rules, there exist open sentences for which the converse statements are false.

2. *Proofs Involving Multiple Quantifiers.* When proving complex statements involving multiple quantifiers, process the quantifiers in the order they are encountered using the proof templates for $\exists$ and $\forall$. To use known statements involving quantifiers, carefully apply Existential Instantiation (EI) and the Inference Rule for ALL (IRA). Be sure to distinguish quantified variables from constants and note whether constants are arbitrary or chosen by the proof writer.

Exercises

1. Is each statement true or false? Explain each answer.
    (a) $\forall y \in \mathbb{R}, y \geq 0$.
    (b) $\forall z \in \mathbb{R}, z < 0$.
    (c) $(\forall y \in \mathbb{R}, y \geq 0) \Leftrightarrow (\forall z \in \mathbb{R}, z < 0)$.
    (d) $x_0 \geq 0 \Leftrightarrow x_0 < 0$, where x_0 is a fixed real number.
    (e) $\forall x \in \mathbb{R}, (x \geq 0 \Leftrightarrow x < 0)$.
    (f) For all open sentences $P(x)$ and $Q(x)$,
    $$\forall x \in \mathbb{R}, (P(x) \Leftrightarrow Q(x)) \quad \text{iff} \quad (\forall y \in \mathbb{R}, P(y)) \Leftrightarrow (\forall z \in \mathbb{R}, Q(z)).$$

2. Deduce part (b) of Theorem 2.74 from part (a) by taking the contrapositive and using the denial rules.

3. Use proof templates to prove part (b) of Theorem 2.73 without using part (a).

4. (a) Prove (2.5) as a consequence of the analogous result for reordering universal quantifiers. (b) Prove (2.5) directly, using proof templates for existential statements and existential instantiation.

5. Prove: for all nonempty universes U and all open sentences $P(x)$,
    $$[\forall x \in U, P(x)] \quad \Rightarrow \quad [\exists x \in U, P(x)].$$

6. (a) Prove: for all universes U and all open sentences $P(x)$ and $Q(x)$,
    $$[\forall x \in U, (P(x) \Rightarrow Q(x))] \quad \Rightarrow \quad [(\forall y \in U, P(y)) \Rightarrow (\forall z \in U, Q(z))].$$

    (b) Disprove: for all universes U and all open sentences $P(x)$ and $Q(x)$,
    $$[(\forall y \in U, P(y)) \Rightarrow (\forall z \in U, Q(z))] \quad \Rightarrow \quad [\forall x \in U, (P(x) \Rightarrow Q(x))].$$

7. Use part (a) of the previous problem and other known theorems to prove:
    $$[\forall x \in U, (P(x) \Leftrightarrow Q(x))] \quad \Rightarrow \quad [(\forall y \in U, P(y)) \Leftrightarrow (\forall z \in U, Q(z))].$$

8. Try to prove the converse of each statement in Theorem 2.74 using the proof templates. Explain what goes wrong.

9. Consider the two statements:
    (i) $\exists x \in U, (P(x) \oplus Q(x))$ (ii) $(\exists y \in U, P(y)) \oplus (\exists z \in U, Q(z))$
    (a) Does (i) imply (ii) in all situations? Prove your answer.
    (b) Does (ii) imply (i) in all situations? Prove your answer.

10. Consider the two statements:
    (i) $\forall x \in U, (P(x) \oplus Q(x))$ (ii) $(\forall y \in U, P(y)) \oplus (\forall z \in U, Q(z))$
    (a) Does (i) imply (ii) in all situations? Prove your answer.
    (b) Does (ii) imply (i) in all situations? Prove your answer.

11. Consider the two statements:
    (i) $\exists x \in U, (P(x) \Leftrightarrow Q(x))$ (ii) $(\exists y \in U, P(y)) \Leftrightarrow (\exists z \in U, Q(z))$
    (a) Does (i) imply (ii) in all situations? Prove your answer.
    (b) Does (ii) imply (i) in all situations? Prove your answer.

12. Consider the two statements:
    (i) $\exists x \in U, (P(x) \Rightarrow Q(x))$ (ii) $(\exists y \in U, P(y)) \Rightarrow (\exists z \in U, Q(z))$
    (a) Does (i) imply (ii) in all situations? Prove your answer.
    (b) Does (ii) imply (i) in all situations? Prove your answer.

13. (a) Prove: for all propositions P and all open sentences $Q(x)$,
    $(P \Rightarrow \forall x, Q(x)) \quad \Leftrightarrow \quad \forall x, (P \Rightarrow Q(x))$.
    (b) Disprove: for all propositions Q and all open sentences $P(x)$,
    $(\forall x, P(x)) \Rightarrow Q \quad \Leftrightarrow \quad \forall x, (P(x) \Rightarrow Q)$.
    (c) Can you find a way to modify the right half of (b) to produce a true statement?
    If so, prove your answer.

14. Fix an arbitrary proposition P and an arbitrary open sentence $Q(x)$. Prove:
    (a) $[\forall x, (P \wedge Q(x))] \quad \Leftrightarrow \quad P \wedge [\forall y, Q(y)]$.
    (b) $[\forall x, (P \vee Q(x))] \quad \Leftrightarrow \quad P \vee [\forall y, Q(y)]$.

15. Fix an arbitrary proposition P and an arbitrary open sentence $Q(x)$. Prove:
    (a) $[\exists x, (P \wedge Q(x))] \quad \Leftrightarrow \quad P \wedge [\exists y, Q(y)]$.
    (b) $[\exists x, (P \vee Q(x))] \quad \Leftrightarrow \quad P \vee [\exists y, Q(y)]$.
    Do this directly, or by using the previous exercise.

16. Let U be a fixed, nonempty universe. Consider the two statements:
    (i) $[\exists x \in U, (P \oplus Q(x))]$ (ii) $P \oplus [\exists y \in U, Q(y)]$.
    (a) Does (i) imply (ii) in all situations? Prove your answer.
    (b) Does (ii) imply (i) in all situations? Prove your answer.

17. Let U be a fixed, nonempty universe. Consider the two statements:
    (i) $[\forall x \in U, (P \Leftrightarrow Q(x))]$ (ii) $P \Leftrightarrow [\forall y \in U, Q(y)]$.
    (a) Does (i) imply (ii) in all situations? Prove your answer.
    (b) Does (ii) imply (i) in all situations? Prove your answer.

18. Prove: for all nonempty universes U and all open sentences $P(x)$,
    $\exists y \in U, (P(y) \Rightarrow \forall x \in U, P(x))$.

19. Prove: for all nonempty universes U and all open sentences $P(x)$,
    $\forall y \in U, (P(y) \Rightarrow \exists x \in U, P(x))$.

20. Does $\exists! x, \forall y, P(x, y)$ always imply $\forall y, \exists! x, P(x, y)$? Explain.

21. Does $\exists! x, (P(x) \vee Q(x))$ always imply $[\exists! y, P(y)] \vee [\exists! z, Q(z)]$? Explain.

22. Does $[\exists! y, P(y)] \vee [\exists! z, Q(z)]$ always imply $\exists! x, (P(x) \vee Q(x))$? Explain.

23. Does $\exists! x, (P(x) \wedge Q(x))$ always imply $[\exists! y, P(y)] \wedge [\exists! z, Q(z)]$? Explain.

24. Does $\exists! x, \exists! y, P(x, y)$ always imply $\exists! y, \exists! x, P(x, y)$? Explain.

25. Which of the answers to the previous five problems change if every $\exists!$ is replaced by $\exists$ in the conclusions?

Review of Logic and Proofs

Logical Symbols

Complicated propositions are built up from simpler ones using the following symbols.

Formal Expression	English Translation
$\sim P$	P is not true.
$P \wedge Q$	P and Q.
$P \vee Q$	P or Q.
$P \Rightarrow Q$	If P, then Q.
$P \Leftrightarrow Q$	P if and only if Q.
$P \oplus Q$	P or Q, but not both.
$\forall x \in U, P(x)$	For every x_0 in U, $P(x_0)$ is true.
$\exists x \in U, P(x)$	There exists at least one x_0 in U for which $P(x_0)$ is true.
$\exists! x \in U, P(x)$	There exists exactly one x_0 in U for which $P(x_0)$ is true.

The meaning of $\sim$, $\wedge$, $\vee$, $\Rightarrow$, $\Leftrightarrow$, and $\oplus$ is given by the following truth tables.

		NOT	AND	OR	IF	IFF	XOR
P	Q	$\sim P$	$P \wedge Q$	$P \vee Q$	$P \Rightarrow Q$	$P \Leftrightarrow Q$	$P \oplus Q$
T	T	F	T	T	T	T	F
T	F	F	F	T	F	F	T
F	T	T	F	T	T	F	T
F	F	T	F	F	T	T	F

Possible translations of $P \Rightarrow Q$ include:

if P, then Q.	P implies Q.	Q if P.	P only if Q.
P is sufficient for Q.	Q is necessary for P.	Q whenever P.	(not P) or Q.

In mathematics, OR always means inclusive-OR ($\vee$).

Denial Rules

To form a useful denial of a given statement, repeatedly apply the following rules.

Statement	Denial of Statement	Symbolic Version of Rule
A and B	(denial of A) or (denial of B)	$\sim(A \wedge B) \equiv (\sim A) \vee (\sim B)$
A or B	(denial of A) and (denial of B)	$\sim(A \vee B) \equiv (\sim A) \wedge (\sim B)$
if A, then B	A and (denial of B)	$\sim(A \Rightarrow B) \equiv A \wedge (\sim B)$
not A	A	$\sim(\sim A) \equiv A$
For all x, $P(x)$	There is x, (denial of $P(x)$)	$\sim \forall x \in U, P(x)$ iff $\exists x \in U, \sim P(x)$
There is x, $P(x)$	For all x, (denial of $P(x)$)	$\sim \exists x \in U, P(x)$ iff $\forall x \in U, \sim P(x)$
A iff B	A or B, not both	$\sim(A \Leftrightarrow B) \equiv (A \oplus B)$
A or B, not both	A iff B	$\sim(A \oplus B) \equiv (A \Leftrightarrow B)$

Note: *the useful denial of an IF-statement is an AND-statement, not an IF-statement.*

Example. A useful denial of
"$\forall \epsilon > 0, \exists \delta > 0, \forall x \in \mathbb{R}, (2 - \delta < x < 2 + \delta) \Rightarrow (4 - \epsilon < x^2 < 4 + \epsilon)$" is
"$\exists \epsilon > 0, \forall \delta > 0, \exists x \in \mathbb{R}, (2 - \delta < x < 2 + \delta)$ and $(4 - \epsilon \geq x^2$ or $x^2 \geq 4 + \epsilon)$."

Propositional Logic

1. *Filling in Truth Tables.* Make a column for each subexpression of the propositional form. Fill in each column using the defining truth tables for the logical operators (given above). Be sure to refer to the correct previous columns. Recall NOT flips true and false; AND outputs true precisely when both inputs are true; OR outputs false precisely when both inputs are false; IF (i.e., $A \Rightarrow B$) outputs false precisely when A is true and B is false; IFF outputs true precisely when the inputs agree; XOR outputs true precisely when the inputs disagree.

2. *Logical Equivalence.* Two propositional forms P and Q are *logically equivalent*, written $P \equiv Q$, iff all rows of the truth tables for P and Q match. Some frequently used logical equivalences appear in the Theorem on Logical Equivalences, the Theorem on IF, and the Theorem on IFF.

3. *Tautologies and Contradictions.*
   A propositional form is a *tautology* iff all rows of its truth table have output true. A propositional form is a *contradiction* iff all rows of its truth table have output false. Every proposition is true or false; but most propositional forms are neither tautologies nor contradictions.

4. *Terminology for IF-statements.*
   (a) $A \Rightarrow B$ is called a *conditional* with *hypothesis* A and *conclusion* B.
   (b) The *converse* of $A \Rightarrow B$ is $B \Rightarrow A$.
   (c) The *contrapositive* of $A \Rightarrow B$ is $(\sim B) \Rightarrow (\sim A)$.
   (d) A *denial* of $A \Rightarrow B$ is $A \wedge (\sim B)$.
   (e) The *inverse* of $A \Rightarrow B$ is $(\sim A) \Rightarrow (\sim B)$.
   (f) "If A then B" is logically equivalent to its contrapositive.
   (g) "If A then B" is not logically equivalent to its converse, its inverse, or its negation (in general).

5. *Eliminating $\Rightarrow$, $\Leftrightarrow$, and $\oplus$.* Use these logical equivalences to rewrite IF, IFF, and XOR in terms of other logical symbols:
   (a) $A \Rightarrow B \quad \equiv \quad (\sim A) \vee B$.
   (b) $A \Leftrightarrow B \quad \equiv \quad (A \Rightarrow B) \wedge (B \Rightarrow A) \quad \equiv \quad (A \wedge B) \vee ((\sim A) \wedge (\sim B))$.
   (c) $A \oplus B \quad \equiv \quad (A \wedge (\sim B)) \vee ((\sim A) \wedge B) \quad \equiv \quad (A \vee B) \wedge (\sim(A \wedge B))$.

6. *Inference Rules for IF.* If $P \Rightarrow Q$ is a known theorem and P is a known theorem, you can deduce Q as a new theorem. If $P \Rightarrow Q$ is a known theorem and $\sim Q$ is a known theorem, you can deduce $\sim P$ as a new theorem.

Quantifiers

Let U be a fixed universe (nonempty set of objects).

1. *Restricted and Unrestricted Quantifiers.* For all open sentences $P(x)$,
   $\forall x \in U, P(x) \quad$ iff $\quad \forall x, (x \in U \Rightarrow P(x))$;
   $\exists x \in U, P(x) \quad$ iff $\quad \exists x, (x \in U \wedge P(x))$.

2. *Quantifiers Rules and Pitfalls.* For all open sentences $P(x)$, $Q(x)$, and $R(x, y)$:

(a) $\qquad\qquad\qquad \forall x, P(x) \qquad \Leftrightarrow \qquad \forall v, P(v)$ where v is a new letter.

(b) $\qquad\qquad\qquad \exists x, P(x) \qquad \Leftrightarrow \qquad \exists v, P(v)$ where v is a new letter.

(c) $\qquad\qquad \forall x, \forall y, R(x, y) \qquad \Leftrightarrow \qquad \forall y, \forall x, R(x, y).$

(d) $\qquad\qquad \exists x, \exists y, R(x, y) \qquad \Leftrightarrow \qquad \exists y, \exists x, R(x, y).$

(e) $\qquad\qquad \exists x, \forall y, R(x, y) \qquad \Rightarrow \qquad \forall y, \exists x, R(x, y).$

(f) $\qquad \forall x \in U, [P(x) \wedge Q(x)] \qquad \Leftrightarrow \qquad (\forall y \in U, P(y)) \wedge (\forall z \in U, Q(z)).$

(g) $\qquad \exists x \in U, [P(x) \vee Q(x)] \qquad \Leftrightarrow \qquad (\exists y \in U, P(y)) \vee (\exists z \in U, Q(z)).$

(h) $\quad [(\forall y \in U, P(y)) \vee (\forall z \in U, Q(z))] \qquad \Rightarrow \qquad [\forall x \in U, (P(x) \vee Q(x))].$

(i) $\qquad [\exists x \in U, (P(x) \wedge Q(x))] \qquad \Rightarrow \qquad [(\exists y \in U, P(y)) \wedge (\exists z \in U, Q(z))].$

The right side of (e) says for all y, there exists an x (which can be chosen differently for different y) such that $R(x, y)$ is true. The left side of (e) is the stronger statement that there is one choice of x (the same choice for all y) such that $R(x, y)$ is true. To see that (h) cannot be strengthened to an IFF-statement, let U be the set of humans, let $P(x)$ mean "x is male," and let $Q(x)$ mean "x is female." To see that (i) cannot be strengthened to an IFF-statement, let $U = \mathbb{R}$, let $P(x)$ mean "$x < 0$," and let $Q(x)$ mean "$x > 0$."

3. *Contrasting Translations.* Let $B(x)$ mean x is black; let $C(x)$ mean x is a crow.
"No crows are black" translates to $\forall x, C(x) \Rightarrow {\sim}B(x)$.
"Not all crows are black" translates to ${\sim}\forall x, C(x) \Rightarrow B(x)$.
$\qquad$ We can transform this translation to $\exists x, C(x) \wedge {\sim}B(x)$.
"All crows are not black" translates to $\forall x, C(x) \Rightarrow {\sim}B(x)$.
"Black crows do not exist" translates to ${\sim}\exists x, C(x) \wedge B(x)$.
$\qquad$ We can transform this translation to $\forall x, C(x) \Rightarrow {\sim}B(x)$.

4. *Inference Rule for ALL.* If "$\forall x \in U, P(x)$" is a known theorem and "$c \in U$" is a known theorem (where c is any expression), then you can deduce the new theorem $P(c)$.

5. *Multiple Quantifier Example.* To *prove* $\forall x, \exists y, \forall z, P(x, y, z)$: fix x_0; choose an appropriate y_0 (which may depend on x_0); fix z_0; prove $P(x_0, y_0, z_0)$. To *disprove* $\forall x, \exists y, \forall z, P(x, y, z)$, prove $\exists x, \forall y, \exists z, {\sim}P(x, y, z)$ as follows. Choose an appropriate specific object x_0; fix y_0; choose an appropriate z_0 (which may depend on x_0 and y_0); prove ${\sim}P(x_0, y_0, z_0)$.

6. *Rule to Eliminate the Uniqueness Symbol.*

$$\exists! x \in U, P(x) \quad \Leftrightarrow \quad (\exists x \in U, P(x)) \wedge \forall x \in U, \forall y \in U, [(P(x) \wedge P(y)) \Rightarrow x = y].$$

Proof Templates

General Comments:

(a) **Memorize the proof templates! Practice the proof templates!**

(b) To decide which template to apply at each stage, look for the outermost logical operator in the current statement to be proved.

(c) In each template, the numbered statements (1), (2), ... should be explicitly written down at the appropriate point in the proof.

(d) After writing these statements, continue the proof by either using more proof templates or expanding definitions in what you have just written.

(e) Remember to introduce each sentence, formula, and new variable with **logical status words** ("assume," "we must prove," "we know," "choose," etc.).

(f) Do not reuse a variable to mean two things.

(g) Do not deduce consequences from an equation (or statement) that is yet to be proved. Instead, work forward from known and assumed statements to gather information leading toward the statement to be proved.

1. *Direct Proof of "if P then Q."* (1) Assume P. (2) Prove Q.

2. *Contrapositive Proof of "if P then Q."* (1) Assume $\sim Q$. (2) Prove $\sim P$. [Use the denial rules to find useful denials of P and Q here.]

3. *Contradiction Proof of "if P then Q."* (1) Assume, to get a contradiction, "P and $\sim Q$." (2) Deduce a contradiction.

4. *Contradiction Proof of Any Statement P.* (1) Assume, to get a contradiction, $\sim P$ [write a useful denial of P here]. (2) Use this assumption, definitions, and known results to produce a contradiction (such as $S \wedge (\sim S)$). (3) Conclude that P is true.

5. *Two-Part Proof of "P iff Q."* (1) Prove "if P then Q" by any method. (2) Prove "if Q then P" by any method.

6. *Constructive Proof of $\exists x \in U, P(x)$.* (1) Describe a specific object x_0 that will make $P(x)$ true (we often say: "choose $x_0 = ...$"). (2) Prove $x_0 \in U$. (3) Prove $P(x_0)$. Note x_0 is allowed to depend on constants introduced earlier in the proof.

7. *Generic-Element Proof of $\forall x \in U, P(x)$.* (1) Fix an arbitrary object $x_0 \in U$. (2) Prove $P(x_0)$. (It is not correct to choose a *specific* object in U here; step (2) must work for *generic* elements of U.)

8. *Proof by Cases.* To prove Q when you already know $P_1 \vee P_2 \vee \cdots \vee P_k$: (0) Say: "We know P_1 or P_2 or ... or P_k, so use cases."
   (1) <u>Case 1.</u> Assume P_1. Prove Q.
   (2) <u>Case 2.</u> Assume P_2. Prove Q.
   ...
   (k) <u>Case k.</u> Assume P_k. Prove Q. (Thus Q must be proved k times, where there is a different assumption each time.)

9. *Proving $P \wedge Q$.* (1) Prove P. (2) Prove Q.

10. *Proving $P \vee Q$.* (1) Assume $\sim P$. (2) Prove Q.

11. *Proving $P \oplus Q$.* Prove $P \Leftrightarrow (\sim Q)$.

12. *Proving $\sim P$.* (1) Use the denial rules to find a useful denial of P not starting with NOT. (2) Prove the denial found in (1).

13. *Disproving P when P is False.* Prove $\sim P$ (see previous item).

14. *Proof by Exhaustion.* If $U = \{z_1, \ldots, z_n\}$ is a finite set, you can prove $\forall x \in U, P(x)$ by proving "$P(z_1)$ and $P(z_2)$ and ... and $P(z_n)$."

Review Problems

1. Complete the following definitions: (a) for a fixed integer c, c divides $c^2 + 7$ iff... (b) A real number r is rational iff... (c) A propositional form $\mathcal{A}$ is a contradiction iff...

2. Give a truth table for the propositional form $((\sim P) \wedge Q) \Leftrightarrow (P \vee Q)$. Show all columns. Is this a tautology?

3. Write seven different English phrases that could be used to translate $P \Rightarrow Q$.

4. Let Q be the statement "If X is compact, then: X is separable or X is not closed." (a) Write the converse of Q. (b) Write the contrapositive of Q. (c) Outline a direct proof of Q. (d) Outline a proof of Q by contradiction.

5. True or false?
   (a) 7 is odd or 7 is odd.
   (b) $1 + 1 = 2$ if $0 = 1$.
   (c) $0 = 0$ whenever $0 \neq 0$.
   (d) $5 \in \mathbb{Z}$ implies $0.37 \in \mathbb{Z}$.
   (e) 0.5 is rational only if 8 is odd.
   (f) $\sqrt{2} \in \mathbb{R}$ is a necessary condition for $\sqrt{2} \in \mathbb{Z}$.
   (g) 2 is odd iff 7 is even.
   (h) $\exists! x \in \mathbb{R}, (x^3 = x)$.
   (i) A useful denial of "$4 < x \leq y$" is "$4 \geq x > y$."
   (j) Every propositional form is either a tautology or a contradiction.
   (k) The propositional form $P \oplus P$ is a contradiction.
   (l) $(P \wedge Q) \vee R$ is logically equivalent to $P \wedge (Q \vee R)$.
   (m) Given that $P \Rightarrow Q$ is a known theorem and Q is a known theorem, we may always deduce P as a new theorem.
   (n) One way to prove $P \oplus Q$ is to assume $\sim P$ and prove Q.

6. (a) Make truth tables for $(P \Leftrightarrow Q) \Rightarrow P$ and $P \Leftrightarrow (Q \Rightarrow P)$.
   (b) Are the two forms in (a) logically equivalent? Explain briefly.
   (c) Is either form in (a) a tautology? Explain briefly.

7. Let Q be the statement "If (X is compact and Hausdorff) or X is metrizable, then X is normal." (a) Write the converse of Q. (b) Write the contrapositive of Q. (c) Outline a direct proof of Q.

8. Consider the statement: $(A \wedge B) \Rightarrow (C \oplus D)$.
   (a) Outline a direct proof of this statement.
   (b) Outline a contrapositive proof of this statement.
   (c) Outline a proof by contradiction of this statement.

9. Let $W(x)$ mean "x is white," $R(x)$ mean "x is a rose," and $P(x)$ mean "x is pretty." (a) Using this notation, logical connectives, and unrestricted quantifiers, write this statement in symbolic form: "Although all white roses are pretty, some pretty roses are not white." (b) Give a useful denial of the previous statement. Do not use symbols in your answer.

10. Prove: for all $x, y \in \mathbb{R}$, if $x \in \mathbb{Q}$ and $y \in \mathbb{Q}$ and $y \neq 0$, then $x/y \in \mathbb{Q}$.

11. (a) Prove: For all integers x, if $x^2 + 4$ is even, then x is even. (b) Is the converse of (a) true?

12. Prove: for all $a, b, c, r, s \in \mathbb{Z}$, if a divides b and a divides c, then a divides $rb + sc$.

13. Outline a direct proof of this statement, assuming P, a, b are fixed objects: "if P is a prime ideal, then $ab \in P$ iff $a \in P$ or $b \in P$."

14. (a) Prove: $\forall x \in \mathbb{R}, \exists y \in \mathbb{R}, y > x + 2$. (b) Disprove: $\exists y \in \mathbb{R}, \forall x \in \mathbb{R}, y > x + 2$.

15. (a) Prove: $\forall x \in \mathbb{R}_{>0}, \exists y \in \mathbb{R}_{>0}, x < y < 2x$.
    (b) Disprove: $\exists y \in \mathbb{R}_{>0}, \forall x \in \mathbb{R}_{>0}, x < y < 2x$.

16. Prove: for all $r \in \mathbb{R}$, r is rational iff $3r + 5$ is rational.

17. Disprove: for all $a, b, c, r, s \in \mathbb{Z}$, if a divides $rb + sc$, then a divides b or a divides c or a divides r or a divides s.

18. (a) Prove: for all $a, b, c \in \mathbb{Z}$, if a divides b or a divides c, then a divides bc^2.
    (b) Prove or disprove the converse of (a).

19. Show that $P \Rightarrow (Q \wedge R)$ is logically equivalent to $(P \Rightarrow Q) \wedge (P \Rightarrow R)$ using known logical equivalences, not truth tables.

20. Give a proof by contradiction of this statement:
    For all $b, c \in \mathbb{Z}$, if bc^2 is even, then b is even or c is even.

3

Sets

3.1 Set Operations and Subset Proofs

At this point, we have introduced nearly all of the fundamental rules for generating proofs of logical statements. (The main rule yet to be discussed is mathematical induction, which we cover in Chapter 4.) Armed with our knowledge of logic and proofs, we can now proceed to study whatever mathematical theories we are interested in: set theory, number theory, graph theory, abstract algebra, analysis, probability theory, complexity theory, and so on. Surprisingly, the theory of sets turns out to be a foundational subject on which every other branch of mathematics can be built. Many aspects of set theory relate very closely to the propositional logic and quantifier logic that we have just studied. So, we continue our journey up the tower of mathematics by examining elementary set theory.

Informal Introduction to Set Theory

Most readers of this text have already seen some aspects of set theory at an intuitive level. We begin by introducing the basic ideas intuitively, to motivate the formal definitions given below. Informally, a *set* is any collection of objects. For example, $\{a, b, c, d\}$ is a set of four letters; $\mathbb{Z}$ is the set of all integers; $[0, 3]$ is the set of all real numbers x satisfying $0 \leq x \leq 3$, and so on. If x is an object and S is a set, we write $x \in S$ to mean that x is a *member* of S; $x \notin S$ means that x is not a member of S. For example, $3 \in \mathbb{Z}$, $b \in \{a, b, c, d\}$, $5 \notin [0, 3]$, and so on. (We already used this notation in our discussion of restricted quantifiers.) The symbol $x \in S$ can be read aloud as "x is a member of S" or "x belongs to S" or "x is in S." Beginners sometimes read the symbol $x \in S$ as "x exists in S," but this phrase should not be used, as it suggests an existential quantifier that is not present.

The members of a set can be any objects whatsoever, including other sets. For example, $\{0, 1, 2, 3, \mathbb{Z}, [0, 3]\}$ is a set with six members: the four integers 0, 1, 2, and 3; the set $\mathbb{Z}$ of all integers; and the closed interval $[0, 3]$, which is a set of real numbers. The following diagram, called a *Venn diagram*, shows two geometric sets A and B; the members of A are the points on this page lying within the left circle.

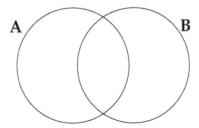

We can combine sets to form new sets using various set operations. Let A and B be any two sets. The *union* of A and B, denoted $A \cup B$, is formed by lumping together all the members of A and B into one big set. Symbolically, $A \cup B = \{x : x \in A \text{ or } x \in B\}$. For example, $\{1, 4, 5, 7\} \cup \{2, 3, 5, 7, 11\} = \{1, 2, 3, 4, 5, 7, 11\}$. For A and B in the Venn diagram above, $A \cup B$ is the shaded region below consisting of all points lying within at least one of the two circles:

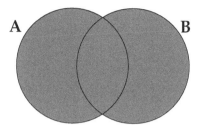

The *intersection* of A and B, denoted $A \cap B$, consists of all objects that are members of *both* A and B. Symbolically, $A \cap B = \{x : x \in A \text{ and } x \in B\}$. Continuing our previous example, $\{1, 4, 5, 7\} \cap \{2, 3, 5, 7, 11\} = \{5, 7\}$. The Venn diagram for $A \cap B$ looks like this:

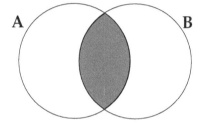

The *set difference* $A - B$ is the set of all members of A that are not members of B. Symbolically, $A - B = \{x : x \in A \text{ and } x \notin B\}$. For example, $\{1, 4, 5, 7\} - \{2, 3, 5, 7, 11\} = \{1, 4\}$ and $\{2, 3, 5, 7, 11\} - \{1, 4, 5, 7\} = \{2, 3, 11\}$. Here is the Venn diagram for $A - B$:

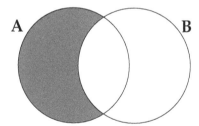

We say that A is a *subset* of B, written $A \subseteq B$, to mean that every member of A is also a member of B. The notation $A \nsubseteq B$ means that A is not a subset of B. For example, $\{1, 4, 5, 7\} \nsubseteq \{2, 3, 5, 7, 11\}$ since 1 is a member of $\{1, 4, 5, 7\}$ that is not a member of $\{2, 3, 5, 7, 11\}$. On the other hand, $\{1, 4, 5, 7\} \subseteq \{1, 2, 3, 4, 5, 6, 7\}$, $\{1, 4, 5, 7\} \subseteq \mathbb{Z}$, and $\mathbb{Z} \subseteq \mathbb{Q} \subseteq \mathbb{R}$. In the Venn diagrams displayed above, $A \nsubseteq B$ and $B \nsubseteq A$, since each of the two circles contains some points not in the other circle. In the following Venn diagram, $C \subseteq D$ but $D \nsubseteq C$:

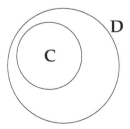

It is possible for a set to have no members whatsoever. Such a set is called an *empty set*, and it is denoted by the symbol $\emptyset$ or $\{\}$ (a set of curly braces with nothing inside). Be warned that these two notations for the empty set must not be combined: $\{\emptyset\}$ is a set with one member (namely, the empty set), so $\{\emptyset\}$ is not the same set as the empty set. We give a more detailed discussion of this point later, after introducing the appropriate definitions.

Formal Definitions of Subsets and Set Operations

We now begin afresh with a formal development of set theory, which is motivated by the informal discussion in the previous section. In our theory, the following words and symbols are *undefined terms* (see §2.1): set, membership in a set, $x \in S$. Every other concept in set theory will be defined from these undefined terms and the logical operators discussed earlier. We begin with the formal definition of subsets.

3.1. Definition: Subsets. For any sets A and B, $\boxed{A \subseteq B}$ iff $\boxed{\forall x, x \in A \Rightarrow x \in B}$.

The sentence in this definition is a new definitional axiom that lets us replace the newly defined symbol $A \subseteq B$ ("A is a subset of B") by the definition text appearing after the word "iff," and vice versa. Since $P \Leftrightarrow Q$ is equivalent to $(\sim P) \Leftrightarrow (\sim Q)$, we can obtain a definition of $A \not\subseteq B$ by applying the denial rules. We get:

For any sets A and B, $\boxed{A \not\subseteq B}$ iff $\boxed{\exists x, x \in A \land x \notin B}$.

Similarly, we can deny both sides of all definitions given below to find out what the denial of a newly defined term means.

The following variants of the subset notation are sometimes needed. For all sets A and B, we say A is a *proper subset* of B, written $\boxed{A \subsetneq B}$, iff $\boxed{A \subseteq B \text{ and } A \neq B}$. Do not confuse this notation with $A \not\subseteq B$, which means that A is not a subset of B. We can also reverse the order of A and B. We say B is a *superset* of A, written $\boxed{B \supseteq A}$, iff $\boxed{A \subseteq B}$. Similarly, B is a *proper superset* of A, written $\boxed{B \supsetneq A}$, iff $\boxed{A \subsetneq B}$.

Next we define union, intersection, set difference, and the empty set. These definitions tacitly require the following idea: *to define a new set, we must specify exactly which objects are members of that set.* (We develop this idea more formally in the next section, when we define set equality.) So, for example, to define the new set $A \cup B$, we need an axiom telling us precisely which objects belong to this set. Here is the definitional axiom we need.

3.2. Definition: Union of Two Sets. For all sets A and B and all objects x:

$\boxed{x \in A \cup B}$ iff $\boxed{x \in A \text{ or } x \in B.}$

3.3. Definition: Intersection of Two Sets. For all sets A and B and all objects x:

$\boxed{x \in A \cap B}$ iff $\boxed{x \in A \text{ and } x \in B.}$

Restating these definitions more symbolically, $x \in A \cup B$ means $(x \in A) \lor (x \in B)$, whereas $x \in A \cap B$ means $(x \in A) \land (x \in B)$. Observe that $\cup$ (the union symbol) and $\lor$ (the OR symbol) both open upwards, whereas $\cap$ (the intersection symbol) and $\land$ (the AND symbol) both open downwards. This observation may help avoid confusing the two symbols $\cup$ and $\cap$. We should also remember that $\cup$ and $\lor$ are not the same symbol; $\cup$ is used to combine sets, whereas $\lor$ is used to combine logical statements. Similar comments apply to $\cap$ and $\land$.

3.4. Definition: Set Difference. For all sets A and B and all objects x:

$\boxed{x \in A - B}$ iff $\boxed{x \in A \text{ and } x \notin B.}$

3.5. Definition: the Empty Set $\emptyset$. $\boxed{\text{For all objects } x, x \notin \emptyset.}$

Subset Proofs

Let A and B be given sets. How can we *prove* that $A \subseteq B$? By definition, we must prove $\forall x, x \in A \Rightarrow x \in B$. We can obtain a proof outline of this statement by combining the generic-element proof template with the direct proof template for IF-statements. The resulting sequence of steps occurs so frequently in proofs that it is worth stating (and memorizing) as its own separate proof template:

3.6. Subset Proof Template to Prove $A \subseteq B$. (A and B are sets.)
Fix an arbitrary object x_0. **Assume** $x_0 \in A$. **Prove** $x_0 \in B$.
(Continue the proof by expanding the definitions of "$x_0 \in A$" and "$x_0 \in B$" and using more proof templates.)

Note that A and B can be any expressions standing for sets, not necessarily individual variables, as seen in the next example.

3.7. Example. Let U and V be sets. Outline a proof that $U \times V \subseteq \mathcal{P}(\mathcal{P}(U \cup V))$.
Solution. Although the given statement uses potentially unfamiliar symbols (to be discussed later), we note that the statement is asserting that one set is a subset of another. So we can generate the proof outline immediately:
Fix an arbitrary object x_0. Assume $x_0 \in U \times V$. Prove $x_0 \in \mathcal{P}(\mathcal{P}(U \cup V))$.

The next theorem lists a host of properties of the subset relation.

3.8. Theorem on Subsets. For all sets A, B, C:

(a) *Reflexivity:* $A \subseteq A$.

(b) *Transitivity:* If $A \subseteq B$ and $B \subseteq C$, then $A \subseteq C$.

(c) *Lower Bound:* $A \cap B \subseteq A$ and $A \cap B \subseteq B$.

(d) *Greatest Lower Bound:* $C \subseteq A \cap B$ iff ($C \subseteq A$ and $C \subseteq B$).

(e) *Upper Bound:* $A \subseteq A \cup B$ and $B \subseteq A \cup B$.

(f) *Least Upper Bound:* $A \cup B \subseteq C$ iff ($A \subseteq C$ and $B \subseteq C$).

(g) *Minimality of Empty Set:* $\emptyset \subseteq A$.

(h) *Difference Property:* $A - B \subseteq A$.

(i) *Monotonicity:* If $A \subseteq B$, then: $A \cap C \subseteq B \cap C$ and $A \cup C \subseteq B \cup C$ and $A - C \subseteq B - C$.

(j) *Inclusion Reversal:* If $A \subseteq B$, then $C - B \subseteq C - A$.

We now use the subset template, combined with other proof templates and definitions, to prove some parts of this theorem. We ask the reader to prove the other parts in later exercises. Throughout all these proofs, we let A, B, and C be fixed, but arbitrary, sets.

3.9. Proof of Reflexivity. We must prove $A \subseteq A$. Fix an arbitrary object x_0. Assume $x_0 \in A$. Prove $x_0 \in A$. Since we just assumed that $x_0 \in A$, the proof is done.

3.10. Proof of Upper Bound. We must prove "$A \subseteq A \cup B$ and $B \subseteq A \cup B$." To prove the AND-statement, we prove each part separately.
$\quad$ *Part 1.* Prove $A \subseteq A \cup B$. Fix an arbitrary object x_0. Assume $x_0 \in A$. Prove $x_0 \in A \cup B$. We must prove "$x_0 \in A$ or $x_0 \in B$." Since $x_0 \in A$ was assumed to be true, the truth table for OR shows that the required OR-statement is also true.
$\quad$ *Part 2.* Prove $B \subseteq A \cup B$. Fix an arbitrary object y_0. Assume $y_0 \in B$. Prove $y_0 \in A \cup B$. We must prove "$y_0 \in A$ or $y_0 \in B$." Since $y_0 \in B$ was assumed to be true, the truth table for OR shows that the required OR-statement is also true.

3.11. Proof of Minimality of Empty Set. We must prove $\emptyset \subseteq A$. [If we mechanically recite the subset proof template, we generate these lines: "Fix x_0. Assume $x_0 \in \emptyset$. Prove $x_0 \in A$." It may be unclear how to proceed from this point. So we try the proof again, starting from the definition of subsets.] We must prove $\forall x, x \in \emptyset \Rightarrow x \in A$. Fix an arbitrary object x_0. We must prove the IF-statement "$x_0 \in \emptyset \Rightarrow x_0 \in A$." Now, the hypothesis "$x_0 \in \emptyset$" is false by definition of the empty set. So the entire IF-statement is true, by the truth table for IF. Thus we have proved $\emptyset \subseteq A$.

You may find the proof just given to be somewhat cryptic. If so, you may prefer the following contrapositive proof of the IF-statement. Assume $x_0 \notin A$; prove $x_0 \notin \emptyset$. The goal is true by definition of the empty set [and we do not need to use the assumption here]. Similarly, you can prove $\emptyset \subseteq A$ by a contradiction proof (see Exercise 10). This example illustrates the following general observation. *It is often easier to prove statements involving the empty set using negative logic, such as a contrapositive proof or a proof by contradiction.*

We continue proving parts of the Theorem on Subsets in the next section.

Section Summary

1. *Subsets.* $\boxed{A \subseteq B}$ means $\boxed{\forall x, x \in A \Rightarrow x \in B}$.
   To **prove** $A \subseteq B$: **fix** an arbitrary object x_0; **assume** $x_0 \in A$; **prove** $x_0 \in B$.

2. *Set Operations.* **Memorize** these definitions: for all sets A, B and all objects x,

   Union: $\boxed{x \in A \cup B}$ iff $\boxed{x \in A \text{ or } x \in B}$.

   Intersection: $\boxed{x \in A \cap B}$ iff $\boxed{x \in A \text{ and } x \in B}$.

   Set Difference: $\boxed{x \in A - B}$ iff $\boxed{x \in A \text{ and } x \notin B}$.

   Empty Set: $\boxed{\forall x, x \notin \emptyset}$.

3. *Theorem on Subsets.* **Memorize** the facts listed in the Theorem on Subsets for later use. In particular, for all sets A, B, C: $\emptyset$ and $A - B$ and A are always subsets of A; $A \subseteq B \subseteq C$ implies $A \subseteq C$; $A \cap B$ is a subset of both A and B; $C \subseteq A \cap B$ iff $C \subseteq A$ and $C \subseteq B$; A and B are both subsets of $A \cup B$; $A \cup B \subseteq C$ iff $A \subseteq C$ and $B \subseteq C$; and $A \subseteq B$ implies $A \cap C \subseteq B \cap C$, $A \cup C \subseteq B \cup C$, $A - C \subseteq B - C$, and $C - B \subseteq C - A$.

Exercises

1. Let $X = \{1, 2, 3, 4, 5, 6, 7, 8, 9, 10\}$, $A = \{1, 3, 5, 7, 9\}$, $B = \{1, 2, 3, 4, 5\}$, and $C = \{2, 4, 6, 8, 10\}$. Informally compute the following sets: (a) $A \cup B$, $A \cap B$, $A \cap C$, $(A \cup B) \cap C$, $A \cup (B \cap C)$. (b) $B - A$, $A - B$, $B - C$, $C - B$, $A - C$, $C - A$, $B - B$. (c) $X - A$, $X - B$, $A - X$, $X - (A \cap B)$, $(X - A) \cup (X - B)$, $(X - A) - B$, $X - (A - B)$.

2. In the previous exercise, consider the following ten sets: $\emptyset$, $A \cap B$, $B \cap C$, A, B, C, $A \cup B$, $B \cup C$, $X - A$, X. Make a 10×10 chart showing, for each pair S, T of sets from this list, whether $S \subseteq T$.

3. Let $S = \{0, 3, 5, 9, 0.6, \pi\}$, $T = \{4, 5, 7, 3/5, \sqrt{2}\}$, and $U = \{\sqrt{2}, \sqrt{4}, \sqrt{9}, \sqrt{0.36}\}$. Informally compute the following sets: (a) $S \cap T$, $S \cap U$, $T \cap U$, $S \cap T \cap U$. (b) $S \cap \mathbb{Z}$, $T \cap \mathbb{Q}$, $U \cap \mathbb{Z}$, $S - \mathbb{Q}$, $U - \mathbb{Q}$, $(U - S) - T$, $U - (S - T)$.

4. In the Venn diagram shown below, express each of the seven numbered regions in terms of the sets A, B, and C. For example, region 1 can be written $(A - B) - C$ or $A - (B \cup C)$.

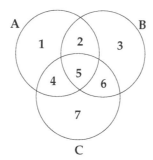

5. Make copies of the Venn diagram shown above with the following regions shaded.
   (a) $A - (B - C)$ (b) $(B - A) \cap C$ (c) $A \cup (B - C)$ (d) $(A \cup C) - B$ (e) $(A \cup B) \cap C$
   (f) $A \cup (B \cap C)$.

6. Draw a Venn diagram showing four sets A, B, C, D with the following properties: $A \cap C = \emptyset$; $A \cap B \neq \emptyset \neq B \cap C$; $D \subseteq B - A$. Then shade in the set $(C \cup D) - (C \cap D)$.

7. Draw a Venn diagram with five sets where the intersection of any two of the sets is not empty, but the intersection of any three of the sets is empty.

8. Create a proof template for each situation below.
   (a) Prove $A \subseteq B$ by expanding the definition and using a contrapositive proof.
   (b) Prove $A \subseteq B$ using proof by contradiction.
   (c) Disprove $A \subseteq B$.

9. Prove parts (c) and (h) of the Theorem on Subsets: for all sets A, B, $A \cap B \subseteq A$ and $A \cap B \subseteq B$ and $A - B \subseteq A$.

10. Fix an arbitrary set A. Prove $\emptyset \subseteq A$ using proof by contradiction.

11. Disprove each statement.
    (a) For all sets A and B, $A \cup B \subseteq B$.
    (b) For all sets A and B, $A \subseteq A - B$.
    (c) For all sets A and B, if $A \subseteq B$ then $C - A \subseteq C - B$.

12. Suppose we replace "For all" by "There exists" in each part of the previous exercise. Prove that the resulting statements are all true.

13. Draw Venn diagrams illustrating (as best you can) the following items from the Theorem on Subsets: (a) lower bound; (b) difference property; (c) least upper bound; (d) monotonicity.

14. Outline the proof of each statement; do not expand unfamiliar definitions.
    (a) For all sets S, $\text{Int}(S) \subseteq \text{Clo}(S)$.
    (b) For all sets F and S, if F is closed and $F \supseteq S$, then $F \supseteq \text{Clo}(S)$.
    (c) For all sets S and all x, if $x \in \text{Int}(S)$, then $\exists r \in \mathbb{R}_{>0}, B(x; r) \subseteq S$.

15. Prove this part of the Theorem on Subsets: for all sets A, B, if $A \subseteq B$, then $A \cap C \subseteq B \cap C$ and $A - C \subseteq B - C$.

16. Give a specific example of sets A, B, C, D, E where $A \subseteq C$, $B \subseteq C$, $C \subseteq D$, and $C \subseteq E$ are all true, but $A \subseteq B$, $B \subseteq A$, $D \subseteq E$, and $E \subseteq D$ are all false.

17. (a) Prove: for all sets A, B, C, D, if $A \subseteq B$ and $B \subseteq C$ and $C \subseteq D$ and $D \subseteq A$ then $C \subseteq B$. (b) Suppose we omit the hypothesis $A \subseteq B$ in (a). Prove or disprove the resulting statement. (c) Suppose we omit the hypothesis $B \subseteq C$ in (a). Prove or disprove the resulting statement.

18. Ten sets created from sets A, B, and C are listed below. Make a 10×10 chart showing, for each pair of sets S, T from this list, whether $S \subseteq T$ must be true for all choices of A, B, and C.

$$(A \cap B) \cap C, \ (A \cap B) - C, \ (A \cap B) \cup C, \ (A \cup B) \cap C, \ (A \cup B) - C, \ (A \cup B) \cup C,$$

$$(A - B) \cup C, \ (A - B) \cap C, \ (A - B) - C, \ A - (B - C).$$

19. For each pair of sets S, T listed in the previous exercise, decide whether $S \cap T$ must be the empty set for all choices of A, B, and C.

3.2   Subset Proofs and Set Equality Proofs

This section continues our discussion of subset proofs, as well as the related idea of using a known statement $A \subseteq B$ to prove other facts. We then define set equality and look at various ways to prove that two sets are equal.

Subset Proofs, Continued

In the last section, we described how to prove statements of the form $A \subseteq B$. Recall the main template used for such proofs goes like this:

Fix x_0. Assume $x_0 \in A$. Prove $x_0 \in B$.

Typically, the proof continues by expanding the definitions of "$x_0 \in A$" and "$x_0 \in B$."

Sometimes, instead of *proving* a statement of the form $A \subseteq B$, we already *know* that this statement is true. The following rule, which we might call a *knowledge template*, tells us how the known statement $A \subseteq B$ can be used to deduce further conclusions.

3.12. When $A \subseteq B$ is Already Known or Assumed:
(a) If you know that $x_0 \in A$, you may deduce that $x_0 \in B$.
(b) If you know that $x_0 \notin B$, you may deduce that $x_0 \notin A$.

To justify this, recall that $A \subseteq B$ means $\forall x, x \in A \Rightarrow x \in B$. By the Inference Rule for ALL, we may take $x = x_0$ here to deduce that "$x_0 \in A \Rightarrow x_0 \in B$" is true. On one hand, if $x_0 \in A$ is known, then $x_0 \in B$ follows by the Inference Rule for IF. On the other hand, if $x_0 \notin B$ is known, then $x_0 \notin A$ follows by taking the contrapositive and using the Inference Rule for IF.

The next few proofs illustrate both the subset proof template and the subset knowledge template. We are continuing to prove parts of the Theorem on Subsets, stated in the previous section. Throughout the discussion, we fix arbitrary sets A, B, and C. Comments about how we are generating the proofs appear in square brackets.

3.13. Proof of Transitivity. We must prove: if $A \subseteq B$ and $B \subseteq C$, then $A \subseteq C$. [Try a direct proof of the IF-statement.] Assume $A \subseteq B$ and $B \subseteq C$. Prove $A \subseteq C$. [We are now proving a subset statement, so use the subset proof template.] Fix an arbitrary object x_0. Assume $x_0 \in A$. Prove $x_0 \in C$. [To use the assumptions, invoke the subset knowledge template twice.] Since $A \subseteq B$ and $x_0 \in A$, we deduce $x_0 \in B$. Since $B \subseteq C$ and $x_0 \in B$, we deduce $x_0 \in C$. The proof is now done.

We can give some intuition for the result just proved using Venn diagrams. Here is a visual version of the transitivity of $\subseteq$:

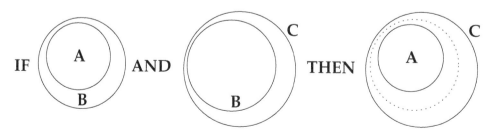

Although these Venn diagrams provide valuable visual insight into why transitivity should be true, the diagrams by themselves are *not* a proof of this result. The preceding picture only shows that transitivity holds for the three *specific* sets shown (namely, the sets of points within the circles labeled A, B, and C). In other words, the Venn diagrams prove the *existential* statement: "There exist three sets A, B, C, such that if $A \subseteq B$ and $B \subseteq C$, then $A \subseteq C$." But the diagrams do not prove the stronger *universal* statement: "for all sets A, B, C, if $A \subseteq B$ and $B \subseteq C$, then $A \subseteq C$" asserted in the theorem. Although the visual intuition conveyed by the diagrams is certainly compelling, we have no guarantee that this intuition extends to *all* abstract sets, which might be sets of numbers, or sets of functions, or sets of cows, or sets of other sets, not merely sets of points that we can draw. To get a rigorous proof of the universal version of the statement, we must rely on the formal definitions and the rules of logic, rather than invoking pictures. Similar comments apply to many other theorems about sets: *Venn diagrams can prove existential statements, and they can supply intuition for other kinds of statements, but they cannot prove universal statements about all sets.*

3.14. Proof of Inclusion Reversal. We must prove: if $A \subseteq B$, then $C - B \subseteq C - A$. [Try a direct proof of the IF-statement.] Assume $A \subseteq B$. Prove $C - B \subseteq C - A$. [We are proving a subset statement, so use that template.] Fix an arbitrary object x_0. Assume $x_0 \in C - B$. Prove $x_0 \in C - A$. We have assumed $x_0 \in C$ and $x_0 \notin B$. We must prove $x_0 \in C$ and $x_0 \notin A$. First, prove $x_0 \in C$; note this statement has already been assumed. Second, prove $x_0 \notin A$. [Use the subset knowledge template, part (b).] Since we have assumed $A \subseteq B$ and $x_0 \notin B$, we can deduce that $x_0 \notin A$, as needed.

The proof reveals why the inclusion of A within B gets reversed to become an inclusion of $C - B$ within $C - A$. The following Venn diagrams provide additional intuition for why this reversal happens. The circle for A is nested within the circle for B, since we are assuming $A \subseteq B$. The shaded part of the left diagram shows $C - B$, whereas the shaded part of the right diagram shows $C - A$. We see that the first shaded shape is contained within the second shaded shape.

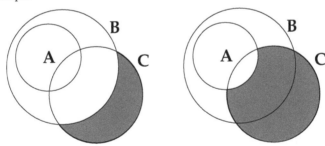

If you still think that these diagrams are enough to prove the result, note that the diagram makes hidden assumptions about how the set C intersects the other two sets. What if the circle for C is completely outside the circle for A, or completely inside the circle for B, or in some other position? What if A happens to be the empty set? We would need new Venn diagrams to cover these and other situations that might arise. On the other hand, the formal proof given earlier works for generic (arbitrary) sets A, B, and C.

Our next proof introduces another key ingredient in proof-writing: using previously proved results to shorten the proofs of new results.

3.15. Proof of Least Upper Bound. We must prove $A \cup B \subseteq C$ iff $A \subseteq C$ and $B \subseteq C$. [This is an IFF-statement, so we need a two-part proof.]
Part 1. Assume $A \cup B \subseteq C$. Prove $A \subseteq C$ and $B \subseteq C$. [We must prove an AND-statement, so we need a two-part proof within part 1.]

Part 1a. Prove $A \subseteq C$. Fix x_0. Assume $x_0 \in A$. Prove $x_0 \in C$. We already proved (Theorem on Subsets, part (e)) that $A \subseteq A \cup B$. Since $x_0 \in A$, it follows that $x_0 \in A \cup B$. Since we assumed $A \cup B \subseteq C$, it follows that $x_0 \in C$, as needed.

Part 1b. Prove $B \subseteq C$. [We could prove this as in part 1a, but here is a faster proof using known theorems.] We have already proved that $B \subseteq A \cup B$, and we have assumed that $A \cup B \subseteq C$ in part 1. By transitivity (Theorem on Subsets, part (b)), $B \subseteq C$ follows. [Part 1a could have been proved in this way, as well. Observe that our proof of Part 1a essentially repeats the proof of transitivity.]

<u>Part 2.</u> Assume $A \subseteq C$ and $B \subseteq C$. Prove $A \cup B \subseteq C$. [Our goal is a subset statement, so we use that template.] Fix z_0. Assume $z_0 \in A \cup B$. Prove $z_0 \in C$. We have assumed $z_0 \in A$ or $z_0 \in B$. [Having assumed an OR-statement, we are led to a proof by cases.]

Case 1. Assume $z_0 \in A$. Prove $z_0 \in C$. Since we have assumed $A \subseteq C$ and $z_0 \in A$, $z_0 \in C$ follows.

Case 2. Assume $z_0 \in B$. Prove $z_0 \in C$. Since we have assumed $B \subseteq C$ and $z_0 \in B$, $z_0 \in C$ follows.

Set Equality

When are two sets equal? Informally, a set is a collection of objects called the members of the set. With this intuition, it seems reasonable to say that two sets should be equal iff they have precisely the same members. This leads to the following definition, which is sometimes called the *Axiom of Extension*.

3.16. Definition: Set Equality. For all sets A and B:

$\boxed{A = B}$ iff $\boxed{\forall x, x \in A \Leftrightarrow x \in B}$.

We can use logical transformations to relate this definition to the definition of subsets. First, $\forall x, x \in A \Leftrightarrow x \in B$ is equivalent to

$$\forall x, ((x \in A \Rightarrow x \in B) \wedge (x \in B \Rightarrow x \in A))$$

by the Theorem on IFF. Second, by the quantifier property linking $\forall$ and $\wedge$, the previous formula is equivalent to

$$(\forall y, y \in A \Rightarrow y \in B) \quad \wedge \quad (\forall z, z \in B \Rightarrow z \in A).$$

We recognize the definition of subset in two places here, so the last formula is equivalent to $A \subseteq B$ and $B \subseteq A$. Chaining together these equivalences, we conclude:

For all sets A and B, $\boxed{A = B}$ iff $\boxed{A \subseteq B \text{ and } B \subseteq A.}$

This leads to the following proof template for proving that two sets are equal.

3.17. Proof Template to Prove a Set Equality $A = B$. (*A* and *B* are sets.)
<u>Part 1.</u> **Prove $A \subseteq B$.**
<u>Part 2.</u> **Prove $B \subseteq A$.**

Typically, we would use the subset proof template to accomplish parts 1 and 2, though other methods can sometimes be used (e.g., invoking previously proved theorems). The next theorem lists many identities asserting the equality of various sets.

3.18. Theorem on Set Equality. For all sets A, B, C, A', B':

(a) *Reflexivity:* $A = A$.

(b) *Symmetry:* If $A = B$, then $B = A$.

(c) *Transitivity:* If $A = B$ and $B = C$, then $A = C$.

(d) *Substitution Properties:* If $A = A'$ and $B = B'$, then: $A \cap B = A' \cap B'$ and $A \cup B = A' \cup B'$ and $A - B = A' - B'$ and ($A \subseteq B$ iff $A' \subseteq B'$).

(e) *Commutativity:* $A \cap B = B \cap A$ and $A \cup B = B \cup A$.

(f) *Associativity:* $(A \cap B) \cap C = A \cap (B \cap C)$ and $(A \cup B) \cup C = A \cup (B \cup C)$.

(g) *Distributive Laws:* $A \cap (B \cup C) = (A \cap B) \cup (A \cap C)$ and $A \cup (B \cap C) = (A \cup B) \cap (A \cup C)$.

(h) *Idempotent Laws:* $A \cap A = A$ and $A \cup A = A$.

(i) *Absorption Laws:* $A \cup (A \cap B) = A$ and $A \cap (A \cup B) = A$.

(j) *Detecting the Empty Set:* $(B = \emptyset$ iff $\forall z, z \notin B)$ and $(B \neq \emptyset$ iff $\exists z, z \in B)$.

(k) *Properties of the Empty Set:* $A \cup \emptyset = A$ and $A \cap \emptyset = \emptyset$ and $A - \emptyset = A$ and $\emptyset - A = \emptyset$ and $A - A = \emptyset$.

(l) *de Morgan Laws for Sets:* $A - (B \cup C) = (A - B) \cap (A - C)$ and $A - (B \cap C) = (A - B) \cup (A - C)$.

(m) *Set Partition of One Set by Another:* $A = (A - B) \cup (A \cap B)$ and $(A - B) \cap (A \cap B) = \emptyset$.

(n) *Set Partition of One Set by a Subset:* If $B \subseteq A$, then $A - (A - B) = B$, $A = B \cup (A - B)$, and $B \cap (A - B) = \emptyset$.

(o) *Characterizations of Subsets:* The following conditions are equivalent: $A \subseteq B$; $A \cap B = A$; $A \cup B = B$; $A - B = \emptyset$.

The first four properties listed here may appear to be obvious facts about equality that do not require proof. In certain theories, where equality is taken as an undefined term, these facts might be adopted as axioms about the equality symbol. But in set theory, set equality is a defined term, so we can prove these facts from more basic principles. As a sample, we prove parts (a) and (c).

3.19. Proof of Reflexivity of Set Equality. Let A be a fixed but arbitrary set. We will prove $A = A$. [Follow the new template, replacing B by A.] <u>Part 1.</u> Prove $A \subseteq A$. This is a known theorem (Theorem on Subsets, part (a)). <u>Part 2.</u> Prove $A \subseteq A$. This is also a known theorem. So $A = A$ holds.

3.20. Proof of Transitivity of Set Equality. Let A, B, and C be fixed but arbitrary sets. We prove: if $A = B$ and $B = C$, then $A = C$. Assume $A = B$ and $B = C$; we must prove $A = C$. [Expand the assumptions using the second version of the definition of set equality.] We have assumed that $A \subseteq B$ and $B \subseteq A$ and $B \subseteq C$ and $C \subseteq B$. [Now attack the goal using the set equality template.]
<u>Part 1.</u> Prove $A \subseteq C$. Since $A \subseteq B$ and $B \subseteq C$ were assumed, $A \subseteq C$ follows by the previously proved transitivity of $\subseteq$ (Theorem on Subsets, part (b)).
<u>Part 2.</u> Prove $C \subseteq A$. Since $C \subseteq B$ and $B \subseteq A$ were assumed, $C \subseteq A$ follows by the known transitivity of $\subseteq$.

3.21. Remark (Optional). Our definition of set equality states that for all sets A and B, $A = B$ iff $\forall x, x \in A \Leftrightarrow x \in B$. The forward direction of this IFF-statement can be viewed as a *substitution property* that allows us to replace a set A appearing on the right side of the membership symbol $\in$ by an equal set B. We need one more axiom stating an analogous substitution property for sets appearing on the left side of the membership symbol. Here is the axiom: for all sets A, B, C, if $A = B$, then $[(A \in C) \Leftrightarrow (B \in C)]$. Various other substitution properties for equal sets (such as those in part (d) of the Theorem on Set Equality) need not be stated as axioms, since they can be proved from the definitions.

Set Equality Proofs

We continue to illustrate the proof template for set equality by proving more items from the Theorem on Set Equality. Remember that it is no longer necessary to prove everything from the definitions alone; you can and should use previously known results such as the Theorem on Subsets. In the following, fix arbitrary sets A, B, C, A', and B'.

3.22. Proof of Idempotent Laws.
<u>Part 1.</u> We prove $A \cap A = A$. *Part 1a.* We prove $A \cap A \subseteq A$. This follows from the known result $A \cap B \subseteq A$ (Theorem on Subsets, part (c)) by replacing B by A. *Part 1b.* We prove $A \subseteq A \cap A$. Fix an arbitrary object x_0. Assume $x_0 \in A$. Prove $x_0 \in A \cap A$. We must prove "$x_0 \in A$ and $x_0 \in A$." To do this, [following the AND-template] first prove $x_0 \in A$ and then prove $x_0 \in A$. In each part, we already assumed the statement to be proved.
<u>Part 2.</u> We prove $A \cup A = A$. *Part 2a.* We prove $A \cup A \subseteq A$. Fix an object y_0. Assume $y_0 \in A \cup A$. Prove $y_0 \in A$. We have assumed the truth of "$y_0 \in A$ or $y_0 \in A$." By the truth table for OR, the truth of $P \vee P$ forces P to be true. So $y_0 \in A$ must be true, as needed. [If we follow the templates mechanically, we could also have finished using a proof by cases. Case 1 assumes $y_0 \in A$ and proves $y_0 \in A$; case 2 assumes $y_0 \in A$ and proves $y_0 \in A$. It seemed easier here to invoke the truth table instead.] *Part 2b.* We prove $A \subseteq A \cup A$. This follows from the known result $A \subseteq A \cup B$ (Theorem on Subsets, part (e)) by replacing B by A.

3.23. Proof that $A \cap \emptyset = \emptyset$. <u>Part 1.</u> We prove $A \cap \emptyset \subseteq \emptyset$. This follows from the lower bound property (Theorem on Subsets, part (c)). <u>Part 2.</u> We prove $\emptyset \subseteq A \cap \emptyset$. This follows from the minimality of $\emptyset$ (Theorem on Subsets, part (g)).

The previous proof showed that a certain set was empty by using the standard set equality proof template. However, there is another way to prove that a set is empty that is much more frequently used. It is based on part (j) of the Theorem on Set Equality, which we now prove.

3.24. Detecting the Empty Set. We prove: $\boxed{B \neq \emptyset}$ iff $\boxed{\exists z, z \in B}$. The other assertion in part (j) of the theorem follows from this one by denying each side of the IFF-statement. [The following proof may be considered optional.]
<u>Part 1.</u> Assume $B \neq \emptyset$. Prove $\exists z, z \in B$. Negating the original definition of set equality, we have assumed that $\exists x, x \in B \oplus x \in \emptyset$. Let x_0 be a particular object that makes "$x_0 \in B$ XOR $x_0 \in \emptyset$" true. By definition of the empty set, $x_0 \in \emptyset$ is false. So, by definition of XOR, $x_0 \in B$ is true. Now choose z to be x_0 to see that $\exists z, z \in B$ is true.
<u>Part 2.</u> Assume $\exists z, z \in B$. Prove $B \neq \emptyset$. We have assumed there is a fixed object z_0 with $z_0 \in B$. We must prove $\exists x, x \in B \oplus x \in \emptyset$. Choose $x = z_0$. As in part 1, $z_0 \in B$ is true and $z_0 \in \emptyset$ is false, making the required XOR-statement true.

The equivalence of the two boxed statements in 3.24 is used constantly in proofs involving the empty set, so it should be memorized carefully. (The equivalence should be intuitively evident as well: a set B is *not* empty iff there exists at least one thing that is a member of B.) We can use this equivalence to prove that a set *is* empty using proof by contradiction, as illustrated next.

3.25. Second Proof that $A \cap \emptyset = \emptyset$. Assume, to get a contradiction, that $A \cap \emptyset \neq \emptyset$. Our result above shows that there is an object $z_0 \in A \cap \emptyset$. By definition of $\cap$, we deduce $z_0 \in A$ and $z_0 \in \emptyset$. But also $z_0 \notin \emptyset$ by definition of $\emptyset$. So we have the contradiction "$z_0 \in \emptyset$ and $z_0 \notin \emptyset$." This contradiction shows that $A \cap \emptyset = \emptyset$ is true.

We introduce the following terminology for sets that do not overlap.

3.26. Definition: Disjoint Sets. We say sets S and T are $\boxed{disjoint}$ iff $\boxed{S \cap T = \emptyset}$. We say sets $S_1, S_2, \ldots, S_n$ are *pairwise disjoint* iff for all $i, j \in \{1, 2, \ldots, n\}$ with $i \neq j$, $S_i \cap S_j = \emptyset$.

For example, part (m) of the Theorem on Set Equality says that $A - B$ and $A \cap B$ are disjoint sets whose union is A.

Section Summary

1. *Proving $A \subseteq B$ vs. Knowing $A \subseteq B$.* To **prove** $A \subseteq B$: fix x_0; assume $x_0 \in A$; prove $x_0 \in B$. When you already **know** $A \subseteq B$, knowing $x_0 \in A$ allows you to deduce $x_0 \in B$; and knowing $y_0 \notin B$ allows you to deduce $y_0 \notin A$.

2. *Set Equality.* For all sets A, B, $\boxed{A = B}$ iff $\boxed{\forall x, x \in A \Leftrightarrow x \in B}$.
   Equivalently, $\boxed{A = B}$ iff $\boxed{A \subseteq B \text{ and } B \subseteq A}$.
   To prove two sets A and B are equal: 1. Prove $A \subseteq B$. 2. Prove $B \subseteq A$.

3. *Theorem on Set Equality.* **Memorize** the facts listed in the Theorem on Set Equality for later use. In particular, set equality is reflexive, symmetric, and transitive; equal sets may be substituted for one another in expressions; set intersection and set union obey the commutative, associative, distributive, idempotent, absorption, and de Morgan laws; any set A is the union of the disjoint sets $A - B$ and $A \cap B$; the four statements $A \subseteq B$, $A \cap B = A$, $A \cup B = B$, $A - B = \emptyset$ are all equivalent.

4. *Proofs Involving the Empty Set.* For any set B, prove $\boxed{B = \emptyset}$ by proving $\boxed{\forall x, x \notin B}$. Prove $\boxed{B \neq \emptyset}$ by proving $\boxed{\exists x, x \in B}$. In general, statements involving the empty set may be easier to prove using negative logic (e.g., proof by contrapositive or proof by contradiction).

Exercises

1. Prove part (d) of the Theorem on Subsets: for all sets A, B, C, $C \subseteq A \cap B$ iff $(C \subseteq A$ and $C \subseteq B)$.

2. Prove part (b) of the Theorem on Set Equality: for all sets A, B, if $A = B$, then $B = A$.

3. Prove part (e) of the Theorem on Set Equality: for all sets A, B, $A \cap B = B \cap A$ and $A \cup B = B \cup A$.

4. Prove: for all sets A, $A - \emptyset = A$.

5. (a) Prove: for all sets A, B, C, if $C \subseteq B$, then $A - B$ and C are disjoint. (b) Use (a) to prove: for all sets A and B, $A - B$ and $B - A$ are disjoint.

6. Prove: for all sets A and B, $A - B = A - (A \cap B)$.

7. Give intuition for parts (l), (m), and (n) in the Theorem on Set Equality by drawing appropriate Venn diagrams. Do these diagrams prove these set identities?

8. Prove part (m) of the Theorem on Set Equality: for all sets A and B, $A = (A - B) \cup (A \cap B)$ and $(A - B) \cap (A \cap B) = \emptyset$.

9. Prove part (n) of the Theorem on Set Equality: for all sets A, B, if $B \subseteq A$, then $A - (A - B) = B$ and $A = B \cup (A - B)$ and $B \cap (A - B) = \emptyset$. (Use other parts of the theorem, such as parts (m) and (o), to obtain a shorter proof.)

10. Prove: for all sets R and S, the sets $R - S$ and $S - R$ and $R \cap S$ are pairwise disjoint, and the union of these sets is $R \cup S$. Illustrate this result with a Venn diagram.

11. Prove: for all sets A, B, C, D, if $A \supseteq B \supseteq C \supseteq D$, then $A - B$, $B - C$, $C - D$, D are pairwise disjoint, and the union of these sets is A. Illustrate this result with a Venn diagram.

12. Use a Venn diagram to express $A \cup B \cup C$ as the union of: (a) two disjoint sets; (b) three disjoint sets; (c) seven disjoint sets.

13. Use the Theorem on Set Equality (*not* definitions or proof templates) to prove the following identities hold for all sets. Justify each step by citing a part of the theorem.
    (a) $(R \cup S) \cup (R \cap S) = R \cup S$.
    (b) $R \cup ((S \cap T) \cap U) = ((R \cup S) \cap (R \cup T)) \cap (R \cup U)$.

14. Use Venn diagrams to disprove each statement. Discuss why Venn diagrams are allowed here, despite earlier claims that set identities could not be proved via Venn diagrams.
    (a) For all sets A and C, $C = A \cup (C - A)$.
    (b) For all sets A, B, C, $A - (B - C) = (A - B) - C$.
    (c) For all sets A, B, C, $A \cup (B \cap C) = (A \cap B) \cup (A \cap C)$.

15. Disprove each statement in the preceding problem by constructing specific examples where every set is a subset of $\{1, 2, 3, 4\}$.

16. Find and prove necessary and sufficient conditions on the sets A and C for $C = A \cup (C - A)$ to be true.

17. Find and prove necessary and sufficient conditions on the sets A, B, and C for $A - (B - C) = (A - B) - C$ to be true.

18. (a) Prove: for all sets A, B, C, $(A - B) \cup (B - C) \supseteq A - C$.
    (b) Disprove: for all sets A, B, C, $(A - B) \cup (B - C) = A - C$.
    (c) Prove: for all sets A, B, C, if $C \subseteq B \subseteq A$, then $(A - B) \cup (B - C) = A - C$.
    (d) Prove or disprove the converse of the statement in (c).

19. (a) Prove: for all sets A, B, and C, if $A - C = B - C$, then $A \cup C = B \cup C$.
    (b) Prove or disprove the converse of part (a).

20. Prove or disprove: for all sets A, B, and C, if $A - C = B - C$ and $A \cap C = B \cap C$, then $A = B$.

21. Prove or disprove: for all sets A, B, and C, if $A - C = B - C$ and $C - A = C - B$, then $A = B$.

22. Find and prove necessary and sufficient conditions on the sets A, B, and C for $A \cup (B \cap C) = (A \cup B) \cap C$ to be true.

3.3 Set Equality Proofs, Circle Proofs, and Chain Proofs

Set equality proofs occur everywhere in mathematics, so we keep studying them in this section. A set equality is a particular kind of IFF-statement. In the second part of the section, we develop two new ways for proving IFF-statements — the circle proof method and the chain proof method — that can help prove set equalities and other biconditional statements.

Set Equality Proofs, Continued

The most common method for proving $A = B$, where A and B are sets, is to use the set equality proof template:
<u>Part 1.</u> **Prove** $A \subseteq B$.
<u>Part 2.</u> **Prove** $B \subseteq A$.
In this section, we continue practicing this proof method, but we also introduce several other ways of proving that two sets are equal. We first recall one such alternate method that is often used to prove that a given set equals the empty set.

3.27. To Prove $B = \emptyset$ (where B is a set):
Assume, to get a contradiction, that $B \neq \emptyset$.
We have **assumed** that there is an object z_0 with $z_0 \in B$.
Expand the definition of "$z_0 \in B$" to derive a contradiction.
Conclude that $B = \emptyset$.

3.28. Example. Prove that for all sets A, $A - A = \emptyset$.
Proof. Assume, to get a contradiction, that $A - A \neq \emptyset$. Then there exists an object $z_0 \in A - A$. By definition of set difference, $z_0 \in A$ and $z_0 \notin A$. This is a contradiction, so $A - A = \emptyset$.

3.29. Practice with Set Proofs. Try to prove the following three statements yourself, before reading the answers given below. Let A, B, and C be fixed, but arbitrary sets. Here, you may use any parts of the Theorem on Subsets except the result being proved.

(a) Prove: If $A \subseteq B$, then $A \cup C \subseteq B \cup C$.

(b) Prove: $\emptyset - A = \emptyset$ and $A = A \cup \emptyset$.

(c) Prove: $A - (B \cup C) = (A - B) \cap (A - C)$.

3.30. Answer to Practice Proof (a). [Use the IF-template to prove an IF-statement.] Assume $A \subseteq B$. Prove $A \cup C \subseteq B \cup C$. [Now use the subset template to prove the new goal.] Fix an arbitrary object x_0. Assume $x_0 \in A \cup C$. Prove $x_0 \in B \cup C$. We have assumed $x_0 \in A$ or $x_0 \in C$, and we must prove $x_0 \in B$ or $x_0 \in C$. [Since an OR-statement was assumed, use cases.]
<u>Case 1.</u> Assume $x_0 \in A$; prove $x_0 \in B$ or $x_0 \in C$. Since $A \subseteq B$ is known, we deduce that $x_0 \in B$, so "$x_0 \in B$ or $x_0 \in C$" is true by definition of OR.
<u>Case 2.</u> Assume $x_0 \in C$; prove $x_0 \in B$ or $x_0 \in C$. This follows by definition of OR.

3.31. Answer to Practice Proof (b). <u>Part 1.</u> Prove $\emptyset - A = \emptyset$. Assume, to get a contradiction, that $\emptyset - A \neq \emptyset$. Then there exists $z_0 \in \emptyset - A$. By definition of set difference, $z_0 \in \emptyset$ and $z_0 \notin A$. But, we also know that $z_0 \notin \emptyset$. The contradiction "$z_0 \in \emptyset$ and $z_0 \notin \emptyset$" proves that $\emptyset - A = \emptyset$.
<u>Part 2.</u> Prove $A = A \cup \emptyset$. *Part 2a.* Prove $A \subseteq A \cup \emptyset$. This is a known theorem (take $B = \emptyset$ in part (e) of the Theorem on Subsets). *Part 2b.* Prove $A \cup \emptyset \subseteq A$. Since $\emptyset$ appears here,

we try a contradiction proof. Assume, to get a contradiction, that $A \cup \emptyset \not\subseteq A$. Then there exists z_0 with $z_0 \in A \cup \emptyset$ and $z_0 \notin A$. We know "$z_0 \in A$ or $z_0 \in \emptyset$" is true. However, $z_0 \in A$ is false by assumption, and $z_0 \in \emptyset$ is false by definition of the empty set. So, "$z_0 \in A$ or $z_0 \in \emptyset$" is false. This contradiction proves part 2b.

3.32. Answer to Practice Proof (c). <u>Part 1.</u> Prove $A - (B \cup C) \subseteq (A - B) \cap (A - C)$. Fix x_0; assume $x_0 \in A - (B \cup C)$; prove $x_0 \in (A - B) \cap (A - C)$. We have assumed $x_0 \in A$ and $x_0 \notin B \cup C$; we must prove $x_0 \in A - B$ and $x_0 \in A - C$. Continuing to expand definitions, we have assumed $x_0 \in A$ and ($x_0 \notin B$ AND $x_0 \notin C$); we must prove $x_0 \in A$ and $x_0 \notin B$ and $x_0 \in A$ and $x_0 \notin C$. All four parts of the new goal have already been assumed, so part 1 is done.

<u>Part 2.</u> Prove $(A - B) \cap (A - C) \subseteq A - (B \cup C)$. Fix y_0; assume $y_0 \in (A - B) \cap (A - C)$; prove $y_0 \in A - (B \cup C)$. Expanding definitions as in part 1, we eventually see that we have assumed $(y_0 \in A \wedge y_0 \notin B) \wedge (y_0 \in A \wedge y_0 \notin C)$, and we must prove $(y_0 \in A \wedge (y_0 \notin B \wedge y_0 \notin C))$. All three parts of this AND-statement have already been assumed, so part 2 is done.

Circle Proof Method

Consider part (o) of the Theorem on Set Equality. This item asserts that four propositions (call them P, Q, R, and S) are all equivalent. More explicitly, part (o) says that

$$(P \Leftrightarrow Q) \wedge (Q \Leftrightarrow R) \wedge (R \Leftrightarrow S) \wedge (S \Leftrightarrow P) \wedge (P \Leftrightarrow R) \wedge (Q \Leftrightarrow S).$$

Using the AND-template and IFF-template to prove this, we would need to perform twelve independent subproofs:

1a. Prove $P \Rightarrow Q$.	1b. Prove $Q \Rightarrow P$.
2a. Prove $Q \Rightarrow R$.	2b. Prove $R \Rightarrow Q$.
3a. Prove $R \Rightarrow S$.	3b. Prove $S \Rightarrow R$.
4a. Prove $S \Rightarrow P$.	4b. Prove $P \Rightarrow S$.
5a. Prove $P \Rightarrow R$.	5b. Prove $R \Rightarrow P$.
6a. Prove $Q \Rightarrow S$.	6b. Prove $S \Rightarrow Q$.

That is a lot of work! It turns out, however, that there is a shorter way. Suppose we give proofs for Steps 1a, 2a, 3a, and 4a, so we know $P \Rightarrow Q$ and $Q \Rightarrow R$ and $R \Rightarrow S$ and $S \Rightarrow P$. You can check that

$$((P \Rightarrow Q) \wedge (Q \Rightarrow R)) \Rightarrow (P \Rightarrow R)$$

is a tautology (called "Transitivity of $\Rightarrow$"). So, once we complete Steps 1a and 2a, the Inference Rule for IF tells us that $P \Rightarrow R$ is true, completing Step 5a. Using this fact and Step 3a, the Inference Rule for IF tells us that $P \Rightarrow S$ is true, completing Step 4b. Repeatedly using Transitivity of IF and the Inference Rule for IF, we see that Steps 2a and 3a and 4a yield $Q \Rightarrow S$ and $Q \Rightarrow P$, which takes care of Step 6a and Step 1b. Continuing in this way, we see that after proving Steps 1a, 2a, 3a, and 4a, the eight remaining statements all follow automatically. Thus it suffices to prove the four statements in 1a, 2a, 3a, and 4a. This is called the *circle method* for proving the equivalence of four statements, because of the following picture:

$$
\begin{array}{ccc}
P & \Rightarrow & Q \\
\Uparrow & & \Downarrow \\
S & \Leftarrow & R
\end{array}
\quad \text{implies} \quad
\begin{array}{ccc}
P & \Leftrightarrow & Q \\
\Updownarrow & & \Updownarrow \\
S & \Leftrightarrow & R
\end{array}.
$$

Following the single arrows (known IF-statements) around the left circle and using Transitivity of IF, we deduce the bidirectional arrows (IFF-statements to be proved) appearing in the right circle, together with the diagonals $P \Leftrightarrow R$ and $Q \Leftrightarrow S$. The circle method generalizes to any number of statements, as follows.

3.33. Circle Method to Prove $P_1, P_2, \ldots, P_n$ are Logically Equivalent.

Part 1. **Prove** $P_1 \Rightarrow P_2$ (by any method).

Part 2. **Prove** $P_2 \Rightarrow P_3$ (by any method).

. . .

Part i. **Prove** $P_i \Rightarrow P_{i+1}$ (by any method).

. . .

Part n. **Prove** $P_n \Rightarrow P_1$ (by any method).

Let us illustrate the circle method by proving part (o) of the Set Equality Theorem.

3.34. Proof of Subset Characterization. Let A and B be fixed, but arbitrary sets.

Part 1. Prove: $A \subseteq B \Rightarrow A \cap B = A$.

Part 2. Prove: $A \cap B = A \Rightarrow A \cup B = B$.

Part 3. Prove: $A \cup B = B \Rightarrow A - B = \emptyset$.

Part 4. Prove: $A - B = \emptyset \Rightarrow A \subseteq B$.

[Stop reading here, and try to fill in the proofs of each part yourself. *Hints:* Use the Theorem on Subsets when you can; try proof by contradiction in part 3; try a contrapositive proof in part 4. Remember that sets already known to be equal have the same members.]

Proof of Part 1. Assume $A \subseteq B$. Prove $A \cap B = A$.

Part 1a. Prove $A \cap B \subseteq A$. This is already known (part (c) of the Theorem on Subsets).

Part 1b. Prove $A \subseteq A \cap B$. Fix an arbitrary object x_0. Assume $x_0 \in A$. Prove $x_0 \in A \cap B$. We must prove $x_0 \in A$ and $x_0 \in B$. First, prove $x_0 \in A$; note we have already assumed this. Second, prove $x_0 \in B$. To do so, recall we have assumed $A \subseteq B$ and $x_0 \in A$, from which $x_0 \in B$ follows [by the knowledge template for subsets].

Proof of Part 2. Assume $A \cap B = A$. Prove $A \cup B = B$.

Part 2a. Prove $A \cup B \subseteq B$. Fix an arbitrary object y_0. Assume $y_0 \in A \cup B$. Prove $y_0 \in B$. We know $y_0 \in A$ or $y_0 \in B$, so use cases within part 2a.

Case 1. Assume $y_0 \in A$; prove $y_0 \in B$. Since $y_0 \in A$ and $A \cap B = A$ is known, we deduce that $y_0 \in A \cap B$. Since $A \cap B \subseteq B$ is known, we deduce that $y_0 \in B$ as needed.

Case 2. Assume $y_0 \in B$; prove $y_0 \in B$. This is immediate.

Part 2b. Prove $B \subseteq A \cup B$. This is known from part (e) of the Theorem on Subsets.

Proof of Part 3. Assume, to get a contradiction, that $A \cup B = B$ and $A - B \neq \emptyset$. The second assumption tells us that there is an object z_0 with $z_0 \in A - B$. This means $z_0 \in A$ and $z_0 \notin B$. On one hand, since $z_0 \in A$ and $A \subseteq A \cup B$, we see that $z_0 \in A \cup B$. We have also assumed $A \cup B = B$, so $z_0 \in B$ follows. We now have the contradiction "$z_0 \in B$ and $z_0 \notin B$," which completes the proof of part 3.

Proof of Part 4. Using the contrapositive method, assume $A \not\subseteq B$; prove $A - B \neq \emptyset$. We have assumed there exists an object z_0 such that $z_0 \in A \wedge z_0 \notin B$. We must prove $\exists w, w \in A - B$, which means $\exists w, w \in A \wedge w \notin B$. Choosing $w = z_0$ completes the proof.

Having proved part (o) of the Theorem on Set Equality, we can quickly deduce the absorption laws (part (i)) as follows.

3.35. Proof of Absorption Laws. Let X and Y be arbitrary sets; we prove $X \cup (X \cap Y) = X$ and $X \cap (X \cup Y) = X$. Since $X \cap Y \subseteq X$ is already known, part (o) of the theorem (with A there replaced by $X \cap Y$, and with B there replaced by X) tells us that $(X \cap Y) \cup X = X$. Assuming commutativity of union has already been proved, we conclude that $X \cup (X \cap Y) = X$. Similarly, since $X \subseteq X \cup Y$ is already known, part (o) gives $X \cap (X \cup Y) = X$, which is the second absorption law.

Chain Proof Method

Suppose we are trying to prove an algebraic identity of the form $a = c$, where a and c are expressions denoting real numbers. A common way to proceed is to write down a sequence of known identities that looks like this:

$$\textbf{We know } a = b_1 \wedge b_1 = b_2 \wedge b_2 = b_3 \wedge b_3 = b_4 \wedge b_4 = c.$$

By repeatedly using the theorem "$\forall x, y, z \in \mathbb{R}$, if $x = y$ and $y = z$, then $x = z$," we can conclude that $a = c$. This theorem is almost never mentioned explicitly; instead, we abbreviate the above discussion by presenting a single *chain* of known equalities

$$a = b_1 = b_2 = b_3 = b_4 = c$$

that begins with a (the left side of the equality to be proved) and ends with c (the right side of the equality to be proved). We state this method as a new proof template.

3.36. Chain Proof Template to Prove $a = c$**.** To **prove** $a = c$, we can present a chain of **known** equalities

$$a = b_1 = b_2 = b_3 = \cdots = b_{n-1} = b_n = c$$

starting at a and ending at c. The letters $a, b_1, \ldots, b_n, c$ represent any expressions denoting objects of the same type (e.g., sets, numbers, functions, etc.).

3.37. Example. We use a chain proof to prove $\forall x, y \in \mathbb{R}, (x - y)(x + y) = x^2 - y^2$. Fix arbitrary $x, y \in \mathbb{R}$. We compute

$$(x-y)(x+y) = x(x+y)-y(x+y) = xx+xy-yx-yy = x^2+xy-xy-y^2 = x^2+0-y^2 = x^2-y^2.$$

Each equality in the chain is true because of a law of algebra; for instance, the first two equalities hold by the distributive law, and the third holds by commutativity of multiplication and the definitions $x^2 = xx$ and $y^2 = yy$. We conclude that $(x - y)(x + y) = x^2 - y^2$, as needed.

Now let us return to the problem of proving IFF-statements. You can check that

$$((P \Leftrightarrow Q) \wedge (Q \Leftrightarrow R)) \Rightarrow (P \Leftrightarrow R)$$

is a tautology (called "Transitivity of $\Leftrightarrow$"). By analogy with what we did above for equality proofs, we can formulate the following chain method for proving IFF-statements.

3.38. Chain Proof Template to Prove $P \Leftrightarrow R$**.** To **prove** the IFF-statement $P \Leftrightarrow R$, we can present a chain of **known** IFF-statements

$$P \Leftrightarrow Q_1 \Leftrightarrow Q_2 \Leftrightarrow Q_3 \Leftrightarrow \cdots \Leftrightarrow Q_{n-1} \Leftrightarrow Q_n \Leftrightarrow R$$

starting at P and ending at R.

Note that the displayed formula is really an abbreviation for

$$(P \Leftrightarrow Q_1) \wedge (Q_1 \Leftrightarrow Q_2) \wedge \cdots \wedge (Q_n \Leftrightarrow R).$$

For instance, $P \Leftrightarrow Q \Leftrightarrow R$ (with no parentheses) is being used to abbreviate $(P \Leftrightarrow Q) \wedge (Q \Leftrightarrow R)$. This abbreviating convention is dangerous, since (for example) the propositional form $(P \Leftrightarrow Q) \Leftrightarrow R$ is not logically equivalent to $(P \Leftrightarrow Q) \wedge (Q \Leftrightarrow R)$. However, expressions such as $(P \Leftrightarrow Q) \Leftrightarrow R$ seldom arise in practice, and context frequently tells us when this abbreviation is being used. To illustrate the chain method, we prove part (f) of the Theorem on Set Equality.

3.39. Proof of Associativity of $\cap$ and $\cup$. Let A, B, C be fixed sets. We must prove $(A \cap B) \cap C = A \cap (B \cap C)$. By the original definition of set equality, we must prove:

$$\forall x, x \in (A \cap B) \cap C \Leftrightarrow x \in A \cap (B \cap C).$$

Let x_0 be a fixed object. To prove the IFF-statement for x_0, we present the following chain of known equivalences:

$$
\begin{aligned}
x_0 \in (A \cap B) \cap C &\Leftrightarrow (x_0 \in A \cap B) \wedge x_0 \in C \\
&\Leftrightarrow (x_0 \in A \wedge x_0 \in B) \wedge x_0 \in C \\
&\Leftrightarrow x_0 \in A \wedge (x_0 \in B \wedge x_0 \in C) \\
&\Leftrightarrow x_0 \in A \wedge (x_0 \in B \cap C) \\
&\Leftrightarrow x_0 \in A \cap (B \cap C).
\end{aligned}
$$

The third equivalence is true by the known associativity of AND (see the Theorem on Logical Equivalences). All the other equivalences are instances of the definition of $\cap$. We can prove associativity of $\cup$ in the same way, by replacing $\cap$ by $\cup$ and $\wedge$ by $\vee$ everywhere in the proof above.

We can give similar chain proofs for several other parts of the Set Equality Theorem (including commutativity, the distributive laws, and the absorption laws), by reducing the identities for sets to corresponding known equivalences from the Theorem on Logical Equivalences. This proof technique vividly reveals how properties of the logical operators lead directly to analogous properties of set operations.

Transitivity and Chain Proofs (Optional)

At this point, we have encountered several mathematical relations that have a property called *transitivity*. We give an abstract definition of this property later, when we study relations (see §6.1). We now list some specific examples of the transitive law occurring in algebra, set theory, and logic.

3.40. Transitivity Properties.

(a) *Transitivity of Equality:* For all objects a, b, c, if $a = b$ and $b = c$, then $a = c$.

(b) *Transitivity of $\leq$:* For all $a, b, c \in \mathbb{R}$, if $a \leq b$ and $b \leq c$, then $a \leq c$.

(c) *Transitivity of $>$:* For all $a, b, c \in \mathbb{R}$, if $a > b$ and $b > c$, then $a > c$.

(d) *Transitivity of Subsets:* For all sets A, B, C, if $A \subseteq B$ and $B \subseteq C$, then $A \subseteq C$.

(e) *Transitivity of IF:* For all propositions P, Q, R, if $P \Rightarrow Q$ and $Q \Rightarrow R$, then $P \Rightarrow R$.

(f) *Transitivity of IFF:* For all propositions P, Q, R, if $P \Leftrightarrow Q$ and $Q \Leftrightarrow R$, then $P \Leftrightarrow R$.

Whenever we have a transitive relation, we can use the chain proof technique to prove that this relation holds between two given objects. We have already discussed chain proofs of equalities and IFF-statements. To give a chain proof of an inequality such as $a > c$, we present a chain of **known** inequalities

$$a > b_1 > b_2 > b_3 > \cdots > b_n > c,$$

which abbreviates "$a > b_1$ and $b_1 > b_2$ and $b_2 > b_3$ and ... and $b_n > c$." Using Transitivity of $>$ repeatedly, we successively deduce $a > b_2$, then $a > b_3$, and so on, until finally we obtain the needed conclusion $a > c$. To prove $A \subseteq C$ by the chain proof method, we find a chain of **known** set inclusions

$$A \subseteq B_1 \subseteq B_2 \subseteq B_3 \subseteq \cdots \subseteq B_n \subseteq C.$$

To prove $P \Rightarrow R$ by the chain proof method, we find a chain of **known** IF-statements

$$P \Rightarrow Q_1 \Rightarrow Q_2 \Rightarrow Q_3 \Rightarrow \cdots \Rightarrow Q_n \Rightarrow R.$$

Beware: the chain proof technique is very convenient when it succeeds, but this proof method cannot be used in all situations. When trying to prove a complicated set inclusion or set equality, it is often safer to use the subset proof template or set equality template rather than attempting a chain proof.

Section Summary

1. *Circle Proof Method.* To prove that statements $P_1, P_2, \ldots, P_n$ are all logically equivalent, we can prove $P_1 \Rightarrow P_2$ and $P_2 \Rightarrow P_3$ and ... and $P_{n-1} \Rightarrow P_n$ and $P_n \Rightarrow P_1$. Reordering the original list of statements is allowed and may make the implications easier to prove.

2. *Chain Proof Method for Equalities.* To prove an equality $a = c$, find a chain of *known* equalities $a = b_1 = b_2 = \cdots = b_n = c$ starting at a and ending at c.

3. *Chain Proof Method for IFF Statements.* To prove an IFF statement $P \Leftrightarrow R$, find a chain of *known* equivalences $P \Leftrightarrow Q_1 \Leftrightarrow Q_2 \Leftrightarrow \cdots \Leftrightarrow Q_n \Leftrightarrow R$ starting at P and ending at R.

4. *Chain Proof Method for Transitive Relations.* Whenever we have a transitive relation (such as $=$, $\leq$, $<$, $\geq$, $>$, $\subseteq$, $\Rightarrow$, or $\Leftrightarrow$), we can prove the relation holds between given objects A and B by finding a finite chain of known relations leading from A to B through intermediate objects. Each step in the chain should be explained by citing known results.

5. *Advice for Set Proofs.* When asked to prove a complicated statement involving sets, steadfastly keep applying the proof templates for subsets, set equality, set operations, and the logical operators as they arise within the statement. Remember the special templates involving the empty set, and be ready to try contrapositive proofs or proof by contradiction. Some set identities can be reduced to logical identities by using a chain proof.

Exercises

1. Outline a proof of the following statement using the circle method. For all sets $X \subseteq \mathbb{R}^n$, the following statements are equivalent: X is compact; X is sequentially compact; X is closed and bounded; X is complete and totally bounded.

2. Given statements P, Q, R, S, T, suppose we have proved $P \Rightarrow Q$ and $Q \Rightarrow S$ and $S \Rightarrow P$ and $T \Rightarrow R$ and $R \Rightarrow P$. Have we proved that all five statements are logically equivalent? If not, find one additional IF-statement that will complete the proof; give all possible answers.

3. Repeat the previous question assuming we have proved $P \Leftrightarrow R$ and $Q \Leftrightarrow R$ and $S \Leftrightarrow R$ and $T \Leftrightarrow S$.

4. Give a chain proof of the commutative laws for set union and intersection (Theorem on Set Equality, part (e)).

5. Give a chain proof of the distributive laws for set union and intersection (Theorem on Set Equality, part (g)).

6. Give chain proofs of the following algebraic identities. (You will need the distributive law and similar facts.)
   (a) $\forall x, y \in \mathbb{R}, (x+y)^2 = x^2 + 2xy + y^2$.
   (b) $\forall x, y \in \mathbb{R}, (x-y)(x^2 + xy + y^2) = x^3 - y^3$.

7. Let x be a fixed positive real number. Prove the following statements are equivalent using a circle proof: $x > 1$; $x^2 > x$; $1/x^2 < 1/x$; $1/x < 1$.

8. Let x be a fixed negative real number. Which pairs of statements in the previous exercise are logically equivalent?

9. Using any method, prove this part of the Theorem on Set Equality: for all sets A, B, C, $A - (B \cap C) = (A - B) \cup (A - C)$.

10. Give direct proofs of the implications $A \subseteq B \Rightarrow A \cup B = B$ and $A \cap B = A \Rightarrow A \subseteq B$ in part (o) of the Theorem on Set Equality.

11. Prove: for all sets A, B, C, $(A - B) - C = A - (B \cup C) = (A - C) - B$.

12. Prove part (d) of the Theorem on Set Equality (substitution properties).

13. Prove: for all sets A and B, the following statements are equivalent: $A \cap B = \emptyset$; $A - B = A$; $B - A = B$; $\forall C \subseteq A, \forall D \subseteq B, C \cap D = \emptyset$.

14. Prove: for all $x \in \mathbb{R}_{\neq 0}$, the following statements are equivalent: x is not rational; $x/3$ is not rational; $3/x$ is not rational; $(3/x) + 2$ is not rational; $2x + 3$ is not rational.

15. Define the *symmetric difference* of two sets A and B as follows:
$$\forall x, \boxed{x \in A \triangle B} \Leftrightarrow \boxed{x \in A \oplus x \in B}.$$

   Informally, $A \triangle B$ consists of all objects that belong to *exactly one* of the sets A and B. Prove the following identities hold for all sets A, B, and C.
   (a) $A \triangle B = B \triangle A$ (commutativity).
   (b) $(A \triangle B) \triangle C = A \triangle (B \triangle C)$ (associativity).
   (c) $A \triangle \emptyset = A = \emptyset \triangle A$ (identity).
   (d) $(A \triangle B) \cap C = (A \cap C) \triangle (B \cap C)$ (distributive law).

16. Illustrate the identities in the previous exercise (as best you can) using Venn diagrams.

17. Let A and B be arbitrary sets.
   (a) Prove: $A \triangle B = (A - B) \cup (B - A)$.
   (b) Prove: $A \triangle B = (A \cup B) - (A \cap B)$.

18. Prove: $\forall A, \exists! B, A \triangle B = \emptyset = B \triangle A$.

19. Prove or disprove: For all A, B, C, x, $x \in (A \triangle B) \triangle C$ iff x is a member of exactly one of the sets A, B, C.

20. (a) Is $\vee$ (OR) transitive? In other words, is it always true that if $A \vee B$ and $B \vee C$, then $A \vee C$? Prove your answer.
   (b) Decide (with proof) whether $\wedge$ (AND) is transitive.
   (c) Decide (with proof) whether $\oplus$ (XOR) is transitive.

3.4   Small Sets and Power Sets

In our initial informal discussion of set theory, we used the notation $\{1, 4, 5, 7\}$ to describe a set whose members were 1, 4, 5, and 7. In this section, we give a precise formal definition of this notation. This enables us to understand several key technical issues involving sets — whether order and repetition matter in sets, how sets may appear as members of other sets, and the distinction between $\in$ (the set membership relation) and $\subseteq$ (the subset relation). We also discuss the power set of a set X, which is the set of all subsets of X.

Small Sets

We have already introduced the informal notation $U = \{z_1, z_2, \ldots, z_n\}$ to represent a finite set whose members are $z_1, z_2, \ldots, z_n$ (and nothing else). We now give more formal definitions of what this notation means, for specific small values of n. To see why this formal definition is necessary, consider the following questions.

(a) Is $\{1, 2\} = \{2, 1\}$? (Does *order* matter in a set?)

(b) Is $\{3\} = \{3, 3, 3\}$? (Does *repetition* matter in a set?)

(c) Is $\{\emptyset\} = \emptyset$? (Is a set containing the empty set the same as the empty set?)

(d) Is $\{1, \{2, 3\}\} = \{1, 2, 3\}$? (Do nested braces change the set?)

To give incontrovertible answers to these questions, we need formal definitions to work with. As before, we define new sets by specifying precisely which objects are members of these sets.

3.41. Definitions of Small Sets.

(a) *Empty Set:* $\boxed{\text{For all } z,\ z \notin \emptyset.}$ Using curly brace notation: $\boxed{\text{for all } z,\ z \notin \{\,\}.}$

(b) *Singleton Sets:* For all a, z, $\boxed{z \in \{a\}}$ iff $\boxed{z = a}$.

(c) *Unordered Pairs:* For all a, b, z, $\boxed{z \in \{a, b\}}$ iff $\boxed{z = a \text{ or } z = b}$.

(d) *Unordered Triples:* For all a, b, c, z, $\boxed{z \in \{a, b, c\}}$ iff $\boxed{z = a \text{ or } z = b \text{ or } z = c}$.

We can give analogous definitions for $\{a, b, c, d\}$, $\{a, b, c, d, e\}$, etc.

We have called $\{a, b\}$ the *unordered pair* with members a and b, suggesting that order does *not* matter in a set. We can prove this fact now.

3.42. Theorem. For all a, b, $\{a, b\} = \{b, a\}$.

Proof. Fix arbitrary objects a, b. We must prove $\forall z, z \in \{a, b\} \Leftrightarrow z \in \{b, a\}$. We use a chain proof. Fix an arbitrary object z_0. We know

$$z_0 \in \{a, b\} \Leftrightarrow (z_0 = a \lor z_0 = b) \Leftrightarrow (z_0 = b \lor z_0 = a) \Leftrightarrow z_0 \in \{b, a\}.$$

The first and third equivalences come from the definition of unordered pair, and the middle equivalence is an instance of the tautology $(P \lor Q) \Leftrightarrow (Q \lor P)$.

An analogous proof shows that for all objects a, b, c,

$$\{a, b, c\} = \{a, c, b\} = \{b, a, c\} = \{b, c, a\} = \{c, a, b\} = \{c, b, a\},$$

and similarly for larger sets. In general, *displaying the members of a set in a different order does not change the set.* Later, we discuss concepts (such as ordered pairs, ordered triples, and sequences) in which the order does make a difference.

Next consider the issue of repetition of elements. The definition of unordered pair allows the possibility that $a = b$. In this case, we can prove that the unordered pair is the same as a singleton set.

3.43. Theorem. For all a, $\{a, a\} = \{a\}$.
Proof. Fix an arbitrary object a. We must prove $\forall z, z \in \{a, a\} \Leftrightarrow z \in \{a\}$. We again use a chain proof. Fix an arbitrary object z_0. We know

$$z_0 \in \{a, a\} \Leftrightarrow (z_0 = a \vee z_0 = a) \Leftrightarrow (z_0 = a) \Leftrightarrow z_0 \in \{a\}.$$

The equivalences follow by definition of unordered pair, by the tautology $(P \vee P) \Leftrightarrow P$, and by definition of singleton sets.

Similarly, we can prove that for all a, b, c, $\{a\} = \{a, a, a, a\}$, $\{a, b, c\} = \{a, b, b, c, c, c\}$, and so on. In general, *listing the members of a set with repetitions does not change the set.* Combining this with the previous observation, we can summarize by saying that

> *Order* and *repetition* do not matter for sets;
> only *membership in the set* matters when deciding set equality.

Next, let us turn to the nuances of the curly brace notation for small sets. Our first result shows that adding new braces to a set can change the set.

3.44. Theorem. $\boxed{\{\emptyset\} \neq \emptyset}$ and $\{\{\emptyset\}\} \neq \{\emptyset\}$.
Proof. We first prove $\{\emptyset\} \neq \emptyset$. To do so, we prove $\exists z, z \in \{\emptyset\}$. Choose $z = \emptyset$; we must prove $\emptyset \in \{\emptyset\}$. By definition of singleton, we must prove $\emptyset = \emptyset$, which is a known fact (reflexivity of set equality).

Next we prove $\{\{\emptyset\}\} \neq \{\emptyset\}$. Note that we can prove a set inequality $A \neq B$ by finding a member of A that is not a member of B. In this case, A is $\{\{\emptyset\}\}$ and B is $\{\emptyset\}$. Consider $z = \{\emptyset\}$. On one hand, by definition of a singleton set, $z \in A$ (i.e., $\{\emptyset\} \in \{\{\emptyset\}\}$) because $\{\emptyset\} = \{\emptyset\}$. On the other hand, using the result proved in the last paragraph, $z \notin B$ (i.e., $\{\emptyset\} \notin \{\emptyset\}$) since $\{\emptyset\} \neq \emptyset$.

If you find the proof just given to be cryptic, here is some intuition that may help. A set (collection of objects) should not be confused with the members of the set. In particular, a singleton set $\{b\}$ containing a single object b is not the same thing as the object b itself. For example, the set {George Washington} is not the same object as George Washington (a set is not a person). We might think of a set as a box that contains its members (in no particular order). Thus, $\{1, 2, 3\}$ is a box containing the three numbers $1, 2, 3$ and nothing else. On the other hand, $S = \{1, \{2, 3\}\}$ is a box containing two objects: the number 1, and a second box that itself contains the numbers 2 and 3. Question: is $2 \in S$? The answer may be unclear using the box analogy, so we refer to the formal definition. We know $2 \in \{1, \{2, 3\}\}$ iff $(2 = 1$ or $2 = \{2, 3\})$. Neither alternative is true, so 2 is not a member of S. However, 2 is a member of one of the members of S.

Similarly, $T = \{\{1, 2, 3\}\}$ is a set with just one member, namely the set $\{1, 2, 3\}$. We have $1 \notin T$, $2 \notin T$, $3 \notin T$, but $\{1, 2, 3\} \in T$. We think of T as a box containing another box containing the numbers 1, 2, and 3. Now, the empty set (which can be denoted by $\emptyset$ or $\{\}$) can be viewed as an empty box — a box that contains nothing. This set is *not* the same as $\{\emptyset\}$, which is a box containing an empty box. Similarly, $\{\{\emptyset\}\} = \{\{\{\}\}\}$ is a box containing a box containing an empty box. The set $\{\emptyset, \{\emptyset\}\}$ is a box containing two things: an empty box, and a box containing an empty box. And so on. In summary, *nested braces are important when determining set membership; only items at the outermost level are considered members of the outermost set. Sets may contain other sets as members.*

3.45. Example. Let $U = \{5.3, \{7\}, \{8, 9, \{10, 11\}\}, \mathbb{Z}\}$. U is a finite set with four members. One member of U is the real number 5.3. Another member of U is a set whose only member is 7. Another member of U is an unordered triple whose three members are 8, 9, and the unordered pair $\{10, 11\}$. Another member of U is the infinite set $\mathbb{Z}$. We could write

$$U = \{\{7\}, \mathbb{Z}, 5.3, \{8, \{11, 10\}, 9\}, \mathbb{Z}, 53/10\}, \text{ but } U \neq \{5.3, \{7\}, \{8, 9, 10, 11\}, \mathbb{Z}\}.$$

Although $8 \in \{8, 9, \{10, 11\}\}$ and $8 \in \mathbb{Z}$, $8 \notin U$.

Power Sets

Informally, for any set X, the *power set* $\mathcal{P}(X)$ is the set of all subsets of X. Formally, we define the power set by specifying its members, as follows.

3.46. Definition: Power Set. For all objects S and all sets X, $\boxed{S \in \mathcal{P}(X)}$ iff $\boxed{S \subseteq X}$. So, the *members* of $\mathcal{P}(X)$ are precisely the *subsets* of X.

3.47. Example. You can check that

$$\mathcal{P}(\{7, 8, 9\}) = \{\emptyset, \{7\}, \{8\}, \{9\}, \{7, 8\}, \{7, 9\}, \{8, 9\}, \{7, 8, 9\}\}.$$

The right side is an eight-element set with members $\emptyset$, $\{7\}$, $\{8\}$, and so on. Changing the braces used here produces an incorrect statement. For instance, we know $\emptyset \subseteq \{7, 8, 9\}$ by the Theorem on Subsets, and so $\emptyset \in \mathcal{P}(\{7, 8, 9\})$. On the other hand, $\{\emptyset\} \not\subseteq \{7, 8, 9\}$, since $\emptyset$ is a member of $\{\emptyset\}$ that is not a member of $\{7, 8, 9\}$. So $\{\emptyset\} \notin \mathcal{P}(\{7, 8, 9\})$. Similarly, $7 \not\subseteq \{7, 8, 9\}$, since the number 7 is not even a set[1]. But $\{7\} \subseteq \{7, 8, 9\}$, since every member of $\{7\}$ (namely, the number 7) is a member of $\{7, 8, 9\}$.

3.48. Example. What is $\mathcal{P}(\{\emptyset, \{\emptyset\}\})$? In general, for any unordered pair $\{a, b\}$, we can see that

$$\mathcal{P}(\{a, b\}) = \{\emptyset, \{a\}, \{b\}, \{a, b\}\}.$$

Making the replacements $a = \emptyset$ and $b = \{\emptyset\}$, we conclude that

$$\mathcal{P}(\{\emptyset, \{\emptyset\}\}) = \{\emptyset, \{\emptyset\}, \{\{\emptyset\}\}, \{\emptyset, \{\emptyset\}\}\}.$$

Similary, since $\mathcal{P}(\{a\}) = \{\emptyset, \{a\}\}$, we see that $\mathcal{P}(\{\emptyset\}) = \{\emptyset, \{\emptyset\}\}$, $\mathcal{P}(\{\{\emptyset\}\}) = \{\emptyset, \{\{\emptyset\}\}\}$, and so on. Finally, what is $\mathcal{P}(\emptyset)$? Answer: $\{\emptyset\}$, *not* $\emptyset$.

We must take care not to confuse the meaning of $T \in S$ and $T \subseteq S$. The statement $T \in S$ says that the object T (which might or might not be a set) is a member of the set S. The statement $T \subseteq S$ says that every member of the set T is also a member of the set S, i.e., $\forall z, z \in T \Rightarrow z \in S$. By definition, $T \subseteq S$ means the same thing as $T \in \mathcal{P}(S)$, but in general, $T \subseteq S$ does not mean the same thing as $T \in S$.

---

[1]Actually, in an axiomatic development of mathematics from set theory, 7 (and every other mathematical object) *is* defined to be a certain set. Without knowing the definition of 7 (and 8 and 9), we cannot decide whether $7 \subseteq \{7, 8, 9\}$. We ignore this technical point in this section.

3.49. Example. Let $U = \{5, 6, 7, \{5, 7\}, \{6\}, \{\{5\}\}, \{\{5, 7\}\}, \{5, \{6\}, 7\}, \{4\}, \emptyset\}$. What is $U \cap \mathcal{P}(U)$? *Solution.* For each fixed z, we know $z \in U \cap \mathcal{P}(U)$ iff ($z \in U$ and $z \in \mathcal{P}(U)$) iff ($z \in U$ and $z \subseteq U$). Thus, we are looking for all objects z that are simultaneously *members* of U and *subsets* of U. Let us test each of the ten members of U to see which of them are also subsets of U. The first three displayed members of U, namely 5, 6 and 7, are numbers, not subsets. The next member of U is $\{5, 7\}$, which is also a subset of U because $5 \in U$ and $7 \in U$. Similarly, $\{6\} \in U$ and $\{6\} \subseteq U$ (because $6 \in U$). On the other hand, $\{\{5\}\} \in U$ but $\{\{5\}\} \not\subseteq U$ because $\{5\}$ is a member of $\{\{5\}\}$ that is not a member of U. Next, $\{\{5, 7\}\} \in U$ and also $\{\{5, 7\}\} \subseteq U$, where the second fact holds because $\{5, 7\} \in U$. We see that $\{5, \{6\}, 7\} \subseteq U$ since each of the three members of the set on the left is a member of U. But $\{4\} \not\subseteq U$ since $4 \notin U$. Finally, $\emptyset \subseteq U$. In summary,

$$U \cap \mathcal{P}(U) = \{\{5, 7\}, \{6\}, \{\{5, 7\}\}, \{5, \{6\}, 7\}, \emptyset\}.$$

3.50. Theorem on Power Sets (Optional). For all sets A and B:

(a) $\emptyset \in \mathcal{P}(A)$ and $A \in \mathcal{P}(A)$.

(b) $A \subseteq B$ iff $\mathcal{P}(A) \subseteq \mathcal{P}(B)$.

(c) $\mathcal{P}(A \cap B) = \mathcal{P}(A) \cap \mathcal{P}(B)$.

(d) $\mathcal{P}(A) \cup \mathcal{P}(B) \subseteq \mathcal{P}(A \cup B)$.

(e) $\mathcal{P}(A - B) - \{\emptyset\} \subseteq \mathcal{P}(A) - \mathcal{P}(B)$.

This theorem can be proved more quickly if we remember facts proved earlier in the Theorem on Subsets. We illustrate by proving parts (a), (b), and (d). Let A and B be fixed, but arbitrary sets. To prove part (a), we must prove $\emptyset \subseteq A$ and $A \subseteq A$. Both parts are already known from the Theorem on Subsets. For part (b), we first prove that $A \subseteq B$ implies $\mathcal{P}(A) \subseteq \mathcal{P}(B)$. Assume $A \subseteq B$; prove $\mathcal{P}(A) \subseteq \mathcal{P}(B)$. [Continue with the subset template.] Fix an arbitrary object S; assume $S \in \mathcal{P}(A)$; prove $S \in \mathcal{P}(B)$. We have assumed that $S \subseteq A$; we must prove $S \subseteq B$. Now, from $S \subseteq A$ and $A \subseteq B$, we deduce that $S \subseteq B$ (by transitivity of $\subseteq$, part of the Theorem on Subsets).

For the converse in part (b), we now assume that $\mathcal{P}(A) \subseteq \mathcal{P}(B)$, and prove that $A \subseteq B$. Fix x; assume $x \in A$; prove $x \in B$. [Here we need a creative step — how can we use our assumption about the power sets to gain information about x? The key is to realize that $\{x\}$ is a subset of A, hence is a member of $\mathcal{P}(A)$.] We claim that $\{x\} \subseteq A$. To prove this claim, fix y, assume $y \in \{x\}$, and prove $y \in A$. Our assumption tells us that $y = x$, so that $y \in A$ by assumption on x. It follows from the claim and the definition of power set that $\{x\} \in \mathcal{P}(A)$. We have assumed $\mathcal{P}(A) \subseteq \mathcal{P}(B)$, so we deduce that $\{x\} \in \mathcal{P}(B)$. This means $\{x\} \subseteq B$. Since we know $x \in \{x\}$ (as $x = x$), we conclude that $x \in B$ as needed.

We continue by proving part (d). Fix an arbitrary object S; assume $S \in \mathcal{P}(A) \cup \mathcal{P}(B)$; prove $S \in \mathcal{P}(A \cup B)$. We have assumed $S \in \mathcal{P}(A)$ or $S \in \mathcal{P}(B)$, so use cases.
<u>Case 1.</u> Assume $S \in \mathcal{P}(A)$; prove $S \in \mathcal{P}(A \cup B)$. Since $A \subseteq A \cup B$ is known, we see from part (b) of the theorem that $\mathcal{P}(A) \subseteq \mathcal{P}(A \cup B)$. Now, because $S \in \mathcal{P}(A)$, we deduce that $S \in \mathcal{P}(A \cup B)$.
<u>Case 2.</u> Assume $S \in \mathcal{P}(B)$; prove $S \in \mathcal{P}(A \cup B)$. Since $B \subseteq A \cup B$ is known, we see from part (b) of the theorem that $\mathcal{P}(B) \subseteq \mathcal{P}(A \cup B)$. Now, because $S \in \mathcal{P}(B)$, we deduce that $S \in \mathcal{P}(A \cup B)$.
This completes the proof of part (d). The next example shows that we cannot replace $\subseteq$ by $=$ in part (d).

3.51. Example. Disprove: for all sets A and B, $\mathcal{P}(A) \cup \mathcal{P}(B) = \mathcal{P}(A \cup B)$.

Solution. We must prove there exist sets A and B with $\mathcal{P}(A) \cup \mathcal{P}(B) \neq \mathcal{P}(A \cup B)$. Choose $A = \{1\}$ and $B = \{2\}$. We know $\mathcal{P}(A) = \{\emptyset, \{1\}\}$ and $\mathcal{P}(B) = \{\emptyset, \{2\}\}$, so

$$\mathcal{P}(A) \cup \mathcal{P}(B) = \{\emptyset, \{1\}, \{2\}\}.$$

On the other hand, $A \cup B = \{1, 2\}$, so

$$\mathcal{P}(A \cup B) = \{\emptyset, \{1\}, \{2\}, \{1, 2\}\}.$$

We see that $\{1, 2\}$ is a member of $\mathcal{P}(A \cup B)$ that is not a member of $\mathcal{P}(A) \cup \mathcal{P}(B)$, so these two sets are unequal.

We conclude with one last example of the distinction between $\in$ and $\subseteq$. We have proved that for all sets X, $X \subseteq X$. Using an axiom of set theory called the *Axiom of Foundation*, it can be proved that for all sets X, $X \notin X$. (For details, see §3.7.) So, although every set is a *subset* of itself, no set is a *member* of itself. On the other hand, for all sets X, $X \in \{X\}$ and $X \in \mathcal{P}(X)$. Similarly, "for all sets X, $\emptyset \subseteq X$" is true, but "for all sets X, $\emptyset \in X$" is false.

Section Summary

1. **Memorize** these definitions:

   Empty Set: $\boxed{\forall z, z \notin \emptyset.}$

   Singleton Sets: For all a, z, $\boxed{z \in \{a\}}$ iff $\boxed{z = a}$.

   Unordered Pairs: For all a, b, z, $\boxed{z \in \{a, b\}}$ iff $\boxed{z = a \text{ or } z = b}$.

   Unordered Triples: For all a, b, c, z, $\boxed{z \in \{a, b, c\}}$ iff $\boxed{z = a \text{ or } z = b \text{ or } z = c}$.

   Power Set: For all S, X, $\boxed{S \in \mathcal{P}(X)}$ iff $\boxed{S \subseteq X}$.

2. *Order* and *repetition* do not matter for sets; only *membership in the set* matters when deciding set equality. For example, $\{1, 2\} = \{2, 1\}$ and $\{3\} = \{3, 3\}$ and $\{4, 4, 4, 5, 5, 6\} = \{6, 6, 5, 4, 4\}$.

3. Sets may contain other sets as members. Nested braces are important when determining set membership; only items at the outermost level are considered members of the outermost set. Note carefully that $\emptyset$ and $\{\}$ denote the empty set, but $\{\emptyset\}$ is a set with one member (namely, the empty set).

4. Take care not to confuse $\in$ (the set membership relation) with $\subseteq$ (the subset relation). Note $x \in B$ means x is a member of B, whereas $A \subseteq B$ means every member of the set A is a member of the set B. In particular, $A \in A$ is always false although $A \subseteq A$ is always true. For the set $B = \{4, 5, \{6, 7\}\}$, we have $4 \in B$ and $5 \in B$ and $\{4, 5\} \subseteq B$ and $\{4, 5\} \notin B$, whereas $\{6, 7\} \in B$ and $6 \notin B$ and $7 \notin B$ and $\{6, 7\} \not\subseteq B$.

Exercises

1. (a) Carefully prove that for all a, b, c, $\{a, b, c\} = \{b, a, c\}$ and $\{a, b, c\} = \{a, c, b\}$.
   (b) Use (a) to deduce that for all x, y, z, $\{x, y, z\} = \{x, z, y\} = \{y, x, z\} = \{y, z, x\} = \{z, x, y\} = \{z, y, x\}$.

2. (a) Prove: for all a, $\{a\} = \{a, a, a, a\}$.
   (b) Prove: for all a, b, c, $\{a, b, c\} = \{a, b, b, c, c, c\}$.

3. Is the set equality $\{a, \{b\}\} = \{\{a\}, b\}$ true for *all* objects a, b, for *some* objects a, b, or for *no* objects a, b? Explain carefully.

4. Let $S = \{\emptyset, 8, 7/5, \sqrt{2}, \mathbb{Z}, \mathbb{Q}, \{\mathbb{R}\}, \{3, 8\}, \mathcal{P}(\mathbb{Z})\}$. (Assume numbers are not sets here.) (a) Find all $x, y \in S$ such that $x \in y$. (b) Find all $x, y \in S$ such that $x \subseteq y$.

5. True or false? Proofs are not required, but expanding definitions may help you reach the correct answer.
    (a) $\{1, 2, 3, 4, 5\} \neq \{2, 4, 5, 3, 1\}$.
    (b) $\{1, 2, 3, 4\} = \{1, 2, 2, 3, 3, 3, 4, 4, 4, 4\}$.
    (c) $5 \in \{5\}$.
    (d) $5 \in \{\{5\}\}$.
    (e) $\{5\} \in \{5\}$.
    (f) $\{5\} \in \{\{5\}\}$.
    (g) $5 \in \{5, \{\{5\}\}\}$.
    (h) $\{5\} \in \{5, \{\{5\}\}\}$.
    (i) $\{\{5\}\} \in \{5, \{\{5\}\}\}$.
    (j) $5 \subseteq \{5, \{\{5\}\}\}$.
    (k) $\{5\} \subseteq \{5, \{\{5\}\}\}$.
    (l) $\{\{5\}\} \subseteq \{5, \{\{5\}\}\}$.
    (m) $\{\{\{5\}\}\} \subseteq \{5, \{\{5\}\}\}$.
    (n) $\{5\} \in \mathcal{P}(\{5, \{\{5\}\}\})$.
    (o) $\{\{5\}\} \in \mathcal{P}(\{5, \{\{5\}\}\})$.
    (p) $\{\{\{5\}\}\} \in \mathcal{P}(\{5, \{\{5\}\}\})$.
    (q) $\{\{5\}\} \subseteq \mathcal{P}(\{5, \{\{5\}\}\})$.
    (r) $\{\{\{5\}\}\} \subseteq \mathcal{P}(\{5, \{\{5\}\}\})$.

6. Compute each of the following power sets; be sure to use braces correctly.
    (a) $\mathcal{P}(\{5, 6\})$ (b) $\mathcal{P}(\{5, \{6\}\})$ (c) $\mathcal{P}(\{\{5, 6\}\})$ (d) $\mathcal{P}(\{5, \{5\}, \{5, \{5\}\}\})$

7. (a) If S has four members, how many members does $\mathcal{P}(S)$ have?
    (b) If S has five members, how many members does $\mathcal{P}(S)$ have?

8. Someone claims that $\{\emptyset, \emptyset\} \neq \{\emptyset\}$ because a box containing two empty boxes is not the same as a box containing one empty box. Is this claim correct? Explain carefully.

9. Let a, b, c be distinct objects that are not sets. Define

    $$S = \{a, \{b\}, c, \{\{b\}, c\}, \{a, c\}, \{\{a, c\}\}, \{\{\{a, c\}\}\}, \{\{a, c\}, b\}, \{\{b\}, a, \{c\}\}, \{a, \{b\}, c\}\}.$$

    (a) How many members does S have? How many members of S are sets?
    (b) List all members of $S \cap \mathcal{P}(S)$.
    (c) List all members of $S \cap \mathcal{P}(\mathcal{P}(S))$.
    (d) How many members does $\mathcal{P}(S) \cap \mathcal{P}(\mathcal{P}(S))$ have?

10. True or false? Explain each answer. Assume the variable X ranges over sets.
    (a) $\exists X, \emptyset \subseteq X$. (b) $\exists X, \emptyset \in X$. (c) $\exists! X, X \subseteq \emptyset$. (d) $\exists! X, X \in \emptyset$.
    (e) $\exists! X, X \in \{\emptyset\}$. (f) $\forall X, X \subseteq \mathcal{P}(X)$. (g) $\exists! X, X \in \mathcal{P}(X)$.
    (h) $\exists X, \exists! y, y \in \mathcal{P}(X)$. (i) $\exists! y, \forall X, y \in \mathcal{P}(X)$.

11. Prove: $\forall X, \sim(\{X\} \subseteq X)$.

12. Prove part (c) of the Theorem on Power Sets: for all sets A and B, $\mathcal{P}(A \cap B) = \mathcal{P}(A) \cap \mathcal{P}(B)$. (This can be done very quickly using part of the Theorem on Subsets.)

13. (a) Prove part (e) of the Theorem on Power Sets:
    for all sets A and B, $\mathcal{P}(A - B) - \{\emptyset\} \subseteq \mathcal{P}(A) - \mathcal{P}(B)$.

(b) Disprove: for all sets A and B, $\mathcal{P}(A - B) - \emptyset \subseteq \mathcal{P}(A) - \mathcal{P}(B)$.

(c) Prove or disprove: for all sets A and B, $\mathcal{P}(A) - \mathcal{P}(B) \subseteq \mathcal{P}(A - B)$.

14. Prove from the definitions: for all sets X and all objects a, $a \in X$ iff $\{a\} \in \mathcal{P}(X)$.

15. Let S be an arbitrary set. Find all members of $\mathcal{P}(S) \cap \mathcal{P}(\{S\})$.

16. (a) Prove $\forall x, \forall y, \{x\} = \{y\} \Leftrightarrow x = y$.

    (b) Prove $\forall x, \forall y, \forall z, \{x, y\} = \{z\} \Leftrightarrow (z = x \wedge z = y)$.

    (c) Prove $\forall x, \forall y, \forall z, \forall w, \{x, y\} = \{z, w\} \Leftrightarrow ((x = z \wedge y = w) \vee (x = w \wedge y = z))$.

17. Prove: $\forall a, b, c, d, \{\{a\}, \{a, b\}\} = \{\{c\}, \{c, d\}\} \Leftrightarrow [(a = c) \wedge (b = d)]$.

18. Give a fully detailed proof of the following fact, used in this section: for all a, b, $\mathcal{P}(\{a, b\}) = \{\emptyset, \{a\}, \{b\}, \{a, b\}\}$. (You may need proof by cases. Note that for any set S, you know $a \in S$ or $a \notin S$; and you know $b \in S$ or $b \notin S$.)

19. (a) Prove:

$$\forall S, \forall x \notin S, \forall T, T \in \mathcal{P}(S \cup \{x\}) \text{ iff } [T \in \mathcal{P}(S) \oplus (x \in T \wedge T - \{x\} \in \mathcal{P}(S))].$$

    (b) Explain informally what the statement proved in part (a) means. In particular, for finite S, how does the size of $\mathcal{P}(S \cup \{x\})$ compare to the size of $\mathcal{P}(S)$?

20. A set S is called $\in$-*transitive* iff for all x, y, if $x \in y$ and $y \in S$, then $x \in S$.

    (a) Prove: For all sets S, S is $\in$-transitive iff $(\forall y, y \in S \Rightarrow y \subseteq S)$ iff $S \subseteq \mathcal{P}(S)$.

    (b) Is $\emptyset$ $\in$-transitive? Explain.

    (c) Is $\{\emptyset\}$ $\in$-transitive? Explain.

    (d) Is $\{\{\emptyset\}\}$ $\in$-transitive? Explain.

21. (a) Prove: for all S, T, if S and T are $\in$-transitive, then $S \cap T$ is $\in$-transitive.

    (b) Prove or disprove: for all S, T, if S and T are $\in$-transitive, then $S \cup T$ is $\in$-transitive.

    (c) Prove: for all $\in$-transitive sets S, $S \cup \{S\}$ is $\in$-transitive.

22. Give an explicit example of an $\in$-transitive set S with exactly five distinct members.

23. Prove or disprove: for all $\in$-transitive sets S, $\mathcal{P}(S)$ is $\in$-transitive.

3.5 Ordered Pairs and Product Sets

This section introduces the concept of an ordered pair (a, b). In contrast to the unordered pairs studied earlier, (a, b) is not the same object as (b, a) when $a \neq b$. Ordered pairs are used in analytic geometry and calculus to represent points in the plane. We also define a product operation for sets to build sets of ordered pairs such as the xy-plane $\mathbb{R}^2 = \mathbb{R} \times \mathbb{R}$.

Ordered Pairs and Intervals

Recall that the unordered pair $\{a, b\}$ is a set with a and b as members. We proved that $\{a, b\}$ is the same set as $\{b, a\}$; order does not matter in a set. In many situations, however, order does matter. We now introduce a new undefined term, called an ordered pair, for such situations.

3.52. Undefined Term: Ordered Pair. For any objects a and b, the notation (a, b) is called the *ordered pair* with *first component a* and *second component b*.

We always use round parentheses for ordered pairs and curly braces for unordered pairs. The following axiom states the main property of ordered pairs.

3.53. Ordered Pair Axiom. For all objects a, b, c, d, $\boxed{(a, b) = (c, d)}$ iff $\boxed{(a = c \text{ and } b = d)}$.

In words, two ordered pairs are equal if and only if their first coordinates agree and their second coordinates agree. Thus, for instance, $(3, 5) \neq (5, 3)$, but $(1 + 1, 2 + 2) = (3 - 1, 2 \times 2)$. The components of an ordered pair can be any objects — integers, real numbers, sets, or even other ordered pairs. For instance, $(\{1, 2\}, (2, 1))$ is an ordered pair whose first component is the set $\{1, 2\}$ and whose second component is the ordered pair $(2, 1)$. For now, we postulate that ordered pairs, sets, and numbers are different kinds of objects. So, for instance, $((1, 2), 3) \neq (1, (2, 3))$ because "$(1, 2) = 1$ and $3 = (2, 3)$" is false.

Remarkably, it is possible to *define* ordered pairs in terms of the previously introduced concepts of unordered pairs and singleton sets. Then ordered pairs are certain sets, and the Ordered Pair Axiom can be proved as a theorem. We give the details of this process in an optional subsection at the end of this section.

We now introduce interval notation for certain subsets of $\mathbb{R}$. Intuitively, for given $a, b \in \mathbb{R}$, the closed interval $[a, b]$ consists of all real numbers x satisfying $a \leq x \leq b$. The open interval (a, b) consists of all real numbers x satisfying $a < x < b$. Half-open intervals and intervals extending to $\pm\infty$ are defined similarly. The formal definition is as follows.

3.54. Definition: Intervals. Let a, b be fixed real numbers.

(a) *Closed Intervals:* For all z, $\boxed{z \in [a, b]}$ iff $\boxed{z \in \mathbb{R} \text{ and } a \leq z \leq b}$.

(b) *Open Intervals:* For all z, $\boxed{z \in (a, b)}$ iff $\boxed{z \in \mathbb{R} \text{ and } a < z < b}$.

(c) *Half-Open Intervals:* For all z, $\boxed{z \in (a, b]}$ iff $\boxed{z \in \mathbb{R} \text{ and } a < z \leq b}$; and $\boxed{z \in [a, b)}$ iff $\boxed{z \in \mathbb{R} \text{ and } a \leq z < b}$.

(d) *Infinite Intervals:* For all z, $\boxed{z \in (-\infty, a]}$ iff $\boxed{z \in \mathbb{R} \text{ and } z \leq a}$. We define $(-\infty, a)$, $[a, \infty)$, and (a, ∞) analogously. Finally, $(-\infty, \infty)$ is the set $\mathbb{R}$. We remark that the symbols ∞ and $-\infty$ are *not* real numbers, and we do not define intervals such as $[-\infty, \infty]$.

Note that *the notation for open intervals conflicts with the notation for ordered pairs*; but this notation conflict seldom causes problems, since context often tells us what is intended. Some authors avoid this conflict by writing $]a,b[$ or $\langle a,b \rangle$ to denote an open interval, but we will not use this convention. We stress that curly braces, round parentheses, and square brackets are not interchangeable in mathematics: $\{a,b\}$ denotes an unordered pair, (a,b) denotes an ordered pair or open interval, and $[a,b]$ denotes a closed interval. The three pictures below illustrate the set $\{1.5,4\}$, the open interval $(1.5,4)$, and the closed interval $[1.5,4]$, respectively. Note that open circles are used to indicate endpoints that are excluded, while closed (filled-in) circles indicate endpoints that are included.

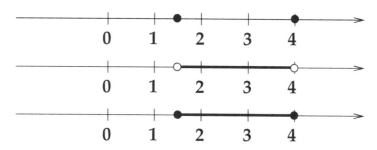

Product Sets

Given two sets A and B, the *product set* $A \times B$ is the set of all ordered pairs (x,y) where $x \in A$ and $y \in B$. We state this formally as follows.

3.55. Definition: Product Set. For all sets A and B and all objects z:

$$\boxed{z \in A \times B} \text{ iff } \boxed{\exists x, \exists y, (x \in A \wedge y \in B) \wedge z = (x,y)}.$$

It can be proved from this definition and the Ordered Pair Axiom that for all sets A and B and all objects x, y,

$$\boxed{(x,y) \in A \times B} \text{ iff } \boxed{x \in A \text{ and } y \in B}.$$

This version of the definition is frequently used in proofs.

3.56. Example. Given $A = \{1,2,3\}$ and $B = \{3,5\}$,

$$A \times B = \{(1,3),(2,3),(3,3),(1,5),(2,5),(3,5)\}.$$

On the other hand,

$$B \times A = \{(3,1),(3,2),(3,3),(5,1),(5,2),(5,3)\} \neq A \times B.$$

We have $(A \times B) \cap (B \times A) = \{(3,3)\}$. Note that $A \cap B = \{3\}$, and $(A \cap B) \times (A \cap B) = \{(3,3)\}$. We have $(A - B) \times (B - A) = \{1,2\} \times \{5\} = \{(1,5),(2,5)\}$. On the other hand, $(A \times B) - (B \times A) = \{(1,3),(2,3),(1,5),(2,5),(3,5)\}$.

3.57. Example. Given $A = \{\{1,2\},(3,4)\}$ and $B = \{\emptyset,\{3\}\}$,

$$A \times B = \{(\{1,2\},\emptyset),(\{1,2\},\{3\}),((3,4),\emptyset),((3,4),\{3\})\}.$$

3.58. Example. Given $C = \{0,1\}$, we have $C \times C = \{(0,0),(0,1),(1,0),(1,1)\}$. We can then compute

$$(C \times C) \times C = \{((0,0),0), ((0,1),0), ((1,0),0), ((1,1),0),$$
$$((0,0),1), ((0,1),1), ((1,0),1), ((1,1),1)\};$$
$$C \times (C \times C) = \{(0,(0,0)), (0,(1,0)), (1,(0,0)), (1,(1,0)),$$
$$(0,(0,1)), (0,(1,1)), (1,(0,1)), (1,(1,1))\}.$$

These two sets are not equal; in fact, $[(C \times C) \times C] \cap [C \times (C \times C)] = \emptyset$.

The product set $\mathbb{R} \times \mathbb{R}$ consists of all ordered pairs (x, y) with $x \in \mathbb{R}$ and $y \in \mathbb{R}$. We write $\mathbb{R}^2 = \mathbb{R} \times \mathbb{R}$; more generally, define $\boxed{A^2 = A \times A}$ for any set A. We visualize $\mathbb{R}^2$ as the xy-plane by drawing two perpendicular axes intersecting at a point called the *origin*, and identifying the ordered pair (x, y) with the point reached by traveling from the origin a directed distance of x horizontally and a directed distance of y vertically. Then, as seen in the next example, the product set $[a, b] \times [c, d]$ is a solid rectangle in the plane.

3.59. Example. Draw the following sets in the xy-plane.
(a) $R = [1, 3] \times [-2, 2]$; (b) $S = (-2, 2) \times (1, 3)$; (c) $R \cup S$; (d) $R \cap S$;
(e) $\{1, 3, 4\} \times \{-1, 0, 2\}$; (f) $((1, 3) \times \{-2, 2\}) \cup (\{1, 3\} \times (-2, 2])$.
Solution. See Figure 3.1. The four edges of the rectangle are part of the set R in (a), but not part of the set S in (b). We use dashed lines and open circles to indicate excluded regions, as in (c) and (d). Part (e) is a finite set of nine ordered pairs, each of which is drawn as a filled dot. In part (f), the graph of $(1, 3) \times \{-2, 2\}$ consists of two horizontal line segments with both endpoints excluded; whereas the graph of $\{1, 3\} \times (-2, 2]$ consists of two vertical line segments with the top endpoints included. The union of these sets is shown in the figure; the interior of the rectangle is not part of the union.

Properties of Product Sets

The next theorem lists general properties of product sets, which can all be proved from the definitions using the proof templates.

3.60. Theorem on Product Sets. For all sets A, B, C, D:

(a) *Monotonicity:* If $A \subseteq C$ and $B \subseteq D$, then $A \times B \subseteq C \times D$.

(b) *Distributive Laws:*
$A \times (B \cup C) = (A \times B) \cup (A \times C)$ and $(A \cup B) \times C = (A \times C) \cup (B \times C)$ and
$A \times (B \cap C) = (A \times B) \cap (A \times C)$ and $(A \cap B) \times C = (A \times C) \cap (B \times C)$ and
$A \times (B - C) = (A \times B) - (A \times C)$ and $(A - B) \times C = (A \times C) - (B \times C)$.

(c) *Intersection of Products:* $(A \times B) \cap (C \times D) = (A \cap C) \times (B \cap D)$.

(d) *Union of Products:* $(A \times B) \cup (C \times D) \subseteq (A \cup C) \times (B \cup D)$,
but equality does not always hold.

(e) *Product of Empty Sets:* $A \times \emptyset = \emptyset$ and $\emptyset \times B = \emptyset$.

(f) *Criterion for Commutativity:* $A \times B = B \times A$ iff $(A = B$ or $A = \emptyset$ or $B = \emptyset)$.

We prove a few parts of this theorem to illustrate the techniques, leaving the remaining parts as exercises.

3.61. Proof of Monotonicity. Let A, B, C, D be fixed, but arbitrary sets. We prove: if $A \subseteq C$ and $B \subseteq D$, then $A \times B \subseteq C \times D$. [Start with the IF-template.] Assume $A \subseteq C$ and $B \subseteq D$. Prove $A \times B \subseteq C \times D$. [Continue with the subset template.] Fix an arbitrary

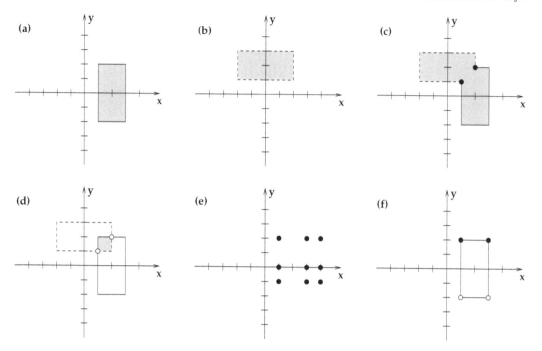

FIGURE 3.1
Solution to Example 3.59.

object z; assume $z \in A \times B$; prove $z \in C \times D$. Our assumption means that $z = (a, b)$ for some $a \in A$ and $b \in B$. We must prove $(a, b) \in C \times D$, which means $a \in C$ and $b \in D$. On one hand, since $a \in A$ and $A \subseteq C$, we deduce $a \in C$. On the other hand, since $b \in B$ and $B \subseteq D$, we deduce $b \in D$.

The following figure gives some intuition for why monotonicity holds, though the figure cannot replace the formal proof.

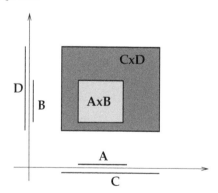

3.62. Proof that $\emptyset \times B = \emptyset$. Let B be an arbitrary set. [To show a set is empty, use proof by contradiction.] Assume, to get a contradiction, that $\emptyset \times B \neq \emptyset$. This assumption means $\exists z, z \in \emptyset \times B$. In turn, this means that $\exists x, \exists y, (x \in \emptyset \wedge y \in B) \wedge z = (x, y)$. In particular, $x \in \emptyset$. But we know $x \notin \emptyset$ also, which is a contradiction. So $\emptyset \times B = \emptyset$, after all.

Before proceeding to the next proof, we describe a modification of the subset proof template that is often convenient when dealing with product sets.

3.63. Subset Proof Template to Prove $A \times B \subseteq T$.
Fix an arbitrary ordered pair (x, y).
Assume $(x, y) \in A \times B$, which means $x \in A$ and $y \in B$.
Prove $(x, y) \in T$.

The template works since the only objects that can belong to the product set $A \times B$ are ordered pairs. So we may as well fix an ordered pair at the outset, rather than fixing an object z and then saying $\exists x, \exists y, (x \in A \wedge y \in B) \wedge z = (x, y)$. We illustrate this template in the next proof.

3.64. Proof that $A \times (B - C) = (A \times B) - (A \times C)$. Let A, B, C be arbitrary sets.
Part 1. Assume (x, y) is an arbitrary member of $A \times (B - C)$; prove $(x, y) \in (A \times B) - (A \times C)$. We have assumed that $x \in A$ and $y \in B - C$, so $y \in B$ and $y \notin C$. We must prove that $(x, y) \in A \times B$ and $(x, y) \notin A \times C$. The first part of what we must prove means that $x \in A$ and $y \in B$, and both statements are known from our current assumptions. The second part of what we must prove is that "$x \notin A$ or $y \notin C$." Since $y \notin C$ was assumed, this OR-statement is true.
Part 2. [We are about to prove $(A \times B) - (A \times C) \subseteq A \times (B - C)$. We observe that every member of the left set belongs to $A \times B$, hence must be an ordered pair. Thus we may begin part 2 by fixing an ordered pair rather than an arbitrary object.] Assume (x, y) is an arbitrary member of $(A \times B) - (A \times C)$; prove $(x, y) \in A \times (B - C)$. We have assumed that $(x, y) \in A \times B$ and $(x, y) \notin A \times C$, which means $(x \in A$ and $y \in B)$ and $(x \notin A$ or $y \notin C)$. We must prove $x \in A$ and $y \in B - C$; i.e., we must prove $x \in A$ and $y \in B$ and $y \notin C$. The first two statements follow from the first parenthesized statement in our assumption. On the other hand, since $x \in A$ is true and "$x \notin A$ or $y \notin C$" is also true, we deduce $y \notin C$ by definition of OR.

3.65. Proof of Commutativity Criterion. Let A and B be arbitrary sets; we prove $A \times B = B \times A$ iff $(A = B$ or $A = \emptyset$ or $B = \emptyset)$.
Part 1. We prove: "if $A \times B = B \times A$, then $A = B$ or $A = \emptyset$ or $B = \emptyset$" using a contrapositive proof. Assume $A \neq B$ and $A \neq \emptyset$ and $B \neq \emptyset$; prove $A \times B \neq B \times A$. Since A and B are nonempty, we know there exist objects $a_0 \in A$ and $b_0 \in B$. We also know $A \not\subseteq B$ or $B \not\subseteq A$, so use cases.
Case 1. Assume $A \not\subseteq B$, which means there exists c with $c \in A$ and $c \notin B$. Consider the ordered pair (c, b_0). On one hand, $(c, b_0) \in A \times B$ because $c \in A$ and $b_0 \in B$. On the other hand, $(c, b_0) \notin B \times A$ since $c \notin B$. Thus, $A \times B \not\subseteq B \times A$ since we have found a member of $A \times B$ that is not a member of $B \times A$.
Case 2. Assume $B \not\subseteq A$. The proof is similar to the proof of Case 1, so we let you fill in the details of Case 2.
Part 2. We prove: "if $A = B$ or $A = \emptyset$ or $B = \emptyset$, then $A \times B = B \times A$." Assume $A = B$ or $A = \emptyset$ or $B = \emptyset$. Prove $A \times B = B \times A$. [We have assumed an OR-statement, so use cases.]
Case 1. Assume $A = B$. Then $A \times B = A \times A = B \times A$.
Case 2. Assume $A = \emptyset$. Then both $A \times B$ and $B \times A$ equal the empty set, by part (e) of the theorem, hence $A \times B = B \times A$.
Case 3. Assume $B = \emptyset$. Then both $A \times B$ and $B \times A$ equal the empty set, by part (e) of the theorem, hence $A \times B = B \times A$.

Formal Definition of Ordered Pairs (Optional)

We mentioned earlier that ordered pairs could be defined in terms of the previously introduced concepts of singleton sets and unordered pairs. Here is the relevant definition.

3.66. Definition: Ordered Pairs. For any objects a, b, define $\boxed{(a, b) = \{\{a\}, \{a, b\}\}}$.

3.67. Example. We have $(3,5) = \{\{3\}, \{3,5\}\}$ and $(5,3) = \{\{5\}, \{5,3\}\} = \{\{5\}, \{3,5\}\}$; note that the set $(3,5)$ does not equal the set $(5,3)$. The ordered pair $(4,4)$ is the set $\{\{4\}, \{4,4\}\}$. Since $\{4,4\} = \{4\}$, we can also write $(4,4) = \{\{4\}, \{4\}\} = \{\{4\}\}$. On the other hand, $(4, \{4\}) = \{\{4\}, \{4, \{4\}\}\}$, whereas $(\{4\}, 4) = \{\{\{4\}\}, \{\{4\}, 4\}\} = \{\{\{4\}\}, \{4, \{4\}\}\}$. Thus, the three sets $(4,4)$ and $(4, \{4\})$ and $(\{4\}, 4)$ are all distinct.

We can now *prove* the Ordered Pair Axiom as a theorem.

3.68. Ordered Pair Theorem. For all objects a, b, c, d, $(a,b) = (c,d)$ iff $(a = c$ and $b = d)$.

Proof. Our proof uses the following three facts, stated in the exercises of §3.4. For all objects x, y, z, w: (1) $\{x\} = \{y\} \Leftrightarrow x = y$; (2) $\{x,y\} = \{z\} \Leftrightarrow (z = x \wedge z = y)$; and (3) $\{x,y\} = \{z,w\} \Leftrightarrow ((x = z \wedge y = w) \vee (x = w \wedge y = z))$. Fix arbitrary objects a, b, c, d.
<u>Part 1.</u> Assume $(a,b) = (c,d)$. Prove $a = c$ and $b = d$. By definition of ordered pairs, we have assumed $\{\{a\}, \{a,b\}\} = \{\{c\}, \{c,d\}\}$. Using Fact (3) with $x = \{a\}$, $y = \{a,b\}$, $z = \{c\}$, and $w = \{c,d\}$, we deduce that $(\{a\} = \{c\}$ and $\{a,b\} = \{c,d\})$ or $(\{a\} = \{c,d\}$ and $\{a,b\} = \{c\})$. [We have just written a known OR-statement, so use proof by cases.]
   Case 1. Assume $\{a\} = \{c\}$ and $\{a,b\} = \{c,d\}$; prove $a = c$ and $b = d$. Since $\{a\} = \{c\}$, Fact (1) gives $a = c$. Since $\{a,b\} = \{c,d\}$, Fact (3) gives $(a = c$ and $b = d)$ or $(a = d$ and $b = c)$. [This is another OR-statement, so use cases within Case 1.]
   Case 1a. Assume $a = c$ and $b = d$; prove $a = c$ and $b = d$. The two statements to be proved have already been assumed.
   Case 1b. Assume $a = d$ and $b = c$; prove $a = c$ and $b = d$. We already know $a = c$. Using the assumptions, we see $b = c = a = d$, so $b = d$.
   Case 2. Assume $\{a\} = \{c,d\}$ and $\{a,b\} = \{c\}$; prove $a = c$ and $b = d$. Since $\{a\} = \{c,d\}$, Fact (2) gives $a = c$ and $a = d$. Since $\{a,b\} = \{c\}$, Fact (2) gives $c = a$ and $c = b$. So, $a = c$ and $b = c = a = d$, as needed. [We might initially think Case 2 cannot occur, since $\{a\}$ is a one-element set and $\{c,d\}$ is a two-element set. But when $c = d$, $\{c,d\} = \{c\}$ is a one-element set after all! Compare to the example $(4,4) = \{\{4\}\}$.]
<u>Part 2.</u> Assume $a = c$ and $b = d$. Prove $(a,b) = (c,d)$. By definition of ordered pairs, we must prove $\{\{a\}, \{a,b\}\} = \{\{c\}, \{c,d\}\}$. Since $a = c$, Fact (1) shows that $\{a\} = \{c\}$. Similarly, since $a = c$ and $b = d$, Fact (3) gives $\{a,b\} = \{c,d\}$. Using Fact (3) again taking $x = \{a\}$, $y = \{a,b\}$, $z = \{c\}$, and $w = \{c,d\}$, we conclude that $\{\{a\}, \{a,b\}\} = \{\{c\}, \{c,d\}\}$, as needed. $\square$

We should not get too distracted by the odd-looking definition of (a,b) or the logical subtleties of the preceding proof. The definition $(a,b) = \{\{a\}, \{a,b\}\}$ has various bizarre consequences, such as $(a,b) \subseteq \mathcal{P}(\{a,b\})$. It is theoretically interesting that we can construct ordered objects from sets, which are intrinsically unordered. However, in applications of set theory to the rest of mathematics, the only important feature of this construction is the criterion for equality of ordered pairs stated in the Ordered Pair Theorem. That is why we can safely ignore the details of this definition, as we do in the main text, treating (a,b) as an undefined concept obeying the Ordered Pair Axiom.

Section Summary

1. *Intervals in $\mathbb{R}$.* For all real numbers a, b and all x:

   Open Interval: $\boxed{x \in (a,b)}$ iff $\boxed{x \in \mathbb{R} \text{ and } a < x < b}$.

   Closed Interval: $\boxed{x \in [a,b]}$ iff $\boxed{x \in \mathbb{R} \text{ and } a \le x \le b}$.

   Half-Open Interval: $\boxed{x \in [a,b)}$ iff $\boxed{x \in \mathbb{R} \text{ and } a \le x < b}$.

We define $(a, b]$ and intervals extending to $\pm\infty$ analogously.

2. *Ordered Pairs.* For all objects a, b, c, d, $(a, b) = (c, d)$ iff ($a = c$ and $b = d$). The round parentheses in (a, b) denote an ordered pair (or an open interval); the curly braces in $\{a, b\}$ denote an unordered pair; the square brackets in $[a, b]$ denote a closed interval.

3. *Product Sets.* For all objects x, y, $\boxed{(x, y) \in A \times B}$ iff $\boxed{x \in A \text{ and } y \in B}$. For all sets A, B and all objects z, $\boxed{z \in A \times B}$ iff $\boxed{\exists x, \exists y, x \in A \wedge y \in B \wedge z = (x, y)}$.
   To prove that $A \times B \subseteq T$, we may fix an ordered pair (x, y) satisfying $x \in A$ and $y \in B$, and prove that $(x, y) \in T$.

4. *Theorem on Product Sets.* The product set construction has properties such as monotonicity, distributive laws, $(A \times B) \cap (C \times D) = (A \cap C) \times (B \cap D)$, $(A \times B) \cup (C \times D) \subseteq (A \cup C) \times (B \cup D)$, and $A \times \emptyset = \emptyset = \emptyset \times A$. However, $A \times B$ does not equal $B \times A$, except when $A = B$ or one of the sets is empty.

5. *Formal Definition of Ordered Pairs (Optional).* For all objects a, b, we can define $(a, b) = \{\{a\}, \{a, b\}\}$ and use this definition to prove the Ordered Pair Axiom from facts about unordered pairs.

Exercises

1. Let $A = \{2, 4, 5\}$, $B = \{4, 5, 6\}$, and $C = \{1, 4, 6\}$. Compute:
   (a) $A \times B$ (b) $B \times A$ (c) $(A \times B) \cap (B \times A)$ (d) $(A \times B) \cap (B \times C)$
   (e) $(A \cap B) \times (B \cap C)$ (f) $(B - C) \times (C - B)$ (g) $(B \times A) - (C \times B)$
   (h) $[(B - A) \times C] \cup [C \times (A - B)]$.

2. Let $A = \{1, 2\}$ and $B = \{3, 4\}$. Compute:
   (a) $A \times B$ (b) $\mathcal{P}(A)$ (c) $\mathcal{P}(B)$ (d) $\mathcal{P}(A \times B)$ (e) $\mathcal{P}(A) \times \mathcal{P}(B)$ (f) $(A \times B) \times A$
   (g) $A \times (B \times A)$ (h) Are the sets found in (d) and (e) equal? Why? (i) Are the sets found in (f) and (g) equal? Why?

3. Given $A = \{[0, 3], \mathbb{Z}, \{2\}\}$ and $B = \{(1, \{3\}), \{1, 3\}\}$, compute $A \times B$ and $B \times A$.

4. Let $A = [-1, 2]$, $B = [-1, 3]$, $C = [1, 4]$, and $D = [2, 5]$.
   (a) Draw pictures of A, B, C, D, $A - C$, and $B - D$ on a number line. Express $A - C$ and $B - D$ in interval notation.
   (b) Draw pictures in $\mathbb{R}^2$ showing $A \times B$, $C \times D$, $(A \times B) - (C \times D)$, and $(A - C) \times (B - D)$.

5. Define intervals $A = (0, 5)$, $B = [2, 3)$, $C = (-1, 2]$, and $D = (1, 7)$.
   (a) Draw pictures of A, B, C, D, $A - B$, $A - C$, $C \cup D$, and $D - B$ on a number line. Express $A - B$ and $D - B$ as unions of intervals.
   (b) Draw pictures in $\mathbb{R}^2$ showing $(A - B) \times (D - B)$, $(D - B) \times (A - B)$, $(A \times D) - (B \times B)$, and $(D \times A) - (C \times C)$.

6. Define $A = [-3, 3]$, $B = (-2.5, 1.5)$, and $C = \{-1, 1, 2\}$. Draw pictures in $\mathbb{R}^2$ of the following sets: (a) $A \times B$ (b) $C \times C$ (c) $(A \times B) \cap (\mathbb{Z} \times \mathbb{Z})$ (d) $(A - C) \times (B - C)$
   (e) $\mathbb{R} \times A$ (f) $B \times \mathbb{Z}$ (g) $C \times \mathbb{R}$ (h) $\mathbb{Z} \times C$.

7. (a) Let S be the boundary of the rectangle in $\mathbb{R}^2$ with corners $(1, 2)$, $(5, 2)$, $(5, 7)$, and $(1, 7)$. Describe S as a union of product sets. (b) Give another description of the set S in part (a) using only interval notation, product sets, and set difference.

8. Let T be the cube $[0, 2] \times [0, 2] \times [1, 3]$.
   (a) What is the geometric significance of the set $\{0, 2\} \times \{0, 2\} \times \{1, 3\}$? What about $[0, 2] \times \{2\} \times [1, 3]$? (b) Use set operations such as $\times$ and $\cup$ to describe the

set of points in $\mathbb{R}^3$ lying on one of the six faces of T. (c) Similarly, describe the set of points in $\mathbb{R}^3$ lying on one of the 12 edges of T.

9. Assume a, b, c, d are fixed real numbers. Prove the following properties of intervals in $\mathbb{R}$, and illustrate with sketches.
   (a) If $a < b$, then $[a, b] - (a, b) = \{a, b\}$.
   (b) If $a < b < c$, then $[a, b] \cup (b, c] = [a, c]$.
   (c) If $a < b < c < d$, then $[a, d] - [b, c] = [a, b) \cup (c, d]$.
   (d) If $a < b$, then $(a, b] = (-\infty, b] \cap (a, \infty)$.

10. (a) Prove: for all sets A, B, C, D, $(A \times B) \cap (C \times D) = (A \cap C) \times (B \cap D)$.
    (b) Draw pictures illustrating the identity in (a), taking the sets A, B, C, D to be specific closed intervals.

11. Prove: for all sets A, $A \times \emptyset = \emptyset$.

12. Fix arbitrary sets A, B, C.
    (a) Prove the distributive law $(A \cup B) \times C = (A \times C) \cup (B \times C)$.
    (b) Prove the distributive law $A \times (B \cap C) = (A \times B) \cap (A \times C)$.
    (c) Prove the distributive law $(A - B) \times C = (A \times C) - (B \times C)$.

13. Draw pictures in $\mathbb{R}^2$ illustrating each identity in the preceding problem.

14. (a) Prove: for all sets A, B, C, D, $(A \times B) \cup (C \times D) \subseteq (A \cup C) \times (B \cup D)$.
    (b) Draw a picture in $\mathbb{R}^2$ to show that equality does not always hold in (a).
    (c) Based on your picture, guess a formula expressing $(A \cup C) \times (B \cup D)$ as a union of other sets. Prove your formula.

15. (a) Let $A = [0, 5]$, $B = [1, 2]$, $C = [0, 6]$, and $D = [3, 4]$. Draw a picture of $(A - B) \times (C - D)$.
    (b) Repeat part (a) with $A = C = [0, 3]$, $B = [2, 4]$, and $D = [-1, 1]$.
    (c) Repeat part (a) with $A = C = \{1, 2, 3, 4, 5\}$, $B = \{1, 2, 4\}$, and $D = \{4, 5, 7\}$.
    (d) For arbitrary sets A, B, C, D, guess a formula expressing $(A - B) \times (C - D)$ in terms of $A \times C$ and other sets. Prove your formula.

16. Carefully sketch the set $A \cup B \cup C$, where: $A = ([1, 8] \cap \mathbb{Z}) \times (1, 3]$, $B = ((4, 5) \cup (7, 8)) \times \{2\}$, and $C = (([1, 6.5] - (3, 4)) - (5, 5.5)) \times \{3\}$.

17. (a) Prove: for any closed intervals I and J in $\mathbb{R}$, $I \cap J$ is a closed interval or the empty set. (b) Prove: for any open intervals I and J in $\mathbb{R}$, if $I \cap J \neq \emptyset$, then $I \cup J$ is an open interval. (c) Prove or disprove the converse of the statement in (b).

18. For any objects a, b, c, define the *ordered triple* (a, b, c) to be $((a, b), c)$. Use the Ordered Pair Axiom to prove:
$$\forall a, b, c, d, e, f, (a, b, c) = (d, e, f) \Leftrightarrow [(a = d) \wedge (b = e) \wedge (c = f)].$$

19. For any objects a, b, c, we give an alternate definition of the ordered triple by letting (a, b, c) be the set $\{(1, a), (2, b), (3, c)\}$. Use this definition and properties of ordered pairs to prove:
$$\forall a, b, c, d, e, f, (a, b, c) = (d, e, f) \Leftrightarrow [(a = d) \wedge (b = e) \wedge (c = f)].$$

20. Use Definition 3.66 to prove: (a) $\forall a, b, (a, b) \subseteq \mathcal{P}(\{a, b\})$.
    (b) $\forall A, B, A \times B \subseteq \mathcal{P}(\mathcal{P}(A \cup B))$. (c) $\forall x, y, x = y \Leftrightarrow \exists! z, z \in (x, y)$.

21. Consider the following alternate definition of ordered pairs: for all objects a, b, define $(a, b) = \{\{1, a\}, \{2, b\}\}$. Prove the Ordered Pair Axiom holds using this definition of (a, b).

22. Suppose we try to define ordered triples by letting $(a, b, c) = \{\{1, a\}, \{2, b\}, \{3, c\}\}$ for all objects a, b, c. Disprove:

$$\forall a, b, c, d, e, f, (a, b, c) = (d, e, f) \Leftrightarrow [(a = d) \wedge (b = e) \wedge (c = f)].$$

23. Suppose we try to define ordered pairs by letting $(a, b) = \{a, b, \{a\}\}$ for all objects a and b. Prove or disprove the Ordered Pair Axiom using this definition of (a, b).

3.6   General Unions and Intersections

Given two sets A and B, we have defined the union $A \cup B$ and the intersection $A \cap B$ of these sets. To take the union or intersection of finitely many sets, one way to proceed is to iterate the binary version of the operation. For instance, the union of five sets A, B, C, D, E could be defined to be $(((A \cup B) \cup C) \cup D) \cup E$. Because of the commutative and associative properties of union and intersection, the order of sets and placement of parentheses in this expression do not affect the answer. However, we cannot use this method to define the union or intersection of infinitely many sets. This section describes how to handle this situation, using the idea of an indexed collection of sets. We also introduce set-builder notation, which gives a concise way to define new sets.

Unions and Intersections of Indexed Collections of Sets

Let I be any nonempty set, called an *index set*. Each member i of I is called an *index*. Suppose that for each index $i \in I$, we have a set A_i. Informally, the *union* of the sets A_i, as i ranges over I, is the set obtained by joining together all the members of all the sets A_i into one giant set. The *intersection* of the sets A_i is the set of all objects z (if any) that belong to all of the A_i. Formally, we have the following definitions.

3.69. Definition: Union and Intersection of an Indexed Family of Sets. Given a nonempty set I and a set A_i for each $i \in I$:

(a) *General Union:* For all z, $\boxed{z \in \bigcup_{i \in I} A_i}$ iff $\boxed{\exists i \in I, z \in A_i}$.

(b) *General Intersection:* For all z, $\boxed{z \in \bigcap_{i \in I} A_i}$ iff $\boxed{\forall i \in I, z \in A_i}$.

For certain index sets I, variations of this notation are used. Specifically, if $I = \{1, 2, \ldots, n\}$ for some positive integer n, we write

$$\bigcup_{i \in I} A_i = \bigcup_{i=1}^{n} A_i = A_1 \cup A_2 \cup \cdots \cup A_n; \qquad \bigcap_{i \in I} A_i = \bigcap_{i=1}^{n} A_i = A_1 \cap A_2 \cap \cdots \cap A_n.$$

If $I = \mathbb{Z}_{>0}$ is the set of all positive integers, we write

$$\bigcup_{i \in I} A_i = \bigcup_{i=1}^{\infty} A_i, \qquad \bigcap_{i \in I} A_i = \bigcap_{i=1}^{\infty} A_i.$$

We use similar notation for $I = \mathbb{Z}_{>a}$, $I = \{b, b+1, \ldots, c\}$, etc. We can also change the name of the index i to any unused letter; for instance, $\bigcup_{i \in I} A_i = \bigcup_{k \in I} A_k = \bigcup_{n \in I} A_n$.

3.70. Example. Let the index set I be $\mathbb{R}_{>0}$, the set of positive real numbers. For each $r \in \mathbb{R}_{>0}$, let $A_r = \{-r, 0, r\}$. Then $\bigcup_{r \in \mathbb{R}_{>0}} A_r = \mathbb{R}$, whereas $\bigcap_{r \in \mathbb{R}_{>0}} A_r = \{0\}$. If we shrink the index set from $\mathbb{R}_{>0}$ to $\mathbb{Z}_{>0}$, then $\bigcup_{n \in \mathbb{Z}_{>0}} A_n = \mathbb{Z}$.

3.71. Example. Suppose $I = \mathbb{Z}_{>0}$, and for each $n \in I$, let $A_n = [-n, 1/n]$. For instance, $A_1 = [-1, 1]$, $A_2 = [-2, 1/2]$, $A_3 = [-3, 1/3]$, and so on, as shown in this picture:

It is visually apparent from this picture that $\bigcup_{n=1}^{\infty} A_n = (-\infty, 1]$ and $\bigcap_{n=1}^{\infty} A_n = [-1, 0]$. In particular, for any positive real number r, there is an integer n_0 with $1/n_0 < r$, so that $r \notin A_{n_0}$ and hence $r \notin \bigcap_{n=1}^{\infty} A_n$.

3.72. Example. Let I be the set of rational numbers larger than π. For each $q \in I$, let $A_q = [q, \infty)$. On one hand, $\bigcap_{q \in I} A_q = \emptyset$, so that the intersection of infinitely many infinite sets can be the empty set. On the other hand, $\bigcup_{q \in I} A_q = (\pi, \infty)$, so that the union of infinitely many closed intervals can be an open interval.

Let us check the claimed set equalities in this example in more detail. First, assume to get a contradiction that $\bigcap_{q \in I} A_q \neq \emptyset$. Then there exists z_0 such that $z_0 \in \bigcap_{q \in I} A_q$. This means that for all $q \in I$, $z_0 \in A_q$. So for every rational $q > \pi$, we have $q \leq z_0 < \infty$. However, if we let q be the smallest integer larger than z_0 (such an integer exists and is a rational number larger than π), we have $z_0 < q$, which is a contradiction. So the intersection is indeed empty.

To check $\bigcup_{q \in I} A_q = (\pi, \infty)$, we prove set inclusions in both directions. To prove $\subseteq$, fix an arbitrary $z_0 \in \bigcup_{q \in I} A_q$, and prove $z_0 \in (\pi, \infty)$. We know there exists $q_0 \in I$ with $z_0 \in A_{q_0}$. This means that $q_0 \in \mathbb{Q}$, $q_0 > \pi$, $z_0 \in \mathbb{R}$, and $q_0 \leq z_0$. Since $\pi < q_0 \leq z_0$, we see that $\pi < z_0$, proving $z_0 \in (\pi, \infty)$. To prove $\supseteq$, fix an arbitrary $y_0 \in (\pi, \infty)$. Then $y_0 \in \mathbb{R}$ and $\pi < y_0$. Assume we already know the theorem that between any two distinct real numbers there exists a rational number. So, we can find $q_0 \in \mathbb{Q}$ with $\pi < q_0 < y_0$. Then $y_0 \in A_{q_0}$, and so $y_0 \in \bigcup_{q \in I} A_q$, as needed.

We remark that if we replace I by a finite subset $I' = \{q_1 < q_2 < \cdots < q_n\}$, then $\bigcup_{q \in I'} A_q = (q_1, \infty) = A_{q_1}$ and $\bigcap_{q \in I'} A_q = (q_n, \infty) = A_{q_n}$.

3.73. Example. Take the index set to be $I = \mathbb{Z}_{>0}$. For fixed $n \in I$, let A_n be the set of real numbers of the form k/n for some $k \in \mathbb{Z}$. Thus, A_n is the set of real numbers that can be written as fractions where the numerator is any integer and the denominator is n. Then $\bigcup_{n=1}^{\infty} A_n = \mathbb{Q}$, whereas $\bigcap_{n=1}^{\infty} A_n = \mathbb{Z}$. When checking the latter equality in detail (see Exercise 4), it helps to first prove that $\mathbb{Z} \subseteq A_n$ for every $n \in I$, and $\mathbb{Z} = A_1$.

Properties of General Unions and Intersections

The next theorem states identities satisfied by the union and intersection operations for indexed collections of sets. Many of these properties are analogous to properties of binary union and binary intersection, which were given earlier in the Theorem on Subsets, the Theorem on Set Equality, and the Theorem on Product Sets.

3.74. Theorem on General Unions and Intersections. For all nonempty index sets I and J and all sets X and A_i and B_i:

(a) *Monotonicity:* If $A_i \subseteq B_i$ for all $i \in I$, then $\bigcup_{i \in I} A_i \subseteq \bigcup_{i \in I} B_i$ and $\bigcap_{i \in I} A_i \subseteq \bigcap_{i \in I} B_i$.

(b) *de Morgan Laws:* $X - \left(\bigcup_{i \in I} A_i \right) = \bigcap_{i \in I} (X - A_i)$ and $X - \left(\bigcap_{i \in I} A_i \right) = \bigcup_{i \in I} (X - A_i)$.

(c) *Distributive Laws:* $X \cap \left(\bigcup_{i \in I} A_i \right) = \bigcup_{i \in I} (X \cap A_i)$ and $X \cup \left(\bigcap_{i \in I} A_i \right) = \bigcap_{i \in I} (X \cup A_i)$ and

$$X \times \left(\bigcup_{i \in I} B_i \right) = \bigcup_{i \in I} (X \times B_i) \text{ and } X \times \left(\bigcap_{i \in I} B_i \right) = \bigcap_{i \in I} (X \times B_i);$$

similarly if X appears as the second input to $\cap$, $\cup$, or $\times$.

(d) *Lower/Upper Bound:* For all $k \in I$, $\bigcap_{i \in I} A_i \subseteq A_k$ and $A_k \subseteq \bigcup_{i \in I} A_i$.

(e) *Greatest Lower Bound:* $X \subseteq \bigcap_{i \in I} A_i$ iff $(\forall i \in I, X \subseteq A_i)$.

(f) *Least Upper Bound:* $\bigcup_{i \in I} A_i \subseteq X$ iff $(\forall i \in I, A_i \subseteq X)$.

(g) *Enlarging the Index Set:* If $I \subseteq J$, then $\bigcup_{i \in I} A_i \subseteq \bigcup_{j \in J} A_j$ and $\bigcap_{j \in J} A_j \subseteq \bigcap_{i \in I} A_i$.

(h) *Combining Index Sets:*

$$\left(\bigcup_{i \in I} A_i \right) \cup \left(\bigcup_{j \in J} A_j \right) = \bigcup_{k \in I \cup J} A_k \quad \text{and} \quad \left(\bigcap_{i \in I} A_i \right) \cap \left(\bigcap_{j \in J} A_j \right) = \bigcap_{k \in I \cup J} A_k.$$

The proofs of these results provide excellent practice with the proof templates involving quantifiers. We prove some parts to illustrate, and leave the other parts as exercises. Fix arbitrary sets $I \neq \emptyset$, $J \neq \emptyset$, X, A_i, and B_i (for $i \in I$ or $i \in J$, as appropriate).

3.75. Proof of Lower/Upper Bound Property. Fix an arbitrary index $k \in I$. We first prove $\bigcap_{i \in I} A_i \subseteq A_k$. Fix an arbitrary z; assume $z \in \bigcap_{i \in I} A_i$; prove $z \in A_k$. We have assumed $\forall i \in I, z \in A_i$. Since $k \in I$, we can let $i = k$ in this assumed universal statement to conclude (by the Inference Rule for ALL) that $z \in A_k$, as needed.

Next we prove $A_k \subseteq \bigcup_{i \in I} A_i$. Fix w; assume $w \in A_k$; prove $w \in \bigcup_{i \in I} A_i$. We must prove $\exists i \in I, w \in A_i$. Choose $i = k$, which is a member of I. Our assumption gives $w \in A_i$, as needed.

3.76. Proof of First de Morgan Law.
<u>Part 1.</u> We prove $X - \left(\bigcup_{i \in I} A_i \right) \subseteq \bigcap_{i \in I} (X - A_i)$. Fix z; assume $z \in X - \left(\bigcup_{i \in I} A_i \right)$; prove $z \in \bigcap_{i \in I} (X - A_i)$. We have assumed that $z \in X$ and $z \notin \bigcup_{i \in I} A_i$; we must prove $z \in \bigcap_{i \in I} (X - A_i)$. Continuing to expand definitions, we have assumed that "$\exists i \in I, z \in A_i$" is false, so that "$\forall i \in I, z \notin A_i$" is true. We must prove $\forall i \in I, z \in (X - A_i)$. Fix an index $i_0 \in I$. Prove $z \in X - A_{i_0}$. We must prove $z \in X$ and $z \notin A_{i_0}$. On one hand, $z \in X$ was already assumed. On the other hand, taking $i = i_0$ in the assumption that $\forall i \in I, z \notin A_i$, we see that $z \notin A_{i_0}$, as needed.
<u>Part 2.</u> We prove $\bigcap_{i \in I} (X - A_i) \subseteq X - \left(\bigcup_{i \in I} A_i \right)$. The proof is similar to part 1; you are asked to supply the details in Exercise 8.

3.77. Proof of Enlarging the Index Set. Assume $I \subseteq J$. First, we prove $\bigcup_{i \in I} A_i \subseteq \bigcup_{j \in J} A_j$. Fix z; assume $z \in \bigcup_{i \in I} A_i$; prove $z \in \bigcup_{j \in J} A_j$. We have assumed $\exists i \in I, z \in A_i$; we must prove $\exists j \in J, z \in A_j$. Let i_0 be a fixed index in I with $z \in A_{i_0}$. Choose $j_0 = i_0$. We know $i_0 \in I$ and $I \subseteq J$, so $j_0 = i_0 \in J$. Also $z \in A_{j_0}$ for this choice of j_0.

Next, we prove $\bigcap_{j \in J} A_j \subseteq \bigcap_{i \in I} A_i$ (note the reversal of the inclusion here). Fix w; assume $w \in \bigcap_{j \in J} A_j$; prove $w \in \bigcap_{i \in I} A_i$. We know $\forall j \in J, w \in A_j$; we must prove $\forall i \in I, w \in A_i$. Fix $i_0 \in I$; prove $w \in A_{i_0}$. Since $i_0 \in I$ and $I \subseteq J$, we know $i_0 \in J$. So we can take $j = i_0$ in the known universal statement to see that $w \in A_{i_0}$, as needed.

3.78. Proof of $\left(\bigcap_{i \in I} A_i \right) \cap \left(\bigcap_{j \in J} A_j \right) = \bigcap_{k \in I \cup J} A_k.$ We prove this set equality using a two-part proof.

<u>Part 1.</u> [proving $\subseteq$] Fix z; assume $z \in (\bigcap_{i \in I} A_i) \cap (\bigcap_{j \in J} A_j)$; prove $z \in \bigcap_{k \in I \cup J} A_k$. We have assumed $z \in \bigcap_{i \in I} A_i$ and $z \in \bigcap_{j \in J} A_j$, which means $\forall i \in I, z \in A_i$ and $\forall j \in J, z \in A_j$. We must prove $\forall k \in I \cup J, z \in A_k$. Fix $k \in I \cup J$; prove $z \in A_k$. We know $k \in I$ or $k \in J$, so consider two cases.

Case 1. Assume $k \in I$; prove $z \in A_k$. This follows by taking $i = k$ in the first universal statement in our earlier assumption.

Case 2. Assume $k \in J$; prove $z \in A_k$. This follows by taking $j = k$ in the second universal statement in our earlier assumption.

<u>Part 2.</u> [proving $\supseteq$] Fix w; assume $w \in \bigcap_{k \in I \cup J} A_k$; prove $w \in (\bigcap_{i \in I} A_i) \cap (\bigcap_{j \in J} A_j)$. We have assumed $\forall k \in I \cup J, w \in A_k$. We must prove $w \in \bigcap_{i \in I} A_i$ and $w \in \bigcap_{j \in J} A_j$. First, we must prove $\forall i \in I, w \in A_i$. Fix $i \in I$; prove $w \in A_i$. Since $I \subseteq I \cup J$, we know $i \in I \cup J$ and hence $w \in A_i$ (by taking $k = i$ in the earlier assumption). Next, we must prove $\forall j \in J, w \in A_j$. Fix $j \in J$; prove $w \in A_j$. Since $J \subseteq I \cup J$, we know $j \in I \cup J$ and hence $w \in A_j$ (by taking $k = j$ in the earlier assumption).

Set-Builder Notation

Before concluding our study of set theory, we need to consider a notational device called *set-builder notation* that is frequently used to introduce new sets. Suppose $P(x)$ is an open sentence with free variable x. Informally, the notation $S = \{x : P(x)\}$ means that S is the set of all objects x such that $P(x)$ is true. Similarly, if B is a given set, the notation $S = \{x \in B : P(x)\}$ means that S is the set of all objects $x \in B$ such that $P(x)$ is true. Many variations of this notation are possible; for example, we could define product sets by writing $A \times B = \{(x, y) : x \in A \wedge y \in B\}$. This formula could be read: "$A \times B$ is the set of all ordered pairs (x, y) such that x is in A and y is in B." Here is a formal definition of the two basic forms of set-builder notation.

3.79. Definition: Set-Builder Notation. For all open sentences $P(x)$, all sets B, and all objects z, $\boxed{z \in \{x : P(x)\}}$ iff $\boxed{P(z) \text{ is true}}$; and $\boxed{z \in \{x \in B : P(x)\}}$ iff $\boxed{z \in B \text{ and } P(z) \text{ is true}}$. The letter x can be replaced by any unused letter. For instance,

$$\{x : P(x)\} = \{y : P(y)\} = \{t : P(t)\}.$$

Table 3.1 illustrates set-builder notation by redefining the main set operations of this chapter using this notation. You *must* become familiar with this notation, as it is used universally in mathematics texts. You might wonder why I have avoided this notation so far in this text (and why I use it only sparingly in later chapters). The reason is this: in many years of teaching proofs, I have observed that set-builder notation is one of the biggest stumbling blocks for beginning students trying to write proofs about sets. In a proof involving set unions, say, one must fluidly transition back and forth between the statement "$z \in A \cup B$" and the equivalent statement "$z \in A$ or $z \in B$" — so I use this equivalence as the very definition of set unions. The set-builder definition $A \cup B = \{z : z \in A \vee z \in B\}$ means exactly the same thing, but the extra level of translation needed to use this definition in a proof seems to cause endless trouble.

$$
\begin{aligned}
A \cup B &= \{x : x \in A \vee x \in B\} \\
A \cap B &= \{x : x \in A \wedge x \in B\} \\
A - B &= \{x : x \in A \wedge x \notin B\} = \{x \in A : x \notin B\} \\
\mathcal{P}(A) &= \{S : S \subseteq A\} \\
A \times B &= \{(x,y) : x \in A \wedge y \in B\} = \{z : \exists x \in A, \exists y \in B, z = (x,y)\} \\
(a,b) &= \{x \in \mathbb{R} : a < x < b\} \\
[a,b] &= \{x \in \mathbb{R} : a \le x \le b\} \\
\bigcup_{i \in I} A_i &= \{x : \exists i \in I, x \in A_i\} \\
\bigcap_{i \in I} A_i &= \{x : \forall i \in I, x \in A_i\}
\end{aligned}
$$

TABLE 3.1
Definitions of Set Operations in Set-Builder Notation.

Section Summary

1. *Definitions.* Suppose $I \neq \emptyset$ is an index set and, for each $i \in I$, A_i is a given set.
   General Union: For all z, $\boxed{z \in \bigcup_{i \in I} A_i}$ iff $\boxed{\exists i \in I, z \in A_i}$.

   General Intersection: For all z, $\boxed{z \in \bigcap_{i \in I} A_i}$ iff $\boxed{\forall i \in I, z \in A_i}$.

2. *Properties of General Unions and Intersections.* The general union and intersection operations satisfy properties such as monotonicity, de Morgan laws, and the distributive laws. Also, $\bigcap_{i \in I} A_i$ is the largest set contained in every A_i, whereas $\bigcup_{i \in I} A_i$ is the smallest set containing every A_i. Replacing I by a larger index set J produces a larger union and a smaller intersection. We can simplify a union of unions or an intersection of intersections by combining the index sets.

3. *Set-Builder Notation.* Informally, $\{x \in B : P(x)\}$ is the set of all objects x in the set B that make $P(x)$ true. Formally, for all open sentences $P(x)$, all sets B, and all objects z, $\boxed{z \in \{x \in B : P(x)\}}$ iff $\boxed{z \in B \wedge P(z)}$. Notation such as $\{x : P(x)\}$ and $\{(x,y) : P(x,y)\}$ is defined similarly. The letter x can be replaced by any unused letter; e.g., $\{x : P(x)\} = \{t : P(t)\}$.

Exercises

1. Let I be the set of odd integers. For $n \in I$, define $A_n = (n-1, n+1)$. Sketch and describe the set $\mathbb{R} - \bigcup_{n \in I} A_n$.

2. For each index set I and sets A_i below, describe $\bigcup_{i \in I} A_i$ and $\bigcap_{i \in I} A_i$.
   Do not prove your answers; it can help to draw pictures in $\mathbb{R}$ and $\mathbb{R}^2$.
   (a) $I = \mathbb{Z}_{>0}$; for each $n \in I$, A_n is the half-open interval $(1 + 1/n, 5 - 1/n]$.
   (b) $I = [0, 3]$; for each $r \in I$, A_r is the open interval $(r - 2, r + 2)$.

(c) $I = \mathbb{R}_{>0}$; for each $r \in I$, A_r is the rectangle $[0, r] \times [0, 1/r]$.

(d) $I = \mathbb{R}_{>0}$; for each $r \in I$, A_r is the set of (x, y) with $(x - r)^2 + y^2 \leq r^2$.

3. Let a, b be fixed real numbers with $a < b$. (a) Express the closed interval $[a, b]$ as an intersection of infinitely many open intervals. (b) Express the open interval (a, b) as a union of infinitely many closed intervals.

4. In Example 3.73, carefully prove the claimed set equalities $\bigcup_{n=1}^{\infty} A_n = \mathbb{Q}$ and $\bigcap_{n=1}^{\infty} A_n = \mathbb{Z}$.

5. For each $n \in \mathbb{Z}_{>0}$, let

$$A_n = \{m \in \mathbb{Z} : m \text{ divides } n\} \text{ and } B_n = \{k \in \mathbb{Z} : n \text{ divides } k\}.$$

(a) Compute $\bigcup_{n=1}^{\infty} A_n$ and $\bigcap_{n=1}^{\infty} A_n$.

(b) Compute $\bigcup_{n=1}^{\infty} B_n$ and $\bigcap_{n=1}^{\infty} B_n$.

(c) Compute $\bigcup_{n=1}^{\infty} (A_n \cap B_n)$.

6. (a) Suppose $A_1, A_2, \ldots, A_n, \ldots$ are sets such that $A_n \subseteq A_{n+1}$ for all $n \in \mathbb{Z}_{>0}$. For $i \leq j$, what are the sets $\bigcup_{n=i}^{j} A_n$ and $\bigcap_{n=i}^{j} A_n$?

(b) Suppose $B_1, B_2, \ldots, B_n, \ldots$ are sets such that $B_n \supseteq B_{n+1}$ for all $n \in \mathbb{Z}_{>0}$. For $i \leq j$, what are the sets $\bigcup_{n=i}^{j} B_n$ and $\bigcap_{n=i}^{j} B_n$?

7. Prove part (a) of the Theorem on General Unions and Intersections (monotonicity).

8. (a) Finish the proof of the de Morgan Law in Proof 3.76.

(b) Prove the second de Morgan Law in part (b) of the Theorem on General Unions and Intersections.

9. (a) Prove part (e) of the Theorem on General Unions and Intersections (greatest lower bound property). (b) Prove part (f) of the Theorem on General Unions and Intersections (least upper bound property).

10. Prove the distributive laws for general unions and intersections (part (c) of the Theorem on General Unions and Intersections).

11. Use set-builder notation to describe the following sets:

(a) $(-\infty, 3]$ (b) $[a, b)$ (c) the set of odd integers   (d) $\mathbb{Q}$.

12. Draw pictures of these sets described in set-builder notation:

(a) $\{x \in \mathbb{R} : x > 3 \lor x < -3\}$

(b) $\{(x, y) \in \mathbb{R} \times \mathbb{R} : 0 \leq y \leq 4 - x^2\}$

(c) $\{(x, y, z) \in \mathbb{R}^3 : \max(|x|, |y|, |z|) = 1\}$

(d) $\{(x, y) \in [0, 3] \times [0, 3] : x + y \in \mathbb{Z}\}$

13. Give specific examples of sets $B_n \subseteq \mathbb{R}$ (for each $n \in \mathbb{Z}_{>0}$) such that $B_n \supsetneq B_{n+1}$ for all n and $\bigcap_{n=1}^{\infty} B_n$ is: (a) the empty set; (b) $\{7\}$; (c) $[-1, 1]$; (d) $\mathbb{Z}$.

14. The goal of this problem is to get practice *using known theorems* to help prove new theorems. Fill in all the blanks in the proof below; the reasons for each step should be appropriate parts of the Theorem on Subsets, the Theorem on Set Equality, and the Theorem on General Unions and Intersections (excluding the first identity in part (h), which we are proving).

Theorem. For all $I, J \neq \emptyset$, A_i, A_j, $\left(\bigcup_{i \in I} A_i \right) \cup \left(\bigcup_{j \in J} A_j \right) = \bigcup_{k \in I \cup J} A_k$.

Proof. Fix arbitrary sets $I, J \neq \emptyset$, A_i, and A_j.

<u>Part 1.</u> We must prove $\left(\bigcup_{i \in I} A_i \right) \cup \left(\bigcup_{j \in J} A_j \right) \subseteq \bigcup_{k \in I \cup J} A_k.$

We know $I \subseteq I \cup J$ (by ___(a)___ from the Theorem on Subsets), and so $\bigcup_{i \in I} A_i \subseteq \bigcup_{k \in I \cup J} A_k$ (by ___(b)___ from the Theorem on General Unions and Intersections). Similarly, we know ___(c)___ (by ___(d)___ from the Theorem on Subsets), and so $\bigcup_{j \in J} A_j \subseteq \bigcup_{k \in I \cup J} A_k$ (by ___(e)___ from the Theorem on General Unions and Intersections). Combining these two facts with ___(f)___ from the Theorem on Subsets, part 1 is proved.

<u>Part 2.</u> We must prove $\bigcup_{k \in I \cup J} A_k \subseteq \left(\bigcup_{i \in I} A_i \right) \cup \left(\bigcup_{j \in J} A_j \right).$

By ___(g)___ from the Theorem on General Unions and Intersections, we may instead prove the equivalent statement:

$$\forall k \in I \cup J, A_k \subseteq \left(\bigcup_{i \in I} A_i \right) \cup \left(\bigcup_{j \in J} A_j \right).$$

Fix $k \in I \cup J$, so we know $k \in I$ or $k \in J$. To prove $A_k \subseteq (\bigcup_{i \in I} A_i) \cup (\bigcup_{j \in J} A_j)$, use cases.

Case 1. Assume $k \in I$. First, $A_k \subseteq \bigcup_{i \in I} A_i$ by ___(h)___ .

   Second, $\bigcup_{i \in I} A_i \subseteq (\bigcup_{i \in I} A_i) \cup (\bigcup_{j \in J} A_j)$ by ___(i)___ .

   Therefore, $A_k \subseteq (\bigcup_{i \in I} A_i) \cup (\bigcup_{j \in J} A_j)$ by ___(j)___ .

Case 2. Assume $k \in J$. First, $A_k \subseteq \bigcup_{j \in J} A_j$ by ___(k)___ .

   Second, $\bigcup_{j \in J} A_j \subseteq (\bigcup_{i \in I} A_i) \cup (\bigcup_{j \in J} A_j)$ by ___(l)___ .

   Therefore, $A_k \subseteq (\bigcup_{i \in I} A_i) \cup (\bigcup_{j \in J} A_j)$ by ___(m)___ .

15. Recall that an indexed collection of sets $\{S_i : i \in I\}$ is called *pairwise disjoint* iff for all $i, j \in I$, if $i \neq j$, then $S_i \cap S_j = \emptyset$. Suppose $A_1, A_2, \ldots, A_n, \ldots$ are sets such that $A_n \subseteq A_{n+1}$ for all $n \in \mathbb{Z}_{>0}$. Define $B_1 = A_1$ and $B_n = A_n - A_{n-1}$ for all $n \in \mathbb{Z}_{>1}$. Prove that $\{B_n : n \in \mathbb{Z}_{>0}\}$ is pairwise disjoint, and $\bigcup_{n=1}^{\infty} B_n = \bigcup_{n=1}^{\infty} A_n$. Illustrate with a Venn diagram. [*Hint:* If $x \in \bigcup_{n=1}^{\infty} A_n$, then there exists a *least* positive integer n_0 with $x \in A_{n_0}$. Show that $x \in B_{n_0}$.]

16. Let $\{A_n : n \in \mathbb{Z}_{>0}\}$ be an indexed family of sets. Define $B_1 = A_1$ and, for each $n > 1$, define $B_n = A_n - \bigcup_{j=1}^{n-1} A_j$. Prove that $\{B_n : n \in \mathbb{Z}_{>0}\}$ is pairwise disjoint, and $\bigcup_{n=1}^{\infty} B_n = \bigcup_{n=1}^{\infty} A_n$. Illustrate A_n and B_n for $1 \leq n \leq 4$ in a Venn diagram.

17. Let $\{A_n : n \in \mathbb{Z}_{>0}\}$ be an indexed family of sets. Define $B_n = \bigcup_{j=1}^{n} A_j$ for each $n \in \mathbb{Z}_{>0}$. Prove that $B_n \subseteq B_{n+1}$ for all $n \in \mathbb{Z}_{>0}$, and $\bigcup_{n=1}^{\infty} B_n = \bigcup_{n=1}^{\infty} A_n$.

18. Let $\{C_n : n \in \mathbb{Z}_{>0}\}$ be an indexed family of subsets of a given set X. Define $D_n = \bigcap_{j=1}^{n} C_j$ for each $n \in \mathbb{Z}_{>0}$. Prove that $D_n \supseteq D_{n+1}$ for all $n \in \mathbb{Z}_{>0}$, and $\bigcap_{n=1}^{\infty} D_n = \bigcap_{n=1}^{\infty} C_n$. Do this by using the result of the previous exercise and other known theorems. [*Hint:* Consider $A_n = X - C_n$.]

19. (a) Prove or disprove: for all $I \neq \emptyset$ and all sets $A_i, B_i,$

$$\bigcup_{i \in I} (A_i \cap B_i) \subseteq \left(\bigcup_{j \in I} A_j \right) \cap \left(\bigcup_{k \in I} B_k \right).$$

(b) Prove or disprove: for all $I \neq \emptyset$ and all sets A_i, B_i,

$$\bigcup_{i \in I}(A_i \cap B_i) \supseteq \left(\bigcup_{j \in I} A_j\right) \cap \left(\bigcup_{k \in I} B_k\right).$$

20. A subset U of $\mathbb{R}$ is called an *open set* iff $\forall x \in U, \exists r \in \mathbb{R}_{>0}, (x - r, x + r) \subseteq U$.
    (a) Prove: $\emptyset$ and $\mathbb{R}$ are open sets.
    (b) Prove: every open interval (a, b) is an open set.
    (c) Prove: for all $U, V \subseteq \mathbb{R}$, if U and V are open sets, then $U \cap V$ is an open set.
    (d) Prove: for any indexed collection $\{U_i : i \in I\}$ of subsets of $\mathbb{R}$, if U_i is an open set for all $i \in I$, then $\bigcup_{i \in I} U_i$ is an open set.

21. A subset C of $\mathbb{R}$ is called a *closed set* iff $\mathbb{R} - C$ is an open set (see Exercise 20).
    (a) Prove: $\emptyset$ and $\mathbb{R}$ are closed sets.
    (b) Prove: every closed interval $[a, b]$ is a closed set.
    (c) Prove: $\mathbb{Z}$ is a closed set.
    (d) Find (with proof) a specific subset of $\mathbb{R}$ that is neither an open set nor a closed set.
    (e) Prove: the union of any two closed sets is a closed set.
    (f) Prove: the intersection of any nonempty family of closed sets is a closed set.

22. An *open disk* is a subset of $\mathbb{R}^2$ of the form $\{(x, y) : (x - h)^2 + (y - k)^2 < r^2\}$, where $h, k, r \in \mathbb{R}$ are fixed real numbers with $r > 0$. Express each of the following subsets of $\mathbb{R}^2$ as a union of an indexed collection of open disks.
    (a) $\mathbb{R}^2$ (b) $\{(x, y) \in \mathbb{R}^2 : x > 0 \wedge y < 0\}$ (c) $(-1, 1) \times \mathbb{R}$
    (d) $(0, 5) \times (2, 4)$ (e) $\{(x, y) \in \mathbb{R}^2 : y > |x|\}$.

23. Show that the intersection of two open disks in $\mathbb{R}^2$ is either empty or is the union of an indexed collection of open disks.

24. Let X be a fixed set, and let $\mathcal{A}$ be a collection of subsets of X with the following properties: (i) $\emptyset \in \mathcal{A}$; (ii) for all $S \in A$, $X - S \in \mathcal{A}$; (iii) given $\{S_n : n \in \mathbb{Z}_{>0}\}$ with every $S_n \in \mathcal{A}$, $\bigcup_{n=1}^{\infty} S_n \in \mathcal{A}$. Prove: (a) $X \in \mathcal{A}$; (b) for any $\{S_n : n \in \mathbb{Z}_{>0}\}$ with every $S_n \in \mathcal{A}$, $\bigcap_{n=1}^{\infty} S_n \in \mathcal{A}$; (c) for all $S, T \in \mathcal{A}$, $S \cup T \in \mathcal{A}$; (d) for all $S, T \in \mathcal{A}$, $S \cap T \in \mathcal{A}$.

25. Let $X = \{1, 2, 3\}$. (a) Give three different examples of collections $\mathcal{A}$ of subsets of X satisfying conditions (i) through (iii) in the previous exercise. (b) How many such collections $\mathcal{A}$ are there for this set X?

3.7   Axiomatic Set Theory (Optional)

So far in this chapter, we have defined several operations on sets and learned how to prove theorems involving these set operations. However, there is a subtle logical omission in our presentation of set theory so far. In many instances, we have introduced a new set (such as the union $A \cup B$, or the intersection $A \cap B$, or the power set $\mathcal{P}(A)$) without giving any formal justification for why the new set actually *exists*. Such justifications may seem unnecessary, but (as we will see below) if we indiscriminately assume the existence of certain sets, logical paradoxes can result.

A completely rigorous logical development of set theory (as opposed to the semi-formal treatment given in earlier sections) is called *axiomatic set theory*. In axiomatic set theory, we must use explicitly stated axioms to justify the existence (and uniqueness) of each set under consideration. This optional section gives a sketch of the framework and axioms for one version of axiomatic set theory called the ZFC (Zermelo–Fraenkel–Choice) system. More detailed introductions to axiomatic set theory can be found in the texts listed in the Suggestions for Further Reading at the end of the book.

Undefined Concepts

Axiomatic set theory begins with two undefined notions: the idea of a *set*, and the concept of *membership* in a set. One aspect of the theory that may seem unusual at first is that *all objects under consideration are sets*. You might think that we would need to allow other types of mathematical objects such as numbers, functions, geometric figures, and so forth. Remarkably, all of these objects can be defined as certain kinds of sets. It follows that all variables appearing in the axioms and formulas below represent sets. As above, we use $\in$ as a symbol for the undefined concept of set membership. Thus, for any two objects (sets) x and y, exactly one of the following alternatives holds: $x \in y$ ("x is a member of y"), or $x \notin y$ ("x is not a member of y"). You must learn not to be distracted by the fact that a lowercase letter is used for the set y. We could have equally well used uppercase letters for both these sets, or a mixture of lowercase and uppercase letters.

Russell's Paradox

Our intuition tells us that, for any logical property, there should exist a set whose members are precisely the objects that satisfy the given property. Formalizing this intuition, we are led to propose the following axiom.

3.80. Axiom of Abstraction (Tentative). Let $P(z)$ be a statement with a free variable z, where the variable x does not occur in $P(z)$. **Axiom:** $\boxed{\exists x, \forall z, (z \in x) \Leftrightarrow P(z).}$

This axiom is really a family of axioms, one for each statement $P(z)$. The members of the set x in the axiom are precisely those objects (sets) z for which $P(z)$ is true. In set-builder notation, we could write $x = \{z : P(z)\}$ (although use of this notation suggests the set x is unique, which we have not proved yet).

3.81. Example. Let $P(z)$ be the statement "$z \neq z$." By the Axiom of Abstraction, there is a set x such that for all z, $z \in x$ iff $z \neq z$. By properties of logical equality, $z \neq z$ is false for any z. It follows that for all z, $z \notin x$. So, we have used the axiom to prove the existence of an *empty set* x.

3.82. Example. On the other hand, letting $P(z)$ be the statement "$z = z$," which holds for every z, we obtain a set x such that $z \in x$ for all sets z. This set x is a *universal set* that contains all objects (sets) in the universe being studied. In particular, since x itself is a set, we have $x \in x$ (x is a member of itself).

3.83. Example. Now let $P(z)$ be the statement "$z \notin z$." The Axiom of Abstraction asserts that there is a set x for which the following statement is true:

$$\forall z, (z \in x) \Leftrightarrow (z \notin z).$$

In words, x is the set of all sets z such that z is not a member of itself. In the universal statement just written, we can choose z to be the particular set x, thereby obtaining the proposition

$$(x \in x) \Leftrightarrow (x \notin x).$$

But this proposition is false, since the statements on the two sides of the biconditional symbol must have opposite truth values. So, the Axiom of Abstraction has led us to a contradiction!

The phenomenon in the last example is called *Russell's Paradox* (named after its discoverer, Bertrand Russell). The only way to avoid this paradox is to give up the Axiom of Abstraction, despite its initial intuitive appeal, and replace it with more delicate existence axioms for sets. We describe these axioms below.

The Axiom of Extension

We now state the axioms of the ZFC theory. For each axiom, we give a brief discussion of the intuitive meaning of the axiom and how it is used.

3.84. Axiom of Extension. $\boxed{\forall x, \forall y, [x = y \Leftrightarrow (\forall z, z \in x \Leftrightarrow z \in y)].}$

Intuitively, this axiom indicates the relationship between the logical equality of sets and the undefined concept of set membership. The axiom says that two sets x and y are equal iff they have exactly the same members. Recall that we defined $x \subseteq y$ to mean $\forall z, z \in x \Rightarrow z \in y$. We also saw earlier that an equivalent way to formulate the Axiom of Extension is to say that $x = y$ iff $x \subseteq y$ and $y \subseteq x$. The Axiom of Extension is often used to verify the uniqueness of a newly constructed set, as we will see in examples below.

We should mention here that the axioms of set theory are overlaid on top of certain logical axioms, which we have not listed explicitly. Some of these logical axioms describe the properties of the equality symbol $=$. For example, we might have axioms asserting that equality is reflexive, symmetric, and transitive (although we have seen that these properties can be proved from the Axiom of Extension). Some other axioms are needed to ensure that equal objects can be substituted for one another wherever they appear in a statement. In the context of set theory, it suffices to adopt these two Substitution Axioms:

$$\forall x, \forall y, [x = y \Rightarrow (\forall z, x \in z \Leftrightarrow y \in z)];$$

$$\forall x, \forall y, [x = y \Rightarrow (\forall z, z \in x \Leftrightarrow z \in y)].$$

With these logical axioms already in place, the new content provided by the Axiom of Extension is the converse of the second Substitution Axiom:

$$\forall x, \forall y, [(\forall z, z \in x \Leftrightarrow z \in y) \Rightarrow x = y].$$

Additional substitution properties involving defined concepts such as union and intersection can then be proved as theorems (see part (d) of the Theorem on Set Equality).

The Axiom of Specification

Our first set construction axiom lets us create subsets of given sets, if we can state a condition that determines which objects belong to the subset.

3.85. Axiom of Specification. Let $P(z)$ be a statement with a free variable z, in which the variable x does not occur. **Axiom:** $\boxed{\forall u, \exists x, \forall z, [z \in x \Leftrightarrow (z \in u \land P(z))].}$

Intuitively, given any logical property $P(z)$ and given a particular set u, we can form a new set x consisting of all objects (sets) *belonging to* u that have the property $P(z)$. In set-builder notation, we could write $x = \{z \in U : P(z)\}$. Note how closely this axiom resembles the forbidden Axiom of Abstraction. The key difference is that we must already have a set u given in advance, and then we use the property $P(z)$ to select or "specify" which elements should belong to the new set x (which is automatically a subset of u).

3.86. Example: Binary Intersections. Let us use this axiom to justify the existence of the intersection of two sets. Earlier, we "defined" $A \cap B$ by saying that for all z, $z \in A \cap B$ iff $z \in A$ and $z \in B$. To obtain this set by invoking the Axiom of Specification, we let $P(z)$ be the statement "$z \in B$" and we let u be the set A. The axiom tells us there is a set x such that for all z, $z \in x$ iff $z \in A$ and $z \in B$. We define this set to be the *intersection* of A and B, denoted $A \cap B$. Before introducing this notation for the intersection, we should also check that the set x (whose *existence* is guaranteed by the Axiom of Specification) is *uniquely* determined by the given sets A and B. This follows from the Axiom of Extension. For suppose x' is another set such that for all z, $z \in x'$ iff $z \in A$ and $z \in B$. We deduce that for all z, $z \in x$ iff $z \in x'$, hence $x = x'$.

Similarly, whenever we use the Axiom of Specification to create a new set x, this set is uniquely determined by $P(z)$ and u. For if another set x' satisfies the conditions in the axiom, then for all z, $z \in x'$ iff ($z \in u$ and $P(z)$) iff $z \in x$. So $x = x'$ by the Axiom of Extension. It is now safe to use the notation $x = \{z \in u : P(z)\}$ to denote the set of all z such that $z \in u$ and $P(z)$. On the other hand, the related notation $\{z : P(z)\}$ must **not** be used.

3.87. Example: Set Difference. Given any sets A and B, we previously "defined" $A - B$ by declaring that for all z, $z \in A - B$ iff $z \in A$ and $z \notin B$. The Axioms of Specification and Extension justify this definition by proving the existence and uniqueness of the set denoted $A - B$. This follows by writing $A - B = \{z \in A : z \notin B\}$.

Let us see what happens if we try to recreate Russell's Paradox using the Axiom of Specification. Let u be a fixed set, and take $P(z)$ to be the statement "$z \notin z$." Our axiom provides a set x satisfying

$$\forall z, [z \in x \Leftrightarrow (z \in u \land z \notin z)].$$

Taking $z = x$ here leads to the statement

$$x \in x \Leftrightarrow (x \in u \land x \notin x).$$

If $x \in u$, the statement just written must be false, as in the original discussion of Russell's Paradox. But now we can escape the contradiction by concluding that $x \notin u$. Since u was a perfectly arbitrary set, we have converted Russell's Paradox into the following theorem.

3.88. No-Universe Theorem. For every set u, there exists a set x with $x \notin u$.

This theorem tells us that, in the ZFC theory, there does not exist a "universal set" that contains every set. Note also that if such a set u did exist, the Axiom of Specification (using this u) would reduce to the Axiom of Abstraction.

The Axioms of Pairs, Power Sets, and Unions

Our next three axioms allow us to combine several sets to produce a larger set.

3.89. Axiom of Pairs. $\boxed{\forall a, \forall b, \exists x, \forall z, [z \in x \Leftrightarrow (z = a \vee z = b)].}$

Informally, given any two sets (objects) a and b, there exists a set x whose members are precisely a and b. The Axiom of Extension shows that this set x is uniquely determined by a and b. So we can introduce the notation $x = \{a, b\}$; as before, we call this set the *unordered pair* with members a and b. Observe that we can take $a = b$ here, in which case we obtain the singleton set $\{a\}$ whose only member is a. By using the Axiom of Pairs on the sets $\{a\}$ and $\{a, b\}$, we obtain the set $\{\{a\}, \{a, b\}\}$, which was our formal definition of the *ordered pair* (a, b) in Definition 3.66. In §3.5, we used this definition to prove that for all objects (sets) a, b, c, d, $(a, b) = (c, d)$ iff $a = c$ and $b = d$.

3.90. Axiom of Power Sets. $\boxed{\forall a, \exists x, \forall z, [z \in x \Leftrightarrow z \subseteq a].}$

Informally, given any set a, there exists a set x such that the *members* of x are precisely the *subsets* of a. By the Axiom of Extension, the set x is uniquely determined by this condition. As before, we write $x = \mathcal{P}(a)$ and call x the *power set of a*.

3.91. Axiom of Unions. $\boxed{\forall a, \exists x, \forall z, [z \in x \Leftrightarrow \exists s \in a, z \in s].}$

Informally, given any set a, there exists a set x such that the members of x are precisely the members of the members of a. By the Axiom of Extension, the set x is uniquely determined by this condition. We write $\boxed{x = \bigcup a}$ and call x the *union of the sets in a*. The following examples may clarify what is going on here.

3.92. Example. Suppose we have a set $a = \{b, c, d, e\}$. Applying the Axiom of Unions to this set, we get a set $x = \bigcup a$ such that for all z, $z \in x$ iff z is a member of some member of a. As the members of a are b, c, d, and e, we see that $z \in x$ iff $z \in b$ or $z \in c$ or $z \in d$ or $z \in e$. Using informal notation from earlier in the chapter, this says that $z \in x$ iff $z \in b \cup c \cup d \cup e$.

3.93. Example: Binary Unions. Given any two sets B and C, define $B \cup C = \bigcup \{B, C\}$. In more detail, we use the Axiom of Pairs to obtain the unordered pair $a = \{B, C\}$, and then use the Axiom of Unions to obtain the set $\bigcup a$. As in the preceding example, it follows that for all z, $z \in B \cup C$ iff $z \in B$ or $z \in C$. Iterating this construction, we can define the union of three sets by letting $B \cup C \cup D = (B \cup C) \cup D$. Similarly, we can define the union of four sets, five sets, or any specific finite number of sets by repeated use of the binary union operation.

3.94. Example. Suppose $a = \{B_0, B_1, B_2, \ldots, B_k, \ldots\} = \{B_k : k \in \mathbb{Z}_{\geq 0}\}$. (This is an informal example, since we have not formally discussed $\mathbb{Z}_{\geq 0}$ yet or defined what is meant by $\{B_k : k \in \mathbb{Z}_{\geq 0}\}$.) Here, $x = \bigcup a$ consists of all z such that $z \in B_k$ for some $k \in \mathbb{Z}_{\geq 0}$. This is the set that we denoted earlier by $\bigcup_{k=0}^{\infty} B_k$.

3.95. Example: Product Sets. Given any two sets A and B, we can now justify the definition of the *product set* $A \times B$, which we defined earlier by the condition

$$\forall z, [z \in A \times B \Leftrightarrow \exists a \in A, \exists b \in B, z = (a, b)].$$

By the Axiom of Extension, there is at most one set satisfying this condition. But, how can we prove that $A \times B$ *exists*? To begin, recall that $(a, b) = \{\{a\}, \{a, b\}\}$. Given $a \in A$ and $b \in B$, we know $\{a\}$ and $\{a, b\}$ are subsets of $A \cup B$, hence $\{a\}$ and $\{a, b\}$ are members of

$\mathcal{P}(A \cup B)$. Then (a, b), which is a set of two members of $\mathcal{P}(A \cup B)$, is a subset of $\mathcal{P}(A \cup B)$ and hence a member of $\mathcal{P}(\mathcal{P}(A \cup B))$. Finally, $A \times B$ is a set of ordered pairs, each of which is a member of $\mathcal{P}(\mathcal{P}(A \cup B))$, so that $A \times B$ is a certain subset of $\mathcal{P}(\mathcal{P}(A \cup B))$. To be precise,

$$A \times B = \{\, z \in \mathcal{P}(\mathcal{P}(A \cup B)) : \exists a \in A, \exists b \in B, z = \{\{a\}, \{a, b\}\} \,\}.$$

We have seen that $A \cup B$ exists using the Axiom of Pairs and the Axiom of Unions. Using the Axiom of Power Sets twice, we see that $\mathcal{P}(\mathcal{P}(A \cup B))$ exists. Finally, the displayed formula and the Axiom of Specification shows that the set $A \times B$ exists. We can iterate this construction to define the product of three sets, of four sets, and so on; for instance, $A \times B \times C = (A \times B) \times C$.

The Axioms of $\emptyset$ and Infinity

Based on the axioms stated so far, it is impossible to prove that there exists any set whatsoever! The reason is that each set-construction axiom can only be invoked if we already have at least one set that is already known to exist (such as the set u in the Axiom of Specification). The following axiom gets us started by postulating the existence of the empty set.

3.96. Axiom of the Empty Set. $\boxed{\exists x, \forall z, z \notin x.}$

Informally, there is a set x having no members. By the Axiom of Extension, if another set x' satisfies $\forall z, z \notin x'$, then $x = x'$. So we can denote the unique set x with no members by $\emptyset$, and we can call x *the empty set*.

Once we have the empty set, we can start applying the other set-construction axioms to build more and more new sets. Intuitively, we can get infinitely many distinct sets in this manner, but we claim that each individual set that we build is still finite. Reasoning informally (as we have not yet rigorously defined "finite"), the preceding claim holds since we can only obtain a given set by finitely many applications of the axioms, and each axiom stated so far produces a finite output set when applied to finite input sets. Thus, with the apparatus developed so far, there is no way to justify the existence of an infinite set such as $\mathbb{Z}$ or $\mathbb{R}$. The next axiom resolves this difficulty.

3.97. Axiom of Infinity. $\boxed{\exists x, [\emptyset \in x \wedge \forall z, (z \in x \Rightarrow z \cup \{z\} \in x)].}$

Although it may not be evident from an initial reading of the axiom, this axiom guarantees the existence of an *infinite* set x. In fact, the implication appearing in the axiom will assure that all the natural numbers $\{0, 1, 2, \ldots\}$ (to be defined later as certain sets) appear as members of x. To give a hint of what is going on here, we mention that the natural number 0 is defined to be the empty set $\emptyset$, and for a natural number z (which is a certain type of set), $z + 1$ is defined to be the set $z \cup \{z\}$. Then the axiom is saying (among other things) that $0 \in x$, and for any natural number $z \in x$, $z + 1$ also belongs to x. It should then be intuitively believable that $\mathbb{Z}_{\geq 0} \subseteq x$ (compare to the discussion of mathematical induction in the next chapter), so that the set x is infinite.

The Axiom of Replacement

Although our discussion of the Axiom of Infinity and $\mathbb{Z}_{\geq 0}$ is not yet complete, let us suppose for a moment that we have constructed the set $\mathbb{Z}_{\geq 0} = \{0, 1, 2, 3, \ldots\}$ within axiomatic set theory. Returning to a previous example, we could now hope to give meaning to the indexed union $\bigcup_{k=0}^{\infty} B_k$, where $B_0, B_1, B_2, \ldots$ are sets indexed by the nonnegative integers. We would like to use the Axiom of Unions to define $\bigcup_{k=0}^{\infty} B_k = \bigcup \{B_k : k \in \mathbb{Z}_{\geq 0}\}$. In order to do this, we need to justify the existence of the *indexed set* $\{B_k : k \in \mathbb{Z}_{\geq 0}\}$. We can use the next axiom to accomplish this task.

3.98. Axiom of Replacement. Let X be a given set, and let $P(x, z)$ be a given statement with free variables x and z. Suppose that for each $x \in X$, there exists a set y (depending on x) such that for all z, $z \in y$ iff $P(x, z)$. By the Axiom of Extension, y is uniquely determined by x, so we denote y as $y(x)$. Then there exists a set Y such that for all w, $w \in Y$ iff there exists $x \in X$ with $w = y(x)$.

Intuitively, the purpose of this axiom is to let us apply a "function" to all members of a given set X to create a new set Y. In this case, a "function" is *not* a set of ordered pairs (as discussed in later chapters), but rather a logical statement that tells us how each input set $x \in X$ is related to the corresponding output set $y \in Y$. In more detail, our "function" F with domain X is defined by letting $F(x) = y(x) = \{z : P(x, z) \text{ is true}\}$ for each $x \in X$. Part of the hypothesis of the Axiom of Replacement is that each output $y(x)$, as x varies over X, is already known to be a set. Given that this hypothesis holds, the axiom asserts the existence of a new *set* $Y = \{F(x) : x \in X\} = \{y(x) : x \in X\}$, which is obtained by "applying F to each member of X" and collecting together all the resulting outputs into a new set. In other words, we obtain a new set Y from the set X by *replacing* each $x \in X$ by the set $F(x)$.

3.99. Example. This example assumes we have already constructed the sets $\mathbb{Z}_{\geq 0}$ and $\mathbb{R}$ within axiomatic set theory. Let $P(x, z)$ be the statement $(x \in \mathbb{Z}_{\geq 0}) \wedge (z \in \mathbb{R}) \wedge (-2x \leq z \leq x)$. By the Axiom of Specification, for each $x \in \mathbb{Z}_{\geq 0}$, the set $y(x) = \{z : P(x, z)\} = \{z \in \mathbb{R} : -2x \leq z \leq x\}$ exists. So, the Axiom of Replacement transforms the known set $\mathbb{Z}_{\geq 0}$ into the set of closed intervals $Y = \{[-2x, x] : x \in \mathbb{Z}_{\geq 0}\}$. We should point out that, for this particular set Y, we can construct Y from other axioms without invoking the Axiom of Replacement.

Generalizing the last example, if we have an index set I and a formula that characterizes a unique set B_i for each $i \in I$, then the Axiom of Replacement converts the index set I into the indexed set $\{B_i : i \in I\}$. Then, for instance, we could use the Axiom of Unions to obtain the set $\bigcup_{i \in I} B_i$. Also, provided I is nonempty, we could use the Axiom of Specification to obtain $\bigcap_{i \in I} B_i$.

The Axiom of Choice

Suppose X is a set of nonempty sets. Then each $x \in X$ is a nonempty set, so there exists at least one object (set) y with $y \in x$. The y appearing here depends on x, and y is usually not unique. In many parts of mathematics, it is technically helpful to have access to a set G of ordered pairs with the property that for each $x \in X$, there is exactly one ordered pair $(x, y) \in G$ with first component x, and for all $(x, y) \in G$ we have $y \in x$. In order to prove the existence of such a set G, we need a new axiom, called the *Axiom of Choice*. The name arises because the set G represents a simultaneous choice of one element y from each set x in the given collection of sets X.

3.100. Axiom of Choice. $\boxed{\forall X, [\emptyset \notin X \Rightarrow \exists F, \forall x \in X, \exists! \, y \in x, (x, y) \in F].}$

The set F provided by the axiom may have "too many" elements. For example, F could have members that are not ordered pairs, or members of the form (a, b) with $a \notin X$, or members (x, z) with $x \in X$ and $z \notin x$. However, using the Axiom of Specification, we can shrink F to a set G having the properties stated above (see Exercise 15).

The Axiom of Choice may seem intuitively obvious, since we can readily imagine forming a collection of ordered pairs (x, y) by considering each nonempty set x in turn and arbitrarily choosing some member $y \in x$. However, being intuitively convinced that a set ought to exist is not the same as being able to prove that such a set exists based on explicitly stated axioms. In some special situations, we can avoid the Axiom of Choice, by using a *specific* choice rule (in the form of a logical statement) together with the Axiom of Specification to build the set of ordered pairs G. For example, if every $x \in X$ is a nonempty subset of $\mathbb{Z}_{\geq 0}$, we could define

$$G = \{z \in X \times \mathbb{Z}_{\geq 0} : \exists x \in X, \exists y \in \mathbb{Z}_{\geq 0}, z = (x, y) \wedge y \in x \wedge \forall w \in x, y \leq w\}.$$

This definition works because of the fact (discussed in more detail later) that every nonempty subset of $\mathbb{Z}_{\geq 0}$ has a unique least element.

It is a famous and difficult result of axiomatic set theory that the Axiom of Choice is *independent* of the other axioms in the ZFC theory, assuming that those axioms are themselves consistent. This means that we could add either the Axiom of Choice, or the negation of this axiom, to the other axioms without introducing new contradictions. In particular, it is impossible to *prove* the Axiom of Choice (or its negation) as a consequence of the other axioms, unless the other axioms are inconsistent (in which case every statement could be proved from those axioms, and the entire theory is worthless).

There are many logically equivalent versions of the Axiom of Choice, some of which are discussed later in this text. Two such versions are: the product of any family of nonempty sets is nonempty; and for any set X, there exists a well-ordering $\leq$ on X. Many key theorems in various areas of mathematics require the Axiom of Choice in their proofs. Some of these theorems include: the union of countably many countable sets if countable (proved in §7.3 when we discuss cardinality); every vector space has a basis (proved in linear algebra); the product of any collection of compact topological spaces is compact (proved in topology); every commutative ring with identity has a maximal ideal (proved in abstract algebra); every consistent set of sentences has a model (proved in formal logic); and not every subset of $\mathbb{R}$ is a Lebesgue-measurable set (proved in real analysis). Most mathematicians today accept the Axiom of Choice in order to have access to these and many other theorems that depend on this axiom.

The Axiom of Foundation

The final axiom of the ZFC theory is the following odd-looking statement.

3.101. Axiom of Foundation. $\boxed{\forall a, [a \neq \emptyset \Rightarrow (\exists b, b \in a \wedge a \cap b = \emptyset)].}$

In words, for any nonempty set a, a has a member b that has empty intersection with a. The hypothesis $a \neq \emptyset$ is not surprising, since the empty set does not have any members. Next we give some intuition for what the conclusion means.

We begin with an informal example involving the set $\mathbb{Z}_{\geq 0} = \{0, 1, 2, 3, \dots\}$. It is intuitively evident that there does not exist an infinite strictly decreasing sequence of nonnegative integers

$$a_0 > a_1 > a_2 > \cdots > a_k > \cdots.$$

(We discuss this point in more detail in later chapters.) Roughly speaking, the goal of the Axiom of Foundation is to rule out analogous strictly decreasing sequences in the world of sets. In this setting, the concept $b < c$ for integers is replaced by the analogous concept $x \in y$ for sets. So, we are trying to rule out the existence of an infinite sequence of sets

$x_0, x_1, x_2, x_3, \ldots$ such that $\cdots \in x_3 \in x_2 \in x_1 \in x_0$, i.e., such that $x_{k+1} \in x_k$ for all $k \in \mathbb{Z}_{\geq 0}$. As special cases of this, we want to forbid a set x such that $x \in x$ (which would lead to the infinite sequence $\cdots x \in x \in x \in x \in x$), as well as a pair of sets x, y such that $x \in y$ and $y \in x$ (which would lead to the sequence $\cdots x \in y \in x \in y \in x \in y$), and so on.

Thus, although the Axiom of Foundation is phrased as an existence statement, its true goal is to force the *non-existence* of certain pathological combinations of set membership relationships. We now sketch how this goal is achieved. Suppose (to get a contradiction) that there did exist specific sets x_k with $x_{k+1} \in x_k$ for all $k \in \mathbb{Z}_{\geq 0}$. By the Axiom of Replacement, we can form the set $a = \{x_k : k \in \mathbb{Z}_{\geq 0}\}$. The set a is nonempty, since $x_0 \in a$, so the Axiom of Foundation provides a set $b \in a$ with $a \cap b = \emptyset$. Now, $b = x_k$ for some $k \in \mathbb{Z}_{\geq 0}$. We deduce $x_{k+1} \in b$ and $x_{k+1} \in a$, so that $x_{k+1} \in a \cap b = \emptyset$, which is a contradiction.

The Axiom of Foundation is also called the *Axiom of Regularity*.

Section Summary

All of mathematics can be formalized within set theory. One development of set theory is based on the following axioms.

3.102. Axioms of ZFC Set Theory.

(a) *Axiom of Extension:* $\forall x, \forall y, [x = y \Leftrightarrow (\forall z, z \in x \Leftrightarrow z \in y)]$.

(b) *Axiom of Specification:* For all open sentences $P(z)$,
$\forall u, \exists x, \forall z, [z \in x \Leftrightarrow (z \in u \wedge P(z))]$.

(c) *Axiom of Pairs:* $\forall a, \forall b, \exists x, \forall z, [z \in x \Leftrightarrow (z = a \vee z = b)]$.

(d) *Axiom of Power Sets:* $\forall a, \exists x, \forall z, [z \in x \Leftrightarrow z \subseteq a]$.

(e) *Axiom of Unions:* $\forall a, \exists x, \forall z, [z \in x \Leftrightarrow \exists s \in a, z \in s]$.

(f) *Axiom of the Empty Set:* $\exists x, \forall z, z \notin x$.

(g) *Axiom of Infinity:* $\exists x, [\emptyset \in x \wedge \forall z, (z \in x \Rightarrow z \cup \{z\} \in x)]$.

(h) *Axiom of Replacement:* For all open sentences $P(x, z)$ and all sets X,
$[\forall x \in X, \exists y, \forall z, (z \in y \Leftrightarrow P(x, z))] \Rightarrow \exists Y, \forall w, [w \in Y \Leftrightarrow \exists x \in X, \forall z, (z \in w \Leftrightarrow P(x, z))]$.

(i) *Axiom of Choice:* $\forall X, [\emptyset \notin X \Rightarrow \exists F, \forall x \in X, \exists! y \in x, (x, y) \in F]$.

(j) *Axiom of Foundation:* $\forall a, [a \neq \emptyset \Rightarrow (\exists b, b \in a \wedge a \cap b = \emptyset)]$.

These axioms can be used to prove the *existence* and *uniqueness* of sets such as $\emptyset$, $\mathbb{Z}_{\geq 0}$, $A \cup B$, $A \cap B$, $A - B$, $\mathcal{P}(A)$, $A \times B$, $\bigcup_{i \in I} A_i$, $\bigcap_{i \in I} A_i$, and so on. The *No-Universe Theorem* states that for every set u, there exists some set $x \notin u$. The Axiom of Foundation rules out infinite descending chains of sets belonging to other sets: i.e., there do not exist sets x_k (for $k \in \mathbb{Z}_{\geq 0}$) with $x_{k+1} \in x_k$ for all $k \geq 0$.

Exercises

1. Prove or disprove: $\forall x, \forall y, [\forall z, x \in z \Leftrightarrow y \in z] \Rightarrow x = y$.

2. Use the axioms to prove: for all sets a, b, c, there exists a unique set x such that for all z, $z \in x$ iff $z = a$ or $z = b$ or $z = c$. (The set x is the unordered triple $\{a, b, c\}$.)

3. Use the axioms to prove: for all sets a, b, c, d, e, there exists a unique set x such that for all z, $z \in x$ iff $z = a$ or $z = b$ or $z = c$ or $z = d$ or $z = e$.

4. Assuming that $\mathbb{R}$ and the order relation $<$ on $\mathbb{R}$ have already been constructed, use the axioms to prove the existence and uniqueness of the intervals (a, b), $[a, b]$, $(a, b]$, and $[a, b)$ for all real numbers $a < b$.

5. Assuming that $\mathbb{Z}$ and the arithmetic operations and order relation on $\mathbb{Z}$ have already been constructed, use the axioms to prove the existence and uniqueness of the following subsets of $\mathbb{Z}$: (a) $\mathbb{Z}_{<0}$; (b) the set of odd integers; (c) the set of prime integers (these are integers $p > 1$ such that the only positive integers dividing p are 1 and p).

6. Assuming that $\mathbb{R}$ and $\mathbb{Z}_{\geq 0}$ have already been constructed, prove the existence of the set $Y = \{[-2x, x] : x \in \mathbb{Z}_{\geq 0}\}$ without using the Axiom of Replacement.

7. Show that the Axiom of the Empty Set follows from the alternate axiom $\exists x, x = x$ and the other axioms of ZFC set theory.

8. (a) Show that the Axiom of Pairs (as stated in this section) follows from the alternate version $\forall a, \forall b, \exists x, \forall z, [(z = a \vee z = b) \Rightarrow z \in x]$ along with the other axioms of ZFC set theory. (b) State and prove similar results for the Axiom of Power Sets and the Axiom of Unions.

9. Compute: (a) $\bigcup \emptyset$ (b) $\bigcup \{\emptyset\}$ (c) $\bigcup \{\emptyset, \{\emptyset\}, \{\{\emptyset\}\}\}$ (d) $\bigcup \mathcal{P}(S)$ (e) $\bigcup (A \times B)$.

10. For $A \neq \emptyset$, informally define the set $\bigcap A$ by the condition: for all objects z, $z \in \bigcap A$ iff for all $y \in A$, $z \in y$. (a) Use the axioms to justify the existence and uniqueness of the set $\bigcap A$. (b) In the case $A = \{B, C\}$, prove that $\bigcap A = B \cap C$. (c) Explain carefully why our informal definition cannot be used to define the set $\bigcap \emptyset$. In contrast, what is $\bigcap \{\emptyset\}$?

11. Prove or disprove each variation of the Axiom of Foundation.
    (a) $\forall a, \exists b, b \in a \wedge a \cap b = \emptyset$.
    (b) $\forall a, [a \neq \emptyset \Rightarrow \exists b, a \cap b = \emptyset]$.
    (c) $\forall a, [a \neq \emptyset \Rightarrow \exists b, b \subseteq a \wedge a \cap b = \emptyset]$.

12. (a) Use the Axiom of Foundation to prove $\forall x, x \notin x$ without referencing $\mathbb{Z}_{\geq 0}$.
    (b) Similarly, prove $\sim \exists x, \exists y, (x \in y \wedge y \in x)$.

13. Suppose we try to define ordered pairs by letting $(a, b) = \{a, \{a, b\}\}$ for all objects a and b. Prove or disprove the Ordered Pair Axiom using this definition of (a, b).

14. Suppose we define the ordered pair (a, b) to be the set $\{\{\emptyset, a\}, \{\{\emptyset\}, b\}\}$ (compare to Exercise 21 in §3.5). (a) Use the axioms to justify the existence and uniqueness of (a, b) for all sets a and b. (b) Use the axioms to justify the existence and uniqueness of $A \times B$ for all sets A and B.

15. In the discussion following the statement of the Axiom of Choice, show how to use the Axiom of Specification to convert the set F in the axiom to the set G described prior to the axiom.

16. Rewrite the Axiom of Infinity without using the defined symbols $\emptyset$, $\cup$, and $\{z\}$; use only logical operators, variables, $=$, and $\in$.

17. Show how the Axiom of the Empty Set can be deduced from the Axiom of Infinity and other axioms.

18. Define $0 = \emptyset$, $1 = \{0\}$, and $2 = \{0, 1\}$. Prove the existence of the set $\mathcal{P}(\{0, 1, 2\})$ without using the Axiom of Power Sets.

19. (a) Without using the Axiom of Pairs, prove the existence of a set S such that there exist exactly two objects a with $a \in S$. (b) Use (a) to show that the Axiom of Pairs can be proved from the Axiom of Replacement.

20. Consider a universe in which there exists only one object (denoted 0), and suppose $0 \in 0$ is false. Assuming all quantified variables range over only the objects in this universe, which of the axioms of ZFC set theory are true?

4

Integers

4.1 Recursive Definitions and Proofs by Induction

We turn now from set theory to a closer examination of facts involving integers, including proofs by induction, integer division with remainder, prime numbers, and the theorem that every integer has a unique factorization into a product of primes. Proofs by induction are often needed when dealing with concepts defined recursively. So we begin by discussing recursive definitions.

Recursive Definitions

Let n be a fixed positive integer, and let $x, a_1, a_2, \ldots, a_n$ be fixed real numbers. Consider the following informal definitions.

(a) *Factorials:* Define $n! = n \cdot (n-1) \cdot (n-2) \cdot \ldots \cdot 3 \cdot 2 \cdot 1$.

(b) *Summation Notation:* Define $\sum_{k=1}^{n} a_k = a_1 + a_2 + \cdots + a_n$.

(c) *Product Notation:* Define $\prod_{k=1}^{n} a_k = a_1 \cdot a_2 \cdot \ldots \cdot a_n$.

(d) *Exponent Notation:* Define $x^n = x \cdot x \cdot \ldots \cdot x$, where the right side has n copies of x.

These definitions may seem perfectly legitimate, but they all contain a subtle flaw. Each definition uses the *ellipsis* symbol "$\ldots$" to represent missing terms in the middle of each formula. The reader is supposed to know what terms are intended here by following the pattern of the terms that are present. But, in a formal development of these concepts, we cannot omit information in this way. One solution is to rephrase these definitions *recursively*. We introduce the idea of a recursive definition with the following formal definition of $n!$.

4.1. Recursive Definition of Factorials. Define $\boxed{0! = 1}$.

For each $n \in \mathbb{Z}_{\geq 0}$, recursively define $\boxed{(n+1)! = (n+1) \cdot n!}$.

This recursive definition begins with an *initial condition*, which provides an ordinary (non-recursive) definition for $0!$. We are told that $0! = 1$. Next, the *recursive part* of the definition gives us a formula for computing $(n+1)!$ that assumes we already know the value of $n!$. For instance, taking $n = 0$ in the recursive formula, we see that $1! = (0+1)! = (0+1) \cdot 0! = 1 \cdot 1 = 1$. Next, taking $n = 1$ in the recursive formula, we get $2! = (1+1)! = (1+1) \cdot 1! = 2 \cdot 1 = 2$. Continuing similarly, we see that $3! = 3 \cdot 2! = 3 \cdot 2 = 6$, $4! = 4 \cdot 3! = 4 \cdot 6 = 24$, $5! = 5 \cdot 4! = 5 \cdot 24 = 120$, and so on. We can also *iterate* the recursive definition to recover the original informal definition; for instance,

$$4! = 4 \cdot 3! = 4 \cdot 3 \cdot 2! = 4 \cdot 3 \cdot 2 \cdot 1! = 4 \cdot 3 \cdot 2 \cdot 1 \cdot 0! = 4 \cdot 3 \cdot 2 \cdot 1 \cdot 1.$$

A basic recursive definition introduces a sequence of quantities $f(0), f(1), \ldots, f(n), \ldots$ by specifying what $f(0)$ is, and then telling us (for each $n \geq 0$) how to compute $f(n+1)$

if we already know the value of $f(n)$. In a more elaborate recursive definition, $f(n+1)$ might depend on any of the values $f(0), f(1), \ldots, f(n)$ preceding it in the sequence, not just $f(n)$. The sequence might start at $n = 1$ instead of $n = 0$, or at any other fixed integer. We illustrate with a few more recursive definitions. As always, **memorize all of these definitions!**

4.2. Definition: Exponent Notation. Let x be a fixed real number. Define $\boxed{x^0 = 1}$. For all $n \in \mathbb{Z}_{\geq 0}$, define $\boxed{x^{n+1} = x^n \cdot x}$.

For example, $x^3 = x^2 \cdot x = x^1 \cdot x \cdot x = x^0 \cdot x \cdot x \cdot x = 1 \cdot x \cdot x \cdot x = x \cdot x \cdot x$, as expected.

4.3. Definition: Summation Notation. Suppose we are given a real number a_n for each $n \in \mathbb{Z}_{\geq 1}$. We define $\boxed{\sum_{k=1}^{1} a_k = a_1}$. For all $n \in \mathbb{Z}_{\geq 1}$, define $\boxed{\sum_{k=1}^{n+1} a_k = \left(\sum_{k=1}^{n} a_k \right) + a_{n+1}}$.

Intuitively, the sum of the first $n + 1$ terms is found by taking the sum of the first n terms and adding a_{n+1}. We can introduce variations of this definition for sums starting at index zero, such as $\sum_{k=0}^{n} a_k$, or sums starting at any other integer. By convention, for all integers $j < i$, we let $\sum_{k=i}^{j} a_k = 0$. The *summation index* k can be replaced by any other letter that does not already have a meaning. For instance, $\sum_{k=1}^{n} a_k = \sum_{j=1}^{n} a_j = \sum_{s=1}^{n} a_s$. But, we could not write $\sum_{n=1}^{n} a_n$, since n is already being used as the upper limit of the sum.

4.4. Definition: Product Notation. Suppose we are given a real number a_n for each $n \in \mathbb{Z}_{\geq 1}$. We define $\boxed{\prod_{k=1}^{1} a_k = a_1}$. For all $n \in \mathbb{Z}_{\geq 1}$, define $\boxed{\prod_{k=1}^{n+1} a_k = \left(\prod_{k=1}^{n} a_k \right) \cdot a_{n+1}}$.

We could have defined powers and factorials using product notation. For any real x and positive integer n, you can check that $x^n = \prod_{k=1}^{n} x$ and $n! = \prod_{j=1}^{n} j$.

The Induction Axiom

In §2.1, we listed axioms describing some algebraic properties of $\mathbb{Z}$ (the set of integers). To prove facts about recursively defined concepts, we need a new axiom giving a fundamental property of the set $\mathbb{Z}_{\geq 1} = \{1, 2, 3, \ldots\}$ of positive integers.

4.5. Induction Axiom for $\mathbb{Z}_{\geq 1}$.
For all open sentences $P(n)$, the following two statements are equivalent:

(a) $\forall n \in \mathbb{Z}_{\geq 1}, P(n)$.

(b) $P(1) \wedge \forall n \in \mathbb{Z}_{\geq 1}, (P(n) \Rightarrow P(n+1))$.

Recall that the axioms of a mathematical theory do not need to be proved. Nevertheless, in the case of the Induction Axiom, we can provide an intuitive justification for the axiom that helps us understand what the axiom is saying. Let $P(n)$ be a fixed open sentence. On one hand, it is a routine exercise using the proof templates to prove that statement (a) in the Induction Axiom implies statement (b). The real content of the axiom is the assertion that statement (b) implies statement (a). Suppose we know that statement (b) is true; why should we believe that statement (a) must follow? By using the Inference Rule for ALL (taking $n = 1, 2, 3, \ldots$ in statement (b)), we see that statement (b) contains the following information:

(1) $P(1)$ is true.
(2) If $P(1)$ is true, then $P(2)$ is true.
(3) If $P(2)$ is true, then $P(3)$ is true.
(4) If $P(3)$ is true, then $P(4)$ is true.
(5) If $P(4)$ is true, then $P(5)$ is true.
... and so on ...

Now, by (1) and (2) and the Inference Rule for IF, we deduce that $P(2)$ is true. It then follows from (3) and the Inference Rule for IF that $P(3)$ is true. It then follows from (4) and the Inference Rule for IF that $P(4)$ is true. It then follows from (5) and the Inference Rule for IF that $P(5)$ is true. Continuing this chain of reasoning forever, we see that $P(n)$ is indeed true for every positive integer n, as asserted in statement (a).

This argument can be informally visualized as follows. We imagine setting up an infinite row of dominos standing on edge, with one domino for each positive integer n.

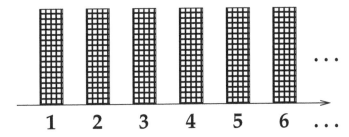

Let $P(n)$ be the statement "domino n falls down." Statement (b) in the Induction Axiom says that domino 1 falls down; and for every positive integer n, if domino n falls down then domino $n + 1$ falls down. The axiom itself tells us that this information is enough to conclude that all the dominos fall down. The argument in the previous paragraph spells out in more detail how the cascade of falling dominos occurs. First domino 1 falls down, which causes domino 2 to fall, which causes domino 3 to fall, and so on.

We hope that this intuitive argument convinces the reader of the truth of the Induction Axiom. The argument is not a *proof* of the Induction Axiom, however. This is because we would need to use the Inference Rule for IF infinitely many times to finish proving the universal statement $\forall n \in \mathbb{Z}_{\geq 1}, P(n)$, and proofs are required to be finite in length. In a similar vein, we can think of the Induction Axiom as indirectly characterizing what the positive integers *are*: they are precisely the objects we get by starting with 1 and then adding 1 again and again, forever.

Now that we have intuition for why the Induction Axiom is true, let us see how to use this axiom in proofs. The key idea is that if we need to prove statement (a) in the axiom (a universal statement about all positive integers), it is sufficient to instead prove the equivalent statement (b). We can use the proof templates for AND, $\forall$, and IF to generate the outline of a proof of statement (b). This leads to the following template for *ordinary induction proofs*, which should be carefully memorized.

4.6. Template for Ordinary Induction Proofs. To prove $\forall n \in \mathbb{Z}_{\geq 1}, P(n)$:
Say "We use induction on n."
<u>Part 1.</u> **Prove** $P(1)$.
<u>Part 2.</u> **Fix** an arbitrary integer $n_0 \in \mathbb{Z}_{\geq 1}$. **Assume** $P(n_0)$ is true. **Prove** $P(n_0 + 1)$ is true.

The two parts of this template arise since statement (b) in the Induction Axiom is an AND-statement. In part 2, we prove $\forall n \in \mathbb{Z}_{\geq 1}, (P(n) \Rightarrow P(n + 1))$ by combining the generic-element proof template with the direct proof template for IF-statements. Part 1 of an induction proof is called the *base case*. Part 2 is called the *induction step*. The assumption $P(n_0)$ in part 2 is called the *induction hypothesis*.

Returning to statement (a) and statement (b) in the Induction Axiom, we can now see that statement (b) will likely be easier to prove than statement (a), even though statement (b) appears to be more complex. The reason is that we are allowed to *assume certain information* in the course of proving statement (b) (namely, the induction hypothesis $P(n_0)$) to help us prove the target statement $P(n_0 + 1)$. We write n_0 here to emphasize that n_0 is a *fixed* integer (though in many proofs below, we drop the subscript to reduce clutter). We are not using circular reasoning when using $P(n_0)$ to prove $P(n_0 + 1)$, since n_0 and $n_0 + 1$ are different integers, and hence $P(n_0)$ and $P(n_0 + 1)$ are different statements.

Examples of Induction Proofs

We now give some examples illustrating the template for proofs by induction. Our first result states that the sum of the first n odd positive integers is n^2.

4.7. Theorem on Sum of Odd Integers. For all $n \in \mathbb{Z}_{\geq 1}$, $\displaystyle\sum_{k=1}^{n}(2k-1) = n^2$.

Proof. We use induction on n. Here, $P(n)$ is the summation formula $\sum_{k=1}^{n}(2k-1) = n^2$.
<u>Base Case.</u> We prove $P(1)$, which says $\sum_{k=1}^{1}(2k-1) = 1^2$. [To prove this, we use the initial condition in the definition of summation notation, together with a chain proof.] We know $\sum_{k=1}^{1}(2k-1) = 2 \cdot 1 - 1 = 2 - 1 = 1 = 1^2$.
<u>Induction Step.</u> Let n be a fixed positive integer. Assume $P(n)$, i.e., assume $\sum_{k=1}^{n}(2k-1) = n^2$. Prove $P(n+1)$, i.e., prove $\sum_{k=1}^{n+1}(2k-1) = (n+1)^2$. [We now use the recursive part of the definition of summation notation, and another chain proof.] We know

$$\sum_{k=1}^{n+1}(2k-1) = \left[\sum_{k=1}^{n}(2k-1)\right] + 2(n+1) - 1 \qquad \text{(by definition of sums)}$$
$$= n^2 + 2(n+1) - 1 \qquad \text{(by induction hypothesis)}$$
$$= n^2 + 2n + 1 = (n+1)^2 \qquad \text{(by algebra)}.$$

This chain of known equations proves the target equation $P(n+1)$.

The following diagram offers a visual justification of the formula $\sum_{k=1}^{n}(2k-1) = n^2$. A square of area 5^2 has been decomposed into L-shaped pieces of areas 1, 3, 5, 7, and 9; hence, $1 + 3 + 5 + 7 + 9 = 25 = 5^2$.

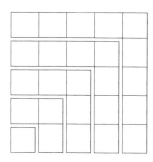

Although the diagram provides compelling visual intuition for why the summation formula is true, it does not *prove* the formula for *all* positive integers n. At best, by considering the full diagram and the smaller squares inside it, the diagram proves the first five instances of the summation formula.

4.8. Theorem on Sum of Squares. For all $n \in \mathbb{Z}_{\geq 1}$, $\displaystyle\sum_{k=1}^{n} k^2 = \frac{n(n+1)(2n+1)}{6}$.

Proof. We use induction on n. Here, $P(n)$ is the statement $\sum_{k=1}^{n} k^2 = n(n+1)(2n+1)/6$.
Base Case. We prove $P(1)$, which says $\sum_{k=1}^{1} k^2 = 1(1+1)(2 \cdot 1+1)/6$. [To prove this, we use the initial condition in the definition of summation notation, together with a chain proof.]
We know $\sum_{k=1}^{1} k^2 = 1^2 = 1 = 1(2)(3)/6 = 1(1+1)(2 \cdot 1 + 1)/6$.
Induction Step. Let n be a fixed positive integer. Assume $P(n)$, i.e., assume

$$\sum_{k=1}^{n} k^2 = \frac{n(n+1)(2n+1)}{6}.$$

Prove $P(n+1)$, i.e., prove $\sum_{k=1}^{n+1} k^2 = (n+1)([n+1]+1)(2[n+1]+1)/6$. [We now use the recursive part of the definition of summation notation and another chain proof.] We know

$$\sum_{k=1}^{n+1} k^2 = \left[\sum_{k=1}^{n} k^2\right] + (n+1)^2 \qquad \text{(by definition of sums)}$$

$$= \frac{n(n+1)(2n+1)}{6} + \frac{6(n+1)^2}{6} \qquad \text{(by induction hypothesis)}$$

$$= \frac{(n+1)(n(2n+1)+6(n+1))}{6} \qquad \text{(by factoring } n+1 \text{ out)}$$

$$= \frac{(n+1)(2n^2+7n+6)}{6} \qquad \text{(by algebra)}$$

$$= \frac{(n+1)(n+2)(2n+3)}{6} \qquad \text{(by factoring)}$$

$$= \frac{(n+1)([n+1]+1)(2[n+1]+1)}{6} \qquad \text{(by algebra)}.$$

This chain of known equalities proves the target formula $P(n+1)$.

Section Summary

1. *Recursive Definitions.* A recursive definition introduces a sequence of quantities $f(0), f(1), \ldots, f(n), \ldots$ by defining $f(0)$ explicitly and giving a formula for $f(n+1)$ involving the previous terms $f(0), f(1), \ldots, f(n)$.

2. *Sums, Products, Powers, and Factorials.* Given $a_k \in \mathbb{R}$ and $n \in \mathbb{Z}_{\geq 1}$, recursively define

$$\sum_{k=1}^{1} a_k = a_1, \quad \sum_{k=1}^{n+1} a_k = \left(\sum_{k=1}^{n} a_k\right) + a_{n+1},$$

$$\prod_{k=1}^{1} a_k = a_1, \quad \prod_{k=1}^{n+1} a_k = \left(\prod_{k=1}^{n} a_k\right) \cdot a_{n+1}.$$

For $x \in \mathbb{R}$ and $n \in \mathbb{Z}_{\geq 0}$, define $x^0 = 1$, $x^{n+1} = x^n \cdot x$, $0! = 1$, and $(n+1)! = (n+1) \cdot n!$.

3. *Induction Axiom.* For any open sentence $P(n)$, the statement "$\forall n \in \mathbb{Z}_{\geq 1}, P(n)$" is equivalent to the statement "$P(1) \wedge \forall n \in \mathbb{Z}_{\geq 1}, (P(n) \Rightarrow P(n+1))$."

4. *Ordinary Induction Proof Template.* To prove "$\forall n \in \mathbb{Z}_{\geq 1}, P(n)$" by induction:
   Say "We use induction on n."
   Step 1. Prove $P(1)$ [the base case].
   Step 2a. Fix $n \in \mathbb{Z}_{\geq 1}$. Assume $P(n)$ is true [the induction hypothesis].
   Step 2b. Aided by this assumption, prove $P(n+1)$ is true [the induction step].

Exercises

1. Compute each quantity by expanding a recursive definition.
   (a) $6!$ (b) 2^5 (c) 5^2 (d) 0^0 (e) $\sum_{k=1}^{4} k^2$ (f) $\prod_{k=1}^{4} k^2$ (g) $\sum_{j=1}^{5} \frac{1}{j}$
   (h) $\prod_{i=0}^{4}(2i+1)$.

2. (a) Prove by induction: for all $c \in \mathbb{R}$ and all $n \in \mathbb{Z}_{\geq 1}$, $\sum_{k=1}^{n} c = cn$.
   (b) Find the value of $\sum_{k=0}^{n} c$.

3. (a) Prove by induction: for all $n \in \mathbb{Z}_{\geq 1}$, $\sum_{k=1}^{n} k = n(n+1)/2$.
   (b) Prove by induction: for all $n \in \mathbb{Z}_{\geq 1}$, $\sum_{k=1}^{n} k^3 = n^2(n+1)^2/4$.
   [Thus the sum of the first n cubes is the square of the sum of the first n positive integers.]

4. Prove by induction: for all $n \in \mathbb{Z}_{\geq 1}$, $\sum_{k=1}^{n} 2 \cdot 3^k = 3^{n+1} - 3$.

5. Prove by induction: for all $n \in \mathbb{Z}_{\geq 1}$, $\sum_{k=1}^{n} k2^k = (n-1)2^{n+1} + 2$.

6. Prove: for all $n \in \mathbb{Z}_{\geq 1}$, $\sum_{k=1}^{n} k \cdot k! = (n+1)! - 1$.

7. Prove: for all $n \in \mathbb{Z}_{\geq 1}$, $\prod_{k=1}^{n}(1 + (3/k)) = (n+1)(n+2)(n+3)/6$.

8. (a) Find $\sum_{k=1}^{n} \frac{1}{k(k+1)}$ for $1 \leq n \leq 5$. (b) Based on your answers to (a), guess the value of this sum for general n. Then prove your guess by induction.

9. Use induction and the recursive definitions to prove:
   (a) for all $a \in \mathbb{R}$ and all $n \in \mathbb{Z}_{\geq 1}$, $a^n = \prod_{k=1}^{n} a$.
   (b) for all $n \in \mathbb{Z}_{\geq 1}$, $n! = \prod_{j=1}^{n} j$.

10. Find the error in the following proposed proof that for all positive integers n, $n = n + 3$. **Proof by induction.** Fix an arbitrary positive integer n; assume $n = n + 3$; prove $n + 1 = (n+1) + 3$. Add 1 to both sides of the induction hypothesis to get $n+1 = (n+3)+1$. By algebra, this becomes $n+1 = (n+1)+3$, completing the induction step.

11. Identify the logical error in the following proposed proof template for an induction proof of $\forall n \in \mathbb{Z}_{\geq 1}, P(n)$. Explain exactly how the proof template in the section differs from the one here.
    Part 1. Prove $P(1)$ is true.
    Part 2. Assume $P(n)$ is true for all positive integers n. Prove $P(n+1)$ is true for all positive integers n.

12. (a) Prove: for all real a, b, c, d, $(a + b) + (c + d) = (a + c) + (b + d)$. Justify every step using one of the axioms from §2.1.
    (b) *Additivity of Sums.* Suppose a_k and b_k are given real numbers for each positive integer k. Prove by induction: for all $n \in \mathbb{Z}_{\geq 1}$, $\sum_{k=1}^{n}(a_k + b_k) = \sum_{k=1}^{n} a_k + \sum_{k=1}^{n} b_k$.
    Explain how part (a) is used in the proof.

13. Let c and a_k (for each positive integer k) be given real numbers. Prove by induction: for all $n \in \mathbb{Z}_{\geq 1}$, $\sum_{k=1}^{n}(ca_k) = c \cdot \sum_{k=1}^{n} a_k$.

14. Without invoking the Induction Axiom, prove that statement (a) in the Induction Axiom implies statement (b).

15. (a) Use the Induction Axiom to prove that the statement "$\forall n \in \mathbb{Z}_{\geq 1}, P(n)$" is equivalent to "$P(1) \wedge \forall m \in \mathbb{Z}_{>1}, P(m-1) \Rightarrow P(m)$." (b) Create a variation of the ordinary induction proof template based on the equivalence in part (a).

16. Use intuitive reasoning (such as a picture of dominos) to create a variation of the Induction Axiom leading to a method for proving a statement of the form $\forall n \in \mathbb{Z}_{<0}, P(n)$. Make a proof template for your proof method.

17. Let A_k be a fixed set for each $k \in \mathbb{Z}_{>0}$. (a) Formulate recursive definitions for $\bigcup_{k=1}^n A_k$ and $\bigcap_{k=1}^n A_k$ similar to the recursive definitions of sums and products. (b) Let $I_n = \{j \in \mathbb{Z} : 1 \leq j \leq n\}$ for each positive integer n. Prove by induction: for all $n \in \mathbb{Z}_{\geq 1}$, $\bigcup_{k=1}^n A_k = \bigcup_{j \in I_n} A_j$ and $\bigcap_{k=1}^n A_k = \bigcap_{j \in I_n} A_j$.

18. We say that a set $S \subseteq \mathbb{R}$ *has a least element* iff $\exists z \in S, \forall x \in S, z \leq x$.
(a) Prove by induction on n: for all $S \subseteq \mathbb{Z}_{\geq 1}$, if $S \cap \{1, 2, \ldots, n\} \neq \emptyset$, then S has a least element. [You can use the fact that for all $n \in \mathbb{Z}_{\geq 1}$, there is no integer k with $n < k < n+1$.] (b) Prove that every nonempty subset of $\mathbb{Z}_{\geq 1}$ has a least element.

4.2   Induction Starting Anywhere and Backwards Induction

This section introduces some variations of the ordinary induction proof template that allow us to apply induction to more general situations. In particular, induction starting anywhere gives us a way to prove statements of the form $\forall n \in \mathbb{Z}_{\geq b}, P(n)$. Backwards induction lets us prove statements of the form $\forall n \in \mathbb{Z}_{\leq b}, P(n)$. Finally, we consider various methods for proving a universal statement $\forall n \in \mathbb{Z}, P(n)$ where n ranges over all integers (positive, negative, and zero).

Induction Starting Anywhere

Sometimes, we need to prove universal statements of the form $\forall n \in \mathbb{Z}_{\geq 0}, P(n)$, which start at $n = 0$ rather than $n = 1$. The Induction Axiom and Induction Proof Template are readily adapted to this situation. The only change is that in the base case, we prove $P(0)$ instead of $P(1)$; and in the induction step, we fix n in the set $\mathbb{Z}_{\geq 0}$ rather than fixing n in the set $\mathbb{Z}_{\geq 1}$. More generally, we have the following proof template for induction proofs starting anywhere.

4.9. Template for Induction Proofs Starting Anywhere. Let b be a fixed integer. To prove $\forall n \in \mathbb{Z}_{\geq b}, P(n)$:
Say "We prove the statement by induction on n."
<u>Part 1</u> (Base Case). **Prove** $P(b)$.
<u>Part 2</u> (Induction Step). **Fix** an arbitrary integer $n_0 \in \mathbb{Z}_{\geq b}$. **Assume** $P(n_0)$ is true. **Prove** $P(n_0 + 1)$ is true.

Exercise 16 outlines how this proof template may be justified using the Induction Axiom.

4.10. Example. Prove: for all $n \in \mathbb{Z}_{\geq 0}$, $\displaystyle\sum_{i=0}^{n} 2^i = 2^{n+1} - 1$.

Proof. We use induction on n.
<u>Base Case.</u> We prove $\sum_{i=0}^{0} 2^i = 2^{0+1} - 1$. We know $\sum_{i=0}^{0} 2^i = 2^0 = 1 = 2 - 1 = 2^{0+1} - 1$.
<u>Induction Step.</u> Fix $n \in \mathbb{Z}_{\geq 0}$. Assume $\sum_{i=0}^{n} 2^i = 2^{n+1} - 1$. Prove $\sum_{i=0}^{n+1} 2^i = 2^{(n+1)+1} - 1$. [Use a chain proof.] We know

$$
\begin{aligned}
\sum_{i=0}^{n+1} 2^i &= \left(\sum_{i=0}^{n} 2^i\right) + 2^{n+1} && \text{(by definition of sums)} \\
&= 2^{n+1} - 1 + 2^{n+1} && \text{(by induction hypothesis)} \\
&= 2^{n+1} \cdot 2 - 1 && \text{(by algebra)} \\
&= 2^{(n+1)+1} - 1 && \text{(by definition of exponents).}
\end{aligned}
$$

All of our examples so far have used induction to prove summation formulas. The next examples show how induction can be used to prove other kinds of statements.

4.11. Example. Prove: $\forall n \in \mathbb{Z}_{\geq 3}$, $\displaystyle\prod_{k=3}^{n} \left(1 - \frac{2}{k}\right) = \frac{2}{n(n-1)}$.

Proof. We use induction on n.
<u>Base Case.</u> We prove $\prod_{k=3}^{3}(1 - \frac{2}{k}) = \frac{2}{3(3-1)}$. By the initial condition in the recursive definition of products and by arithmetic, we know

$$
\prod_{k=3}^{3} \left(1 - \frac{2}{k}\right) = 1 - \frac{2}{3} = \frac{1}{3} = \frac{2}{6} = \frac{2}{3(3-1)}.
$$

Induction Step. Fix an arbitrary integer $n \in \mathbb{Z}_{\geq 3}$. Assume $\prod_{k=3}^{n}(1 - \frac{2}{k}) = \frac{2}{n(n-1)}$.
Prove $\prod_{k=3}^{n+1}(1 - \frac{2}{k}) = \frac{2}{(n+1)(n+1-1)}$. We know

$$\prod_{k=3}^{n+1}\left(1 - \frac{2}{k}\right) = \left[\prod_{k=3}^{n}\left(1 - \frac{2}{k}\right)\right] \cdot \left(1 - \frac{2}{n+1}\right) \qquad \text{(by definition of products)}$$

$$= \frac{2}{n(n-1)} \cdot \left(1 - \frac{2}{n+1}\right) \qquad \text{(by induction hypothesis)}$$

$$= \frac{2}{n(n-1)} \cdot \frac{n-1}{n+1} \qquad \text{(by algebra)}$$

$$= \frac{2}{(n+1)n} = \frac{2}{(n+1)(n+1-1)} \qquad \text{(by algebra)}.$$

4.12. Example. Prove: for all integers $n \geq 4$, $n! > 2^n$.
Proof. We use induction on n. [Here, $P(n)$ is the statement "$n! > 2^n$."]
Base Case. We must prove $4! > 2^4$. By the recursive definitions, we know $4! = 24 > 16 = 2^4$.
Induction Step. Fix an arbitrary integer $n \geq 4$; assume $n! > 2^n$; prove $(n+1)! > 2^{n+1}$. [We use a chain proof for inequalities.] We know

$$(n+1)! = (n+1)n! \qquad \text{(by definition of factorials)}$$

$$> (n+1)2^n \qquad \text{(by induction hypothesis; note } n+1 > 0)$$

$$> 2 \cdot 2^n \qquad \text{(since } n+1 > 2 \text{ and } 2^n > 0)$$

$$= 2^{n+1} \qquad \text{(by definition of exponents)}.$$

[To explain the two inequalities in more detail, recall this theorem: $\forall a \in \mathbb{R}, \forall b \in \mathbb{R}, \forall c \in \mathbb{R}$, if $a > b$ and $c > 0$, then $ca > cb$ and $ac > bc$. The first inequality in the chain follows from this theorem by taking $a = n!$, $b = 2^n$, and $c = n+1$; note that the hypothesis $a > b$ is our assumption in the induction proof, and $c > 0$ since $n+1$ must be at least 5. Similarly, the second inequality follows from this theorem by taking $a = n+1$, $b = 2$ and $c = 2^n$, which is allowed since we know $n+1 > 2$ and $2^n > 0$. Finally, by transitivity of $>$, the displayed chain of steps lets us deduce that $(n+1)! > 2^{n+1}$, which is the goal of the induction step.]

The next example shows why we started the induction at $n = 4$ in the last proof.

4.13. Example. Disprove: for all integers $n \geq 1$, $n! > 2^n$.
Disproof. We must prove $\exists n \in \mathbb{Z}_{\geq 1}, n! \leq 2^n$. [Induction is not required here; we prove the existence statement by giving a single example.] Choose $n = 3$, which is in $\mathbb{Z}_{\geq 1}$. We compute $3! = 6 \leq 8 = 2^3$.

More Examples of Induction Proofs

4.14. Example. Prove that for all $n \in \mathbb{Z}_{\geq 0}$, 8 divides $5^{2n} - 1$.
Proof. We use induction on n. [Here, $P(n)$ is the statement "8 divides $5^{2n} - 1$."]
Base Case. We must prove 8 divides $5^{2 \cdot 0} - 1$, i.e., that 8 divides 0. We must prove $\exists k \in \mathbb{Z}, 0 = 8k$. Choose $k = 0$; we know $0 \in \mathbb{Z}$ and $0 = 8 \cdot 0$.
Induction Step. Fix $n \in \mathbb{Z}_{\geq 0}$; assume 8 divides $5^{2n} - 1$; prove 8 divides $5^{2(n+1)} - 1$. We know $\exists a \in \mathbb{Z}, 5^{2n} - 1 = 8a$; we must prove $\exists b \in \mathbb{Z}, 5^{2n+2} - 1 = 8b$. [How can we choose b? The key is to rewrite 5^{2n+2} in a way that lets us use the induction hypothesis.] On one hand, we know $5^{2n+2} - 1 = 5^{2n} \cdot 5 \cdot 5 - 1 = 25 \cdot 5^{2n} - 1$. Now, from the induction hypothesis, we know $5^{2n} = 8a + 1$. Substituting, we get $5^{2n+2} - 1 = 25(8a + 1) - 1 = 25(8a) + 25 - 1 = 8(25a) + 24 = 8(25a + 3)$. Choose $b = 25a + 3$, which is an integer by closure. By the above algebra, we have $5^{2n+2} - 1 = 8b$, as needed.

The next example introduces the idea of a *recursively defined sequence*. The example also illustrates how we can gather data to guess a theorem that might be true, and then use induction to prove that theorem.

4.15. Example. Define numbers $a_1, a_2, \ldots, a_n, \ldots$ recursively by setting $a_1 = 5$ and $a_{n+1} = 3a_n - 4$ for all integers $n \geq 1$. (a) Calculate a_1, a_2, a_3, a_4, a_5. (b) Use the data in (a) to guess a non-recursive formula for a_n. (c) Prove the formula in (b) by induction.
Solution. (a) We know $a_1 = 5$. Taking $n = 1$ in the recursive formula, we find $a_2 = 3a_1 - 4 = 3(5) - 4 = 11$. Taking $n = 2$ in the recursive formula, we get $a_3 = 3a_2 - 4 = 3(11) - 4 = 29$. Continuing similarly, $a_4 = 3a_3 - 4 = 3(29) - 4 = 83$, and $a_5 = 3a_4 - 4 = 3(83) - 4 = 245$.
(b) What is the pattern in the numbers $5, 11, 29, 83, 245$? We might eventually notice that these numbers are close to powers of 3, which are $3, 9, 27, 81, 243$. In fact, we see that $a_n = 3^n + 2$ for $1 \leq n \leq 5$. We can verify that this formula continues to work for $n = 6, 7, 8, \ldots$ by calculating more values from the recursion. So we guess that $a_n = 3^n + 2$ for all integers $n \geq 1$. (Notice that we cannot fully confirm this guess by calculating finitely many a_n from the recursion.)
(c) We now prove $\forall n \in \mathbb{Z}_{\geq 1}, a_n = 3^n + 2$ by induction on n. For the base case, we must prove $a_1 = 3^1 + 2$. We know $a_1 = 5$ and $3^1 + 2 = 5$, so the base case is true. For the induction step, fix an arbitrary integer $n \in \mathbb{Z}_{\geq 1}$, assume $a_n = 3^n + 2$, and prove $a_{n+1} = 3^{n+1} + 2$. We have the chain of known equalities:

$$
\begin{aligned}
a_{n+1} &= 3a_n - 4 && \text{(by recursive definition of } a_{n+1}) \\
&= 3(3^n + 2) - 4 && \text{(by induction hypothesis)} \\
&= 3(3^n) + 3 \cdot 2 - 4 && \text{(by the distributive law)} \\
&= 3^{n+1} + 2 && \text{(by definition of exponents).}
\end{aligned}
$$

The goal of the next example is to prove one of the *laws of exponents* from algebra. This statement contains multiple universal quantifiers; we prove it by combining a generic-element proof with an induction proof starting at 0.

4.16. Example. Prove: for all real numbers x and all integers $m, n \in \mathbb{Z}_{\geq 0}$, $x^{m+n} = x^m x^n$.
Proof. We must prove $\forall x \in \mathbb{R}, \forall m \in \mathbb{Z}_{\geq 0}, \forall n \in \mathbb{Z}_{\geq 0}, x^{m+n} = x^m x^n$. [We begin by using the generic-element proof template twice.] Let x be a fixed, but arbitrary real number. Let m be a fixed, but arbitrary nonnegative integer. [For the universal statement involving n, we use induction.] We prove by induction on n that $x^{m+n} = x^m x^n$.
Base Case. For $n = 0$, we must prove $x^{m+0} = x^m x^0$. On one hand, $x^{m+0} = x^m$ since $m + 0 = m$. On the other hand, $x^m x^0 = x^m \cdot 1 = x^m$ by the initial condition in the definition of exponents and the multiplicative identity axiom. So the base case holds.
Induction Step. Fix an arbitrary integer $n \geq 0$. Assume $x^{m+n} = x^m x^n$. Prove $x^{m+(n+1)} = x^m x^{n+1}$. We know

$$
\begin{aligned}
x^{m+(n+1)} &= x^{(m+n)+1} && \text{(by associativity of addition)} \\
&= x^{m+n} x && \text{(by recursive definition of exponents)} \\
&= (x^m x^n) x && \text{(by induction hypothesis)} \\
&= x^m (x^n x) && \text{(by associativity of multiplication)} \\
&= x^m x^{n+1} && \text{(by recursive definition of exponents).}
\end{aligned}
$$

Backwards Induction and Induction on $\mathbb{Z}$ (Optional)

So far, we have talked about induction proof templates for proving statements of the form $\forall n \in \mathbb{Z}_{\geq b}, P(n)$. What if we need to prove that $P(n)$ holds for *all* integers n? We discuss two ways to do this in a moment. First, we introduce *backwards induction*, which can be used to prove statements of the form $\forall n \in \mathbb{Z}_{\leq b}, P(n)$.

4.17. Template for Backwards Induction Proofs. Let b be a fixed integer.
To prove $\forall n \in \mathbb{Z}_{\leq b}, P(n)$:
Say "We prove the statement by backwards induction on n."
<u>Part 1</u> (Base Case). **Prove** $P(b)$.
<u>Part 2</u> (Induction Step). **Fix** an arbitrary integer $n_0 \in \mathbb{Z}_{\leq b}$. **Assume** $P(n_0)$ is true.
Prove $P(n_0 - 1)$ is true.

See the exercises for intuitive and formal justifications of why this proof template works. With this template in hand, we can now describe the first method for proving a universal statement about all integers.

4.18. First Template for Proving $\forall n \in \mathbb{Z}, P(n)$.
<u>Part 1</u> (Base Case). **Prove** $P(0)$.
<u>Part 2</u> (Forward Induction Step). **Fix** an arbitrary integer $n_0 \in \mathbb{Z}_{\geq 0}$. **Assume** $P(n_0)$ is true. **Prove** $P(n_0 + 1)$ is true.
<u>Part 3</u> (Backward Induction Step). **Fix** an arbitrary integer $m_0 \in \mathbb{Z}_{\leq 0}$. **Assume** $P(m_0)$ is true. **Prove** $P(m_0 - 1)$ is true.

This template works because "$\forall n \in \mathbb{Z}, P(n)$" is logically equivalent to "$\forall n \in \mathbb{Z}_{\geq 0}, P(n) \wedge \forall m \in \mathbb{Z}_{\leq 0}, P(m)$." We prove the latter statement by combining forwards induction (starting at 0) with backwards induction (also starting at 0). We could replace each occurrence of 0, here and in the proof template above, by any fixed integer b.

Our second method for proving $\forall n \in \mathbb{Z}, P(n)$ first handles nonnegative integers by ordinary induction, and then proves the statement for each negative integer by reducing to the positive case dealt with earlier. This technique is useful when the statement $P(-n)$ is closely related to the statement $P(n)$.

4.19. Second Template for Proving $\forall n \in \mathbb{Z}, P(n)$.
<u>Part 1</u> (Base Case). **Prove** $P(0)$.
<u>Part 2</u> (Forward Induction Step). **Fix** an integer $n_0 \in \mathbb{Z}_{\geq 0}$. **Assume** $P(n_0)$ is true. **Prove** $P(n_0 + 1)$ is true.
<u>Part 3</u> (Proof for Negative Integers). **Fix** a positive integer $n_0 \in \mathbb{Z}_{> 0}$. **We have proved** $P(n_0)$ is true. **Prove** $P(-n_0)$ is true.

Refer to the exercises for a justification of this proof template.

4.20. Example. We outline a proof that *for all integers n, n is even or n is odd*, asking the reader to supply the missing details.
<u>Part 1.</u> Prove 0 is even or 0 is odd. Since $0 = 2 \cdot 0$, 0 is even, so the OR-statement is true.
<u>Part 2.</u> Fix an integer $n \geq 0$. Assume n is even or n is odd. Prove $n + 1$ is even or $n + 1$ is odd. This part is completed by considering two cases (see the exercises).
<u>Part 3.</u> Now fix an integer $n > 0$. By part 2, we already know that n is even or n is odd. We must prove $-n$ is even or $-n$ is odd. Again this part can be completed by considering two cases. [The fastest way to finish is to use the previously proved results that n is even iff $-n$ is even, and n is odd iff $-n$ is odd.]

Combining this example with Example 2.35 (where we showed that no integer is both even and odd), we have finally proved the theorem "every integer is even or odd, but not both." This theorem was announced without proof in §2.3. We give another proof of this theorem a little later, using the Integer Division Theorem.

Section Summary

1. *Induction Starting Anywhere.* To prove $\forall n \in \mathbb{Z}_{\geq b}, P(n)$ by ordinary induction:
   Step 1. Prove $P(b)$ [the base case].
   Step 2a. Fix $n \in \mathbb{Z}_{\geq b}$, and assume $P(n)$ is true [the induction hypothesis].
   Step 2b. Aided by this assumption, prove $P(n+1)$ is true [the induction step].

2. *Backwards Induction.* To prove $\forall n \in \mathbb{Z}_{\leq b}, P(n)$ by backwards induction:
   Step 1. Prove $P(b)$ [the base case].
   Step 2a. Fix $n \in \mathbb{Z}_{\leq b}$, and assume $P(n)$ is true [the induction hypothesis].
   Step 2b. Aided by this assumption, prove $P(n-1)$ is true [the induction step].

3. *First Method to Prove $\forall n \in \mathbb{Z}, P(n)$.*
   Step 1. [Base case] Prove $P(0)$.
   Step 2. [Forward induction step] Fix $n \in \mathbb{Z}_{\geq 0}$; assume $P(n)$; prove $P(n+1)$.
   Step 3. [Backward induction step] Fix $m \in \mathbb{Z}_{\leq 0}$; assume $P(m)$; prove $P(m-1)$.

4. *Second Method to Prove $\forall n \in \mathbb{Z}, P(n)$.*
   Step 1. [Base case] Prove $P(0)$.
   Step 2. [Forward induction step] Fix $n \in \mathbb{Z}_{\geq 0}$; assume $P(n)$; prove $P(n+1)$.
   Step 3. [Negative case] Fix $n \in \mathbb{Z}_{>0}$; note $P(n)$ has been proved; prove $P(-n)$.

Exercises

1. Prove by induction: for all $n \in \mathbb{Z}_{\geq 2}$, $\displaystyle\prod_{k=2}^{n}\left(1 - \frac{1}{k^2}\right) = \frac{n+1}{2n}$.

2. (a) Compute the product $\displaystyle\prod_{k=4}^{n}\left(1 - \frac{3}{k}\right)$ for $4 \leq n \leq 7$. (b) Based on the data in (a), guess the value of this product for general $n \geq 4$, and prove your guess by induction.

3. (a) Prove by induction: for all real $r \neq 1$ and all $n \in \mathbb{Z}_{\geq 0}$,
$$\sum_{k=0}^{n} r^k = \frac{1 - r^{n+1}}{1 - r}.$$
   (b) Compute $\sum_{k=0}^{n} r^k$ when $r = 1$.

4. Prove: $\forall x \in \mathbb{R}_{\geq 0}, \forall n \in \mathbb{Z}_{\geq 0}, (1+x)^n \geq 1 + nx$.

5. (a) Prove: for all $n \in \mathbb{Z}_{\geq 5}$, $n^2 < 2^n$. [To finish, you may need to prove a similar inequality involving a smaller power of n.] (b) Disprove: for all $n \in \mathbb{Z}_{\geq 1}$, $n^2 < 2^n$.

6. Prove by induction: for all $n \in \mathbb{Z}_{\geq 0}$, 6 divides $7^n - 1$.

7. *Telescoping Sums.* For each $k \in \mathbb{Z}$, let a_k be a fixed real number. Prove: for all $i, j \in \mathbb{Z}$, if $i \leq j$, then $\sum_{k=i}^{j}(a_{k+1} - a_k) = a_{j+1} - a_i$.

8. *Laws of Exponents.* Prove: $\forall x \in \mathbb{R}, \forall m \in \mathbb{Z}_{\geq 0}, \forall n \in \mathbb{Z}_{\geq 0}, (x^m)^n = x^{mn}$. [Fix x and m and use induction on n. You will need the law of exponents proved in Example 4.16.]

9. Prove: $\forall x \in \mathbb{R}, \forall y \in \mathbb{R}, \forall n \in \mathbb{Z}_{\geq 0}, (xy)^n = x^n y^n$.

10. Define $a_0 = 1$ and $a_{n+1} = 2a_n - 1$ for all $n \in \mathbb{Z}_{\geq 0}$. (a) Compute a_n for $0 \leq n \leq 5$. (b) Based on the data in (a), guess a general formula for a_n and prove it by induction.

11. Define $a_0 = 4$ and $a_{n+1} = 2a_n + 3 \cdot 2^n$ for all $n \in \mathbb{Z}_{\geq 0}$. Prove:
    $\forall n \in \mathbb{Z}_{\geq 0},\ a_n = (3n+4)2^n$.

12. Let x and y be fixed, unequal real numbers. (a) Simplify $\frac{x^n - y^n}{x - y}$ for $1 \leq n \leq 4$.
    (b) For general $n \in \mathbb{Z}_{\geq 1}$, guess a summation formula that evaluates to $\frac{x^n - y^n}{x - y}$,
    and prove it by induction on n.

13. (a) Prove by induction: for all $n \in \mathbb{Z}_{\geq 1}$, $\displaystyle\sum_{j=1}^{n} j^4 \leq \frac{(n+1)^5}{5}$.
    (b) Prove the inequality in (a) using integral calculus.

14. Fill in the missing details of the proof in Example 4.20.

15. *Negative Exponents.* The goal of this problem is to prove the laws of exponents where the powers involved can be any integers (possibly negative). Given any nonzero real number x, the axioms for $\mathbb{R}$ tell us that there exists a unique real number x' (the multiplicative inverse of x) such that $x \cdot x' = x' \cdot x = 1$. We have already defined x^n when n is a *nonnegative* integer. For a positive integer n, we now define the negative power x^{-n} to be $(x')^n$. In particular, notice that $x^{-1} = (x')^1 = x'$ is the multiplicative inverse of x.
    (a) Prove: $\forall x \in \mathbb{R}_{\neq 0}, \forall n \in \mathbb{Z}_{>0}, (x')^n = (x^n)'$. [So $(x^{-1})^n = (x^n)^{-1}$.]
    (b) Prove: $\forall x \in \mathbb{R}_{\neq 0}, \forall m \in \mathbb{Z}, \forall n \in \mathbb{Z}, x^{m+n} = x^m x^n$.
    (c) Prove: $\forall x \in \mathbb{R}_{\neq 0}, \forall m \in \mathbb{Z}, \forall n \in \mathbb{Z}, (x^m)^n = x^{mn}$.
    (d) Prove: $\forall x \in \mathbb{R}_{\neq 0}, \forall y \in \mathbb{R}_{\neq 0}, \forall n \in \mathbb{Z}, (xy)^n = x^n y^n$.

16. Proof Template 4.9 gives a method to prove statements of the form $\forall n \in \mathbb{Z}_{\geq b}, P(n)$ by induction. Use the Induction Axiom 4.5 to justify this proof template. [*Hint:* Given $P(n)$ and b, let $Q(n)$ be the statement $P(b + n - 1)$ for each positive integer n.]

17. Give an intuitive justification for the Backwards Induction Proof Template 4.17 by imitating our informal explanation of the Induction Axiom, based on repeated use of the Inference Rules for IF and ALL. Illustrate your answer by drawing a row of dominos.

18. Use induction starting anywhere (see Proof Template 4.9) to give a formal justification of backwards induction (see Proof Template 4.17).

19. It is known that for all $m \in \mathbb{Z}$, either $m \geq 0$ or $-m > 0$. Use this theorem to give a formal justification of Proof Template 4.19 for proving statements of the form $\forall n \in \mathbb{Z}, P(n)$.

20. Let $P(n)$ be a fixed open sentence. Suppose we have proved
    "$\forall n \in \mathbb{Z},\ P(n) \Rightarrow P(n+1)$" and "$\forall n \in \mathbb{Z},\ P(n) \Rightarrow P(n-1)$."
    (a) Give a specific example to show that $\forall n \in \mathbb{Z}, P(n)$ can be false.
    (b) Prove, however, that "$\forall n \in \mathbb{Z}, P(n)$ or $\forall m \in \mathbb{Z}, \sim P(m)$" must be true.

21. *Induction Stepping by* 2. Here is a new template for proving $\forall n \in \mathbb{Z}_{\geq 0}, P(n)$:
    Step 1. Prove $P(0)$. Prove $P(1)$.
    Step 2. Fix $n_0 \in \mathbb{Z}_{\geq 0}$. Assume $P(n_0)$ is true. Prove $P(n_0 + 2)$ is true.
    Show that this template is a correct method for proving the given statement.
    [*Hint:* Let $Q(n)$ be the statement "$P(n)$ and $P(n+1)$." Prove $\forall n \in \mathbb{Z}_{\geq 0}, Q(n)$ by ordinary induction.]

22. Define $a_0 = 1$, $a_1 = 4$, and $a_n = 9a_{n-2} - 8$ for all $n \in \mathbb{Z}_{\geq 2}$. Guess a non-recursive formula for a_n, and then prove it using the template in Exercise 21.

23. Prove: for every odd integer n, 8 divides $n^2 - 1$. [*Suggestion:* Write $n = 2k + 1$ and use induction on k.]

24. Let $s \geq 1$ be a fixed integer (the *step size*). State and prove a generalization of Exercise 21 where part 2 of the template fixes $n_0 \in \mathbb{Z}_{\geq 0}$, assumes $P(n_0)$ is true, and proves $P(n_0 + s)$ is true.

4.3 Strong Induction

In an ordinary induction proof of $\forall n \in \mathbb{Z}_{\geq 1}, P(n)$, we are allowed to assume $P(n)$ in the induction step (where n is fixed) to help us prove $P(n+1)$. This method works well in many situations, but sometimes the assumption is not powerful enough to enable us to finish proving $P(n+1)$. Strong induction lets us assume even more information as part of the induction hypothesis. Instead of assuming only $P(n)$ when proving $P(n+1)$, we are now allowed to assume that all of the preceding statements $P(1)$, $P(2)$, ..., $P(n)$ are true. This additional information gives us more to work with as we try to prove $P(n+1)$. Strong induction is often needed to handle recursive definitions where the recursively defined quantity $f(n+1)$ depends on several of the preceding objects $f(1), f(2), \ldots, f(n)$.

The Strong Induction Proof Template

We begin with an example of an attempted proof by ordinary induction where the induction hypothesis does not give us enough information to finish the proof.

4.21. Example. Define a sequence recursively by setting $F_0 = 0$, $F_1 = 1$, and $F_n = F_{n-1} + F_{n-2}$ for all integers $n \geq 2$. (This is the famous *Fibonacci sequence*.)
Prove: $\forall n \in \mathbb{Z}_{\geq 0}, F_n \leq 2^n$.
Proof. We try a proof by induction on n. For the base case, we must prove $F_0 \leq 2^0$. Since $F_0 = 0$ and $2^0 = 1$ and $0 \leq 1$, the base case is proved. Now fix an arbitrary integer $n \geq 0$, assume $F_n \leq 2^n$, and try to prove $F_{n+1} \leq 2^{n+1}$. On one hand, if $n = 0$, we are proving $F_1 \leq 2^1$, which holds since $F_1 = 1$ and $2^1 = 2$. On the other hand, if $n > 0$, then $n+1 \geq 2$, so the recursive definition of F_{n+1} tells us that $F_{n+1} = F_n + F_{n-1}$. Here is where the ordinary induction proof gets stuck. We can use the induction hypothesis to bound the first term on the right side by 2^n. Similarly, we would like to bound the second term on the right side by writing $F_{n-1} \leq 2^{n-1}$. Unfortunately, our induction proof (as we have currently set it up) has not provided us with any assumptions about F_{n-1}. So the proof by induction cannot be completed.

In the previous example, we could have finished the proof if we had been allowed to assume that $P(n-1)$ and $P(n)$ were both true, before trying to prove $P(n+1)$. We now present a variation of induction, called *strong induction* or *complete induction*, that allows us to assume even more information as part of the induction hypothesis. These extra assumptions can make it easier to complete the induction step.

4.22. Strong Induction Theorem. Let $P(n)$ be any open sentence, and let b be a fixed integer. To prove $\forall n \in \mathbb{Z}_{\geq b}, P(n)$, we may instead prove this statement:

> For all integers $n \geq b$, if (for all integers m such that $b \leq m < n$, $P(m)$ is true), then $P(n)$ is true.

We can recast this theorem as the following proof template for a strong induction proof.

4.23. Strong Induction Proof Template. To prove $\forall n \in \mathbb{Z}_{\geq b}, P(n)$ by strong induction:
Fix an arbitrary integer $n_0 \geq b$.
Assume: for all integers m in the range $b \leq m < n_0$, $P(m)$ is true.
Prove $P(n_0)$ is true.

Five Remarks on Strong Induction

(1) The Strong Induction Theorem can be proved from the (ordinary) Induction Axiom; we give a proof in an optional section later.

(2) Here is intuition for why strong induction works. Intuitively, any induction proof starting at b proves the statements $P(b)$, $P(b+1)$, $P(b+2)$, ..., in this order. Let $n_0 \geq b$ be a fixed integer. When the proof reaches $P(n_0)$ and tries to prove it, we will have already successfully proved all preceding statements $P(b)$, $P(b+1)$, ..., $P(n_0-1)$. So we might as well assume all those statements are true when attempting to prove $P(n_0)$. This is exactly the assumption in the proof template above; we call this assumption the *(strong) induction hypothesis*. The word "strong" (or "complete") refers to the fact that we are allowed to assume a stronger statement in this proof template compared to the ordinary induction proof template.

This remark can be informally visualized using the domino analogy.

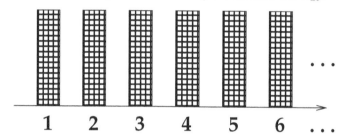

Here $P(n)$ is the statement "domino n falls down." Suppose we are proving that all dominos fall down by induction. By the time we reach domino 6, we already know that the first 5 dominos have fallen down. So it is allowable to assume $P(1) \wedge P(2) \wedge P(3) \wedge P(4) \wedge P(5)$, not just $P(5)$, to help us prove $P(6)$.

(3) Now that strong induction is available, we could always use this proof technique instead of ordinary induction. However, ordinary induction proofs are conceptually simpler than strong induction proofs, so many people prefer ordinary induction for problems where this method succeeds. As a general rule, ordinary induction is applicable to recursive definitions where $f(n+1)$ depends only on $f(n)$; whereas strong induction is needed for more complicated recursive definitions.

(4) The strong induction proof template does not specifically mention a base case, unlike the ordinary induction proof template. However, there is a hidden base case in strong induction. Consider the situation where the fixed integer n_0 happens to equal b. Then there are no integers m satisfying $b \leq m < n_0$, and so our strong induction hypothesis is not actually assuming anything for this value of n_0. It follows that we need to prove $P(b)$ without any prior knowledge, which is exactly what we must do in the base case of an ordinary induction proof.

(5) In strong induction proofs, it often happens that several small values of n_0 (not just $n_0 = b$) need to be treated in separate cases, which could be called the "base cases" of the strong induction proof. These special cases often correspond to the initial conditions in a recursive definition, as illustrated in the following examples.

Examples of Strong Induction Proofs

We begin by revisiting the failed proof above.

4.24. Example. Define $F_0 = 0$, $F_1 = 1$, and $F_n = F_{n-1} + F_{n-2}$ for all integers $n \geq 2$. Prove: for all integers $n \geq 0$, $F_n \leq 2^n$.
Proof. We use strong induction on n. Fix an integer $n \geq 0$. Assume that for all integers m

in the range $0 \le m < n$, $F_m \le 2^m$. Prove $F_n \le 2^n$. We know $n = 0$ or $n = 1$ or $n \ge 2$, so use cases. [The previous statement was motivated by the recursive definition of F_n, which has different cases for $n = 0$, $n = 1$, and $n \ge 2$.]

Case 0. Assume $n = 0$; prove $F_0 \le 2^0$. We know $F_0 = 0$ and $2^0 = 1$ and $0 \le 1$, so $F_0 \le 2^0$.

Case 1. Assume $n = 1$; prove $F_1 \le 2^1$. We know $F_1 = 1$ and $2^1 = 2$ and $1 \le 2$, so $F_1 \le 2^1$.

Case 2. Assume $n \ge 2$; prove $F_n \le 2^n$. To start, we use the recursive definition to write $F_n = F_{n-1} + F_{n-2}$. Now, $0 \le n-1 < n$, so the strong induction hypothesis (taking $m = n-1$) tells us $F_{n-1} \le 2^{n-1}$. Furthermore $0 \le n-2 < n$, so we also know $F_{n-2} \le 2^{n-2}$ by taking $m = n-2$ in the induction hypothesis. Hence,

$$F_n = F_{n-1} + F_{n-2} \le 2^{n-1} + 2^{n-2} = 2^n(2^{-1} + 2^{-2}) = 2^n(3/4) < 2^n,$$

completing the proof of this case. The induction proof is now finished.

4.25. Example. Define $a_0 = 3$, $a_1 = 8$, and $a_n = 4(a_{n-1} - a_{n-2})$ for all integers $n \ge 2$. Prove: for all integers $n \ge 0$, $a_n = (n+3)2^n$.

Proof. We use strong induction on n. Fix an arbitrary integer $n \ge 0$. Assume that for all m in the range $0 \le m < n$, $a_m = (m+3)2^m$. Prove $a_n = (n+3)2^n$. We know $n = 0$ or $n = 1$ or $n \ge 2$, so use cases. [This step was suggested by the cases in the recursive definition of a_n.]

Case 0. Assume $n = 0$; prove $a_0 = (0+3)2^0$. We know $a_0 = 3 = (0+3)2^0$, so this case is proved.

Case 1. Assume $n = 1$; prove $a_1 = (1+3)2^1$. We know $a_1 = 8 = 4 \cdot 2 = (1+3)2^1$, so this case is proved.

Case 2. Assume $n \ge 2$; prove $a_n = (n+3)2^n$. In this case, we have $0 \le n-1 < n$ and $0 \le n-2 < n$, so the induction hypothesis tells us that $a_{n-1} = (n-1+3)2^{n-1}$ and $a_{n-2} = (n-2+3)2^{n-2}$. Using the recursive definition and this information, we deduce

$$a_n = 4(a_{n-1} - a_{n-2}) = 4((n+2)2^{n-1} - (n+1)2^{n-2}) = 2^n(2(n+2) - (n+1)) = 2^n(n+3).$$

So the result holds in all cases, completing the induction.

4.26. Example. Define $a_0 = 1$ and $a_n = (\sum_{k=0}^{n-1} a_k) + 1$ for all $n \ge 1$. Guess an explicit (non-recursive) formula for a_n and then prove it.

Solution. We use the recursive definition to compute $a_0 = 1$, $a_1 = a_0 + 1 = 1 + 1 = 2$, $a_2 = a_0 + a_1 + 1 = 1 + 2 + 1 = 4$, $a_3 = a_0 + a_1 + a_2 + 1 = 1 + 2 + 4 + 1 = 8$, $a_4 = a_0 + a_1 + a_2 + a_3 + 1 = 1 + 2 + 4 + 8 + 1 = 16$, and so on. It looks like $a_n = 2^n$ for all integers $n \ge 0$. Since the recursive formula for a_n involves all preceding values a_k, we prove our guess by strong induction on n. Fix an arbitrary integer $n \ge 0$. Assume: for all integers k in the range $0 \le k < n$, $a_k = 2^k$. Prove $a_n = 2^n$. [The definition of a_n breaks into two parts, suggesting the following two cases.] We know $n = 0$ or $n > 0$, so use cases.

Case 1. Assume $n = 0$; prove $a_0 = 2^0$. We know $a_0 = 1 = 2^0$.

Case 2. Assume $n > 0$; prove $a_n = 2^n$. In this case, we know $a_n = (\sum_{k=0}^{n-1} a_k) + 1$. Each term a_k appearing in the sum has an index k in the range $0 \le k < n$. So the strong induction hypothesis allows us to replace every term a_k by 2^k in the sum defining a_n. We now know $a_n = (\sum_{k=0}^{n-1} 2^k) + 1$. By a previous theorem (proved in Example 4.10 by ordinary induction), the sum appearing here evaluates to $2^{n-1+1} - 1 = 2^n - 1$. So $a_n = (2^n - 1) + 1 = 2^n$, completing the induction proof.

4.27. Example. Prove: $\forall n \in \mathbb{Z}_{\ge 18}, \exists a \in \mathbb{Z}_{\ge 0}, \exists b \in \mathbb{Z}_{\ge 0}, n = 7a + 4b$. [This statement can be concretely interpreted as follows: for all $n \ge 18$, we can pay n cents postage using a combination of 7-cent stamps and 4-cent stamps.]

Proof. We use strong induction on n. Fix an integer $n \ge 18$. Assume: for all integers m in the range $18 \leq m < n$, $\exists c \in \mathbb{Z}_{\geq 0}, \exists d \in \mathbb{Z}_{\geq 0}, m = 7c + 4d$. Prove: $\exists a \in \mathbb{Z}_{\geq 0}, \exists b \in \mathbb{Z}_{\geq 0}, n = 7a + 4b$. [Note the use of non-conflicting letters for the existentially quantified variables in the induction hypothesis and the goal.] We know $n = 18$ or $n = 19$ or $n = 20$ or $n = 21$ or $n \geq 22$, so use cases. [This is the tricky part — how did we know to introduce these particular cases? The reason will retroactively emerge at the end of the proof.]

Case 1. Assume $n = 18$. Choose $a = 2$ and $b = 1$, which are in $\mathbb{Z}_{\geq 0}$. By arithmetic, $7a + 4b = 7(2) + 4(1) = 14 + 4 = 18 = n$.

Case 2. Assume $n = 19$. Choose $a = 1$ and $b = 3$, which are in $\mathbb{Z}_{\geq 0}$. By arithmetic, $7a + 4b = 7(1) + 4(3) = 7 + 12 = 19 = n$.

Case 3. Assume $n = 20$. Choose $a = 0$ and $b = 5$, which are in $\mathbb{Z}_{\geq 0}$. By arithmetic, $7a + 4b = 7(0) + 4(5) = 0 + 20 = 20 = n$.

Case 4. Assume $n = 21$. Choose $a = 3$ and $b = 0$, which are in $\mathbb{Z}_{\geq 0}$. By arithmetic, $7a + 4b = 7(3) + 4(0) = 21 + 0 = 21 = n$. [Note that the induction hypothesis was not needed to complete the first four cases.]

Case 5. Assume $n \geq 22$. The key observation is that for such an n, the integer $m = n - 4$ satisfies $18 \leq m < n$. So we can take m in the induction hypothesis to be $n - 4$. By definition of ALL, we know there exist $c_0 \in \mathbb{Z}_{\geq 0}$ and $d_0 \in \mathbb{Z}_{\geq 0}$ with $n - 4 = 7c_0 + 4d_0$. Adding 4 to both sides, we get $n = 7c_0 + 4d_0 + 4 = 7c_0 + 4(d_0 + 1)$. Choose $a = c_0$ and $b = d_0 + 1$, which are both nonnegative integers by closure (since c_0 and d_0 are in $\mathbb{Z}_{\geq 0}$). Then $n = 7a + 4b$ holds, completing the proof of Case 5. [Now, at the end of the proof, we understand why the given cases were used. The point is that we can go from a solution for $n - 4$ to a solution for n by adding one more 4-cent stamp. But $n - 4$ falls outside the allowable range of m's when n is 18, 19, 20, or 21. So these values of n must be treated in separate base cases.]

Proof of the Strong Induction Theorem (Optional)

We now prove the Strong Induction Theorem. Fix an integer b and an arbitrary open sentence $P(n)$. Assume that we have proved the following statement:

For all integers $n \geq b$, if (for all integers m such that $b \leq m < n$, $P(m)$ is true), then $P(n)$ is true.

We must prove the statement $\forall n \in \mathbb{Z}_{\geq b}, P(n)$. The idea is to begin by proving a related statement by *ordinary* induction starting at b. Define a new open sentence $Q(n)$, which is the statement "for all integers m in the range $b \leq m < n$, $P(m)$ is true." We will prove $\forall n \in \mathbb{Z}_{\geq b}, Q(n)$ by induction on n.

For the base case, we must prove the statement $Q(b)$. This statement says "for all integers m in the range $b \leq m < b$, $P(m)$ is true." There are no integers m satisfying $b \leq m < b$, so this statement is vacuously true.

For the induction step, fix an integer $n \geq b$, assume $Q(n)$ is true, and prove $Q(n+1)$ is true. We have assumed the following induction hypothesis:

for all integers m in the range $b \leq m < n$, $P(m)$ is true.

We must prove:

for all integers m in the range $b \leq m < n + 1$, $P(m)$ is true.

Fix an integer m_0, and assume $b \leq m_0 < n + 1$. Since m_0 is an integer, we must have $b \leq m_0 < n$ or $m_0 = n$. In the case where $b \leq m_0 < n$, the induction hypothesis and the Inference Rule for ALL tell us that $P(m_0)$ is true. In the case where $m_0 = n$, the induction hypothesis coincides with the hypothesis of the IF-statement that we assumed at the start of the proof. By the Inference Rule for IF, we see that $P(n)$ is true. Because $m_0 = n$, $P(m_0)$ is true in this case as well.

We have completed the induction proof of the statement $\forall n \in \mathbb{Z}_{\geq b}, Q(n)$. We use this to prove $\forall n \in \mathbb{Z}_{\geq b}, P(n)$. Fix an arbitrary integer $n \geq b$. Then $n + 1$ is also an integer in $\mathbb{Z}_{\geq b}$, so $Q(n + 1)$ is true. So for all integers m in the range $b \leq m < n + 1$, $P(m)$ is true. Since n is an integer satisfying $b \leq n < n + 1$, we can take $m = n$ to conclude that $P(n)$ is true, as needed.

Section Summary

1. *Strong Induction.* To prove $\forall n \in \mathbb{Z}_{\geq b}, P(n)$ by strong induction:
   Step 1. Fix an arbitrary integer $n \geq b$.
   Step 2a. Assume that for all integers m in the range $b \leq m < n$, $P(m)$ is true.
   Step 2b. Aided by this assumption, prove $P(n)$ is true.

2. *Remarks on Induction Proofs.* Ordinary induction can be used when the truth of $P(n + 1)$ can be deduced assuming only that the previous statement $P(n)$ is true. Strong induction is needed when the truth of $P(n+1)$ depends on several of the preceding statements $P(n)$, $P(n - 1)$, ... Both types of induction proofs can be justified using the ordinary Induction Axiom. Although the strong induction proof template does not specifically mention a base case, we often need to treat small values of n as separate cases. The required cases in the proof are often based on corresponding cases in the recursive definitions of concepts appearing in the statement $P(n)$.

Exercises

1. Recursively define $a_0 = -1$, $a_1 = 1$, and $a_n = 8a_{n-1} - 15a_{n-2}$ for all $n \in \mathbb{Z}_{\geq 2}$. Prove: for all $n \in \mathbb{Z}_{\geq 0}$, $a_n = 2 \cdot 5^n - 3^{n+1}$.

2. Define a sequence recursively by setting $a_1 = 2$, $a_2 = 1$, and $a_n = 2a_{n-1} - a_{n-2}$ for all integers $n \geq 3$. (a) Compute a_n for $1 \leq a_n \leq 6$, and then guess an explicit (non-recursive) formula for a_n. (b) Use strong induction to prove your formula for a_n holds for all integers $n \geq 1$.

3. Define $a_0 = 4$, $a_1 = 32$, and $a_n = 12a_{n-1} - 20a_{n-2}$ for all integers $n \geq 2$. Guess a non-recursive formula for a_n (valid for all $n \geq 0$), and prove your guess by strong induction.

4. Use ordinary induction or strong induction to prove the following summation formulas for Fibonacci numbers (see Example 4.21), valid for all integers $n \geq 0$.
   (a) $\sum_{k=0}^{n} F_k = F_{n+2} - 1$.   (b) $\sum_{k=0}^{n-1} F_{2k+1} = F_{2n}$.   (c) $\sum_{k=0}^{n} F_{2k} = F_{2n+1} - 1$.

5. Prove by strong induction: $\forall n \in \mathbb{Z}_{\geq 11}, \exists a \in \mathbb{Z}_{>0}, \exists b \in \mathbb{Z}_{>0}, n = 2a + 5b$.

6. Consider this statement: $\forall n \in \mathbb{Z}_{\geq n_0}, \exists a \in \mathbb{Z}_{\geq 0}, \exists b \in \mathbb{Z}_{\geq 0}, n = 12a + 5b$. Find the smallest value of n_0 for which the statement is true, and then prove it.

7. Define $a_0 = 1$, $a_1 = 3$, $a_2 = 33$, and $a_n = 7a_{n-1} - 8a_{n-2} - 16a_{n-3}$ for all $n \in \mathbb{Z}_{\geq 3}$. Prove: for all $n \in \mathbb{Z}_{\geq 0}$, $a_n = (-1)^n + n4^n$.

8. Define $a_0 = 1$, $a_1 = 2$, $a_2 = 5$, and $a_n = 3a_{n-1} - 3a_{n-2} + a_{n-3}$ for all $n \in \mathbb{Z}_{\geq 3}$. Guess an explicit formula for a_n and prove your formula holds for all $n \in \mathbb{Z}_{\geq 0}$.

9. Define $b_1 = 1$, $b_2 = 2$, $b_3 = 3$, and $b_n = b_{n-1} + b_{n-2} + b_{n-3}$ for all $n \geq 4$. Prove by strong induction: for all $n \in \mathbb{Z}_{\geq 1}$, $b_n < 2^n$.

10. Let F_n be the Fibonacci sequence defined in Example 4.21. Find and prove a formula for $F_n^2 - F_{n+1}F_{n-1}$, valid for all $n \in \mathbb{Z}_{\geq 1}$.

11. Prove the following exact formula for the Fibonacci numbers F_n (see Example 4.21): for all $n \in \mathbb{Z}_{\geq 0}$, $F_n = [(1 + \sqrt{5})^n - (1 - \sqrt{5})^n]/(2^n\sqrt{5})$.

12. In Example 4.26, prove $a_n = 2^n$ for all integers $n \geq 0$ by ordinary induction (not strong induction).

13. Define $c_0 = 1$ and $c_{n+1} = 3\prod_{k=0}^{n} c_k$ for all integers $n \geq 0$. Guess an explicit formula for c_n and then prove it.

14. Define $d_0 = 1$ and $d_{n+1} = \sum_{k=0}^{n} 2^k d_k$ for all $n \in \mathbb{Z}_{\geq 0}$. Prove: for all integers $n \geq 2$, $d_n = \prod_{k=1}^{n-1}(2^k + 1)$.

15. Give just the structural outline of a proof by strong induction of this statement:

$$\forall n \in \mathbb{Z}_{\geq 1}, \exists a \in \mathbb{Z}_{\geq 0}, \exists b \in \mathbb{Z}_{\geq 0}, \exists c \in \mathbb{Z}_{\geq 0}, \exists d \in \mathbb{Z}_{\geq 0}, n = a^2 + b^2 + c^2 + d^2.$$

16. (a) Prove that for every integer $n \geq 0$, there exists $k \in \mathbb{Z}$ such that $n = 4k$ or $n = 4k+1$ or $n = 4k+2$ or $n = 4k+3$. (b) Prove that the result in (a) also holds for all negative integers n.

17. Using the theorem that every integer is even or odd, prove by strong induction: for all $n \in \mathbb{Z}_{\geq 1}$, there exist $k \in \mathbb{Z}_{\geq 1}$ and distinct exponents $e_1, e_2, \ldots, e_k \in \mathbb{Z}_{\geq 0}$ such that $n = 2^{e_1} + 2^{e_2} + \cdots + 2^{e_k}$.

18. (a) Prove: for all $n \in \mathbb{Z}_{\geq 3}, F_{n+1} < 2F_n$, where F_n is a Fibonacci number (Example 4.21). (b) Use strong induction to prove that every positive integer n can be expressed as a sum of one or more distinct Fibonacci numbers. *Suggestion:* Consider the largest Fibonacci number not exceeding n.

19. Consider the following template for proving $\forall n \in \mathbb{Z}_{\geq 0}, P(n)$:
    Step 1. Prove $P(0)$. Prove $P(1)$.
    Step 2. Fix $n \in \mathbb{Z}_{\geq 0}$. Assume $P(n)$ and $P(n+1)$. Prove $P(n+2)$.
    Prove the correctness of this template using ordinary induction (not strong induction).

20. (a) Create a strong induction proof template to prove statements of the form $\forall n \in \mathbb{Z}_{\leq c}, Q(n)$. (b) Use the Strong Induction Theorem 4.22 to prove that your template in (a) works.

21. Let $P(x)$ be the open sentence "$0 \leq x \leq 1$," where x is a real variable. (a) Disprove: $\forall x \in \mathbb{R}_{\geq 0}, P(x)$. (b) Prove: For all $x \in \mathbb{R}_{\geq 0}$, if (for all real numbers y in the range $0 \leq y < x$, $P(y)$ is true) then $P(x)$ is true. (c) Conclude that the Strong Induction Theorem 4.22 is *not* valid if we replace integers by real numbers.

22. Suppose we know: "for all $n \in \mathbb{Z}$, if (for all integers $m < n$, $P(m)$ is true), then $P(n)$ is true." Must it follow that "$\forall n \in \mathbb{Z}, P(n)$" is true? Explain.

23. Suppose we know that every nonempty subset of $\mathbb{Z}_{\geq 1}$ has a least element (see Exercise 18 in §4.1). Use this fact to prove the Strong Induction Theorem 4.22, taking $b = 1$. [*Suggestion:* Let S be the set of positive integers n such that $P(n)$ is false; prove $S = \emptyset$.]

4.4   Prime Numbers and Integer Division

Now that we have strong induction, we can investigate the divisibility properties of the positive integers more closely. We begin by defining prime numbers and showing that every integer $n > 1$ can be written as a product of primes. Then we study integer division, which is a process for dividing an integer a by a nonzero integer b to produce a unique quotient q and remainder r.

Primes and Prime Factorizations

A composite number is a positive integer that can be expressed as the product of two smaller positive integers, whereas a prime number cannot be written in this way. Formally, we have the following definition.

4.28. Definition: Prime and Composite Integers.

(a) For all $n \in \mathbb{Z}_{>1}$, $\boxed{n \text{ is composite}}$ iff $\boxed{\exists a \in \mathbb{Z}, \exists b \in \mathbb{Z}, (1 < a < n) \wedge (1 < b < n) \wedge n = ab}$.

(b) For all $n \in \mathbb{Z}_{>1}$, $\boxed{n \text{ is prime}}$ iff $\boxed{n \text{ is not composite}}$.

Note that we give these definitions only for integers $n > 1$. By convention, 1 *is not prime* and 1 *is not composite*. Similarly, 0 is neither prime nor composite. It can be proved from the above definition that *an integer $p > 1$ is prime iff the only positive divisors of p are 1 and p*. The first few primes are

$$2, 3, 5, 7, 11, 13, 17, 19, 23, 29, 31, 37, \ldots.$$

Our next theorem states that every integer in $\mathbb{Z}_{>1}$ can be factored into primes.

4.29. Theorem on Existence of Prime Factorizations. Every integer $n > 1$ can be written as a product of one or more primes. More precisely, for all $n \in \mathbb{Z}_{>1}$, there exist $k \in \mathbb{Z}_{\geq 1}$ and primes $p_1, p_2, \ldots, p_k$ with $n = p_1 p_2 \cdots p_k$. (The primes appearing here need not be distinct.)

Proof. We use strong induction on n. Fix an arbitrary integer $n \in \mathbb{Z}_{>1}$. Assume: for all integers m in the range $1 < m < n$, m can be written as a product of one or more primes. Prove n can be written as a product of one or more primes. We know n is prime or n is not prime, so use cases.

Case 1. Assume n is prime. Then n is a product of one prime, namely itself. (In other words, we can choose $k = 1$ and $p_1 = n$ above, to obtain $n = p_1$.)

Case 2. Assume n is not prime. Then n is composite, so there exist $a, b \in \mathbb{Z}$ with $1 < a < n$ and $1 < b < n$ and $n = ab$. Since a and b are both larger than 1 and less than n, we can apply the induction hypothesis to both a and b to see that a is a product of one or more primes, and b is a product of one or more primes. Say $a = p_1 p_2 \cdots p_k$ and $b = q_1 q_2 \cdots q_m$ where every p_i and q_j is prime. Then $n = ab = p_1 p_2 \cdots p_k q_1 q_2 \cdots q_m$, which expresses n as a product of $k + m$ primes.

Later, we strengthen this theorem by proving that the prime factorization of n is unique in a certain respect. More specifically, if we have written $n = p_1 p_2 \cdots p_k$ and $n = q_1 q_2 \cdots q_m$ where all p_i and q_j are prime, then we must have $k = m$ and $q_1, \ldots, q_k$ is a reordering of $p_1, \ldots, p_k$. For now, we use the existence of prime factorizations to prove a famous theorem of Euclid about prime numbers.

4.30. Theorem on Infinitude of Primes. There are infinitely many prime integers. More precisely, for any $k \in \mathbb{Z}_{\geq 1}$ and any finite list of primes $p_1, p_2, \ldots, p_k$, there exists a prime q different from every p_i.

Proof. Fix $k \in \mathbb{Z}_{\geq 1}$ and primes $p_1, p_2, \ldots, p_k$. Consider the integer $n = 1 + \prod_{r=1}^{k} p_r$. We know n is prime or n is not prime.

Case 1. Assume n is prime. Since n is larger than every p_i, n is a prime number distinct from all p_i.

Case 2. Assume n is not prime. Then we can write n as a product of primes, say $n = q_1 q_2 \cdots q_k$ with all q_j prime. We prove q_1 is a prime different from every p_i, using proof by contradiction. Assume, to get a contradiction, that there exists i with $q_1 = p_i$. Then p_i divides n since $n = q_1 q_2 \cdots q_k = p_i(q_2 \cdots q_k)$. Also p_i divides $\prod_{r=1}^{k} p_r$. It follows that p_i divides $1 = n - \prod_{r=1}^{k} p_r$. This is impossible, since $p_i > 1$. This contradiction completes the proof of Case 2.

Integer Division

In grade school, we learn to divide one integer by another, producing a *quotient* and a *remainder*. For example, if we divide 100 by 7, the quotient is 14 and the remainder is 2. One way to present this information is by writing $100/7 = 14 + (2/7)$. When studying integers, it is convenient to avoid fractions by multiplying this equation by the denominator 7. We obtain the integer equation $100 = 7 \cdot 14 + 2$.

In general, we want to be able to divide any nonnegative integer a by any positive integer b to produce a quotient q and a remainder r. In other words, given $a \geq 0$ and $b > 0$, we seek integers $q \geq 0$ and $r \geq 0$ satisfying $a/b = q + (r/b)$, or equivalently, $a = bq + r$. The remainder r should be small in the sense that r/b should be less than 1, which holds iff $0 \leq r < b$. Our next result shows that this division process is always possible. To motivate the proof, consider the following example.

4.31. Example. Let $a = 25$ and $b = 7$; find the quotient and remainder for this a and b.

Solution. The answer can be found by the long division algorithm learned in school. But here is a more basic solution relying on the idea that we can perform a division by repeated subtraction. Start with 25 and subtract 7 repeatedly. First, $25 - 7 = 18$. Second, $18 - 7 = 11$. Third, $11 - 7 = 4$. The current value 4 is less than 7, so we have found the remainder $r = 4$. We subtracted 7 three times to reach the remainder, so the quotient is $q = 3$.

The idea of the induction proof below is to recast this iterative computation in a recursive way. Specifically, in a single subtraction, we pass from the initial input 25 to $25 - 7 = 18$. We could recursively calculate the quotient and remainder when 18 is divided by 7 (namely, $q_1 = 2$ and $r_1 = 4$), and then obtain the quotient and remainder for 25 by adding 1 to the quotient (giving $q = 3$) and keeping the old remainder (giving $r = 4$). The initial condition for the recursive calculation occurs when the initial number a is already less than b. Then we can take $q = 0$ and $r = a$. For example, 4 divided by 7 has quotient 0 and remainder 4.

Let us see how the intuition in the preceding example translates into a formal proof.

4.32. Integer Division Theorem (Preliminary Version). For all integers $a \geq 0$ and $b > 0$, there exist integers q, r with $a = bq + r$ and $0 \leq r < b$.

Proof. To better understand how the induction proof handles quantifiers, we write the statement to be proved in formal symbols as follows:

$$\forall b \in \mathbb{Z}_{>0}, \forall a \in \mathbb{Z}_{\geq 0}, \exists q \in \mathbb{Z}, \exists r \in \mathbb{Z}, a = bq + r \wedge (0 \leq r < b).$$

Fix $b \in \mathbb{Z}_{>0}$. We prove the inner statement by strong induction on a. Fix an arbitrary integer $a \in \mathbb{Z}_{\geq 0}$. Assume: for all integers c in the range $0 \leq c < a$,

$$\exists q_1 \in \mathbb{Z}, \exists r_1 \in \mathbb{Z}, c = bq_1 + r_1 \text{ and } 0 \leq r_1 < b.$$

We must prove:

$$\exists q \in \mathbb{Z}, \exists r \in \mathbb{Z}, a = bq + r \text{ and } 0 \leq r < b.$$

We know $a < b$ or $a \geq b$, so use cases.

Case 1. Assume $a < b$. Choose $q = 0$ and $r = a$. We know $q, r \in \mathbb{Z}_{\geq 0}$, $a = bq + r$ (since $a = b \cdot 0 + a$) and $0 \leq a < b$, so our choice proves the existence statement.

Case 2. Assume $a \geq b$. In the induction hypothesis, let us take $c = a - b$. This is a legitimate value of c, since $c = a - b \geq 0$ and $c = a - b < a$ (because $b > 0$). Now, the induction hypothesis gives us integers $q_1, r_1 \geq 0$ with $c = bq_1 + r_1$ and $0 \leq r_1 < b$. Adding b to both sides, $a = c + b = bq_1 + r_1 + b = b(q_1 + 1) + r_1$. So, choosing $q = q_1 + 1$ and $r = r_1$ (which are in $\mathbb{Z}_{\geq 0}$), we have $a = bq + r$ and $0 \leq r < b$.

Integer Division for Negative Integers (Optional)

We now improve the Integer Division Theorem in two ways. First, we extend the result to the case where a and b might be negative integers (although b can never be zero). Second, we prove that the quotient and remainder are uniquely determined by a and b. The next several lemmas deal with these issues. We begin by allowing the dividend a to be negative; in this case, the quotient q might be negative.

4.33. Lemma: Extension of Integer Division to Negative a.

$$\forall b \in \mathbb{Z}_{>0}, \boxed{\forall a \in \mathbb{Z}}, \exists q \in \mathbb{Z}, \exists r \in \mathbb{Z}, a = bq + r \text{ and } 0 \leq r < b.$$

Proof. Fix $b \in \mathbb{Z}_{>0}$. Fix $a \in \mathbb{Z}$. We know $a \geq 0$ or $a < 0$, so use cases.

Case 1. Assume $a \geq 0$. Then the needed result follows from the previous theorem.

Case 2. Assume $a < 0$. Then $-a > 0$, so the previous theorem says there exist $q_2, r_2 \in \mathbb{Z}$ with $-a = bq_2 + r_2$ and $0 \leq r_2 < b$. Now, $a = -bq_2 - r_2$. [We are tempted to choose $q = -q_2$ and $r = -r_2$, but this produces a negative remainder most of the time. This suggests the following subcases.] We know $r_2 = 0$ or $r_2 > 0$, so use cases.

Case 2a. Assume $r_2 = 0$. Then $a = b(-q_2) + 0$, so we can choose $q = -q_2$ and $r = 0$.

Case 2b. Assume $r_2 > 0$. Then $-b < -r_2 < 0$, so (adding b) $0 < b - r_2 < b$. So we can manipulate $a = -bq_2 - r_2$ as follows: we know

$$a = -bq_2 - r_2 = -bq_2 - b + b - r_2 = b(-q_2 - 1) + (b - r_2).$$

Now choose $q = -q_2 - 1 \in \mathbb{Z}$ and $r = b - r_2 \in \mathbb{Z}$; observe that $0 \leq r < b$ for this choice of r.

4.34. Example. To illustrate the above proof, first suppose $a = -21$ and $b = 7$. Then $-a = 21 = 7 \cdot 3 + 0$ (so $q_2 = 3$, $r_2 = 0$), and we see that $a = -21 = 7 \cdot (-3) + 0$ (so $q = -3$, $r = 0$). On the other hand, suppose $a = -26$ and $b = 7$. Then $-a = 26 = 7 \cdot 3 + 5$ (so $q_2 = 3$, $r_2 = 5$). Negating gives $a = -26 = 7 \cdot (-3) - 5$, which has a forbidden negative remainder. We fix this by writing $a = -26 = 7 \cdot (-3) - 7 + 7 - 5 = 7 \cdot (-4) + 2$, so $q = -4 = -q_2 - 1$ and $r = 2 = b - r_2$, in accordance with the proof above.

For the next extension, we allow the divisor b to be negative. We still want nonnegative remainders, so we now require r to satisfy $0 \leq r < |b|$.

4.35. Lemma: Extension of Integer Division to Negative b.

$$\boxed{\forall b \in \mathbb{Z}_{\neq 0}}, \forall a \in \mathbb{Z}, \exists q \in \mathbb{Z}, \exists r \in \mathbb{Z}, a = bq + r \text{ and } 0 \leq r < |b|.$$

Proof. Fix $b \in \mathbb{Z}_{\neq 0}$. We know $b > 0$ or $b < 0$, so use cases.

<u>Case 1.</u> Assume $b > 0$. In this case, we already proved the needed result in the previous lemma.

<u>Case 2.</u> Assume $b < 0$. Then $b = -|b|$ where $|b| \in \mathbb{Z}_{>0}$. Applying the previous results to divide a by $|b|$, we see there exist integers $q_3, r_3 \in \mathbb{Z}_{\geq 0}$ with

$$a = |b|q_3 + r_3 \text{ and } 0 \leq r_3 < |b|.$$

Now, $|b| = -b$, so $a = (-b)q_3 + r_3 = b(-q_3) + r_3$. Choose $q = -q_3$ and $r_3 = r$ to prove the existence assertion in the lemma.

For example, dividing $a = -26$ by $b = -7$ gives quotient $q = 4$ and remainder $r = 2$.

Uniqueness of the Quotient and Remainder

The final, crucial improvement to our results so far is to prove that the quotient and remainder are *uniquely* determined by a and b. Here is the full-fledged statement of the Integer Division Theorem. The proof assumes some basic facts about absolute value are known; these facts are discussed in more detail in §8.4.

4.36. Integer Division Theorem (Final Version). For all integers a and all nonzero integers b, there exist unique integers q, r with $a = bq + r$ and $0 \leq r < |b|$.

Proof. First we restate the theorem in formal symbols to better understand the uniqueness assertion. We must prove:

$$\forall b \in \mathbb{Z}_{\neq 0}, \forall a \in \mathbb{Z}, \exists! \, (q, r) \in \mathbb{Z} \times \mathbb{Z}, a = bq + r \wedge (0 \leq r < |b|).$$

Fix arbitrary $b \in \mathbb{Z}_{\neq 0}$ and $a \in \mathbb{Z}$. We have already proved the existence of (q, r) above. To prove uniqueness, we use the uniqueness proof template (recall this arises from the rule for eliminating the uniqueness symbol). Fix arbitrary $(q_1, r_1) \in \mathbb{Z} \times \mathbb{Z}$ and arbitrary $(q_2, r_2) \in \mathbb{Z} \times \mathbb{Z}$. Assume $a = bq_1 + r_1$ and $0 \leq r_1 < |b|$ and $a = bq_2 + r_2$ and $0 \leq r_2 < |b|$. We must prove $(q_1, r_1) = (q_2, r_2)$, which means $q_1 = q_2$ and $r_1 = r_2$ (by the Ordered Pair Axiom). We first prove $q_1 = q_2$ by contradiction. Assume $q_1 \neq q_2$, and derive a contradiction. Combining the two equations for a assumed earlier, we get $bq_1 + r_1 = bq_2 + r_2$. Some algebra transforms this to $b(q_1 - q_2) = r_2 - r_1$. Since q_1 and q_2 are unequal integers, $|q_1 - q_2| \geq 1$, and hence $|r_2 - r_1| = |b||q_1 - q_2| \geq |b| \cdot 1 = |b|$. So $|r_2 - r_1| \geq |b|$. On the other hand, the assumed inequalities for r_1 and r_2 combine to show that $r_1 - r_2 < |b|$ and $r_2 - r_1 < |b|$. Since $|r_2 - r_1|$ equals $r_2 - r_1$ or $r_1 - r_2$, we therefore have $|r_2 - r_1| < |b|$. We have reached the contradiction "$|r_2 - r_1| \geq |b|$ and $|r_2 - r_1| < |b|$." Therefore, $q_1 = q_2$. Now, to prove $r_1 = r_2$, rewrite the assumed equations for a to get $r_1 = a - bq_1 = a - bq_2 = r_2$.

4.37. Remark. Consider the special case of the Division Theorem where $b = 2$. On one hand, the existence part of the theorem says that any $a \in \mathbb{Z}$ can be written in the form $a = 2q + r$ where $0 \leq r < 2$. In other words, for every $a \in \mathbb{Z}$, either there exists $q \in \mathbb{Z}$ with $a = 2q + 0$, or there exists $q \in \mathbb{Z}$ with $a = 2q + 1$. This says that every integer a is even or odd. On the other hand, one consequence of the uniqueness of (q, r) for a is that the remainders 0 and 1 cannot both work for the same fixed integer a. So, no integer a is both even and odd. This gives another proof of the theorem that $\boxed{\text{every integer is either even or odd, but not both}}$, which we used frequently in earlier sections.

We can obtain similar results for other specific choices of b. For example, letting $b = 3$, we see that *every integer a can be written in exactly one of the forms $a = 3q$, $a = 3q + 1$, or $a = 3q + 2$ for some integer q*. Similarly, taking $b = 10$, we see that *every integer a has the form $a = 10q + r$ for unique $q, r \in \mathbb{Z}$ with $r \in \{0, 1, 2, 3, 4, 5, 6, 7, 8, 9\}$*. This observation is the first step in rigorously proving the existence and uniqueness of the decimal expansions (or more generally, the base-b expansions) of integers. More details on this process appear in some of the exercises below.

Section Summary

1. *Prime and Composite Numbers.* An integer $n > 1$ is *composite* iff $n = ab$ for some integers a, b with $1 < a < n$ and $1 < b < n$. An integer $p > 1$ is *prime* iff p is not composite iff the only positive divisors of p are 1 and p.

2. *Facts about Primes.* Every integer $n > 1$ can be written as a product of one or more primes. There are infinitely many primes.

3. *Integer Division Theorem.* For all integers a and b with $b \neq 0$, there exist unique integers q and r such that $a = bq + r$ and $0 \leq r < |b|$. We call q the *quotient* and r the *remainder* when a is divided by b. Taking $b = 2$, we see that every integer is either even or odd, but not both.

Exercises

1. (a) List all primes less than 100. (b) Write each integer as a product of primes: 91, 96, 111, 3003, and 8000000.

2. For each a and b below, find the quotient and remainder when a is divided by b.
   (a) $a = 58, b = 11$ (b) $a = 58, b = -11$ (c) $a = -58, b = 11$ (d) $a = -58, b = -11$
   (e) $a = 0, b = 11$ (f) $a = 6, b = 11$ (g) $a = -6, b = 11$ (h) $a = 6, b = -11$.

3. For each a and b below, find the quotient and remainder when a is divided by b.
   (a) $a = 300, b = 27$ (b) $a = 9103, b = 11$ (c) $a = 4574, b = 19$ (d) $a = 100000, b = 123$.

4. Use Definition 4.28 to prove that an integer $p > 1$ is prime iff the only positive divisors of p are 1 and p.

5. Prove: $\forall n \in \mathbb{Z}_{>0}, \forall m \in \mathbb{Z}_{>0}$, if m is composite, then nm is composite.

6. True or false? Prove your answers.
   (a) The product of any two positive integers is always composite.
   (b) The sum of two prime integers is never prime.
   (c) The sum of any two composite integers is composite.
   (d) The product of any two composite integers is composite.
   (e) Every composite integer has an even number of positive divisors.

7. Suppose $k \in \mathbb{Z}_{>0}$ and $p_1, \ldots, p_k$ are integers. Give a careful proof that for all i between 1 and k, p_i divides $\prod_{r=1}^{k} p_r$. (This fact was used in the proof of Theorem 4.30.) *Suggestion:* Prove p_i divides $\prod_{r=1}^{s} p_r$ for all s in the range $i \leq s \leq k$ by induction.

8. (a) Prove that every prime integer other than 2 is odd. (b) Prove that every positive integer is either a power of 2 or is divisible by an odd prime.

9. Use the uniqueness assertion in the Integer Division Theorem to prove carefully that 5 does not divide 22, without assuming in advance that $22/5$ is not an integer.

10. Use the technique in the previous exercise to give a fully rigorous proof that 7 is prime. (Use proof by exhaustion to show that for all $b \in \{2, 3, 4, 5, 6\}$, b does not divide 7.)

11. The following statements resemble the Integer Division Theorem, but they are all false. Negate and disprove each statement.
    (a) $\forall a \in \mathbb{Z}, \forall b \in \mathbb{Z}, \exists! (q,r) \in \mathbb{Z} \times \mathbb{Z}, (a = bq + r \wedge 0 \leq r < |b|)$.
    (b) $\forall a \in \mathbb{Z}, \forall b \in \mathbb{Z}_{\neq 0}, \exists! (q,r) \in \mathbb{Z} \times \mathbb{Z}, (a = bq + r \wedge 0 \leq r < b)$.
    (c) $\forall a \in \mathbb{Z}, \forall b \in \mathbb{Z}_{\neq 0}, \exists! (q,r) \in \mathbb{Z} \times \mathbb{Z}, (a = bq + r \wedge 0 \leq r \leq |b|)$.
    (d) $\forall a \in \mathbb{Z}, \forall b \in \mathbb{Z}_{\neq 0}, \exists! (q,r) \in \mathbb{Z}_{\geq 0} \times \mathbb{Z}_{\geq 0}, (a = bq + r \wedge 0 \leq r < |b|)$.

12. What are the possible remainders when a prime integer $p > 10$ is divided by 10? Prove your answer.

13. What are the possible remainders when a square integer n^2 is divided by 8? Prove your answer.

14. (a) Prove or disprove: for all open sentences $P(q,r)$,
$$[\exists! (q,r) \in \mathbb{Z} \times \mathbb{Z}, P(q,r)] \quad \Rightarrow \quad [\exists! q \in \mathbb{Z}, \exists! r \in \mathbb{Z}, P(q,r)].$$
    (b) Prove or disprove: for all open sentences $P(q,r)$,
$$[\exists! q \in \mathbb{Z}, \exists! r \in \mathbb{Z}, P(q,r)] \quad \Rightarrow \quad [\exists! (q,r) \in \mathbb{Z} \times \mathbb{Z}, P(q,r)].$$

15. (a) For $a, b \in \mathbb{Z}_{>0}$, how are the quotient and remainder when a is divided by b related to the quotient and remainder when $a + 1$ is divided by b? (b) Prove the preliminary version of the Integer Division Theorem by fixing $b > 0$ and using ordinary induction (not strong induction) on a.

16. Prove this variation of the Integer Division Theorem: for all $a, b \in \mathbb{Z}$ with $b \neq 0$, there exist unique integers q, r with $a = bq + r$ and $-|b|/2 < r \leq |b|/2$.

17. Fix $r_0 \in \mathbb{Z}$. State and prove a version of the Integer Division Theorem where the remainder r is required to satisfy $r_0 \leq r < r_0 + |b|$.

18. For $a \in \mathbb{Z}$, let $a \bmod 4 \in \{0,1,2,3\}$ be the remainder when a is divided by 4. (a) Prove by induction: for all $n, a_1, a_2, \ldots, a_n \in \mathbb{Z}_{>0}$, if $a \bmod 4 = 1$ for all i, then $(\prod_{i=1}^n a_i) \bmod 4 = 1$. (b) Prove there exist infinitely many primes p such that $p \bmod 4 = 3$. [Adapt the proof of Theorem 4.30.]

19. *Decimal Expansions of Integers.* (a) Prove by induction that every positive integer n can be written in the form $n = \sum_{k=0}^m d_k 10^k$ where $m \geq 0$, each $d_k \in \{0,1,2,\ldots,9\}$, and $d_m \neq 0$. (b) Prove that m and $d_0, d_1, \ldots, d_m$ in part (a) are uniquely determined by n.

20. State and prove a generalization of the previous exercise in which 10 is replaced by an arbitrary *base* $b \in \mathbb{Z}_{>1}$.

21. (a) Prove: for all $x, c, d \in \mathbb{Z}_{>0}$, $x^{cd} - 1 = (x^c - 1)\sum_{k=0}^{d-1} x^{ck}$.
    (b) Suppose $a \geq 0$, $b > 0$, and dividing a by b yields quotient q and remainder r. For fixed $x \in \mathbb{Z}_{>1}$, find (with proof) the quotient and remainder when $x^a - 1$ is divided by $x^b - 1$.

4.5 Greatest Common Divisors

In this section, we study the greatest common divisor (gcd) of two integers a and b, denoted $\gcd(a, b)$. We present Euclid's algorithm for computing $\gcd(a, b)$ based on repeated division. One consequence of this algorithm is the fact that $\gcd(a, b)$ can always be written in the form $ax + by$ for some integers x and y. This fact has important applications in modern cryptography.

Greatest Common Divisors

Before discussing greatest common divisors, we recall the definition of divisibility.

4.38. Definition: Divisors. For all integers a and c, $\boxed{c \text{ is a divisor of } a}$ iff $\boxed{\exists u \in \mathbb{Z}, a = cu}$. When this condition holds, we also say that $\boxed{c \text{ divides } a}$ and $\boxed{a \text{ is a multiple of } c}$. The notation $\boxed{c|a}$ means that c divides a.

Note that $c|a$ stands for the *statement* "c divides a," whereas c/a denotes the *rational number* obtained when c is divided by a. Take care not to confuse these two symbols!

4.39. Definition: Greatest Common Divisors. Given integers a, b that are not both zero, the *greatest common divisor* $\gcd(a, b)$ is the largest integer d such that d divides a and d divides b. We also define $\gcd(0, 0) = 0$.

As a technical aside, it can be shown that $\gcd(a, b)$ exists and is unique, using the fact that all divisors d of a nonzero integer a satisfy $d \leq |a|$; see Theorem 8.58(b).

4.40. Example. Let $a = 88$ and $b = 60$. The positive divisors of a are $1, 2, 4, 8, 11, 22, 44, 88$. The positive divisors of b are $1, 2, 3, 4, 5, 6, 10, 12, 15, 20, 30, 60$. Comparing these lists, we see that the positive integers d dividing both a and b are $1, 2,$ and 4. So $\gcd(88, 60) = 4$. Another way to compute greatest common divisors is to compare prime factorizations. Here, $88 = 2^3 \cdot 11$ and $60 = 2^2 \cdot 3 \cdot 5$. Both 88 and 60 are divisible by two copies of the prime 2 but have no other common prime factors, so we conclude that $\gcd(88, 60) = 2^2 = 4$. (See Theorem 4.60 for a justification of this technique.)

It turns out that the two methods of finding greatest common divisors just discussed — listing all the divisors, or comparing prime factorizations — are horribly inefficient for large integers a and b. A much faster algorithm was discovered thousands of years ago by Euclid, the author of the famous geometry textbook *Elements* (which is also a text on number theory). The basis for this algorithm is the following result relating integer division to greatest common divisors.

4.41. Theorem on Greatest Common Divisors. (a) For all $a \in \mathbb{Z}$, $\gcd(a, 0) = |a|$. (b) For all $a, b, q, r \in \mathbb{Z}$, if $b \neq 0$ and $a = bq + r$, then $\gcd(a, b) = \gcd(b, r)$.
Proof. (a) Fix $a \in \mathbb{Z}$. We know $a = 0$ or $a \neq 0$, so consider cases.
Case 1. Assume $a = 0$. Here, $\gcd(a, 0) = 0 = |a|$ by definition.
Case 2. Assume $a \neq 0$. We know $d|0$ for all integers d, since $0 = d \cdot 0$. Therefore, the *common* divisors of a and 0 are exactly the same integers as the *divisors* of a itself. We know the largest divisor of a is $|a|$, so $\gcd(a, 0) = |a|$ in this case.
 (b) Fix $a, b, q, r \in \mathbb{Z}$. Assume $b \neq 0$ and $a = bq + r$. Prove $\gcd(a, b) = \gcd(b, r)$. Our strategy is to prove that the set of common divisors of a and b coincides with the set of

common divisors of b and r; since it then follows that the largest element of the first set (namely $\gcd(a, b)$) must equal the largest element of the second set (namely $\gcd(b, r)$).

To show these two sets are equal, fix an integer d; we prove that $d|a$ and $d|b$ iff $d|b$ and $d|r$. For one direction, assume $d|a$ and $d|b$, so $a = ds$ and $b = dt$ for some $s, t \in \mathbb{Z}$. We must prove $d|b$ and $d|r$. We already assumed $d|b$. Also, $r = a - bq = ds - dtq = d(s - tq)$ where $s - tq$ is an integer, so $d|r$. For the other direction, assume $d|b$ and $d|r$, so $b = dt$ and $r = du$ for some $t, u \in \mathbb{Z}$. We must prove $d|a$ and $d|b$. We already assumed $d|b$. Also, $a = bq + r = dtq + du = d(tq + u)$ where $tq + u$ is an integer, so $d|a$.

Euclid's GCD Algorithm

The previous theorem justifies the following recursive algorithm for computing $\gcd(a, b)$, called *Euclid's Algorithm*.

4.42. Euclid's Recursive GCD Algorithm.
Input: $a, b \in \mathbb{Z}_{\geq 0}$.
Output: $\gcd(a, b)$.
Procedure: (a) If $b = 0$, return a.
(b) If $b > 0$, use integer division to write $a = bq + r$ for some $q, r \in \mathbb{Z}$ with $0 \leq r < b$; return $\gcd(b, r)$.

It can be proved, using strong induction on b, that this algorithm terminates with the correct answer after finitely many steps. The key observation is that the recursive call in step (b) replaces the second input b by the smaller nonnegative integer r.

4.43. Example. Let us recompute $\gcd(88, 60)$ using Euclid's algorithm. Initially $a = 88$ and $b = 60$, so we are in case (b) of the algorithm. We divide a by b, getting:

$$88 = 60 \cdot 1 + 28; \qquad \gcd(88, 60) = \gcd(60, 28).$$

In the first recursive call, the inputs are 60 and 28. Dividing again gives

$$60 = 28 \cdot 2 + 4; \qquad \gcd(60, 28) = \gcd(28, 4).$$

The new inputs are 28 and 4; dividing gives

$$28 = 4 \cdot 7 + 0; \qquad \gcd(28, 4) = \gcd(4, 0).$$

By case (a) in the algorithm, $\gcd(4, 0) = 4$. Working our way back up through the recursive calls, we see $\gcd(88, 60) = 4$.

Although we described the gcd algorithm recursively, the previous example shows how to unravel this recursion into an iterative procedure consisting of a series of integer divisions. Specifically, assuming we start with positive inputs a and b, the algorithm performs the following divisions:

$$
\begin{aligned}
a &= bq_1 + r_1, & 0 &< r_1 < b; \\
b &= r_1 q_2 + r_2, & 0 &< r_2 < r_1; \\
r_1 &= r_2 q_3 + r_3, & 0 &< r_3 < r_2; \\
r_2 &= r_3 q_4 + r_4, & 0 &< r_4 < r_3; \qquad\qquad (4.1) \\
&\;\cdots & &\;\cdots \\
r_{k-2} &= r_{k-1} q_k + r_k, & 0 &< r_k < r_{k-1}; \\
r_{k-1} &= r_k q_{k+1} + 0, & &\text{(remainder } r_{k+1} \text{ is zero).}
\end{aligned}
$$

When we finally reach a remainder $r_{k+1} = 0$, we return $\gcd(a, b) = r_k$, the last nonzero remainder. This must eventually occur, since $b > r_1 > r_2 > r_3 > \cdots$ and all remainders are nonnegative integers.

4.44. Example. Let us use the algorithm to compute $\gcd(171, 51)$:

$$171 \ = \ 51 \cdot 3 + 18, \tag{4.2}$$
$$51 \ = \ 18 \cdot 2 + 15, \tag{4.3}$$
$$18 \ = \ 15 \cdot 1 + 3, \tag{4.4}$$
$$15 \ = \ 3 \cdot 5 + 0. \tag{4.5}$$

So $\gcd(171, 51) = 3$.

Linear Combination Property of GCDs

One of the most important properties of greatest common divisors is that for all $a, b \in \mathbb{Z}$, $\gcd(a, b)$ can be written in the form $ax + by$ for some integers x and y. In fact, we can enhance Euclid's algorithm with some extra computations that find x and y explicitly. The idea is to work backwards through the chain of divisions to express $\gcd(a, b)$ (which is the last nonzero remainder) in terms of quantities appearing earlier. We illustrate this idea by revisiting the two previous examples.

4.45. Example. We computed $\gcd(88, 60) = 4$; let us now find $x, y \in \mathbb{Z}$ with $4 = 88x + 60y$. Recall the divisions executed by the recursive algorithm to find $\gcd(88, 60)$:

$$88 = 60 \cdot 1 + 28; \qquad 60 = 28 \cdot 2 + 4; \qquad 28 = 4 \cdot 7 + 0.$$

The gcd 4 appears as the remainder in the second equation. Isolating the 4, we see that $4 = 60 + 28 \cdot (-2)$. Now, from the first equation above, $28 = 88 - 60$. Substituting this expression for 28 into our expression for 4, we find that

$$4 = 60 + 28 \cdot (-2) = 60 + (88 - 60) \cdot (-2) = 88 \cdot (-2) + 60 \cdot 3.$$

The last step used the distributive law to collect terms; note that $(88 - 60) \cdot (-2)$ contributes $(-1)(-2) = 2$ copies of 60, and there is one more copy of 60 added in front, for a total of 3. We now see that $4 = 88x + 60y$ holds if we choose $x = -2$ and $y = 3$.

Before looking at the next example, let us introduce some terminology.

4.46. Definition: Linear Combinations. For all $d, s, t \in \mathbb{Z}$,

$\boxed{d \text{ is a linear combination of } s \text{ and } t}$ iff $\boxed{\exists x, y \in \mathbb{Z}, d = sx + ty}$.

4.47. Example. Let us find $x, y \in \mathbb{Z}$ with $3 = \gcd(171, 51) = 171x + 51y$. Our strategy is to use the divisions (4.4), (4.3), (4.2) to express 3 as a linear combination of 18 and 15, then as a linear combination of 51 and 18, and finally as a linear combination of 171 and 51. First, from (4.4),

$$3 = 18 \cdot 1 + 15 \cdot (-1).$$

Now from (4.3), $15 = 51 + 18 \cdot (-2)$. Substituting into the previous equation and collecting terms,

$$3 = 18 \cdot 1 + [51 + 18 \cdot (-2)] \cdot (-1) = 18 \cdot 3 + 51 \cdot (-1).$$

Note that the new coefficient of 18 was computed from the distributive law as $1 + (-2)(-1) = 3$. Now from (4.2), $18 = 171 + 51 \cdot (-3)$. Substituting into the previous equation and collecting terms,

$$3 = [171 + 51 \cdot (-3)] \cdot 3 + 51 \cdot (-1) = 171 \cdot 3 + 51 \cdot (-10).$$

Note that the new coefficient of 51 was computed from the distributive law as $(-3) \cdot 3 + (-1) = -10$. So $3 = 171x + 51y$ holds if we choose $x = 3$ and $y = -10$.

In summary, *to find x and y with $\gcd(a, b) = ax + by$:* first find the gcd using repeated divisions as in (4.1); then work backwards to express the gcd r_k as a linear combination of r_{k-2} and r_{k-1}, then as a linear combination of r_{k-3} and r_{k-2}, and so on, until finally r_k has been expressed as a linear combination of the original inputs a and b. We do one more complete example to illustrate the full process.

4.48. Example. Find $\gcd(999, 101)$, and find $x, y \in \mathbb{Z}$ with $\gcd(999, 101) = 999x + 101y$.
Solution. First find the gcd by repeated divisions:

$$999 = 101 \cdot 9 + 90,$$
$$101 = 90 \cdot 1 + 11,$$
$$90 = 11 \cdot 8 + 2,$$
$$11 = 2 \cdot 5 + 1,$$
$$2 = 2 \cdot 1 + 0.$$

So $\gcd(999, 101) = 1$. Next, work backwards through the equations, taking care to *collect common terms at each stage*, as shown in the far right column below:

$$
\begin{aligned}
1 \quad &= 11 + 2 \cdot (-5) \\
&= 11 + [90 + 11 \cdot (-8)] \cdot (-5) && = 11 \cdot 41 + 90 \cdot (-5) \\
&= [101 + 90 \cdot (-1)] \cdot 41 + 90 \cdot (-5) && = 101 \cdot 41 + 90 \cdot (-46) \\
&= 101 \cdot 41 + (999 + 101 \cdot (-9)) \cdot (-46) && = 101 \cdot 455 + 999 \cdot (-46).
\end{aligned}
$$

So $1 = 999x + 101y$ holds with $x = -46$ and $y = 455$.

We prove the linear combination property of gcds in §4.6.

Matrix Reduction Algorithm for Computing GCDs (Optional)

In this optional section, we present another algorithm for computing $d = \gcd(a, b)$ and finding integers x, y such that $d = ax + by$. This algorithm proceeds by performing certain *elementary row operations* on matrices with integer entries. There are three allowable row operations: (i) switch two rows in a matrix; (ii) multiply one row in a matrix by -1; (iii) add any *integer* multiple of one row of a matrix to a different row.

You may have studied similar elementary row operations for real-valued matrices in linear algebra. In the real case, we can add any *real* multiple of one row to another row in operation (iii), and we can multiply a given row by any nonzero real number in operation (ii). These row operations are used in the Gaussian elimination algorithm to solve a system of linear equations by row-reducing the augmented matrix for the system.

Our new algorithm for computing $\gcd(a, b)$ is analogous to Gaussian elimination, but we are only allowed to use the integer versions of the elementary row operations. Given nonzero integers a and b as input, we begin with the matrix

$$\left[\begin{array}{cc|c} 1 & 0 & a \\ 0 & 1 & b \end{array} \right].$$

Next we perform a sequence of row operations whose goal is to produce a zero in the rightmost column. For example, we could start by dividing a by b, obtaining a quotient q_1 and a remainder r_1 such that $a = bq_1 + r_1$ and $0 \le r_1 < |b|$. Adding $-q_1$ times row 2 to row 1 produces the new matrix

$$\left[\begin{array}{cc|c} 1 & -q_1 & r_1 \\ 0 & 1 & b \end{array} \right].$$

If r_1 is nonzero, we could continue by dividing b by r_1, obtaining q_2 and r_2 with $b = r_1 q_2 + r_2$ and $0 \leq r_2 < |r_1|$. Adding $-q_2$ times row 1 to row 2 leads to the matrix

$$\begin{bmatrix} 1 & -q_1 & \big| & r_1 \\ -q_2 & 1 + q_1 q_2 & \big| & r_2 \end{bmatrix}.$$

We can continue in this way, following the same sequence of division steps used in (4.1) to row-reduce the matrix. We ultimately arrive at a matrix of the form

$$\begin{bmatrix} x & y & \big| & r_k \\ u & v & \big| & 0 \end{bmatrix},$$

where a zero has appeared in the rightmost column. Remarkably, *the entries x, y, and r_k in the other row now satisfy $r_k = \gcd(a,b) = ax + by$.* A proof of this assertion is outlined in the exercises.

4.49. Example. We illustrate the matrix reduction algorithm and its connection to the previous algorithm by solving Example 4.48 once again. The row-reduction steps are shown here:

$$\begin{bmatrix} 1 & 0 & \big| & 999 \\ 0 & 1 & \big| & 101 \end{bmatrix} \xrightarrow{R_1 - 9R_2} \begin{bmatrix} 1 & -9 & \big| & 90 \\ 0 & 1 & \big| & 101 \end{bmatrix} \xrightarrow{R_2 - R_1} \begin{bmatrix} 1 & -9 & \big| & 90 \\ -1 & 10 & \big| & 11 \end{bmatrix} \xrightarrow{R_1 - 8R_2} \begin{bmatrix} 9 & -89 & \big| & 2 \\ -1 & 10 & \big| & 11 \end{bmatrix}$$

$$\xrightarrow{R_2 - 5R_1} \begin{bmatrix} 9 & -89 & \big| & 2 \\ -46 & 455 & \big| & 1 \end{bmatrix} \xrightarrow{R_1 - 2R_2} \begin{bmatrix} 101 & -999 & \big| & 0 \\ -46 & 455 & \big| & 1 \end{bmatrix}$$

Looking at row 2, we read off $1 = \gcd(999, 101) = 999 \cdot (-46) + 101 \cdot 455$.

The matrix reduction algorithm is really the same as Euclid's algorithm based on repeated division, but the matrix formulation has several advantages. First, the algorithm both computes the gcd and writes it as a linear combination of the original inputs in a single computation; there is no need to backtrack through a chain of divisions to find x and y. Second, when executing the original algorithm by hand, it is easy to make arithmetic errors when substituting expressions into other expressions and using the distributive law. The matrix algorithm executes the same arithmetic via row operations, where mistakes are less likely. Third, we are free to use any row operations we like, not just those based on the divisions in (4.1). By choosing row operations wisely, we can sometimes reach the goal in fewer steps, as seen in the next example.

4.50. Example. We compute $\gcd(119, 21)$ as follows:

$$\begin{bmatrix} 1 & 0 & \big| & 119 \\ 0 & 1 & \big| & 21 \end{bmatrix} \xrightarrow{R_1 - 6R_2} \begin{bmatrix} 1 & -6 & \big| & -7 \\ 0 & 1 & \big| & 21 \end{bmatrix} \xrightarrow{R_2 + 3R_1} \begin{bmatrix} 1 & -6 & \big| & -7 \\ 3 & -17 & \big| & 0 \end{bmatrix} \xrightarrow{R_1 \times -1} \begin{bmatrix} -1 & 6 & \big| & 7 \\ 3 & -17 & \big| & 0 \end{bmatrix}.$$

We see that $\gcd(119, 21) = 7 = 119 \cdot (-1) + 21 \cdot 6$. By subtracting six copies of R_2 at stage 1 instead of five copies, we made the gcd appear more quickly (up to a sign, which is easily removed). As additional savings, the row operation adding three copies of row 1 to row 2 is not really needed, once we notice that -7 divides 21 in the last column.

Section Summary

1. *Facts about Divisors.* For integers a and c, the notation $c|a$ means c divides a (i.e., $\exists u \in \mathbb{Z}, a = cu$). For all $a, b, c, x, y \in \mathbb{Z}$, c divides a iff $|c|$ divides $|a|$; if c divides a and $a > 0$, then $c \leq a$; if $c|a$ and $c|b$, then $c|(ax + by)$.

2. *Greatest Common Divisors.* For integers a and b (not both zero), $\gcd(a, b)$ is the largest integer d such that $d|a$ and $d|b$. We have $\gcd(a, b) = \gcd(|a|, |b|)$; $\gcd(a, 0) = |a|$; and if $a = bq + r$ for some integers q, r, then $\gcd(a, b) = \gcd(b, r)$. We can always write $\gcd(a, b) = ax + by$ for some integers x, y; we say that $\gcd(a, b)$ is a linear combination of a and b.

3. *Euclid's GCD Algorithm.* To find the gcd of positive integers a and b, divide a by b to get a remainder r_1; divide b by r_1 to get a remainder r_2; divide r_1 by r_2 to get a remainder r_3; and so on. The last nonzero remainder r_k is $\gcd(a, b)$. Working backwards through the divisions, we can express r_k as a linear combination of r_{k-1} and r_{k-2}, then as a linear combination of r_{k-2} and r_{k-3}, until eventually r_k has been written as a linear combination of the original inputs a and b.

4. *Matrix Reduction Algorithm for GCDs.* To find the gcd of positive integers a and b, start with the matrix $\begin{bmatrix} 1 & 0 & a \\ 0 & 1 & b \end{bmatrix}$. Repeatedly add integer multiples of one row to the other row (or multiply one row by -1) until a zero appears in the last column. If the other row has entries x, y, d with $d > 0$, then $d = \gcd(a, b) = ax + by$.

Exercises

1. For each pair a, b below, compute $\gcd(a, b)$ using repeated division.
   (a) $a = 98$, $b = 21$ (b) $a = 228$, $b = 168$ (c) $a = 100$, $b = 39$ (d) $a = 513$, $b = 252$.

2. For each pair a, b in the previous problem, find integers x, y with $\gcd(a, b) = ax + by$ by working backwards through the chain of divisions.

3. For each pair a, b below, compute $d = \gcd(a, b)$ and integers x, y with $d = ax + by$ using the matrix reduction algorithm.
   (a) $a = 144$, $b = 89$ (b) $a = 516$, $b = 215$ (c) $a = 111111$, $b = 117845$.

4. Prove: for all $a, b \in \mathbb{Z}$, $\gcd(a, b) = \gcd(|a|, |b|)$.

5. Explain how Euclid's algorithm computes $\gcd(a, b)$ in each of these special cases:
   (a) $a = b = 0$; (b) $a = 0 \neq b$; (c) b divides a; (d) a divides b.

6. Prove by strong induction on b that the recursive Algorithm 4.42 always terminates in finitely many steps with the correct answer.

7. Use induction on n to prove: for all integers $n > 0, d, x_1, \ldots, x_n, a_1, \ldots, a_n$, if $d|x_i$ for all $i \in \{1, 2, \ldots, n\}$, then $d|(a_1x_1 + \cdots + a_nx_n)$.

8. (a) Prove: for all positive integers a, b, $\gcd(b, a) = \gcd(a, b) = \gcd(a - b, b)$.
   (b) Use part (a) to formulate a new recursive algorithm for computing gcds. Compare this algorithm to Euclid's gcd algorithm.

9. Prove or disprove: for all positive integers a, b, $\gcd(a, b) = \gcd(a + b, a - b)$.

10. Suppose p is prime and $a \in \mathbb{Z}$. What are the possible values of $\gcd(a, p)$? Under what conditions on a does each value occur?

11. Prove: for all $p, a, b \in \mathbb{Z}$, if p is prime and p divides ab, then p divides a or p divides b. [Use the previous exercise and the fact that $\gcd(a, p)$ is a linear combination of a and p.]

12. For fixed $n \in \mathbb{Z}_{>0}$, find all possible values of $\gcd(8n + 1, 5n - 3)$ and determine when each value occurs. Also write the gcd as a linear combination of $8n + 1$ and $5n - 3$.

13. (a) Prove: for all $n \in \mathbb{Z}_{>0}$, $\gcd(n^3, n^2 + n + 1) = 1$. (b) Write 1 as an explicit linear combination of n^3 and $n^2 + n + 1$.

14. Prove: for all positive integers a, b, c, $\gcd(ca, cb) = c \cdot \gcd(a, b)$. [*Hint:* Examine how Euclid's Algorithm acts on inputs ca, cb compared to inputs a, b by multiplying each line of (4.1) by c.]

15. Define the Fibonacci numbers by $F_0 = 0$, $F_1 = 1$, and $F_n = F_{n-1} + F_{n-2}$ for all $n \geq 2$. (a) Prove by induction: for all $n \geq 1$, $\gcd(F_n, F_{n-1}) = 1$. (b) For $1 \leq n \leq 6$, express 1 as a linear combination of F_n and F_{n-1}. (c) Based on your answers to (b), conjecture and then prove a formula (valid for all $n \geq 1$) expressing 1 as a linear combination of F_n and F_{n-1}.

16. Suppose a and b are fixed integers, $A = \begin{bmatrix} x & y & | & z \\ u & v & | & w \end{bmatrix}$ is a matrix with integer entries such that $ax + by = z$ and $au + bv = w$, and $B = \begin{bmatrix} x' & y' & | & z' \\ u' & v' & | & w' \end{bmatrix}$ is obtained from A by performing one of the elementary row operations (i), (ii), or (iii) described on page 188. Prove that the entries of B are all integers, $ax' + by' = z'$, $au' + bv' = w'$, and $\gcd(z, w) = \gcd(z', w')$.

17. Suppose a and b are fixed integers, $A_0 = \begin{bmatrix} 1 & 0 & | & a \\ 0 & 1 & | & b \end{bmatrix}$, and $A_0, A_1, A_2, \ldots, A_k$ is any finite sequence of matrices such that for $0 \leq i < k$, A_{i+1} is obtained from A_i by an elementary row operation for integer matrices. Write $A_i = \begin{bmatrix} x_i & y_i & | & z_i \\ u_i & v_i & | & w_i \end{bmatrix}$. Use induction on k and the previous exercise to prove that for all i between 0 and k, $ax_i + by_i = z_i$, $au_i + bv_i = w_i$ and $\gcd(z_i, w_i) = \gcd(a, b)$.

18. Prove that the matrix reduction algorithm for finding $\gcd(a, b)$ always terminates in finitely many steps with the correct answer, assuming that at each step we perform any elementary row operation that reduces the maximum absolute value n of the two entries in the third column. (Use strong induction on n.) Also explain why, at each step before the end, at least one such row operation must be possible.

19. Given $a, b \in \mathbb{Z}_{>0}$, write $a = bq + r$ where $q, r \in \mathbb{Z}$ and $0 \leq r < b$. Prove: for all $x \in \mathbb{Z}_{>1}$, $\gcd(x^a - 1, x^b - 1) = \gcd(x^b - 1, x^r - 1)$. (*Hint:* See Exercise 21 in §4.4.)

20. Use the previous exercise to prove: for all positive integers a, b, x with $x > 1$, $\gcd(x^a - 1, x^b - 1) = x^{\gcd(a,b)} - 1$.

4.6 GCDs and Uniqueness of Prime Factorizations

In the last section, we looked at specific examples where $\gcd(a, b)$ could be expressed in the form $ax + by$ for certain integers x and y. Here we prove that such expressions always exist. We then use this result to obtain further facts about prime integers. In particular, we prove that the factorization of a positive integer into a product of primes is unique except for rearranging the order of the factors.

Proof of the Linear Combination Property of GCDs

We now prove the linear combination property of gcds for all integers $a, b \geq 0$; an exercise asks you to generalize to the case where a or b could be negative.

4.51. Theorem on Writing GCDs as Linear Combinations. For all integers $a, b \geq 0$, there are integers x and y such that $\gcd(a, b) = ax + by$.

Proof. To clarify the structure of the induction proof, we write the theorem statement in formal symbols as follows:

$$\forall b \in \mathbb{Z}_{\geq 0}, \forall a \in \mathbb{Z}_{\geq 0}, \exists x \in \mathbb{Z}, \exists y \in \mathbb{Z}, \gcd(a, b) = ax + by.$$

We prove this statement by strong induction on b. Fix an arbitrary integer $b \in \mathbb{Z}_{\geq 0}$. [Next we must state the strong induction hypothesis. We change to new letters here to avoid a notation conflict between the assumed statement and the statement to be proved.] Assume: for all integers m in the range $0 \leq m < b$,

$$\forall c \in \mathbb{Z}_{\geq 0}, \exists s \in \mathbb{Z}, \exists t \in \mathbb{Z}, \gcd(c, m) = cs + mt. \tag{4.6}$$

We must prove

$$\forall a \in \mathbb{Z}_{\geq 0}, \exists x \in \mathbb{Z}, \exists y \in \mathbb{Z}, \gcd(a, b) = ax + by. \tag{4.7}$$

We know $b = 0$ or $b > 0$, so consider two cases.

Case 1. Assume $b = 0$; prove (4.7). Fix an arbitrary integer $a \in \mathbb{Z}_{\geq 0}$; prove $\exists x \in \mathbb{Z}, \exists y \in \mathbb{Z}, \gcd(a, 0) = ax + 0y$. We know $\gcd(a, 0) = a$, so choose $x = 1$ and $y = 0$. Note 1 and 0 are in $\mathbb{Z}$, and $\gcd(a, 0) = a = a \cdot 1 + 0 \cdot 0$.

Case 2. Assume $b > 0$; prove (4.7). Fix an arbitrary integer $a \in \mathbb{Z}_{\geq 0}$; prove $\exists x \in \mathbb{Z}, \exists y \in \mathbb{Z}, \gcd(a, b) = ax + by$. Because $b > 0$, we can divide a by b, obtaining integers q, r with $a = bq + r$ and $0 \leq r < b$. By the Theorem on Greatest Common Divisors, we know $\gcd(a, b) = \gcd(b, r)$. The key observation is that $0 \leq r < b$, so that we are allowed to choose $m = r$ in the induction hypothesis. By the Inference Rule for ALL, we can take $c = b$ in (4.6), so that the induction hypothesis tells us there are integers s and t with $\gcd(b, r) = bs + rt$. We also know $r = a - bq$, and hence

$$\gcd(a, b) = \gcd(b, r) = bs + rt = bs + (a - bq)t = at + b(s - qt).$$

So, choosing $x = t$ and $y = s - qt$ (which are integers since $s, q, t \in \mathbb{Z}$), we do have $\gcd(a, b) = ax + by$, as needed. This completes the induction proof.

We can translate this proof into a recursive algorithm that takes inputs $a, b \in \mathbb{Z}_{\geq 0}$ and returns $d, x, y \in \mathbb{Z}$ such that $d = \gcd(a, b) = ax + by$. The base case of the algorithm occurs when $b = 0$; then we return $d = a$, $x = 1$, and $y = 0$. Otherwise, when $b > 0$, divide a by b to get $q, r \in \mathbb{Z}$ with $a = bq + r$ and $0 \leq r < b$. Recursively compute d, s, t with $d = \gcd(b, r) = bs + rt$; and return d as the gcd, $x = t$, and $y = s - qt$. Tracing through all the recursive calls, you can check that the net effect of this recursive algorithm agrees with the iterative computation presented in the last section, where we worked backwards through the chain of divisions to find x and y.

Properties of Prime Numbers

In §4.4, we proved that for all $n \in \mathbb{Z}_{>1}$, there exist factorizations of n into products of one or more primes. Our next major goal is to prove a uniqueness theorem about prime factorizations. The uniqueness of the factorizations is harder to prove than their existence, so we must first develop some facts about prime numbers. Recall that $p \in \mathbb{Z}_{>1}$ is prime iff the only positive divisors of p are 1 and p.

4.52. Euclid's Lemma on Divisibility by Primes. For all $p, a, b \in \mathbb{Z}$, if p is prime and p divides ab, then p divides a or p divides b.

Proof. Fix $p, a, b \in \mathbb{Z}$. Assume p is prime and p divides ab. Prove p divides a or p divides b. [Use the OR-template to continue.] Assume p does not divide a; prove p divides b. We have assumed p divides ab, which means $ab = pc$ for some integer c. We must prove $\exists d \in \mathbb{Z}, b = pd$. [How do we proceed? We need to use our other assumptions: p is prime and p does not divide a. A key idea is that these assumptions allow us to compute $\gcd(p, a)$.] Since p is prime, the only positive divisors of p are 1 and p. Thus the greatest common divisor of p and a could only be 1 or p. Since we assumed p is not a divisor of a, we must have $\gcd(p, a) = 1$. We now invoke Theorem 4.51 to conclude there exist integers x and y with $1 = ax + py$. [We need one more trick that does not come from the proof templates: multiply both sides by b, to make ab appear on the right side.] Multiplying by b, using algebra, and recalling that $ab = pc$, we get

$$b = (ax + py)b = axb + pyb = (ab)x + pyb = (pc)x + pyb = p(cx + yb).$$

Choosing d to be the integer $cx + yb$, we have $b = pd$, as needed.

4.53. Theorem on Divisibility by Primes. For all $k \in \mathbb{Z}_{\geq 1}$ and all $p, a_1, a_2, \ldots, a_k \in \mathbb{Z}$, if p is prime and p divides $a_1 a_2 \cdots a_k$, then for some $i \in \{1, 2, \ldots, k\}$, p divides a_i.

Proof. We use ordinary induction on k.

Base Case. Prove that for all $p, a_1 \in \mathbb{Z}$, if p is prime and p divides a_1, then for some $i \in \{1\}$, p divides a_i. This statement is readily proved (choose $i = 1$).

Induction Step. Fix an arbitrary $k \in \mathbb{Z}_{\geq 1}$; assume the theorem statement holds for this k; prove the theorem statement holds for $k + 1$. Fix $p, a_1, \ldots, a_k, a_{k+1} \in \mathbb{Z}$; assume p is prime and p divides $a_1 \cdots a_k a_{k+1}$; prove there exists $i \in \{1, 2, \ldots, k + 1\}$ such that p divides a_i. The key realization is that the number $a_1 \cdots a_k a_{k+1}$ can be regarded as the product of the two integers $a = a_1 \cdots a_k$ and $b = a_{k+1}$. Since p divides ab by assumption, the previous lemma applies to show that p divides a or p divides b. Now use cases.

Case 1. Assume p divides a. This means p divides $a_1 a_2 \cdots a_k$. By induction hypothesis, there exists $i \in \{1, 2, \ldots, k\}$ such that p divides a_i, as needed.

Case 2. Assume p divides b, so p divides a_{k+1}. Choose $i = k + 1$; then p divides a_i, as needed.

Uniqueness of Prime Factorizations

Now we are ready to prove the uniqueness of prime factorizations. The next result is sometimes called the *Fundamental Theorem of Arithmetic*.

4.54. Theorem on Existence and Uniqueness of Prime Factorizations.

(a) Every integer $n > 1$ can be written as the product of one or more primes.

(b) For all $n \in \mathbb{Z}_{>1}$, all $s, t \in \mathbb{Z}_{\geq 1}$, and all primes $p_1, \ldots, p_s, q_1, \ldots, q_t$, if $n = p_1 p_2 \cdots p_s = q_1 q_2 \cdots q_t$, then $s = t$ and we can reorder the q_j so that for all $i \in \{1, 2, \ldots, s\}$, $p_i = q_i$.

Proof. Part (a) was proved in §4.4. We prove part (b) by strong induction on n. Fix an integer $n \in \mathbb{Z}_{>1}$, and assume part (b) holds for all n' in the range $1 < n' < n$. Prove the statement in part (b) holds for n. Fix $s, t \in \mathbb{Z}_{\geq 1}$ and primes $p_1, \ldots, p_s, q_1, \ldots, q_t$, and assume

$$n = p_1 p_2 \cdots p_s = q_1 q_2 \cdots q_t.$$

We know $s = 1$ or $t = 1$ or ($s > 1$ and $t > 1$), so use cases.

<u>Case 1.</u> Assume $s = 1$. Then $n = p_1 = q_1 \cdots q_t$. In this case, t must be 1, since otherwise $p_1 = q_1(q_2 \cdots q_t)$ expresses p_1 as a product of the integers q_1 and $q_2 \cdots q_t$, which are both strictly between 1 and p_1. This is impossible, since p_1 is prime. Knowing that $t = 1$, we now have $s = t = 1$ and $p_1 = n = q_1$.

<u>Case 2.</u> Assume $t = 1$. As in case 1, we see that $s = 1$ (because q_1 is prime), so $s = t$ and $p_1 = n = q_1$.

<u>Case 3.</u> Assume $s > 1$ and $t > 1$. On one hand, since $n = p_1(p_2 \cdots p_s)$, we see that p_1 divides n. On the other hand, $n = q_1 \cdots q_t$, so the prime p_1 divides the product $q_1 \cdots q_t$. By the Theorem on Divisibility by Primes, p_1 divides some q_j where $1 \leq j \leq t$. By reordering the q_i if needed, we can arrange that p_1 divides q_1. But q_1 is prime, so its only positive divisors are 1 and q_1. As $p_1 > 1$, we must have $p_1 = q_1$. Now we know

$$p_1(p_2 \cdots p_s) = n = q_1 q_2 \cdots q_t = p_1(q_2 \cdots q_t).$$

Dividing both sides by the nonzero integer p_1, we obtain a new integer

$$n' = p_2 \cdots p_s = q_2 \cdots q_t.$$

Note $n' < n$ since $p_1 > 1$, whereas $n' > 1$ since $s > 1$ and $p_2 > 1$. Thus, n' is an integer in the range $1 < n' < n$, so we can apply the induction hypothesis to n'. We have written n' as the product of $s - 1$ primes $p_2 \cdots p_s$ and also as the product of $t - 1$ primes $q_2 \cdots q_t$. By the induction hypothesis, we conclude that $s - 1 = t - 1$ and $q_2, \ldots, q_t$ is a rearrangement of $p_2, \ldots, p_s$. Since also $p_1 = q_1$, we see that $s = t$ and $q_1, \ldots, q_t$ is a rearrangement of $p_1, \ldots, p_s$. This completes the induction proof.

The Fundamental Theorem of Arithmetic can be rephrased as follows.

4.55. Theorem on Prime Factorization of Integers. Let $p_1 < p_2 < p_3 < \cdots$ be a list of all the distinct primes. For every nonzero integer n, there exist unique integers $s \in \{+1, -1\}$ and $e_1, e_2, \ldots, e_j, \ldots \in \mathbb{Z}_{\geq 0}$ such that all but finitely many e_j are zero, and

$$n = s p_1^{e_1} p_2^{e_2} \cdots p_j^{e_j} \cdots = s \prod_{j \geq 1} p_j^{e_j}.$$

Proof. Let n be a fixed nonzero integer. We know $n > 1$ or $n = 1$ or $n < 0$, so use cases.

<u>Case 1.</u> Assume $n > 1$. To prove existence of the factorization, use the earlier theorem to write n as a product of one or more primes, say $n = q_1 q_2 \cdots q_k$. Now choose $s = +1$, and for all $j \geq 1$, let e_j be the number of times the prime p_j appears in the list $q_1, q_2, \ldots, q_k$. Here, all but finitely many e_j are zero, and $n = s \prod_{j \geq 1} p_j^{e_j}$ since multiplication of integers is commutative. To prove uniqueness, suppose we also had $n = s' \prod_{j \geq 1} p_j^{e_j'}$ where s' and e_j' satisfy the conditions in the theorem statement. We must have $s' = +1 = s$ since $n > 0$ and every $p_j > 0$. If some e_j' were unequal to e_j, then we would have two prime factorizations $n = \prod_{j \geq 1} p_j^{e_j} = \prod_{j \geq 1} p_j^{e_j'}$ in which the two lists of primes were not rearrangements of each other, because p_j occurs a different number of times in the two lists. This violates the uniqueness property in the earlier theorem. So $e_j' = e_j$ for all j, completing the proof of Case 1.

<u>Case 2.</u> Assume $n = 1$. For existence, choose $s = 1$ and every $e_j = 0$. For uniqueness, note that this is the only choice of s and e_j that produces a positive product less than 2.

<u>Case 3.</u> Assume $n < 0$. This case follows by applying the existence and uniqueness results already proved to $-n > 0$; the only required modification is to take $s = -1$ instead of $s = +1$.

Section Summary

1. *Linear Combination Property of GCDs.* For all $a, b \in \mathbb{Z}$, there are integers $x, y \in \mathbb{Z}$ with $d = \gcd(a, b) = ax + by$. To find d, x, y when $a > 0$ and $b = 0$, let $d = a$, $x = 1$, $y = 0$. When $a, b > 0$, write $a = bq + r$ for $q, r \in \mathbb{Z}$ with $0 \le r < b$; recursively find d, s, t with $d = \gcd(b, r) = bs + rt$; then $\gcd(a, b) = d$, $x = t$, and $y = s - qt$.

2. *Properties of Primes.* For prime p and $a \in \mathbb{Z}$, $\gcd(a, p)$ is p if p divides a and is 1 if p does not divide a. Whenever a prime p divides a product $a_1 a_2 \cdots a_k$ of integers, p must divide some a_i. Every integer $n > 1$ can be written as a product of prime integers. The prime factorization is unique in the following sense: if $n = p_1 \cdots p_s = q_1 \cdots q_t$ with all p_i and q_j prime, then $s = t$ and $q_1, \ldots, q_t$ is a rearrangement of $p_1, \ldots, p_s$.

3. *Fundamental Theorem of Arithmetic.* Existence and uniqueness of prime factorizations can be reformulated as follows. Any nonzero integer n can be expressed uniquely in the form $s \prod_{j \ge 1} p_j^{e_j}$, where s is $+1$ or -1, the e_j are nonnegative integers, all but finitely many e_j are zero, and $p_1 < p_2 < \cdots < p_j < \cdots$ is a complete list of prime integers.

Exercises

1. For each a, b, find $d = \gcd(a, b)$ and integers x, y with $d = ax + by$ using the recursive algorithm from this section. (a) $a = 693$, $b = 525$ (b) $a = 999$, $b = 629$ (c) $a = 34$, $b = 21$.

2. Extend Theorem 4.51 to the case where a or b could be negative.

3. (a) Use the known identity $\gcd(a, b) = \gcd(a - b, b)$ and strong induction on $a + b$ to give another proof of Theorem 4.51. (b) Translate this proof into a recursive algorithm for finding d, x, y with $d = \gcd(a, b) = ax + by$, and illustrate this algorithm on inputs $a = 30$, $b = 8$.

4. Prove: For all $a, b \in \mathbb{Z}$, $\gcd(a, b) = 1$ if and only if there exist $x, y \in \mathbb{Z}$ with $ax + by = 1$.

5. For all $a, b, d \in \mathbb{Z}$, we proved: if $\gcd(a, b) = d$, then $\exists x, y \in \mathbb{Z}$, $ax + by = d$. Is the converse of this statement always true? Prove your answer, and compare to the previous exercise.

6. Prove: for all $a, b, c \in \mathbb{Z}$, c is a common divisor of a and b iff c divides $\gcd(a, b)$. [This conclusion strengthens the original definition of $\gcd(a, b)$, which initially tells us only that c is less than or equal to $\gcd(a, b)$.]

7. Suppose $p \in \mathbb{Z}_{>1}$ is a fixed integer with the following property: for all $a, b \in \mathbb{Z}$, if p divides ab, then p divides a or p divides b. Must p be prime? Prove your answer.

8. Prove: for all $a, b, c \in \mathbb{Z}$, if $\gcd(b, c) = 1$ and $c | ab$, then $c | a$.

9. Prove: for all $q \in \mathbb{Q}$, there exists unique $(m, n) \in \mathbb{Z} \times \mathbb{Z}$ with $n > 0$ and $q = m/n$ and $\gcd(m, n) = 1$. [This proves that every rational number has a unique representation as a fraction in *lowest terms*.]

10. (a) Prove: for all $r, s, t \in \mathbb{Z}$, if $\gcd(r, t) = 1 = \gcd(s, t)$, then $\gcd(rs, t) = 1$. (b) Prove by induction on $n > 0$: for all integers $r_1, r_2, \ldots, r_n, t$, if $\gcd(r_i, t) = 1$ for all $i \in \{1, 2, \ldots, n\}$, then $\gcd(\prod_{i=1}^{n} r_i, t) = 1$.

11. Give a new proof of Theorem 4.51 based on Exercise 18 of §4.1, as follows. Fix $a, b \in \mathbb{Z}_{>0}$, and define $S = \{ax + by : x, y \in \mathbb{Z}\} \cap \mathbb{Z}_{>0}$. Show that S has a least element d, and $d = \gcd(a, b)$.

12. (a) Given $a, b \in \mathbb{Z}$, show that the integers $x, y \in \mathbb{Z}$ such that $\gcd(a, b) = ax + by$ are *not* uniquely determined by the pair (a, b). In fact, show that for each pair $(a, b) \in \mathbb{Z} \times \mathbb{Z}$, there are infinitely many pairs $(x, y) \in \mathbb{Z} \times \mathbb{Z}$ with $d = \gcd(a, b) = ax + by$.
    (b) Given fixed integers a, b, x_0, y_0 with $1 = \gcd(a, b) = ax_0 + by_0$, explicitly describe (with proof) the set of all pairs of integers (x, y) solving $1 = ax + by$.

13. Prove that the iterative method for finding x, y with $\gcd(a, b) = ax + by$ (based on working backwards through the divisions in (4.1)) always gives the same x, y as the recursive algorithm described after the proof of Theorem 4.51.

14. The gcd of a finite list of positive integers $a_1, \ldots, a_n$ is the largest integer d such that $d|a_i$ for all i between 1 and n. (It can be shown that such a d exists.) Use induction to prove: for all $n \in \mathbb{Z}_{\geq 2}$ and all $a_1, \ldots, a_n \in \mathbb{Z}_{>0}$, $\gcd(a_1, \ldots, a_n) = \gcd(\gcd(a_1, \ldots, a_{n-1}), a_n)$.

15. Continuing the previous exercise, prove that for all positive integers $n, a_1, \ldots, a_n$, there exist integers $x_1, \ldots, x_n$ with $\gcd(a_1, \ldots, a_n) = a_1 x_1 + \cdots + a_n x_n$.

16. Describe an algorithm that takes $a_1, \ldots, a_n \in \mathbb{Z}_{>0}$ as input and computes $d, x_1, \ldots, x_n \in \mathbb{Z}$ with $d = \gcd(a_1, \ldots, a_n) = a_1 x_1 + \cdots + a_n x_n$. Prove that your algorithm terminates with the correct answer.

4.7 Consequences of Prime Factorization (Optional)

This optional section explores some consequences of the Fundamental Theorem of Arithmetic. Recall that this theorem expresses each nonzero integer n in the form $n = s \prod_{j \geq 1} p_j^{e_j}$, where $p_1 < p_2 < \cdots < p_j < \cdots$ is the list of all prime numbers, and the integers $s \in \{+1, -1\}$ and $e_1, e_2, \ldots, e_j, \ldots \in \mathbb{Z}_{\geq 0}$ are uniquely determined by n. Furthermore, only finitely many e_j are nonzero. To emphasize that s and all e_j depend on n and are uniquely determined by n, we write $s = s(n)$ and $e_j = e_j(n)$. So for all nonzero integers n,

$$n = s(n) \prod_{j \geq 1} p_j^{e_j(n)}.$$

Multiplication Rule for Prime Factorizations

The following result shows how the prime factorization of the product of two nonzero integers is related to the factorizations of each separate integer. Intuitively, the result reduces multiplication of nonzero integers to the addition of the vectors of exponents in the prime factorizations.

4.56. Theorem on Products of Integers. For all nonzero $n, m \in \mathbb{Z}$, $s(nm) = s(n)s(m)$ and $e_j(nm) = e_j(n) + e_j(m)$ for all $j \in \mathbb{Z}_{\geq 1}$.
Proof. Fix nonzero $n, m \in \mathbb{Z}$. We have the factorizations

$$n = s(n) \prod_{j \geq 1} p_j^{e_j(n)}, \qquad m = s(m) \prod_{j \geq 1} p_j^{e_j(m)}.$$

Multiplying these together, we obtain

$$nm = s(n)s(m) \prod_{j \geq 1} p_j^{e_j(n) + e_j(m)}. \tag{4.8}$$

On the other hand, we know the nonzero integer nm has a *unique* expression

$$nm = s(nm) \prod_{j \geq 1} p_j^{e_j(nm)} \tag{4.9}$$

where $s(nm) \in \{+1, -1\}$, each $e_j(nm) \in \mathbb{Z}_{\geq 0}$, and only finitely many values $e_j(nm)$ are nonzero. Since $s(n)s(m) \in \{+1, -1\}$, every $e_j(n) + e_j(m)$ is a nonnegative integer, and at most finitely many numbers $e_j(n) + e_j(m)$ are nonzero, (4.8) satisfies the conditions required in (4.9). By uniqueness, we must therefore have $s(nm) = s(n)s(m)$ and $e_j(nm) = e_j(n) + e_j(m)$ for all integers $j \geq 1$.

Characterization of Divisors

The next theorem shows how to use prime factorizations to detect when one integer is a divisor (or multiple) of another integer.

4.57. Theorem on Factorization of Divisors. For all nonzero integers a and b, a is a divisor of b iff for all $j \in \mathbb{Z}_{\geq 1}$, $e_j(a) \leq e_j(b)$.
Proof. Fix $a, b \in \mathbb{Z}_{\neq 0}$. First assume a is a divisor of b. Then there exists an integer c with $b = ac$; we have $c \neq 0$ since $b \neq 0$. Fix $j \in \mathbb{Z}_{\geq 1}$. By the multiplication rule for prime factorizations, $e_j(b) = e_j(a) + e_j(c) \geq e_j(a) + 0$, so $e_j(a) \leq e_j(b)$.

Conversely, now assume that for all $j \in \mathbb{Z}_{\geq 1}$, $e_j(a) \leq e_j(b)$. Prove $a|b$, which means $\exists d \in \mathbb{Z}, b = ad$. Choose

$$d = s \prod_{j \geq 1} p_j^{e_j(b) - e_j(a)},$$

where $s = 1$ if a and b have the same sign, and $s = -1$ otherwise. Note that d is an integer, since the exponent of each p_j is the nonnegative integer $e_j(b) - e_j(a)$, and at most finitely many of these exponents can be nonzero. On one hand, the choice of s shows that b and ad have the same sign. On the other hand, for all $j \in \mathbb{Z}_{\geq 1}$, $e_j(ad) = e_j(a) + e_j(d) = e_j(a) + (e_j(b) - e_j(a)) = e_j(b)$. So $b = ad$, as needed.

Prime Factorization of GCDs and LCMs

Our next result shows how to compute the greatest common divisor of two nonzero integers using their prime factorizations. First, we need one new definition.

4.58. Definition: Minimum. For all $x, y \in \mathbb{R}$, define $\min(x, y) = x$ if $x \leq y$, and $\min(x, y) = y$ if $y < x$.

This definition says that $\min(x, y)$ is the smaller of the two numbers x and y. You are asked to prove the next lemma in the exercises.

4.59. Lemma on the Minimum. For all $x, y, z \in \mathbb{R}$, $z \leq \min(x, y)$ iff $z \leq x$ and $z \leq y$.

The next theorem shows that we can calculate greatest common divisors by comparing prime factorizations and taking the minimum exponent of each prime.

4.60. Theorem on Prime Factorization of GCDs. For all nonzero integers a and b, $\gcd(a, b) = \prod_{j \geq 1} p_j^{\min(e_j(a), e_j(b))}$.

Proof. Fix nonzero integers a and b, and let $d = \prod_{j \geq 1} p_j^{\min(e_j(a), e_j(b))}$. First we show d is a common divisor of a and b. For every $j \in \mathbb{Z}_{\geq 1}$, $e_j(d) = \min(e_j(a), e_j(b)) \leq e_j(a)$, so $d|a$ by the Theorem on Factorization of Divisors. Similarly, since $e_j(d) \leq e_j(b)$ for all j, $d|b$ follows. To finish the proof, we show that d is the largest common divisor of a and b. Fix any integer f such that $f|a$ and $f|b$. We prove that $f \leq d$ by showing the stronger statement $f|d$. Because $f|a$ and $f|b$, $e_j(f) \leq e_j(a)$ and $e_j(f) \leq e_j(b)$ for all $j \geq 1$. By the Lemma on the Minimum, $e_j(f) \leq \min(e_j(a), e_j(b)) = e_j(d)$ for all j. This implies $f|d$ by the Theorem on Factorization of Divisors. We have now proved that $d = \gcd(a, b)$.

We can extend the definitions of gcd and min to apply to k inputs instead of two inputs (see Exercise 14 of §4.6). The proof above extends to show

$$\gcd(a_1, \ldots, a_k) = \prod_{j \geq 1} p_j^{\min(e_j(a_1), \ldots, e_j(a_k))}.$$

A similar analysis is possible for the *least common multiple* (lcm) of two or more integers. By definition, $\text{lcm}(a_1, \ldots, a_k)$ is the least positive integer d such that every a_i divides d (we assume every a_i is nonzero here). The proof for gcds can be adapted to show

$$\text{lcm}(a_1, \ldots, a_k) = \prod_{j \geq 1} p_j^{\max(e_j(a_1), \ldots, e_j(a_k))},$$

where $\max(z_1, \ldots, z_k)$ is the largest of the real numbers $z_1, \ldots, z_k$.

Prime Factorization of Rational Numbers

The prime factorization theorem for nonzero integers can be extended to a prime factorization theorem for nonzero rational numbers by allowing negative exponents. We continue to assume that $p_1 < p_2 < \cdots < p_j < \cdots$ is the list of all prime integers.

4.61. Theorem on Unique Prime Factorization of Rational Numbers. For every nonzero $q \in \mathbb{Q}$, there exist unique integers $s(q) \in \{+1, -1\}$ and $e_j(q)$ for $j \in \mathbb{Z}_{\geq 1}$, such that all but finitely many $e_j(q)$ are zero, and

$$q = s(q) \prod_{j \geq 1} p_j^{e_j(q)}.$$

Proof. Fix nonzero $q \in \mathbb{Q}$. First we prove existence of the claimed factorization. Since q is rational, we know there are integers n and m such that $m \neq 0$ and $q = n/m$. Note $n \neq 0$, since $q \neq 0$. By the Theorem on Prime Factorization of Integers, we know $n = s(n) \prod_{j \geq 1} p_j^{e_j(n)}$ and $m = s(m) \prod_{j \geq 1} p_j^{e_j(m)}$. Dividing n by m,

$$q = n/m = [s(n)/s(m)] \prod_{j \geq 1} p_j^{e_j(n) - e_j(m)}.$$

Choosing $s(q) = s(n)/s(m)$, which is in $\{+1, -1\}$, and (for each j) choosing $e_j(q) = e_j(n) - e_j(m)$, which is in $\mathbb{Z}$ since $e_j(n), e_j(m) \in \mathbb{Z}_{\geq 0}$, we have found a factorization of q of the required type.

Proving uniqueness of the factorization is a little tricky, since n and m are not uniquely determined by q. Assume

$$q = s \prod_{j \geq 1} p_j^{e_j} = s' \prod_{j \geq 1} p_j^{e'_j} \tag{4.10}$$

where the integers s and e_j as well as the integers s' and e'_j satisfy all the conditions in the theorem statement. We must prove $s = s'$ and $e_j = e'_j$ for all $j \in \mathbb{Z}_{\geq 1}$. If q is positive, we must have $s = +1 = s'$; while if q is negative, we must have $s = -1 = s'$. To deal with the exponents, we divide the index set $\mathbb{Z}_{\geq 1}$ into three subsets. Let $A = \{j \in \mathbb{Z}_{\geq 1} : e_j = e'_j\}$, let $B = \{j \in \mathbb{Z}_{\geq 1} : e_j < e'_j\}$, and let $C = \{j \in \mathbb{Z}_{\geq 1} : e_j > e'_j\}$. Our goal is to prove $B = C = \emptyset$. To begin, we can divide both expressions for q in (4.10) by the common factor $s \prod_{j \in A} p_j^{e_j}$, which leaves

$$\prod_{j \in B} p_j^{e_j} \prod_{j \in C} p_j^{e_j} = \prod_{j \in B} p_j^{e'_j} \prod_{j \in C} p_j^{e'_j}.$$

(Here and below, we interpret a product over an empty set of indices as 1.) Next, divide both sides by $\prod_{j \in B} p_j^{e_j} \prod_{j \in C} p_j^{e'_j}$, to get

$$\prod_{j \in C} p_j^{e_j - e'_j} = \prod_{j \in B} p_j^{e'_j - e_j}. \tag{4.11}$$

By the definitions of B and C, the left side is either 1 or a product of strictly positive powers of primes p_j with $j \in C$; and the right side is either 1 or a product of strictly positive powers of primes p_j with $j \in B$. Consider various cases that might occur.

<u>Case 1.</u> Assume $B \neq \emptyset$ and $C \neq \emptyset$. Note that the primes p_j for $j \in B$ are all different from the primes p_j for $j \in C$, since B and C cannot overlap. Now (4.11) displays two different prime factorizations of a positive integer, violating uniqueness of prime factorization of integers.

<u>Case 2.</u> Assume $B = \emptyset$ and $C \neq \emptyset$. Then the right side of (4.11) is 1, whereas the left side is larger than 1, which is impossible.

Case 3. Assume $B \neq \emptyset$ and $C = \emptyset$. Then the left side of (4.11) is 1, whereas the right side is larger than 1, which is impossible.

Case 4. Assume $B = \emptyset$ and $C = \emptyset$. This is the only possible case, and in this case we must have $e_j = e'_j$ for all $j \in \mathbb{Z}_{\geq 1}$, as needed. This completes the uniqueness proof.

Proving Irrationality of nth Roots

In §2.4, we proved that $\sqrt{2}$ is irrational using proof by contradiction. Now that we have the unique factorization theorem for rational numbers, we can quickly prove a much more general result that tells us exactly which numbers $\sqrt[n]{q}$ (where $q \in \mathbb{Q}_{>0}$ and $n \in \mathbb{Z}_{>1}$) are rational. In using the notation $\sqrt[n]{q}$, we are tacitly assuming the following fact about real numbers: for every positive real number x and every positive integer n, there exists a unique positive real number w such that $w^n = x$; by definition, $\sqrt[n]{x}$ is the unique such w solving $w^n = x$. This can be proved using the Intermediate Value Theorem from calculus; the case of square roots ($n = 2$) is proved in Theorem 8.61. Informally, our main result says that the only positive rational numbers q with rational nth roots are those for which the exponents of all primes in the factorization of q are multiples of n.

4.62. Theorem on Rationality of Roots. For all $q \in \mathbb{Q}_{>0}$ and all $n \in \mathbb{Z}_{>1}$, $\sqrt[n]{q}$ is rational iff for all $j \in \mathbb{Z}_{\geq 1}$, n divides $e_j(q)$; and in this case,

$$\sqrt[n]{q} = \prod_{j \geq 1} p_j^{e_j(q)/n}.$$

Proof. Fix $q \in \mathbb{Q}_{>0}$ and $n \in \mathbb{Z}_{>1}$. By the fact quoted above, let $w \in \mathbb{R}_{>0}$ be the unique real number such that $w^n = q$. Part 1. Assume $w \in \mathbb{Q}$, and prove n divides every $e_j(q)$. Since w is positive and rational, it has a prime factorization $w = \prod_{j \geq 1} p_j^{e_j(w)}$. Raising this expression to the power n,

$$\prod_{j \geq 1} p_j^{n e_j(w)} = w^n = q = \prod_{j \geq 1} p_j^{e_j(q)}.$$

Since the prime factorization of q is unique, we conclude that $e_j(q) = n e_j(w)$ for every $j \geq 1$. Since $e_j(w) \in \mathbb{Z}$, we see that n divides every $e_j(q)$. Furthermore, $e_j(w) = e_j(q)/n$, so the last formula in the theorem statement holds.

Part 2. Assume n divides every $e_j(q)$; prove $w \in \mathbb{Q}$. Let $w_1 = \prod_{j \geq 1} p_j^{e_j(q)/n}$, which is a well-defined positive rational number since every $e_j(q)/n$ is an integer. By direct computation, $w_1^n = \prod_{j \geq 1} p_j^{e_j(q)} = q = w^n$. By the fact quoted above, w is the *unique* positive real number whose nth power is q, so $w = w_1 \in \mathbb{Q}$, as needed.

4.63. Example. The rational number $q = 0.729 = 2^{-3}3^6 5^{-3}$ has rational cube root $\sqrt[3]{q} = 0.9 = 2^{-1}3^2 5^{-1}$. But $\sqrt{q}$ is not rational since $e_1(q) = -3$ is not divisible by 2. More generally, for all $n \in \mathbb{Z}_{>1} - \{3\}$, $\sqrt[n]{q}$ is not rational.

Similarly, it follows from the theorem that the only positive integers k such that $\sqrt{k}$ is rational are integers of the form $k = j^2$ for some positive integer j; the only $k \in \mathbb{Z}_{>0}$ such that $\sqrt[3]{k} \in \mathbb{Q}$ are those of the form $k = j^3$ for some $j \in \mathbb{Z}_{>0}$; and so on. In particular, for every prime integer p, the numbers $\sqrt{p}$, $\sqrt[3]{p}$, etc., are all irrational.

Section Summary

Let $p_1 < p_2 < \cdots < p_j < \cdots$ be the list of all prime integers.

1. *Computing with Prime Factorizations.* Given positive integers n and m with prime factorizations $n = \prod_{j \geq 1} p_j^{e_j(n)}$ and $m = \prod_{j \geq 1} p_j^{e_j(m)}$, we have:

$$nm = \prod_{j \geq 1} p_j^{e_j(n)+e_j(m)}; \quad n \text{ divides } m \text{ iff } e_j(n) \leq e_j(m) \text{ for all } j \geq 1;$$

$$\gcd(n, m) = \prod_{j \geq 1} p_j^{\min(e_j(n),e_j(m))}; \quad \text{and } \operatorname{lcm}(n, m) = \prod_{j \geq 1} p_j^{\max(e_j(n),e_j(m))}.$$

Similar formulas hold for the product, gcd, and lcm of a finite list of nonzero integers.

2. *Prime Factorization of Rational Numbers.* For every nonzero rational number q, there exist unique $s \in \{+1, -1\}$ and unique integers $e_j(q)$ for $j \geq 1$, such that all but finitely many $e_j(q)$ are zero, and $q = s(q) \prod_{j \geq 1} p_j^{e_j(q)}$.

3. *nth Roots of Rational Numbers.* For any positive rational $q = \prod_{j \geq 1} p_j^{e_j(q)}$ and $n \in \mathbb{Z}_{>1}$, $\sqrt[n]{q}$ is rational iff n divides $e_j(q)$ for all $j \geq 1$.

Exercises

1. Let $n = 2^5 3^3 7^4$, $m = 2^3 3^5 5^2 7^1$, and $s = 3^4 5^4 7^2$.
   (a) Find $\gcd(n, m)$, $\gcd(m, s)$, $\gcd(n, s)$, and $\gcd(n, m, s)$.
   (b) Find $\operatorname{lcm}(n, m)$, $\operatorname{lcm}(m, s)$, $\operatorname{lcm}(n, s)$, and $\operatorname{lcm}(n, m, s)$.
   (c) Are any of $\sqrt{n}$, $\sqrt{m}$, or $\sqrt{s}$ rational?

2. Find the prime factorizations of each rational number.
   (a) 0.625   (b) 212/121   (c) 2.64   (d) 3300/77077.

3. (a) Prove the Lemma on Minimums. (b) State and prove an analogous result for $\max(x, y)$.

4. Suppose $x_1, \ldots, x_k$ are fixed real numbers. (a) Give a recursive definition of $\min(x_1, \ldots, x_k)$ and $\max(x_1, \ldots, x_k)$. (b) Generalize the results of the previous exercise to apply to this situation.

5. Given integers $a_1, \ldots, a_k > 0$, prove $\gcd(a_1, \ldots, a_k) = \prod_{j \geq 1} p_j^{\min(e_j(a_1),\ldots,e_j(a_k))}$.

6. Given integers $a_1, \ldots, a_k > 0$, prove $\operatorname{lcm}(a_1, \ldots, a_k) = \prod_{j \geq 1} p_j^{\max(e_j(a_1),\ldots,e_j(a_k))}$.

7. Extend Theorem 4.56 to the case where n and m are nonzero rational numbers.

8. Given nonzero rational numbers q and r, find and prove a formula for the prime factorization of q/r.

9. Use prime factorizations to prove: for all $r, s, t \in \mathbb{Z}_{>0}$, if $\gcd(r, t) = 1 = \gcd(s, t)$, then $\gcd(rs, t) = 1$.

10. Use prime factorizations to prove: for all $a, b, c \in \mathbb{Z}_{>0}$, if $\gcd(a, b) = 1$ and a divides bc, then a divides c.

11. (a) Prove or disprove: for all $m, n \in \mathbb{Z}_{>0}$, if $m \geq n$, then $e_j(m) \geq e_j(n)$ for all $j \geq 1$. (b) Prove or disprove: for all $m, n \in \mathbb{Z}_{>0}$, if $e_j(m) \geq e_j(n)$ for all $j \geq 1$ then $m \geq n$.

12. Prove: for all nonzero $a, b \in \mathbb{Q}$, there exists $n \in \mathbb{Z}$ with $b = na$ iff for all $j \in \mathbb{Z}_{\geq 1}$, $e_j(a) \leq e_j(b)$.

13. Prove: for all positive integers a, b, c, $\gcd(ca, cb) = c \cdot \gcd(a, b)$ using prime factorizations. (See Exercise 14 in §4.5 for a different proof.)

14. Prove: for all positive integers a, b, d, $\gcd(a, b) = d$ iff $d|a$ and $d|b$ and $\gcd(a/d, b/d) = 1$.

15. Suppose we define divisibility in $\mathbb{Q}$ as follows: for all $q, r \in \mathbb{Q}$, q divides r iff $\exists s \in \mathbb{Q}, r = sq$. Characterize all pairs (q, r) such that q divides r under this definition.

16. (a) Prove: for all $x, y \in \mathbb{R}$, $x + y = \min(x, y) + \max(x, y)$. (b) Deduce that for all $a, b \in \mathbb{Z}_{\neq 0}$, $ab = \gcd(a, b) \cdot \operatorname{lcm}(a, b)$.

17. Prove: for all $m, n \in \mathbb{Z}_{>0}$, if there exists $q \in \mathbb{Q}$ with $q^n = m$, then there exists $k \in \mathbb{Z}$ with $k^n = m$. [Thus if an integer has a rational nth root, then the root must actually be an integer.]

18. Given a nonzero $q \in \mathbb{Q}$, let $P(q)$ be the set of primes p_j such that $e_j(q) \neq 0$. (a) Prove: for all nonzero $q \in \mathbb{Q}$, $P(-q) = P(q)$. (b) Prove: for all nonzero $q, r \in \mathbb{Q}$, $P(qr) \subseteq P(q) \cup P(r)$. Show that equality holds if $P(q) \cap P(r) = \emptyset$.

19. (a) Given nonzero rational numbers $q_1, \ldots, q_k$, define rational versions of $\gcd(q_1, \ldots, q_k)$ and $\operatorname{lcm}(q_1, \ldots, q_k)$. (b) For nonzero $q, r \in \mathbb{Q}$, can we always write $\gcd(q, r) = qx + ry$ for some integers x and y?

Review of Set Theory and Integers

Tables 4.1–4.6 on the following pages review the main definitions, proof templates, and theorems covered in the preceding chapters on set theory and integers. All unquantified variables in the tables represent arbitrary objects, sets, real numbers, or integers (as required by context).

Set Theory Pitfalls

(a) *Order and repetition do not matter in sets.* For example, $\{1, 2\} = \{2, 1\} = \{1, 1, 2, 2\}$.

(b) *Distinguishing $\in$ and $\subseteq$.* $A \in B$ means the object A is a member of the set B. $A \subseteq B$ means every member of the set A is a member of the set B. Do not confuse the set membership relation ($A \in B$) with the subset relation ($A \subseteq B$). Note that sets are allowed to be members of other sets.

(c) *Types of Brackets.* There are three kinds of brackets used in set theory: round parentheses $(\cdots)$, square brackets $[\cdots]$, and curly braces $\{\ldots\}$. Each shape of bracket has a different meaning: use round parentheses for ordered pairs or open intervals, use square brackets for closed intervals, and use curly braces for sets. Adding more braces changes the meaning; for example, $\emptyset$ and $\{\emptyset\}$ and $\{\{\emptyset\}\}$ are all different sets. Similarly, $S = \{1, \{2, 3, 4\}, 5, 6, \{\{7\}, 8\}, [3, 5]\}$ is a set with six members: the integers 1, 5, and 6; the set containing 2 and 3 and 4; another set with members 8 and $\{7\}$; and the closed interval $[3, 5]$.

Theorems on Divisibility, Primes, and GCDs

1. *Integer Division Theorem.* For all integers a and b with $b \neq 0$, there exist unique integers q, r with $a = bq + r$ and $0 \leq r < |b|$. We call q the quotient and r the remainder when a is divided by b.

2. *Theorem on Computing GCDs.* For all $a \in \mathbb{Z}$, $\gcd(a, 0) = |a|$. For all $a, b, q, r \in \mathbb{Z}$, if $a = bq + r$ then $\gcd(a, b) = \gcd(b, r)$.

3. *Linear Combination Property of GCDs.* For all $a, b \in \mathbb{Z}$, there exist $x, y \in \mathbb{Z}$ such that $\gcd(a, b) = ax + by$. This result extends to the gcd of any finite list of integers.

4. *Euclid's Recursive GCD Algorithm.*
   Input: $a, b \in \mathbb{Z}_{\geq 0}$.
   Output: $d = \gcd(a, b)$, and $x, y \in \mathbb{Z}$ with $d = ax + by$.
   Procedure: 1. If $b = 0$, return $d = a$, $x = 1$, $y = 0$.
   2. If $b > 0$, divide a by b to get $a = bq + r$ with $0 \leq r < b$.
   3. Recursively compute $d' = \gcd(b, r)$ and $s, t \in \mathbb{Z}$ with $d' = bs + rt$.
   4. Return $d = d'$, $x = t$, and $y = s - qt$.
   The iterative version of this algorithm uses repeated division to find $\gcd(a, b)$ (the last nonzero remainder), then works backwards through the divisions to find x and y.

TABLE 4.1
Set Definitions.

Concept	Defined Term	Definition Text
Set Membership	$x \in A$ [undefined term]	(informal) x is a member of A.
	$x \notin A$	$\sim(x \in A)$.
Subset	$A \subseteq B$	$\forall x, x \in A \Rightarrow x \in B$.
	$A \nsubseteq B$	$\exists x, x \in A \wedge x \notin B$.
Proper Subset	$A \subsetneq B$	$A \subseteq B$ and $A \neq B$.
Set Equality	$A = B$	$\forall x, x \in A \Leftrightarrow x \in B$
		(equivalently: $A \subseteq B$ and $B \subseteq A$).
Union	$x \in A \cup B$	$x \in A$ or $x \in B$.
Intersection	$x \in A \cap B$	$x \in A$ and $x \in B$.
Set Difference	$x \in A - B$	$x \in A$ and $x \notin B$.
Empty Set	$x \in \emptyset$	$x \in \emptyset$ is **false** for all x.
	$B = \emptyset$	$\forall x, x \notin B$.
	$B \neq \emptyset$	$\exists x, x \in B$.
Singleton Set	$z \in \{a\}$	$z = a$.
Unordered Pair	$z \in \{a, b\}$	$z = a$ or $z = b$.
Unordered Triple	$z \in \{a, b, c\}$	$z = a$ or $z = b$ or $z = c$.
Ordered Pair	(a, b) [undefined term]	**Axiom:** $\forall a, \forall b, \forall c, \forall d,$
		$(a, b) = (c, d) \Leftrightarrow (a = c \wedge b = d)$.
		[or, **define** $(a, b) = \{\{a\}, \{a, b\}\}$]
Open Interval	$x \in (a, b)$	$x \in \mathbb{R}$ and $a < x < b$.
Closed Interval	$x \in [a, b]$	$x \in \mathbb{R}$ and $a \leq x \leq b$.
Half-Open Interval	$x \in [a, b)$	$x \in \mathbb{R}$ and $a \leq x < b$.
Half-Open Interval	$x \in (a, b]$	$x \in \mathbb{R}$ and $a < x \leq b$.
Power Set	$S \in \mathcal{P}(X)$	$S \subseteq X$.
Product Set	$z \in X \times Y$	$\exists x, \exists y, (x \in X \wedge y \in Y) \wedge z = (x, y)$.
	$(a, b) \in X \times Y$	$a \in X$ and $b \in Y$.
Indexed Union	$z \in \bigcup_{i \in I} A_i$	$\exists i \in I, z \in A_i$.
Indexed Intersection	$z \in \bigcap_{i \in I} A_i$	$\forall i \in I, z \in A_i$ (need $I \neq \emptyset$).

TABLE 4.2

Proof Templates Involving Sets and Induction.

Statement	Proof Template to Prove Statement
$A \subseteq B$	**Fix** an arbitrary object x_0. **Assume** $x_0 \in A$. **Prove** $x_0 \in B$. [To continue, expand definitions of "$x_0 \in A$" and "$x_0 \in B$."]
$S \times T \subseteq C$	**Fix** an arbitrary ordered pair (a, b). **Assume** $(a, b) \in S \times T$, which means $a \in S$ and $b \in T$. **Prove** $(a, b) \in C$ [by expanding the definitions of membership in S, T, and C].
$A = B$ (set equality)	Part 1. **Prove** $A \subseteq B$. Part 2. **Prove** $B \subseteq A$.
$A = B$ (set equality)	**Fix** an arbitrary object x_0. **Prove** $x_0 \in A \Leftrightarrow x_0 \in B$ by a chain proof. [If A and B contain **only** ordered pairs, replace x_0 by (x, y).]
$A = \emptyset$	Assume, to get a contradiction, that $A \neq \emptyset$. We have assumed $\exists z, z \in A$. Expand the definition of "$z \in A$," and try to find a contradiction.
$\forall n \in \mathbb{Z}_{\geq b}, P(n)$	We use **(ordinary) induction** on n. 1. <u>Base Case.</u> **Prove** $P(b)$. 2a. <u>Induction Step.</u> **Fix** an arbitrary integer $n \in \mathbb{Z}_{\geq b}$. 2b. <u>Induction Hypothesis.</u> **Assume** $P(n)$. 2c. **Prove** $P(n+1)$. [Use $P(n)$ to help prove $P(n+1)$.]

$\forall n \in \mathbb{Z}_{\geq b}, P(n)$	We use **strong induction** on n.
	1. **Fix** an arbitrary integer $n \in \mathbb{Z}_{\geq b}$.
	2a. <u>Strong Induction Hypothesis.</u> **Assume** that for all
	    integers m in the range $b \leq m < n$, $P(m)$ is true.
	2b. **Prove** $P(n)$.
	[Small values of n may require separate cases. Use $P(m)$,
	    for well-chosen values of $m < n$, to help prove $P(n)$.]

TABLE 4.3
Properties of Subsets.

Reflexivity	$A \subseteq A$.
Antisymmetry	$(A \subseteq B \wedge B \subseteq A) \Rightarrow A = B$.
Transitivity	$(A \subseteq B \wedge B \subseteq C) \Rightarrow A \subseteq C$.
Lower Bound	$A \cap B \subseteq A$ and $A \cap B \subseteq B$.
Greatest Lower Bound	$C \subseteq A \cap B$ iff $(C \subseteq A$ and $C \subseteq B)$.
Upper Bound	$A \subseteq A \cup B$ and $B \subseteq A \cup B$.
Least Upper Bound	$A \cup B \subseteq C$ iff $(A \subseteq C$ and $B \subseteq C)$.
Least Element	$\emptyset \subseteq A$.
Difference Property	$A - B \subseteq A$.
Monotonicity of $\cap$	$A \subseteq B \Rightarrow A \cap C \subseteq B \cap C$.
Monotonicity of $\cup$	$A \subseteq B \Rightarrow A \cup C \subseteq B \cup C$.
Monotonicity of $-$	$A \subseteq B \Rightarrow A - C \subseteq B - C$.
Inclusion Reversal of $-$	$A \subseteq B \Rightarrow C - B \subseteq C - A$.
Subset Characterizations	$A \subseteq B \Leftrightarrow A \cap B = A \Leftrightarrow A \cup B = B \Leftrightarrow A - B = \emptyset$.

TABLE 4.4
Set Equality Theorems.

Reflexivity	$A = A$.
Symmetry	$A = B \Rightarrow B = A$.
Transitivity	$(A = B \wedge B = C) \Rightarrow A = C$.
Commutativity	$A \cup B = B \cup A$, $\quad A \cap B = B \cap A$.
Associativity	$(A \cap B) \cap C = A \cap (B \cap C)$, $\quad (A \cup B) \cup C = A \cup (B \cup C)$.
Distributive Law 1	$A \cap (B \cup C) = (A \cap B) \cup (A \cap C)$.
Distributive Law 2	$A \cup (B \cap C) = (A \cup B) \cap (A \cup C)$.
Idempotent Laws	$A \cap A = A$, $\quad A \cup A = A$.
Absorption Laws	$A \cup (A \cap B) = A$, $\quad A \cap (A \cup B) = A$.
Properties of $\emptyset$	$A \cup \emptyset = A$, $\quad A \cap \emptyset = \emptyset$, $\quad A \times \emptyset = \emptyset$, $\quad \emptyset \times A = \emptyset$,
	$A - \emptyset = A$, $\quad \emptyset - A = \emptyset$, $\quad A - A = \emptyset$.
De Morgan Law 1	$A - (B \cup C) = (A - B) \cap (A - C)$.
De Morgan Law 2	$A - (B \cap C) = (A - B) \cup (A - C)$.
Set Decomposition	$A = (A - B) \cup (A \cap B)$, $\quad (A - B) \cap (A \cap B) = \emptyset$.
Difference Properties	$A - (A \cap B) = A - B$, $\quad A - (A - B) = A \cap B$.

TABLE 4.5

Properties of Power Sets, Product Sets, and General Unions and Intersections.

Power Sets:	
Min/Max Subsets	$\emptyset \in \mathcal{P}(A)$, $A \in \mathcal{P}(A)$.
Monotonicity	$A \subseteq B$ iff $\mathcal{P}(A) \subseteq \mathcal{P}(B)$.
Intersection Property	$\mathcal{P}(A \cap B) = \mathcal{P}(A) \cap \mathcal{P}(B)$.
Union Property	$\mathcal{P}(A) \cup \mathcal{P}(B) \subseteq \mathcal{P}(A \cup B)$.
Difference Property	$\mathcal{P}(A - B) - \{\emptyset\} \subseteq \mathcal{P}(A) - \mathcal{P}(B)$.
Product Sets:	
Monotonicity	$(A \subseteq C \wedge B \subseteq D) \Rightarrow A \times B \subseteq C \times D$.
Distributive Laws	$(A \cup B) \times C = (A \times C) \cup (B \times C)$,
	$(A \cap B) \times C = (A \times C) \cap (B \times C)$,
	$(A - B) \times C = (A \times C) - (B \times C)$,
	similarly if C appears in the first factor.
Intersection Property	$(A \times B) \cap (C \times D) = (A \cap C) \times (B \cap D)$.
Union Property	$(A \times B) \cup (C \times D) \subseteq (A \cup C) \times (B \cup D)$.
Unions and Intersections:	
Monotonicity of $\bigcup$	$(\forall i \in I, A_i \subseteq B_i) \Rightarrow \bigcup_{i \in I} A_i \subseteq \bigcup_{i \in I} B_i$.
Monotonicity of $\bigcap$	$(\forall i \in I, A_i \subseteq B_i) \Rightarrow \bigcap_{i \in I} A_i \subseteq \bigcap_{i \in I} B_i$.
De Morgan Law 1	$X - \bigcup_{i \in I} A_i = \bigcap_{i \in I}(X - A_i)$.
De Morgan Law 2	$X - \bigcap_{i \in I} A_i = \bigcup_{i \in I}(X - A_i)$.
Distributive Laws	$X \cap \bigcup_{i \in I} A_i = \bigcup_{i \in I}(X \cap A_i)$,
	$X \cup \bigcap_{i \in I} A_i = \bigcap_{i \in I}(X \cup A_i)$,
	$X \times \bigcup_{i \in I} B_i = \bigcup_{i \in I}(X \times B_i)$,
	$X \times \bigcap_{i \in I} B_i = \bigcap_{i \in I}(X \times B_i)$,
	similarly if X appears as the second factor.
Lower and Upper Bound	$\forall k \in I, \bigcap_{i \in I} A_i \subseteq A_k \subseteq \bigcup_{i \in I} A_i$.
Greatest Lower Bound	$X \subseteq \bigcap_{i \in I} A_i$ iff $(\forall i \in I, X \subseteq A_i)$.
Least Upper Bound	$\bigcup_{i \in I} A_i \subseteq X$ iff $(\forall i \in I, A_i \subseteq X)$.
Enlarging the Index Set	$I \subseteq J \Rightarrow \bigcup_{i \in I} A_i \subseteq \bigcup_{j \in J} A_j$,
	$I \subseteq J \Rightarrow \bigcap_{j \in J} A_j \subseteq \bigcap_{i \in I} A_i$.
Combining Index Sets	$(\bigcup_{i \in I} A_i) \cup (\bigcup_{j \in J} A_j) = \bigcup_{k \in I \cup J} A_k$,
	$(\bigcap_{i \in I} A_i) \cap (\bigcap_{j \in J} A_j) = \bigcap_{k \in I \cup J} A_k$.

TABLE 4.6

Definitions Involving Recursion or Divisibility.

Concept	Defined Term	Definition Text
Sum	$\sum_{k=b}^{n} x_k$	$\sum_{k=b}^{b} x_k = x_b,$
	(for fixed $b \leq n$)	$\sum_{k=b}^{n+1} x_k = (\sum_{k=b}^{n} x_k) + x_{n+1}.$
Product	$\prod_{k=b}^{n} x_k$	$\prod_{k=b}^{b} x_k = x_b,$
	(for fixed $b \leq n$)	$\prod_{k=b}^{n+1} x_k = (\prod_{k=b}^{n} x_k) \cdot x_{n+1}.$
Factorials	$n!\ (n \in \mathbb{Z}_{\geq 0})$	$0! = 1,\ (n+1)! = (n+1) \cdot n!.$
Exponents	$x^n\ (x \in \mathbb{R}, n \in \mathbb{Z}_{\geq 0})$	$x^0 = 1,\ x^{n+1} = x^n \cdot x.$
Fibonacci sequence	$F_n\ (n \in \mathbb{Z}_{\geq 0})$	$F_0 = 0,\ F_1 = 1,$
		$F_n = F_{n-1} + F_{n-2}$ for all $n \in \mathbb{Z}_{\geq 2}.$
Divisor	$a|b$ (a divides b)	$\exists u \in \mathbb{Z}, b = au.$
Common Divisor	$d|a$ and $d|b$	$\exists s, t \in \mathbb{Z}, a = ds$ and $b = dt.$
Greatest Common Divisor	$d = \gcd(a,b)$	$d|a \wedge d|b \wedge \forall c \in \mathbb{Z}_{>0}, (c|a \wedge c|b \Rightarrow c \leq d).$
Least Common Multiple	$\ell = \mathrm{lcm}(a,b)$	$a|\ell \wedge b|\ell \wedge \forall e \in \mathbb{Z}_{>0}, (a|e \wedge b|e \Rightarrow \ell \leq e).$
Composite	n is composite	$n \in \mathbb{Z}_{>1}$ and $\exists a, (1 < a < n \wedge a|n).$
Prime	n is prime	$n \in \mathbb{Z}_{>1}$ and $\sim\exists a, (1 < a < n \wedge a|n).$
Rational	$x \in \mathbb{Q}$	$\exists m \in \mathbb{Z}, \exists n \in \mathbb{Z}, n \neq 0 \wedge x = m/n.$

5. *Theorem on Divisibility by Primes.* For all primes p and all $a_1, \ldots, a_k \in \mathbb{Z}$, if p divides $a_1 a_2 \cdots a_k$, then p divides some a_i.

6. *Infinitude of Primes.* For every integer n, there exists a prime $p > n$. So there are infinitely many primes.

7. *Unique Prime Factorization of Integers.* Let $p_1 < p_2 < \cdots < p_j < \cdots$ be the list of all prime integers. For all nonzero $n \in \mathbb{Z}$, there exist unique integers $s(n) \in \{+1, -1\}$ and $e_j(n) \geq 0$ (for $j = 1, 2, 3, \ldots$) such that all but finitely many $e_j(n)$ are zero, and $n = s(n) \prod_{j \geq 1} p_j^{e_j(n)}$. In words, every nonzero integer n can be written as a product of zero or more primes, times a sign factor. The sign and primes occurring in the factorization (including the number of times a given prime appears) are uniquely determined by n.

8. *Unique Prime Factorization of Rational Numbers.* Let $p_1 < p_2 < \cdots < p_j < \cdots$ be the list of all prime integers. For all nonzero $q \in \mathbb{Q}$, there exist unique integers $s(q) \in \{+1, -1\}$ and $e_j(q) \in \mathbb{Z}$ (for $j = 1, 2, 3, \ldots$) such that all but finitely many $e_j(q)$ are zero, and $q = s(q) \prod_{j \geq 1} p_j^{e_j(q)}$.

9. *Divisibility Concepts and Prime Factorizations.* For all nonzero $x, y \in \mathbb{Q}$ and all $j \in \mathbb{Z}_{\geq 1}$, $s(xy) = s(x)s(y)$ and $e_j(xy) = e_j(x) + e_j(y)$. For all nonzero $a, b \in \mathbb{Z}$, a divides b iff for all $j \in \mathbb{Z}_{\geq 1}$, $e_j(a) \leq e_j(b)$. For all nonzero $a_1, \ldots, a_k \in \mathbb{Z}$,

$$\gcd(a_1, \ldots, a_k) = \prod_{j \geq 1} p_j^{\min(e_j(a_1), \ldots, e_j(a_k))},$$

$$\mathrm{lcm}(a_1, \ldots, a_k) = \prod_{j \geq 1} p_j^{\max(e_j(a_1), \ldots, e_j(a_k))}.$$

For all $q \in \mathbb{Q}_{>0}$ and $n \in \mathbb{Z}_{\geq 1}$, $\sqrt[n]{q}$ is rational iff for all $j \in \mathbb{Z}_{\geq 1}$, n divides $e_j(q)$; and in this case, $\sqrt[n]{q} = \prod_{j \geq 1} p_j^{e_j(q)/n}$.

10. *Summation Formulas.* (These can be proved by induction.)

$\sum_{k=b}^{n} 1 = n - b + 1$.

$\sum_{k=1}^{n} k = n(n+1)/2$.

$\sum_{k=1}^{n} k^2 = n(n+1)(2n+1)/6$.

$\sum_{k=1}^{n} k^3 = n^2(n+1)^2/4$.

$\sum_{k=0}^{n} r^k = (r^{n+1} - 1)/(r - 1)$ for all real $r \neq 1$

$\sum_{k=b}^{n}(x_k + y_k) = \sum_{k=b}^{n} x_k + \sum_{k=b}^{n} y_k$ for all $x_k, y_k \in \mathbb{R}$ and all $n \geq b$.

$\sum_{k=b}^{n}(ax_k) = a \sum_{k=b}^{n} x_k$ for all $a, x_k \in \mathbb{R}$ and all $n \geq b$.

Review Problems

1. Complete the following definitions and theorem statement.
   (a) $x \in A \cup B$ iff ...
   (b) $x \in \bigcap_{i \in I} A_i$ iff ...
   (c) $(y, z) \in C \times D$ iff ...
   (d) *Integer Division Theorem:* ...

2. Use Euclid's algorithm to compute $\gcd(210, 51)$, and find integers x and y such that $\gcd(210, 51) = 210x + 51y$.

3. Prove this statement by ordinary induction: for all $n \in \mathbb{Z}_{\geq 2}$, $\sum_{k=2}^{n} \frac{1}{k(k-1)} = \frac{n-1}{n}$.

4. Let A, B, C be fixed sets. Prove using only the definitions: if $B \cap A \subseteq C$ and $B - A \subseteq C$, then $B \subseteq C$.

5. Define a sequence recursively by setting $a_0 = 8$, $a_1 = 13$, and $a_n = 7a_{n-1} - 6a_{n-2}$ for all integers $n \geq 2$. Prove by strong induction: for all $n \in \mathbb{Z}_{\geq 0}$, $a_n = 6^n + 7$.

6. (a) Let A be a fixed set. Which of the following sets **must** equal the empty set? $A \cup \emptyset$, $A \cap \emptyset$, $A - \emptyset$, $A \times \emptyset$, $\mathcal{P}(\emptyset)$, $\{\emptyset\}$.
   (b) Disprove: for all sets B, $B \cap \mathcal{P}(B) = \emptyset$.

7. True or false? Explain briefly.
   (a) $\{3, 5, 8\} = \{5, 8, 3\}$.
   (b) $\{1, 2, 2, 4, 4, 4, 4\} = \{1, 2, 4\}$.
   (c) $\{2\} \in \{\{\{2\}\}\}$.
   (d) $\{7, 8\} \in \{\{7, 8\}, 9\}$.
   (e) $\{7, 8\} \subseteq \{\{7, 8\}, 9\}$.
   (f) For all sets A and B, $A \cup B = B \cup A$.
   (g) For all sets A and B, $A \times B = B \times A$.
   (h) For all sets A and B, $A \times B \neq B \times A$.
   (i) For all sets A, $\emptyset \in A$.
   (j) For all sets A, $\emptyset \subseteq A$.

8. For every positive integer n, define a set $A_n = [1/n, n + 2] \cup \{-n\}$. Draw pictures of each set (on a number line or in $\mathbb{R}^2$, as appropriate).
   (a) $A_1 \times A_2$ (b) $\bigcup_{n=1}^{\infty} A_n$ (c) $A_2 - A_1$ (d) $A_2 \times A_3$

9. Use Euclid's Algorithm to compute $\gcd(528, 209)$, and find integers x and y such that $\gcd(528, 209) = 528x + 209y$.

10. Prove using only the definitions: for all sets A, B, C,

$$A \times (B \cup C) = (A \times B) \cup (A \times C).$$

11. Define a sequence recursively by $a_0 = 0$ and $a_{n+1} = 3a_n + 2$ for all $n \in \mathbb{Z}_{\geq 0}$. Compute a_1, a_2, a_3, and a_4, guess a closed (non-recursive) formula for a_n, and then prove your guess by ordinary induction.

12. Prove using only the definitions: for all sets A and B, $A - (A - B) = A \cap B$.

13. Write *just the outline* of a proof by **strong induction** of the following statement: "Every odd integer $n \geq 7$ can be written as the sum of three primes."

14. For each $n \in \mathbb{Z}_{\geq 1}$, define a set $A_n = [1/n, n+3) - \{n\}$. Draw each of the following sets on a number line. (a) $\bigcup_{n=1}^{\infty} A_n$ (b) $\bigcap_{n=1}^{\infty} A_n$ (c) (c)$A_2 - A_1$ (d) $A_2 \times A_3$.

15. Prove: for all sets A, B, C, $(A - B) \times C = (A \times C) - (B \times C)$.

16. Prove: for all $n \in \mathbb{Z}$, the following statements are equivalent: 4 divides $n + 2$; 4 divides $n - 2$; n is even and $n/2$ is odd; 2 divides n but 4 does not divide n.

17. Prove: for all sets A, B, C, if $A \cap C = B \cap C$ and $A \cup C = B \cup C$, then $A = B$.

18. Let I and J be fixed, nonempty index sets. Let A_i (for each $i \in I$) and A_j (for each $j \in J$) be fixed sets. Give two proofs of this fact: $\left(\bigcap_{i \in I} A_i\right) \cap \left(\bigcap_{j \in J} A_j\right) = \bigcap_{k \in I \cup J} A_k$. In the first proof, use only the definitions. In the second proof, try to use previously known theorems wherever possible (except the result being proved).

19. Compute $d = \gcd(1702, 483)$, and find integers x, y with $d = 1702x + 483y$.

20. Prove: $\forall a \in \mathbb{R}, \forall n \in \mathbb{Z}_{\geq 0}, \sum_{k=0}^{n} a5^k = a(5^{n+1} - 1)/4$.

21. Define $c_1 = c_2 = 1$, and $c_n = 2c_{n-1} + 3c_{n-2}$ for all $n \geq 3$. (a) Prove that $c_n > 3^{n-2}$ for all $n \geq 3$. (b) Prove that $c_n < 2 \cdot 3^{n-2}$ for all $n \geq 3$.

22. (a) Prove: for all sets A and B, $\mathcal{P}(A - B) - \{\emptyset\} \subseteq \mathcal{P}(A) - \mathcal{P}(B)$.
    (b) Disprove: for all sets A and B, $\mathcal{P}(A - B) - \emptyset \subseteq \mathcal{P}(A) - \mathcal{P}(B)$.

23. Recursively define a sequence of sets T_n as follows. Define $T_0 = \emptyset$. For all integers $n \geq 0$, define $T_{n+1} = T_n \cup \{T_n\}$. Prove:

$$\forall n \in \mathbb{Z}_{\geq 0}, \forall x, (x \in T_n \Rightarrow x \subseteq T_n).$$

5

Relations and Functions

5.1 Relations

In this chapter, we study relations and functions, which are two of the most central ideas in mathematics. Informally, a function f is a rule that takes each object x in some set of inputs and transforms it into a new object $y = f(x)$ in some set of outputs. A key point is that for every allowable input x, there must exist exactly one output y assigned to this x by the function. One of our main goals is to translate this informal description into a formal definition within set theory. To do so, we first define and study relations, which can be regarded as generalizations of functions in which each input x can be associated with more than one output, or perhaps no outputs at all. Many important mathematical concepts — including functions, partial orders, and equivalence relations — can be developed within the unifying general framework of relation theory.

Relations

In set theory, a *relation* is any set of ordered pairs. If R is a relation and $(x, y) \in R$, we can think of x as an *input* that is associated with the *output* y. In this setting, it is important to keep track of which objects might appear as potential inputs and outputs. This leads to the following definition.

5.1. Definition: Relations. For all sets R, X, Y:

$$\boxed{R \text{ is a relation from } X \text{ to } Y} \text{ iff } \boxed{R \subseteq X \times Y}.$$

This means that every member of R has the form (x, y), for some $x \in X$ and some $y \in Y$.

5.2. Example: Arrow Diagram of a Relation. Let $X = \{1, 2, 3, 4, 5, 6\}$ and $Y = \{a, b, c, d, e, f\}$. Then $R = \{(1, c), (1, e), (1, f), (3, f), (4, a), (4, b), (4, f), (6, c)\}$ is a relation from X to Y. We can visualize the relation R by the following *arrow diagram*.

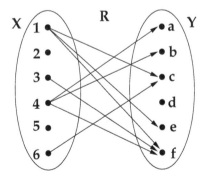

The arrow diagram consists of an oval with members of X (potential inputs for R) on the left side and an oval with members of Y (potential outputs for R) on the right side. For all $x \in X$ and $y \in Y$, we draw an arrow pointing from x to y iff $(x, y) \in R$. When X and Y are finite sets, this arrow diagram contains exactly the same information as the set R.

In the current example, we see that the input 1 is related (under R) to the outputs c and e and f. Input 3 is related to output f only. Inputs 2 and 5 are not related to any outputs under R. Similarly, d is a possible output that does not get associated with any input by the relation R. In particular, R is also a relation from $\{1, 3, 4, 6\}$ to $\{a, b, c, e, f\}$, or a relation from $\mathbb{Z}$ to $\{a, b, c, \ldots, z\}$, and so on. However, R is not a relation from $X' = \{1, 2, 3, 4\}$ to Y since $(6, c) \in R$ but $(6, c) \notin X' \times Y$.

5.3. Example: Graph of a Relation. If X and Y are subsets of $\mathbb{R}$, we can draw the *graph* of a relation S from X to Y by drawing a dot at each point (x, y) in the plane $\mathbb{R}^2$ such that $(x, y) \in S$. For example, suppose $X = [-3, 3]$, $Y = [-2, 2]$, and S is defined by letting $(x, y) \in S$ iff $x \in \mathbb{R}$ and $y \in \mathbb{R}$ and $(x^2/4) + y^2 = 1$. The graph of the relation S is the ellipse shown below.

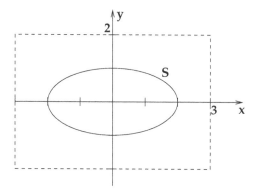

Since this ellipse is completely contained in the solid rectangle $X \times Y$ (whose boundary is indicated by dashed lines in the figure), S is a relation from $X = [-3, 3]$ to $Y = [-2, 2]$. Since $(-2, 0) \in S$, the input $x = -2$ has associated output $y = 0$. The input 0 is related to outputs 1 and -1. The input $3 \in X$ is not related to any output, and the possible output $2 \in Y$ is not related to any input. S is also a relation from $\mathbb{R}$ to $\mathbb{R}$, and S is a relation from $[-2, 2]$ to $[-1, 1]$. But S is not a relation from $[0, 1]$ to $[0, 1]$, because the graph of S does not lie completely within the square $[0, 1] \times [0, 1]$.

5.4. Example. A relation can be any set of ordered pairs, not necessarily a set determined by an explicit equation involving x and y. For example, the next figure shows the graph of a relation T from $\mathbb{R}$ to $\mathbb{R}$.

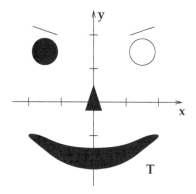

5.5. Example. Suppose X and Y are any sets. Is $\emptyset$ a relation from X to Y? Yes, since $\emptyset \subseteq X \times Y$ by an earlier theorem. When we are thinking of $\emptyset$ as a relation, we could call it the *empty relation*. For finite sets X and Y, the arrow diagram of $\emptyset$ contains no arrows. When X and Y are subsets of $\mathbb{R}$, the graph of $\emptyset$ has nothing in it.

At the other extreme, the full product set $X \times Y$ is a relation from X to Y, since $X \times Y \subseteq X \times Y$ by an earlier theorem. In this relation, every possible input $x \in X$ is related to every possible output $y \in Y$. The arrow diagram for $X \times Y$ contains every possible arrow leading from a point in X to a point in Y. When X and Y are closed intervals, the graph of $X \times Y$ (as we have seen before) is a solid rectangle of points with x-coordinates coming from X and y-coordinates coming from Y.

Images of Sets under Relations

Suppose R is a relation from X to Y. We have seen that each individual member x of X is related by R to zero or more members y of Y, namely those y for which $(x, y) \in R$. More generally, if we are given a set C of possible inputs to R, we could examine the ordered pairs in R to find the set of all possible outputs associated with the inputs coming from C. This new set is called the *image of C under the relation R* and is denoted $R[C]$. Formally, we make the following definition.

5.6. Definition: Image of a Set under a Relation. For all relations R, all sets C, and all objects y, $\boxed{y \in R[C]}$ iff $\boxed{\exists x \in C, (x, y) \in R}$.

Note that this definition places no restriction on the set C, although we usually take C to be a subset of the input space X. To get used to this definition, let us see how to compute images of sets for relations given by arrow diagrams and graphs.

5.7. Example. Consider the relation

$$R = \{(1, c), (1, e), (1, f), (3, f), (4, a), (4, b), (4, f), (6, c)\}$$

from Example 5.2. Given $C = \{1, 3, 5\}$, we compute $R[C] = R[\{1, 3, 5\}]$ by scanning the ordered pairs in R and listing the second component of each ordered pair whose first component is in C. This produces the set $R[C] = \{c, e, f, f\} = \{c, e, f\}$. Note that $R[C] = R[\{1, 3\}]$ since the input 5 is not related to any output. In fact, $R[C] = R[\{1\}]$ since the output f (related to input 3) is already related to input 1. Image computations can be performed conveniently using the arrow diagram of R (shown on page 211). For example, to compute $R[\{2, 4, 6\}]$ from the arrow diagram, start at the dots 2 and 4 and 6 on the left side and follow all possible outgoing arrows. The set of outputs that we reach, namely $\{a, b, f, c\} = \{a, b, c, f\}$, is the image $R[\{2, 4, 6\}]$. The following generic arrow diagram shows how $R[C]$ is obtained from C.

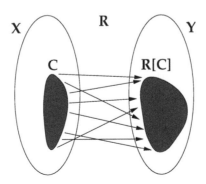

If we think of the arrows as rays of light leading from C to a screen represented by the set Y, then $R[C]$ is the image cast by the object C on this screen. This picture is the source of the term *image*.

5.8. Example. Define a relation S from $\mathbb{R}$ to $\mathbb{R}$ by letting S consist of all $(x,y) \in \mathbb{R}^2$ with $x = y^2$. In the following pictures, we use the graph of S to compute $S[\{4\}]$, $S[(1,4)]$, $S[\mathbb{R}_{>0}]$, and $S[[-1,1/4]]$.

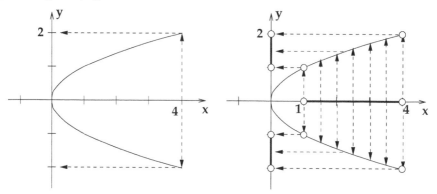

First, $S[\{4\}] = \{2, -2\}$ since $(4, 2) \in S$ and $(4, -2) \in S$ and no output other than 2 or -2 is related to the input 4 (as $4 = y^2$ is solved only by 2 and -2). To obtain this answer from the graph of S, we move up and down from the input 4 on the x-axis until we hit the graph of S. We hit the graph twice, at heights $y = 2$ and $y = -2$, so the image is $\{2, -2\}$.

We compute the image of the open interval $(1, 4)$ similarly, but now we imagine moving up and down from all points on the x-axis in the range $1 < x < 4$. We hit two arcs of the graph, which occupy heights between 1 and 2 (exclusive) and between -2 and -1 (exclusive). In interval notation, we have $S[(1, 4)] = (-2, -1) \cup (1, 2)$.

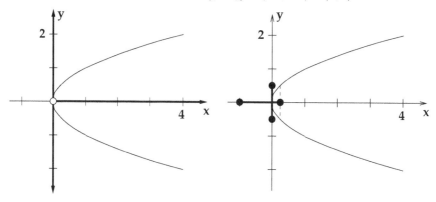

Next, if we allow all strictly positive real numbers as inputs to S, we see from the graph (or the formula $x = y^2$) that we obtain all nonzero real numbers as the related outputs. Thus, $S[\mathbb{R}_{>0}] = (-\infty, 0) \cup (0, \infty) = \mathbb{R}_{\neq 0}$. Starting with the closed interval $[-1, 1/4]$, we find $S[[-1, 1/4]] = [-1/2, 1/2]$. On the other hand, $S[[-1, -1/4]] = \emptyset$.

5.9. Example. Consider the relation T whose graph was drawn in Example 5.4. Use the graph to find (approximately) $T[\{0\}]$, $T[\{1.5\}]$, and $T[\mathbb{R}_{<0}]$.
Solution. To find $T[\{0\}]$, draw a vertical line through the graph of T at $x = 0$, and look at the heights of the points on the graph that meet this line. We estimate $T[\{0\}] = [-2, -4/3] \cup [-1/4, 1/2]$. Similarly, we estimate

$$T[\{1.5\}] = [-1.5, -1.1] \cup \{9/8, 15/8, 17/8\}.$$

Finally, $T[\mathbb{R}_{<0}] = (-2, -1) \cup [-1/4, 1/2) \cup [9/8, 15/8] \cup [2, 9/4]$.

Properties of Images

The next theorem lists properties of the image construction for relations.

5.10. Theorem on Images. For all sets $X, Y, A, B, I \neq \emptyset$, A_i (for $i \in I$) and all relations R and S:

(a) *Monotonicity:* If $A \subseteq B$, then $R[A] \subseteq R[B]$.

(b) *Empty Set Images:* $R[\emptyset] = \emptyset$ and $\emptyset[A] = \emptyset$.

(c) *Image of Unions:* $R[A \cup B] = R[A] \cup R[B]$ and $R[\bigcup_{i \in I} A_i] = \bigcup_{i \in I} R[A_i]$
    and $(R \cup S)[A] = R[A] \cup S[A]$.

(d) *Image of Intersections:* $R[A \cap B] \subseteq R[A] \cap R[B]$ and $R[\bigcap_{i \in I} A_i] \subseteq \bigcap_{i \in I} R[A_i]$
    and $(R \cap S)[A] \subseteq R[A] \cap S[A]$. (Equality is not always true.)

(e) *Image of Set Difference:* $R[A] - R[B] \subseteq R[A - B]$. (Equality is not always true.)

(f) *Input/Output Property:* If R is a relation from X to Y, then $R[A] \subseteq Y$
    and $R[A] = R[A \cap X]$.

We prove some representative parts of this theorem, asking you to prove other parts in the exercises. Fix arbitrary relations R and S, and sets A, B, etc., as in the theorem statement.

5.11. Proof of Monotonicity. Assume $A \subseteq B$. Prove $R[A] \subseteq R[B]$. [Continue with the subset template.] Fix an arbitrary object y; assume $y \in R[A]$; prove $y \in R[B]$. We have assumed there exists $x \in A$ with $(x, y) \in R$. We must prove $\exists z \in B, (z, y) \in R$. Choose $z = x$, so that $(z, y) = (x, y) \in R$. Since $x \in A$ and $A \subseteq B$, we have $z \in B$, as needed.

5.12. Proof that $R[\emptyset] = \emptyset$. Assume, to get a contradiction, that $R[\emptyset] \neq \emptyset$. This assumption means there exists an object $y \in R[\emptyset]$. Expanding the definition of image, we have assumed $\exists x \in \emptyset, (x, y) \in R$. But we know $x \notin \emptyset$. The contradiction "$x \in \emptyset$ and $x \notin \emptyset$" proves $R[\emptyset] = \emptyset$.

5.13. Proof that $R[\bigcup_{i \in I} A_i] = \bigcup_{i \in I} R[A_i]$. [We try a chain proof.] Fix an arbitrary object y. We know

$$y \in R\left[\bigcup_{i \in I} A_i\right] \Leftrightarrow \exists x \in \bigcup_{i \in I} A_i, (x, y) \in R \qquad \text{(by definition of image)}$$

$$\Leftrightarrow \exists x, \left(x \in \bigcup_{i \in I} A_i \wedge (x, y) \in R\right) \qquad \text{(by quantifier conversion rule)}$$

$$\Leftrightarrow \exists x, ((\exists i \in I, x \in A_i) \wedge (x, y) \in R) \qquad \text{(by definition of union)}$$

$$\Leftrightarrow \exists x, \exists i \in I, (x \in A_i \wedge (x, y) \in R) \qquad \text{(by a quantifier property)}$$

$$\Leftrightarrow \exists i \in I, \exists x, (x \in A_i \wedge (x, y) \in R) \qquad \text{(by reordering } \exists \text{ quantifiers)}$$

$$\Leftrightarrow \exists i \in I, \exists x \in A_i, (x, y) \in R \qquad \text{(by quantifier conversion rule)}$$

$$\Leftrightarrow \exists i \in I, y \in R[A_i] \qquad \text{(by definition of image)}$$

$$\Leftrightarrow y \in \bigcup_{i \in I} R[A_i] \qquad \text{(by definition of union)}.$$

5.14. Proof of Input/Output Property. Assume R is a relation from X to Y, which means $R \subseteq X \times Y$.

<u>Part 1.</u> Prove $R[A] \subseteq Y$. Fix an object z; assume $z \in R[A]$; prove $z \in Y$. We have assumed $\exists x \in A, (x, z) \in R$. Since $R \subseteq X \times Y$, we know $(x, z) \in X \times Y$. Thus $x \in X$ and $z \in Y$, so $z \in Y$.

<u>Part 2.</u> Prove $R[A] = R[A \cap X]$. *Part 2a.* Prove $R[A] \subseteq R[A \cap X]$. Fix z; assume $z \in R[A]$; prove $z \in R[A \cap X]$. We assumed $\exists x \in A, (x, z) \in R$. We must prove $\exists w \in A \cap X, (w, z) \in R$. Choose $w = x$, so $(w, z) = (x, z) \in R$. We must check that $w \in A \cap X$, which means $w \in A$ and $w \in X$. On one hand, since $w = x$ and $x \in A$, $w \in A$. On the other hand, since $(w, z) \in R$ and $R \subseteq X \times Y$, we deduce $(w, z) \in X \times Y$ and hence $w \in X$. *Part 2b.* Prove $R[A \cap X] \subseteq R[A]$. Since $A \cap X \subseteq A$ by an earlier theorem, we know $R[A \cap X] \subseteq R[A]$ by the previously proved monotonicity property of images.

Section Summary

1. *Relations.* R is a relation from X to Y iff $R \subseteq X \times Y$. To make the arrow diagram of a relation R, draw an arrow from the input $x \in X$ to the output $y \in Y$ for each ordered pair $(x, y) \in R$. When X and Y are subsets of $\mathbb{R}$, make the graph of R by drawing a dot at each point in the plane with coordinates $(x, y) \in R$.

2. *Images of Sets under Relations.* For any set C, relation R, and object y, $y \in R[C]$ iff $\exists x \in C, (x, y) \in R$. In an arrow diagram for R, $R[C]$ is the set of all objects reachable by following arrows starting at points in C. In the graph of R, we find $R[C]$ by drawing C on the x-axis, moving up and down from C into the graph, and finding the set of all y-coordinates reachable in this way.

3. *Properties of Images.* For all relations R and all sets A and B, $A \subseteq B$ implies $R[A] \subseteq R[B]$; $R[A \cup B] = R[A] \cup R[B]$; and $R[\emptyset] = \emptyset$. So the image construction preserves inclusions, unions, and the empty set. For intersections and differences, we have the weaker properties $R[A \cap B] \subseteq R[A] \cap R[B]$ and $R[A] - R[B] \subseteq R[A - B]$. Similar results hold for indexed unions and intersections.

Exercises

1. Let $X = \{1, 2, 3, 4\}$, $Y = \{1, 2, 3, 4, 5\}$, and $R = \{(1, 4), (1, 1), (2, 3), (2, 4), (4, 3)\}$.
   (a) Draw an arrow diagram for R.
   (b) Draw the graph of R.
   (c) Compute $R[\{1, 2\}]$, $R[\{3\}]$, $R[\{2, 4\}]$, and $R[X]$.
   (d) Find all sets $A, B \subseteq X$ with $R[A] = R[B]$.

2. Let R be the relation with the following arrow diagram.

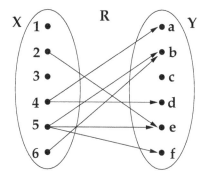

   (a) Describe R as a set of ordered pairs.
   (b) Find $R[\{1, 2, 3\}]$, $R[\{2, 4, 5\}]$, $R[\{3, 6\}]$, and $R[X]$.
   (c) Find a relation S such that $S[X] = Y$ and $R \cap S = \emptyset$.

3. Let $X = \{a, b, c, d, e\}$, $Y = \{v, w, x, y, z\}$, $R = \{(a, z), (b, y), (c, x), (d, y), (e, z)\}$, and $S = \{(a, v), (b, w), (c, x), (d, y), (e, v)\}$.
   (a) Draw arrow diagrams for R, S, $R \cap S$, and $R \cup S$.
   (b) Find the image of $\{a, c, d\}$ under each relation in (a).
   (c) Which subsets of Y have the form $R[A]$ for some set A?

4. Let $R = \{(x, y) \in \mathbb{R} \times \mathbb{R} : x^2 + y^2 = 25\}$. Draw the graph of R. Find $R[\{1\}]$, $R[\{-3\}]$, $R[[6, 9]]$, $R[(-5, 0)]$, and $R[\mathbb{Z}]$.

5. Let $X = \{1, 2, 3, 4, 5\}$, and let $R = \{(x, y) \in X \times X : xy \text{ is even}\}$. Draw an arrow diagram and graph for R. Compute $R[\{1, 3\}]$ and $R[\{2, 4\}]$.

6. Let $R = \{(x, y) \in \mathbb{R} : |xy| = 1\}$.
   (a) Draw the graph of R.
   (b) Find $R[\{2\}]$, $R[\{0\}]$, $R[(3, 4]]$, and $R[\mathbb{R}_{<0}]$.
   (c) Find $R \cap (\mathbb{Z} \times \mathbb{Z})$.

7. Define a relation $R = \{(x, y) : x \in [-3, 3] \text{ and } (y = 2x \text{ or } y = -x)\}$.
   (a) Draw the graph of R in the xy-plane.
   (b) Given $A = [1, 2]$ and $B = (-3, -1)$, compute the sets $R[A]$, $R[B]$, $R[A - B]$, $R[A] - R[B]$, $R[A \cap B]$, and $R[A] \cap R[B]$.

8. Let $R = \{(x, y) \in \mathbb{R} \times \mathbb{R} : y = \sin x\}$ and $S = \{(x, y) \in \mathbb{R} \times \mathbb{R} : x = \sin y\}$. Compute the image of the following sets under R and S: (a) $\{0\}$ (b) $\{\pi\}$ (c) $\{-1/2\}$ (d) $[-1, 1]$ (e) $\mathbb{R}$.

9. Let $X = \{1, 2, 3, 4, 5\}$, $Y = \{i, j, k, m, n\}$, $R = \{(1, n), (1, i), (2, m), (4, i), (4, m), (5, j), (5, k), (5, n)\}$, and $S = \{(1, n), (2, j), (3, k), (3, m), (3, n), (4, m), (5, n)\}$.
   (a) Draw arrow diagrams for R, S, $R \cap S$, and $R - S$, viewed as relations from X to Y. (b) Let $A = \{2, 3, 5\}$. Compute the sets $R[A]$, $S[A]$, $(R \cap S)[A]$, $R[A] \cap S[A]$, $(R - S)[A]$, and $R[A] - S[A]$.

10. Let S be the relation from $\mathbb{R}$ to $\mathbb{R}$ with the graph shown here.

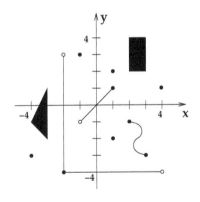

Find $S[\{1, 2, 3, 4\}]$, $S[\{-4, -1\}]$, $S[(-2, 1]]$, $S[[-1, 2)]$, $S[(2, 3)]$, $S[\mathbb{R}_{<0}]$, and $S[\mathbb{R}_{\geq 0}]$.

11. (a) Prove: for all relations R and S and all sets A, if $R \subseteq S$, then $R[A] \subseteq S[A]$.
    (b) Disprove: for all relations R and all sets A and B, if $R[A] \subseteq R[B]$, then $A \subseteq B$.
    (c) Prove or disprove: for all relations R and S and all sets A, if $R[A] \subseteq S[A]$, then $R \subseteq S$.
    (d) Prove or disprove: for all relations R and S, if $R[A] \subseteq S[A]$ for every set A, then $R \subseteq S$.

12. Prove: for all sets A, $\emptyset[A] = \emptyset$.

13. (a) Prove: for all relations R and S and all sets A, $(R \cup S)[A] = R[A] \cup S[A]$.
    (b) State and prove a generalization of (a) involving the image of an indexed union of relations.

14. (a) Prove part (d) of the Theorem on Images. (b) For each set inclusion proved in (a), give a specific example where set equality does not hold.

15. Prove part (e) of the Theorem on Images, and give a specific example where equality does not hold.

16. For any sets A, B, C, find (with proof) the set $(A \times B)[C]$.

17. Suppose R is a relation having the following property:
    for all w, x, y, if $(w, y) \in R$ and $(x, y) \in R$, then $w = x$.
    (a) What does this property say about the arrow diagram of R?
    (b) What does this property say about the graph of R?
    (c) Prove: for all sets A and B, $R[A \cap B] = R[A] \cap R[B]$.
    (d) Prove: for all sets A and B, $R[A] - R[B] = R[A - B]$.

5.2 Inverses, Identity, and Composition of Relations

This section introduces some concepts for relations — inverses, composition, and identity relations — that foreshadow and generalize the corresponding concepts for functions. One advantage of working with relations is that (unlike functions) every relation has an inverse relation.

Inverse of a Relation

Recall that a relation R from X to Y is a set of ordered pairs (x, y), where x is a member of the input set X and y is a member of the output set Y. The relation R associates zero or more outputs $y \in Y$ with each input $x \in X$. We could reverse the action of R by interchanging the roles of inputs and outputs. This produces a new relation R^{-1} from Y to X, called the *inverse of R*, which is found by replacing each (x, y) in R by (y, x). Formally, we have the following definition.

5.15. Definition: Inverse of a Relation. For all relations R and all objects a and b:

R^{-1} is the relation such that $\boxed{(a, b) \in R^{-1}}$ iff $\boxed{(b, a) \in R}$.

This definition states that R^{-1} *is a relation*, so the only possible elements of R^{-1} are ordered pairs. We could be more explicit by saying: for all objects z, $z \in R^{-1}$ iff $\exists a, \exists b, z = (a, b) \wedge (b, a) \in R$. A similar comment applies to other relations defined below.

5.16. Example: Inverses via Arrow Diagrams. Suppose $X = \{1, 2, 3, (4, 5), (5, 4)\}$, $Y = \{a, b, c, d, \emptyset\}$, and

$$R = \{(1, a), (1, c), (3, \emptyset), ((4, 5), a), ((5, 4), c), ((5, 4), d)\}.$$

By reversing all the ordered pairs, we see that

$$R^{-1} = \{(a, 1), (c, 1), (\emptyset, 3), (a, (4, 5)), (c, (5, 4)), (d, (5, 4))\}.$$

The arrow diagrams for R and R^{-1} are shown below. We get the arrow diagram for R^{-1} by reversing all the arrows in the arrow diagram for R.

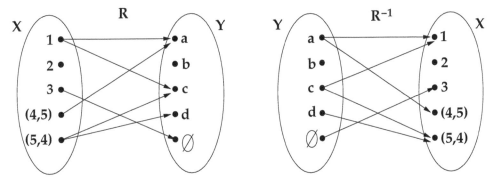

Let us compute the images of some sets under R and R^{-1}. First,

$$R[\{1, 2, 3\}] = \{a, c, \emptyset\}, \quad R^{-1}[\{a, c, \emptyset\}] = \{1, (4, 5), (5, 4), 3\},$$

$$\text{so } R^{-1}[R[\{1, 2, 3\}]] = \{1, (4, 5), (5, 4), 3\} \neq \{1, 2, 3\}.$$

Similarly, we find

$$R^{-1}[\{b,c\}] = \{1,(5,4)\}, \ R[\{1,(5,4)\}] = \{a,c,d\}, \text{ so } R[R^{-1}[\{b,c\}]] = \{a,c,d\} \neq \{b,c\}.$$

Note $R^{-1}[\emptyset] = \emptyset$, but $R^{-1}[\{\emptyset\}] = \{3\}$.

5.17. Example: Inverses via Graphs. The left side of the figure below shows the graph of a relation S from $[-2,2]$ to $[-2,2]$. The right side of the figure shows the graph of the inverse relation S^{-1}.

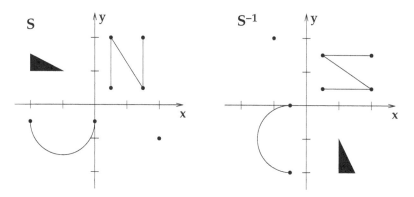

To see how the graph of the inverse is found, first consider the isolated point $(2,-1) \in S$. Switching the coordinates, we see that $(-1,2)$ is a point in S^{-1}. Next, the part of the graph of S that looks like the letter N consists of line segments from $(1/2,1/2)$ to $(1/2,2)$ to $(3/2,1/2)$ to $(3/2,2)$. Reversing the coordinates, we obtain three line segments from $(1/2,1/2)$ to $(2,1/2)$ to $(1/2,3/2)$ to $(2,3/2)$. The counterparts in S^{-1} of the solid triangle and circular arc in S are computed similarly. Geometrically, the graph of S^{-1} is obtained from the graph of S by reflecting the picture through the line $y = x$, since this reflection has the effect of replacing each (a,b) in the original graph by (b,a).

Note that $S[\{0,1/2\}] = \{-1/2\} \cup [1/2,2]$, whereas $S^{-1}[\{0,1/2\}] = \{1/2,3/2\}$. The latter image can be found either by looking at vertical lines through $x = 0$ and $x = 1/2$ in the graph of S^{-1}, or by looking at horizontal lines through $y = 0$ and $y = 1/2$ in the graph of S. We have $S^{-1}[[0,2]] = [-2,-1] \cup [1/2,3/2] = S^{-1}[[1/2,\infty)]$. Next, $S[S^{-1}[\mathbb{R}_{<0}]] = S[[-2,0] \cup \{2\}] = [-3/2,-1/2] \cup [1,3/2]$.

5.18. Example. We know $\emptyset$ is a relation containing no ordered pairs. It follows that $\emptyset^{-1} = \emptyset$. Drawing a picture, you can check that the inverse of the relation $R = [0,3] \times [1,2]$ is the relation $R^{-1} = [1,2] \times [0,3]$. More generally, for any sets X and Y, we claim that $(X \times Y)^{-1} = Y \times X$. Here is a quick chain proof: for any ordered pair (a,b), $(a,b) \in (X \times Y)^{-1}$ iff $(b,a) \in X \times Y$ iff $b \in X$ and $a \in Y$ iff $a \in Y$ and $b \in X$ iff $(a,b) \in Y \times X$.

Identity Relation on a Set

For any set X, we can create an especially simple relation from X to X called the *identity relation on X* and denoted I_X. This relation consists of all ordered pairs (x,x) with $x \in X$. The name "identity" is used since each $x \in X$ is related to itself and nothing else. The formal definition follows.

5.19. Definition: Identity Relation on a Set. For all sets X and all objects a and b:

I_X is the relation such that $\boxed{(a,b) \in I_X}$ iff $\boxed{a \in X \text{ and } a = b}$.

5.20. Example: Arrow Diagram of an Identity Relation.
Let X be the set $\{1, 2, 3, (4, 5), (5, 4)\}$. The identity relation on X is

$$I_X = \{(1, 1), (2, 2), (3, 3), ((4, 5), (4, 5)), ((5, 4), (5, 4))\}.$$

The arrow diagram for I_X is shown below. For each $x \in X$, there is an arrow pointing from x to itself. We see that $I_X^{-1} = I_X$. For all $A \subseteq X$, $I_X[A] = A$. These properties hold in general.

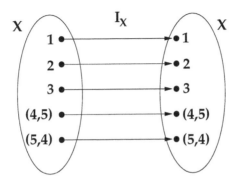

5.21. Example: Graph of an Identity Relation. Let $A = [-2, 0) \cup (1, 2]$ and $B = \{-2, -1, 0, 1/2, 1\}$. The next figure shows the graphs of the identity relations I_A and I_B. The graph I_A consists of the portion of the line $y = x$ consisting of points (c, c) with $c \in A$; similarly for I_B. The entire line $y = x$ is the graph of $I_\mathbb{R}$. Note $I_\emptyset = \emptyset$, so the graph of $I_\emptyset$ is empty.

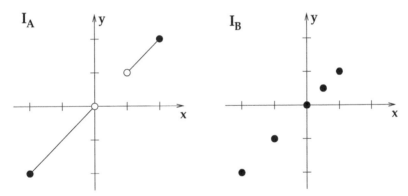

Composition of Relations

Informally, given two real-valued functions f and g, we can form a new function $g \circ f$ (the *composition* of g and f) as follows. Starting with an input $x \in \mathbb{R}$, we feed this number into f to obtain an output $y = f(x)$. Now we regard this output y as the input to g, obtaining a new output $z = g(y)$. The net effect is that $z = g(y) = g(f(x)) = (g \circ f)(x)$. We can generalize this idea to relations, as follows.

5.22. Definition: Composition of Relations. For all relations R and S and all ordered pairs (a, b), $S \circ R$ is the relation such that $\boxed{(a, b) \in S \circ R}$ iff $\boxed{\exists c, (a, c) \in R \wedge (c, b) \in S}$.

The relation $S \circ R$ is the *composition* of the relations S and R, in this order.

Note carefully the order of the relations in the expression $S \circ R$, which is motivated by the analogy with the function composition $g \circ f$. Observe that when computing $(g \circ f)(x)$, we start with the input x, apply the function f, and then apply the function g. Analogously, given an ordered pair $(a, b) \in S \circ R$, we know there exists c with $(a, c) \in R$ and $(c, b) \in S$. This means that the original input a is related (under R) to the intermediate output c, and then c (viewed as an input to S) is related to the final output b. The net effect is that input a is related to output b under $S \circ R$.

5.23. Example: Composition of Relations via Arrow Diagrams. Let $X = \{1, 2, 3, 4\}$ and $Y = \{a, b, c, d, e\}$. Define a relation R from X to Y and a relation S from Y to X by

$$R = \{(1, d), (3, d), (4, a), (4, e)\}, \quad S = \{(a, 2), (b, 1), (c, 3), (c, 4), (d, 2), (e, 1)\}.$$

The arrow diagrams for R and S are shown below.

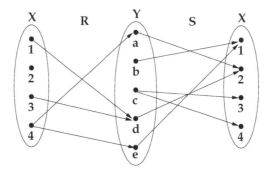

We can compute $S \circ R$, which is a relation from X to X, by starting at points in X and seeing where we can go by traveling along an R-arrow followed by an S-arrow. By examination of the figure, we arrive at this arrow diagram for $S \circ R$:

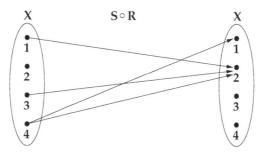

Symbolically, $S \circ R = \{(1, 2), (3, 2), (4, 2), (4, 1)\}$. For example, $(4, 1) \in S \circ R$ because $(4, e) \in R$ and $(e, 1) \in S$. Similarly, we can compute $R \circ S$, which is a relation from Y to Y, by combining the arrow diagrams for R and S in the other order, as shown here:

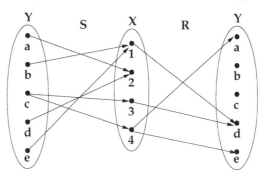

The arrow diagram for $R \circ S$ looks like this:

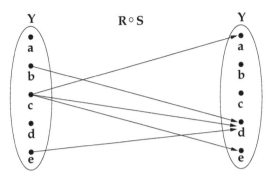

So $R \circ S = \{(b,d),(c,a),(c,d),(c,e),(e,d)\}$. Note carefully that $\boxed{S \circ R \neq R \circ S}$ in this example (and in most examples). This means that *composition of relations is not commutative.* Indeed, in our example, $(R \circ S) \cap (S \circ R) = \emptyset$, which says that these two sets of ordered pairs have no members in common.

5.24. Example. Let $X = \{1,2,3\}$, $R = \{(1,2),(2,3),(3,1)\}$, and $S = \{(1,3),(2,1),(3,2)\}$. You can check that $R \circ S = \{(1,1),(2,2),(3,3)\} = S \circ R = I_X$. Thus, it does sometimes happen that $R \circ S = S \circ R$. In this example, it is also true that $R^{-1} = S = R \circ R$ and $S^{-1} = R = S \circ S$.

5.25. Example. Let $R = [0,2] \times [1,3]$ and $S = I_{\mathbb{R}} \cup \{(x, 2-x) : x \in [0,2]\}$. Trying to read off the compositions $S \circ R$ and $R \circ S$ from the graphs of R and S can be tricky, so let us return to the definition. Which ordered pairs (a,b) belong to $S \circ R$? First, the original input a must belong to $[0,2]$, and all such inputs are related to all numbers $c \in [1,3]$ (and nothing else). Proceeding to S, a given input $c \in [1,3]$ is always related (under S) to c; furthermore, each number c in the range $[0,2]$ is also related to $2-c$. Thus, the final output b can be any number in $[1,3]$, and (by considering $c \in [1,2]$) b can also be any number in $[0,1]$. In summary, $(a,b) \in S \circ R$ iff $a \in [0,2]$ and $b \in [0,3]$, so that $S \circ R = [0,2] \times [0,3]$.

On the other hand, when is $(a,b) \in R \circ S$? Now we need to find c with $(a,c) \in S$ and $(c,b) \in R$. One possibility is that $(a,c) \in I_{\mathbb{R}}$, which means $a \in \mathbb{R}$ and $c = a$. Then (c,b) is in R iff $c = a$ is in $[0,2]$ and b is in $[1,3]$. So every element of $[0,2] \times [1,3]$ is in $R \circ S$. The other possibility for $(a,c) \in S$ is that $a \in [0,2]$ and $c = 2-a \in [0,2]$. In this case, $(c,b) \in R$ holds iff $b \in [1,3]$. We conclude $R \circ S = [0,2] \times [1,3] = R \neq S \circ R$.

Section Summary

1. *Inverse of a Relation.* For any relation R, the inverse of R is the relation R^{-1} such that for all a and b, $(a,b) \in R^{-1}$ iff $(b,a) \in R$. To go from the arrow diagram of R to the arrow diagram of R^{-1}, reverse all the arrows. To go from the graph of R to the graph of R^{-1}, reflect the xy-plane through the line $y = x$.

2. *Identity Relation.* For any set X, the identity relation on X is the relation I_X such that for all a and b, $(a,b) \in I_X$ iff $a \in X$ and $a = b$. The arrow diagram for I_X has an arrow from x to x for each $x \in X$. The graph of I_X is the part of the line $y = x$ consisting of those points (c,c) with $c \in X$.

3. *Composition of Relations.* For any relations R and S, the composition of R followed by S is the relation $S \circ R$ such that for all a and b, $(a,b) \in S \circ R$ iff $\exists c, (a,c) \in R \wedge (c,b) \in S$. The arrow diagram for $S \circ R$ is found by drawing an arrow from a to b whenever there is a path from a to b consisting of an arrow for

R followed by an arrow for S. In most cases, $S \circ R \neq R \circ S$. So composition of relations is not commutative.

Exercises

1. Let $X = \{1, 2, 3, 4, 5, 6\}$. Define a relation R from X to X as follows:

$$R = \{(1, 2), (1, 4), (1, 6), (2, 4), (4, 5), (4, 6), (6, 3)\}.$$

(a) Draw an arrow diagram for R and the graph of R in $\mathbb{R}^2$.
(b) Describe R^{-1} as a set of ordered pairs, as an arrow diagram, and as a graph.
(c) Describe $R \circ R$ as a set of ordered pairs, as an arrow diagram, and as a graph.
(d) Compute the following images of sets under the relations R and R^{-1}:

$$R[\{2, 3, 4\}], \quad R[\{3, 5\}], \quad R[\{1\}], \quad R^{-1}[\{2, 3, 4\}], \quad R^{-1}[\{3, 5\}],$$

$$R^{-1}[\{1\}], \quad R[R^{-1}[\{1, 2, 3\}]], \quad R^{-1}[R[\{1, 2, 3\}]].$$

2. Define X and R as in the previous problem. Define

$$S = \{(2, 1), (2, 2), (2, 3), (2, 4), (1, 5), (1, 6), (3, 3), (4, 6)\}.$$

Draw arrow diagrams for each relation:
(a) $R \circ S$ (b) $S \circ R$ (c) $(R \circ S)^{-1}$ (d) $R^{-1} \circ S^{-1}$ (e) $S^{-1} \circ R^{-1}$ (f) $(R \circ S) \circ R$.

3. Let $X = \{1, 2, 3, 4\}$, $R = \{(a, b) \in X \times X : a < b\}$ and $S = \{(a, b) \in X \times X : a \leq b\}$.
(a) Draw arrow diagrams for R and S.
(b) Compute R^{-1}, S^{-1}, $R \circ S$, $S \circ R$, $R \circ R$, $S \circ S$, $R \cup S$, and $R \cap S$.

4. Define a relation T from $\mathbb{R}$ to $\mathbb{R}$ by the following graph.

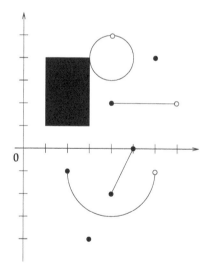

(a) Carefully sketch the graph of T^{-1}.
(b) What is the set $T \cap I_{\mathbb{Z}}$?
(c) Compute $T[\{2\}]$, $T^{-1}[\{2\}]$, $T[[6, 7]]$, and $T^{-1}[[0, 1)]$.

5. Let R be the relation in Example 5.16.
(a) Find all subsets A of X such that $R^{-1}[R[A]] = A$.
(b) Find all subsets B of Y such that $R[R^{-1}[B]] = B$.

6. Let S be the relation in Example 5.17. (a) Find $S[\{-2,-1\}]$. (b) Find $S^{-1}[\{-2,-1\}]$. (c) Find $S[[1,2]]$. (d) Find $S^{-1}[[1,2]]$. (e) Find $S^{-1}[S[\mathbb{R}_{<0}]]$.

7. (a) Draw an arrow diagram and a graph for $I_{\{-1,0,2\}}$. (b) Draw the graph of $I_{\mathbb{R}-\mathbb{Z}}$. (c) Let $X = \{0,1\} \times \{1,2,3\}$. Draw an arrow diagram for I_X.

8. Let $R = \{(x,x^2) : x \in \mathbb{R}\}$, $S = \{(x,3x) : x \in \mathbb{R}\}$, and $T = \{(x,x-1) : x \in \mathbb{R}\}$. Draw graphs of the following relations: (a) $R \circ S$ (b) $S \circ R$ (c) $R \circ T$ (d) $T \circ R$ (e) $S \circ T$ (f) $T \circ R$ (g) $R \circ R^{-1}$ (h) $R^{-1} \circ R$.

9. Let $R = \{(x,y) \in \mathbb{R}^2 : x^2 + y^2 = 4\}$ and $S = \{(x,y) \in \mathbb{R}^2 : y^2 = x^2\}$. True or false? Explain each answer. (a) $R^{-1} = R$. (b) $S^{-1} = S$. (c) $R \circ R = S$. (d) $S \circ S = S$. (e) $R \circ S = R$. (f) $S \circ R = R$.

10. Repeat the previous exercise taking $R = \{(x,y) \in \mathbb{R}^2 : x^2 + y = 4\}$ and $S = \{(x,y) \in \mathbb{R}^2 : y = -x\}$.

11. Let A, B, C, and D be arbitrary sets. Describe the following relations as explicitly as possible: (a) $(A \times B) \circ (C \times D)$ (b) $(A \times B)^{-1}$ (c) $I_A \cap I_B$ (d) $I_A \cup I_B$ (e) $(A \times B) \circ \emptyset$ (f) $(B \times A) \circ R$, where R is a relation from A to B.

12. Let $X = \{1,2,3\}$. Give four examples of relations R from X to X satisfying $R \circ R = R = R^{-1}$. In one of your examples, find a relation R consisting of five ordered pairs.

13. Define $X = \{1,2,3,4\}$, $Y = \{a,b,c\}$, and $R = \{(1,b),(4,a),(3,c)\}$.
    (a) Find a relation S from Y to X such that $S \circ R = I_X$, or explain why no such S exists. (b) Find a relation S from Y to X such that $R \circ S = I_Y$, or explain why no such S exists.

14. Repeat Problem 13 taking $X = \mathbb{R}$, $Y = \mathbb{R}_{\geq 0}$, and $R = \{(x,y) \in \mathbb{R}^2 : y = x^2\}$.

15. Repeat Problem 13 taking $X = Y = [-1,1]$ and $R = \{(x,y) \in \mathbb{R}^2 : x^2 + y^2 = 1\}$.

16. Define a relation R by the following arrow diagram.

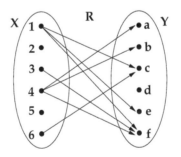

(a) Compute R^{-1}, $R \circ R^{-1}$, and $R^{-1} \circ R$.
(b) Find three relations S from Y to X such that $S \circ R = I_{\{1,4\}}$.
(c) Find all subsets $B \subseteq Y$ for which there exists a relation S from Y to X with $R \circ S = I_B$.

17. Let R be a relation from X to Y.
    (a) Find a condition on R that guarantees the existence of a relation S from Y to X with $S \circ R = X \times X$.
    (b) Find a condition on R that guarantees the existence of a relation T from Y to X with $R \circ T = Y \times Y$.

5.3 Properties of Relations

In this section, we state and prove many properties of inverses, identity relations, and compositions. Before starting, you are encouraged to review the relevant proof templates from set theory and the definitions from the last two sections.

Closure Theorem for Relations

We begin with a preliminary theorem indicating what kind of relations are produced by various constructions.

5.26. Closure Theorem for Relations. For all sets X, Y, Z, $I \neq \emptyset$ and all relations R, S, R_i (for $i \in I$):

(a) If R and S are relations from X to Y, then $R \cup S$ and $R \cap S$ and $R - S$ are relations from X to Y.

(b) If R_i is a relation from X to Y for each $i \in I$, then $\bigcup_{i \in I} R_i$ and $\bigcap_{i \in I} R_i$ are relations from X to Y.

(c) R is a relation from X to Y iff R^{-1} is a relation from Y to X.

(d) I_X is a relation from X to X.

(e) If R is a relation from X to Y and S is a relation from Y to Z, then $S \circ R$ is a relation from X to Z.

We prove (a) and (e) as illustrations. Fix arbitrary sets X, Y, and Z and arbitrary relations R and S. For (a), assume R and S are relations from X to Y, which means $R \subseteq X \times Y$ and $S \subseteq X \times Y$. We must show $R \cup S$ and $R \cap S$ and $R - S$ are relations from X to Y. <u>Part 1.</u> Prove $R \cup S \subseteq X \times Y$. Since we assumed $R \subseteq X \times Y$ and $S \subseteq X \times Y$, this follows from the least upper bound property (part (f) of the Theorem on Subsets; see page 106). <u>Part 2.</u> Prove $R \cap S \subseteq X \times Y$. We know $R \cap S \subseteq R$ by part (c) of the Theorem on Subsets, and we assumed $R \subseteq X \times Y$, so $R \cap S \subseteq X \times Y$ follows by transitivity of $\subseteq$ (part (b) of the Theorem on Subsets). <u>Part 3.</u> Prove $R - S \subseteq X \times Y$. As in part 2, we know $R - S \subseteq R$ and $R \subseteq X \times Y$, so $R - S \subseteq X \times Y$ follows.

For (e), assume R is a relation from X to Y and S is a relation from Y to Z. Prove $S \circ R$ is a relation from X to Z. We have assumed $R \subseteq X \times Y$ and $S \subseteq Y \times Z$. We must prove $S \circ R \subseteq X \times Z$. To prove this, fix an arbitrary ordered pair (a, b), assume $(a, b) \in S \circ R$, and prove $(a, b) \in X \times Z$. We have assumed $\exists c, (a, c) \in R \wedge (c, b) \in S$. We must prove $a \in X$ and $b \in Z$. First, since $(a, c) \in R$ and $R \subseteq X \times Y$, we see that $(a, c) \in X \times Y$. So $a \in X$ and $c \in Y$, hence $a \in X$. Second, since $(c, b) \in S$ and $S \subseteq Y \times Z$, we see that $(c, b) \in Y \times Z$. So $c \in Y$ and $b \in Z$, hence $b \in Z$.

Properties of Inverses, Identity Relations, and Composition

Our next theorem lists some general properties of inverse relations, identity relations, and composition of relations.

5.27. Theorem on Relations. For all sets X, Y, C and all relations R, S, T:

(a) *Inverse of Identity:* $I_X^{-1} = I_X$.

(b) *Double Inverse:* $(R^{-1})^{-1} = R$.

(c) *Inverse of Union:* $(S \cup T)^{-1} = S^{-1} \cup T^{-1} = T^{-1} \cup S^{-1}$.

(d) *Inverse of Intersection:* $(S \cap T)^{-1} = S^{-1} \cap T^{-1} = T^{-1} \cap S^{-1}$.

(e) *Composition with Identity:* If R is a relation from X to Y, then $R \circ I_X = R$ and $I_Y \circ R = R$.

(f) *Associativity of Composition:* $T \circ (S \circ R) = (T \circ S) \circ R$.

(g) *Inverse of a Composition:* $(R \circ S)^{-1} = S^{-1} \circ R^{-1}$.

(h) *Image of a Composition:* $(S \circ R)[C] = S[R[C]]$.

(i) *Monotonicity Properties:* If $X \subseteq Y$, then $I_X \subseteq I_Y$. If $R \subseteq S$, then $R^{-1} \subseteq S^{-1}$ and $T \circ R \subseteq T \circ S$ and $R \circ T \subseteq S \circ T$.

(j) *Empty Set Properties:* $I_{\emptyset} = \emptyset$ and $\emptyset^{-1} = \emptyset$ and $R \circ \emptyset = \emptyset$ and $\emptyset \circ R = \emptyset$.

(k) *Distributive Laws for $\cup$ and $\circ$:*

$$(S \cup T) \circ R = (S \circ R) \cup (T \circ R) \text{ and } T \circ (S \cup R) = (T \circ S) \cup (T \circ R).$$

(l) *Partial Distributive Laws for $\cap$ and $\circ$:*

$$(S \cap T) \circ R \subseteq (S \circ R) \cap (T \circ R) \text{ and } T \circ (S \cap R) \subseteq (T \circ S) \cap (T \circ R).$$

Equality does not always hold.

Items involving $\cup$ and $\cap$ generalize to indexed unions and intersections. As noted earlier, composition of relations is *not commutative*: in most cases, $R \circ S \neq S \circ R$.

We prove some representative parts of the theorem, leaving other parts to the exercises. You are strongly encouraged to work out proofs of all parts of this theorem now, before reading further. [*Hints:* When proving subset statements or set equality statements involving relations, you should fix arbitrary ordered pairs (a, b) rather than fixing arbitrary objects. Chain proofs may work well for sufficiently simple set equalities, but two-part set equality proofs are preferred for more complex statements. When proving a set is empty, contradiction proofs are often the fastest way.]

Have you proved the theorem yet? To see our proofs, read on. Fix arbitrary sets X, Y, C and relations R, S, and T.

5.28. Proof that $(R^{-1})^{-1} = R$. We give a chain proof. Let (a, b) be an arbitrary ordered pair. Using the definition of inverse relation twice, we see that $(a, b) \in (R^{-1})^{-1}$ iff $(b, a) \in R^{-1}$ iff $(a, b) \in R$. So $(R^{-1})^{-1} = R$ since these two relations contain exactly the same ordered pairs.

5.29. Proof of Composition with Identity. Assume R is a relation from X to Y, which means $R \subseteq X \times Y$. We prove $R \circ I_X = R$ (letting you prove $I_Y \circ R = R$).
Part 1. Prove $R \circ I_X \subseteq R$. Fix (a, b); assume $(a, b) \in R \circ I_X$; prove $(a, b) \in R$. By definition of composition, we know there exists c with $(a, c) \in I_X$ and $(c, b) \in R$. Next, by definition of I_X, we know $a \in X$ and $a = c$. Since $(c, b) \in R$, we deduce $(a, b) \in R$, as needed.
Part 2. Prove $R \subseteq R \circ I_X$. Fix (s, t); assume $(s, t) \in R$; prove $(s, t) \in R \circ I_X$. We must prove $\exists w, (s, w) \in I_X \wedge (w, t) \in R$. Choose $w = s$. On one hand, since $(s, t) \in R$ and $R \subseteq X \times Y$, we see that $(s, t) \in X \times Y$. So $s \in X$ and $t \in Y$, which shows that $w = s$ is a member of X. Then $(s, w) = (s, s) \in I_X$. Next, $(w, t) = (s, t) \in R$ by assumption.

5.30. Proof of Distributive Law for $\cup$ and $\circ$. We give a chain proof of $(S \cup T) \circ R = (S \circ R) \cup (T \circ R)$, leaving the proof of the companion identity to you. Fix an arbitrary ordered pair (a, b). We know

$$(a, b) \in (S \cup T) \circ R \Leftrightarrow \exists c, (a, c) \in R \wedge (c, b) \in S \cup T$$
$$\Leftrightarrow \exists c, (a, c) \in R \wedge [(c, b) \in S \vee (c, b) \in T]$$
$$\Leftrightarrow \exists c, ([(a, c) \in R \wedge (c, b) \in S] \vee [(a, c) \in R \wedge (c, b) \in T])$$
$$\Leftrightarrow (\exists c, [(a, c) \in R \wedge (c, b) \in S]) \vee (\exists c, [(a, c) \in R \wedge (c, b) \in T])$$
$$\Leftrightarrow (a, b) \in S \circ R \vee (a, b) \in T \circ R$$
$$\Leftrightarrow (a, b) \in (S \circ R) \cup (T \circ R).$$

The steps follow by definition of composition, definition of union, the distributive law for $\wedge$ and $\vee$, the distributive law for $\exists$ and $\vee$, the definition of composition, and the definition of union.

5.31. Proof of Empty Set Properties. We prove that $\emptyset^{-1} = \emptyset$. Assume, to get a contradiction, that $\emptyset^{-1} \neq \emptyset$. Then there exists an object $z \in \emptyset^{-1}$; in fact, there exists an ordered pair $(a, b) \in \emptyset^{-1}$. By definition of inverse, $(b, a) \in \emptyset$. But we also know $(b, a) \notin \emptyset$, which gives us a contradiction.

Next we prove that $R \circ \emptyset = \emptyset$. Assume, to get a contradiction, that $R \circ \emptyset \neq \emptyset$. Then there exists an ordered pair $(a, b) \in R \circ \emptyset$. By definition of composition, we know there exists c with $(a, c) \in \emptyset$ and $(c, b) \in R$. But $(a, c) \notin \emptyset$, so we have found a contradiction.

5.32. Proof of Partial Distributive Law. We prove $T \circ (S \cap R) \subseteq (T \circ S) \cap (T \circ R)$. Fix (a, b); assume $(a, b) \in T \circ (S \cap R)$; prove $(a, b) \in (T \circ S) \cap (T \circ R)$. Our assumption means there exists w with $(a, w) \in S \cap R$ and $(w, b) \in T$; so there exists w with $(a, w) \in S$ and $(a, w) \in R$ and $(w, b) \in T$. We must prove $(a, b) \in T \circ S$ and $(a, b) \in T \circ R$. First, to prove $(a, b) \in T \circ S$, we must prove $\exists x, (a, x) \in S \wedge (x, b) \in T$. Choose $x = w$; note that $(a, w) \in S$ and $(w, b) \in T$ have been assumed. Second, to prove $(a, b) \in T \circ R$, we must prove $\exists y, (a, y) \in R \wedge (y, b) \in T$. Choose $y = w$; note that $(a, w) \in R$ and $(w, b) \in T$ have been assumed.

We give an example to show that set equality may not hold. Let $R = \{(1, 2)\}$, $S = \{(1, 3)\}$, and $T = \{(2, 4), (3, 4)\}$. By drawing arrow diagrams or looking at the ordered pairs, we find that $T \circ S = \{(1, 4)\}$, $T \circ R = \{(1, 4)\}$, and so $(T \circ S) \cap (T \circ R) = \{(1, 4)\}$. But $S \cap R = \emptyset$, so $T \circ (S \cap R) = \emptyset$. Thus $T \circ (S \cap R)$ is a proper subset of $(T \circ S) \cap (T \circ R)$ in this example.

5.33. Partial Proof of Associativity of Composition. We prove $T \circ (S \circ R) \subseteq (T \circ S) \circ R$; we let you prove $(T \circ S) \circ R \subseteq T \circ (S \circ R)$. Fix (a, b); assume $(a, b) \in T \circ (S \circ R)$; prove $(a, b) \in (T \circ S) \circ R$. We have assumed there exists w with $(a, w) \in S \circ R$ and $(w, b) \in T$. Expanding this again, we have assumed there exists w such that there exists x with $(a, x) \in R$ and $(x, w) \in S$ and $(w, b) \in T$. Now, since $(x, w) \in S$ and $(w, b) \in T$, we see that $(x, b) \in T \circ S$. Since $(a, x) \in R$ and $(x, b) \in T \circ S$, it follows that $(a, b) \in (T \circ S) \circ R$, as needed.

5.34. Proof of $(R \circ S)^{-1} = S^{-1} \circ R^{-1}$. We try a chain proof. Fix an ordered pair (a, b). We know

$$(a, b) \in (R \circ S)^{-1} \Leftrightarrow (b, a) \in R \circ S \qquad \text{(by definition of inverse)}$$
$$\Leftrightarrow \exists c, (b, c) \in S \wedge (c, a) \in R \qquad \text{(by definition of composition)}$$
$$\Leftrightarrow \exists c, (c, a) \in R \wedge (b, c) \in S \qquad \text{(by commutativity of AND)}$$
$$\Leftrightarrow \exists c, (a, c) \in R^{-1} \wedge (c, b) \in S^{-1} \qquad \text{(by definition of inverse)}$$
$$\Leftrightarrow (a, b) \in S^{-1} \circ R^{-1} \qquad \text{(by definition of composition)}.$$

Section Summary

1. *Closure Properties of Relations.* The union, intersection, and set difference of relations from X to Y are also relations from X to Y. The inverse of a relation from X to Y is a relation from Y to X. The identity relation I_X is a relation from X to X. If R is a relation from X to Y and S is a relation from Y to Z, then $S \circ R$ is a relation from X to Z.

2. *Properties of Inverse Relations.* For all sets X and relations R and S, $I_X^{-1} = I_X$, $(R^{-1})^{-1} = R$, $(R \cup S)^{-1} = R^{-1} \cup S^{-1}$, $(R \cap S)^{-1} = R^{-1} \cap S^{-1}$, $(R \circ S)^{-1} = S^{-1} \circ R^{-1}$, $R \subseteq S$ implies $R^{-1} \subseteq S^{-1}$, and $\emptyset^{-1} = \emptyset$.

3. *Properties of Composition of Relations.* For all relations R from X to Y, $R \circ I_X = R = I_Y \circ R$. Composition of relations is associative but not commutative: $T \circ (S \circ R) = (T \circ S) \circ R$ always holds, but $S \circ R = R \circ S$ does not always hold. We have distributive laws such as $(S \cup T) \circ R = (S \circ R) \cup (T \circ R)$ and $(S \cap T) \circ R \subseteq (S \circ R) \cap (T \circ R)$.

Exercises

1. Prove parts (b), (c), and (d) of the Closure Theorem for Relations.

2. Prove part (a) of the Theorem on Relations.

3. Prove parts (c) and (d) of the Theorem on Relations.

4. Finish the proof of part (e) of the Theorem on Relations.

5. Finish the proof of part (f) of the Theorem on Relations.

6. Prove part (h) of the Theorem on Relations.

7. Prove part (i) of the Theorem on Relations.

8. Finish the proof of part (j) of the Theorem on Relations.

9. Finish the proof of part (k) of the Theorem on Relations.

10. Finish the proof of part (l) of the Theorem on Relations.

11. Let A and B be arbitrary sets.
    (a) Prove $I_{A \cup B} = I_A \cup I_B$. (b) Prove $I_{A \cap B} = I_A \cap I_B = I_A \circ I_B$.

12. Prove: for all relations R and S, $R \circ S = S \circ R$ iff $R^{-1} \circ S^{-1} = S^{-1} \circ R^{-1}$.

13. (a) Prove: for all sets X and all relations R, S, T on X, if $R^{-1} \circ R = I_X$ and $R \circ S = R \circ T$, then $S = T$. (b) Show that the result in (a) need not hold if we drop the hypothesis $R^{-1} \circ R = I_X$.

14. Prove: for all relations R, if $R \subseteq R^{-1}$, then $R = R^{-1}$.

15. (a) Prove: for all sets X and all relations R on X, if $I_X \subseteq R$, then $R \subseteq R \circ R$. (b) Prove or disprove the converse of part (a).

16. Prove or disprove: for all sets X and all relations R on X, if $R = R^{-1}$, then $R \circ R = I_X$.

17. Prove or disprove: for all sets X and all relations R on X, if $R \circ R = I_X$, then $R = R^{-1}$.

18. Prove: for all relations R, S, and T, $(S \circ R) \cap T = \emptyset$ iff $(T \circ R^{-1}) \cap S = \emptyset$.

19. Prove: for all relations R, S, and T, $T \cap (S \circ R) \subseteq T \circ (R^{-1} \circ R)$. Give an example where equality does not hold.

20. Prove: for all relations R, S, and T, $R \cap S \cap T \subseteq (T \circ S^{-1}) \circ R$. Give an example where equality does not hold.

21. Prove: for all relations R, S, and T, if $T \circ S \subseteq S$ and $T^{-1} \circ S \subseteq S$, then $S \cap (T \circ R) = T \circ (S \cap R)$.

22. Prove: for all relations R, S, T, U, $(S \circ R) \cap (U \circ T) \subseteq [U \circ ((U^{-1} \circ S) \cap (T \circ R^{-1}))] \circ R$. Does equality always hold?

5.4 Definition of Functions

One of the most fundamental and pervasive concepts in mathematics is the notion of a function. In calculus, we study real-valued functions of a real variable such as the exponential function, the natural logarithm function, polynomial functions, and trigonometric functions. In advanced calculus and other parts of mathematics, we need to consider more general kinds of functions. In this section, we give a formal definition of functions within the framework of set theory. We illustrate the logical nuances of this definition with many examples of functions and non-functions, presented using arrow diagrams, graphs, and formulas.

Arrow Diagrams of Functions and Non-Functions

We begin with an informal discussion of functions, which motivates the formal definition below. Informally, a *function* g from a set A into a set B is a rule that assigns to each object x in A a unique object $g(x)$ in B. The set A is called the *domain* of g; the set B is called the *codomain* of g; and $g(x)$ is called the *value* of g at the *input* x. The crucial point is that each input x in A has *exactly one* output $g(x)$ associated to it, and this output is required to belong to the codomain B. We use the notation $\boxed{g : A \to B}$ to mean that $\boxed{g \text{ is a function with domain } A \text{ and codomain } B.}$

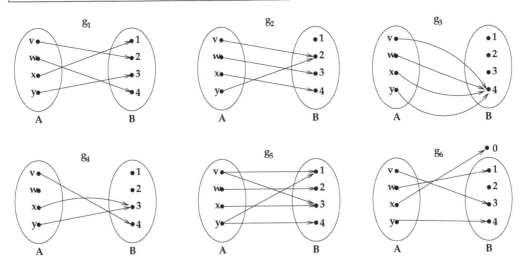

FIGURE 5.1
Arrow diagrams illustrating functions and non-functions.

These requirements can be conveniently visualized using arrow diagrams. Consider the six arrow diagrams in Figure 5.1. The three arrow diagrams in the top row all represent valid functions from the set $A = \{v, w, x, y\}$ to the set $B = \{1, 2, 3, 4\}$. In each case, there is exactly one arrow emanating from each point in the domain A, and each such arrow leads into the codomain B. We have, for example, $g_1(w) = 4$, $g_2(y) = 2 = g_2(v)$, and $g_3(a) = 4$ for all $a \in A$. Note that it is permissible for several arrows to hit the same element of B. For example, in the diagram for g_3, four different arrows hit the same point $4 \in B$. Similarly, there may exist points in B that are not hit by any arrows. For example, in the diagram for g_2, no arrow leads to $1 \in B$, and yet g_2 is still considered to be a function mapping A into B.

On the other hand, g_4 is not a function from A to B because there is no arrow leaving the input point $w \in A$. The arrow diagram g_5 is not a function because there are multiple arrows leaving v (and also y). Finally, g_6 is not a function from A to B because $g(x) = 0$, which is not an element of the codomain B. However, g_6 can be regarded as a function from A to the set $\{0, 1, 2, 3, 4\}$.

Graphs of Functions and Non-Functions

We continue to build intuition for the function concept by looking at the graphs of some functions and non-functions with domain $\mathbb{R}$ and codomain $\mathbb{R}$. Informally, given a function $f : \mathbb{R} \to \mathbb{R}$, the *graph* of f is the set of ordered pairs $G_f = \{(x, f(x)) : x \in \mathbb{R}\}$, which is a certain relation from $\mathbb{R}$ to $\mathbb{R}$. The graph G_f encodes exactly the information needed to understand how the function f transforms inputs to outputs. Specifically, for each input $x \in \mathbb{R}$, there must exist a unique $y \in \mathbb{R}$ such that (x, y) belongs to the graph G_f, and we then know that $y = f(x)$ is the output associated with the input x by the function f.

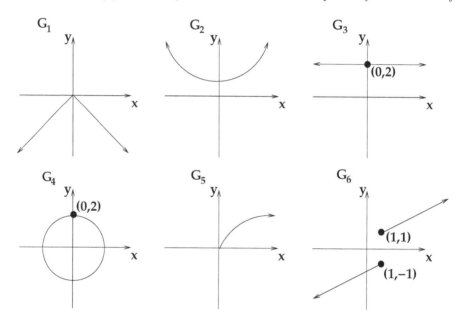

FIGURE 5.2
Graphs of functions and non-functions.

For example, consider the graphs shown in Figure 5.2. Each graph illustrates a relation on $\mathbb{R} \times \mathbb{R}$ consisting of certain ordered pairs (x, y). The first three pictures are graphs of functions from $\mathbb{R}$ to $\mathbb{R}$. Specifically, G_1 is the graph of the function $g_1 : \mathbb{R} \to \mathbb{R}$ defined by $g_1(x) = -|x|$ for each real x. G_2 is the graph of the function $g_2 : \mathbb{R} \to \mathbb{R}$ defined by $g_2(x) = (x^2/4) + 1$ for each real x. G_3 is the graph of the constant function $g_3 : \mathbb{R} \to \mathbb{R}$ defined by $g_3(x) = 2$ for all real x. In each case, every particular real input x has exactly one associated output y. Some of the *potential* outputs $y \in \mathbb{R}$ do not appear as actual outputs of the functions, but this does not matter. Similarly, some outputs have more than one associated input, for instance $2 = g_3(0) = g_3(1) = g_3(3)$, but this does not matter either.

On the other hand, $G_4 = \{(x, y) \in \mathbb{R} \times \mathbb{R} : x^2 + y^2 = 4\}$ is not the graph of any function from $\mathbb{R}$ to $\mathbb{R}$. One reason is that some inputs x in the domain $\mathbb{R}$ do not have any associated output y in the codomain. For example, there is no $y \in \mathbb{R}$ such that $(3, y) \in G_4$. We could try to circumvent this difficulty by shrinking the domain of the function from all of $\mathbb{R}$ to the closed interval $[-2, 2]$. But we still do not get a function, because there are some inputs that have more than one associated output based on the graph. For example, $(0, 2) \in G_4$ and $(0, -2) \in G_4$. In terms of formulas, the condition defining the relation G_4 can be rewritten $y = \pm\sqrt{4 - x^2}$ for $-2 \le x \le 2$. This is a double-valued expression, but we insist that all functions be single-valued. One possible way to extract a function from this example would be to define $g_4 : [-2, 2] \to \mathbb{R}$ by $g_4(x) = -\sqrt{4 - x^2}$ for $-2 \le x \le 2$. The graph of this function would be just the lower half of the circle of radius 2 centered at the origin.

Next, G_5 is not the graph of a function with domain $\mathbb{R}$, since each negative real input x has no associated output using this graph. If we shrink the domain to $\mathbb{R}_{\ge 0}$, we would obtain a function $g_5 : \mathbb{R}_{\ge 0} \to \mathbb{R}$ given by $g_5(x) = \sqrt{x}$ for all $x \ge 0$. Here we could also shrink the codomain (the set of allowable outputs) to get a function $g_5^* : \mathbb{R}_{\ge 0} \to \mathbb{R}_{\ge 0}$ given by $g_5^*(x) = \sqrt{x}$ for all $x \ge 0$. Are g_5 and g_5^* the same function? Based on the formal definition below, the answer turns out to be no, because g_5 and g_5^* have different codomains. Here, g_5 and g_5^* are different functions with the same graph and the same domain.

Finally, G_6 is not the graph of a function because of the ordered pairs $(1, 1) \in G_6$ and $(1, -1) \in G_6$. If exactly one of these ordered pairs were removed from G_6, we would get the graph of a function $g_6 : \mathbb{R} \to \mathbb{R}$. We could make additional functions with the same domain and graph by changing the codomain to be any set containing the image $G_6[\mathbb{R}]$ of the relation G_6.

Recall the key condition in the informal definition of a function: g is a function from the domain $A \subseteq \mathbb{R}$ to the codomain $\mathbb{R}$ iff for each $x \in A$ there exists exactly one y with $y = g(x)$, and moreover y must belong to $\mathbb{R}$. A given relation $G \subseteq \mathbb{R}^2$ will be the graph of such a function iff the following geometric property holds: for every $x_0 \in A$, the vertical line $x = x_0$ intersects G in exactly one point; and for every $x_0 \in \mathbb{R} - A$, the vertical line $x = x_0$ does not intersect G at all. In beginning calculus, this is often called the *vertical line test* for determining whether a given graph is a function. One point not often stressed in elementary treatments is that we must know the intended domain A in advance in order to see if a graph passes this test. Often A is tacitly assumed to be the largest possible domain that might work for the given graph, namely $A = \{x \in \mathbb{R} : \exists y \in \mathbb{R}, (x, y) \in G\}$.

Formal Definition of a Function

We are now ready to present the formal definition of a function. This definition needs to capture all the information contained in the arrow diagrams and graphs we have been discussing: the input set A, the output set B, the rule by which inputs $x \in A$ are associated with outputs $y \in B$, and the key requirement that each input have exactly one associated output. The crucial insight needed to formalize these ideas is that the vague idea of a "rule" for transforming inputs to outputs can be captured precisely by forming the relation consisting of all ordered pairs (x, y) such that input x is mapped to output y by the function. By analogy with the case where $A = B = \mathbb{R}$, this relation is called the *graph* of the function. For instance, the function g_1 shown in Figure 5.1 has graph $G_1 = \{(v, 2), (w, 4), (x, 1), (y, 3)\}$. These considerations motivate the following formal definition.

5.35. Definition: Functions. A *function* g is an ordered triple $g = (A, B, G)$ such that: A and B are sets, G is a relation from A to B, and $\forall x \in A, \exists! y, (x, y) \in G$. A is called the *domain* of g; B is called the *codomain* of g; and G is called the *graph* of g. The notation $\boxed{g : A \to B}$ means that $\boxed{g \text{ is a function with domain } A \text{ and codomain } B.}$

For all objects x, y, we write $\boxed{y = g(x) \text{ iff } (x, y) \in G}$.

So we can describe the graph of g as $\boxed{G = \{(x, g(x)) : x \in A\}}$.

We must emphasize at the outset that *the formal ordered triple notation $g = (A, B, G)$ is almost never used*; instead, we write $g : A \to B$ to introduce a new function with domain A and codomain B. Similarly, only in rare cases do we need to mention the graph by writing $(x, y) \in G$; instead, we use the equivalent *function notation $y = g(x)$*. If we say "let $g : A \to B$ be defined by $g(x) = (\text{expr})$," where (expr) is some expression involving x, it is understood that the graph of g is the relation G consisting of all ordered pairs (x, y) where $x \in A$ and y is the value of the expression for this choice of x.

Let us spell out the three conditions that the relation G must satisfy in order to be the graph of a function $g : A \to B$.

(a) *Valid Inputs and Outputs:* Every member of G is an ordered pair (x, y) with $x \in A$ and $y \in B$.

(b) *Existence of Outputs:* For all $x \in A$, there exists y with $(x, y) \in G$.

(c) *Uniqueness of Outputs:* For all x, y, z, if $(x, y) \in G$ and $(x, z) \in G$, then $y = z$.

Here are the same three conditions for $g : A \to B$ to be a function, written in function notation.

(a) *Valid Inputs and Outputs:* For all x, y, if $y = g(x)$, then $x \in A$ and $y \in B$.

(b) *Existence of Outputs:* For all $x \in A$, there exists y such that $y = g(x)$.

(c) *Uniqueness of Outputs:* For all x, y, z, if $y = g(x)$ and $z = g(x)$, then $y = z$.

The uniqueness condition can be stated in a way that does not explicitly mention y and z.

(c′) *Uniqueness of Outputs:* For all $x_1, x_2 \in A$, if $x_1 = x_2$, then $g(x_1) = g(x_2)$.

The main technical drawback of the function notation $g(x)$ is that the very use of the notation suggests that the existence and uniqueness of the output $g(x)$ associated with input x are already known. But in fact, when introducing a new function for the first time, we really need to *prove* these assertions before the function notation can be legitimately used. Some of the dangers of using this notation prematurely are illustrated in the following subsection.

5.36. Example. Consider the arrow diagram labeled g_2 in Figure 5.1. We could formally define the function g_2 by setting

$$g_2 = (\{v, w, x, y\}, \{1, 2, 3, 4\}, \{(v, 2), (w, 3), (x, 4), (y, 2)\}).$$

But we would usually introduce this function by saying "define $g_2 : \{v, w, x, y\} \to \{1, 2, 3, 4\}$ by setting $g(v) = 2$, $g(w) = 3$, $g(x) = 4$, and $g(y) = 2$."

5.37. Example. Consider the graph G_2 in Figure 5.2. One possible function having this graph could be formally defined as

$$g = (\mathbb{R}, \mathbb{R}_{\geq 0}, G_2) = (\mathbb{R}, \mathbb{R}_{\geq 0}, \{(x, x^2/4 + 1) : x \in \mathbb{R}\}).$$

But we would usually introduce this function by saying "define $g : \mathbb{R} \to \mathbb{R}_{\geq 0}$ by setting $g(x) = x^2/4 + 1$ for all $x \in \mathbb{R}$."

Formulas for Functions and Non-Functions

Here are more examples of function notation using some familiar functions from calculus. We defer to the next section the task of proving that these formulas really do define functions.

5.38. Example. Define the *cube function* $C : \mathbb{R} \to \mathbb{R}$ by $C(x) = x^3$ for $x \in \mathbb{R}$. Define the *absolute value* function $A : \mathbb{R} \to \mathbb{R}$ by $A(x) = x$ for real $x \geq 0$, and $A(x) = -x$ for $x < 0$; we often write $A(x) = |x|$. Define the *sine function* $S : \mathbb{R} \to \mathbb{R}$ by $S(x) = \sin x$ for all real x. If we shrink the domain and codomain of S, we obtain a related function $S_1 : [-\pi/2, \pi/2] \to [-1, 1]$ given by $S_1(x) = \sin x$ for $-\pi/2 \leq x \leq \pi/2$. This function is *not* the same as the function S. Finally, define the *natural logarithm* function $L : (0, \infty) \to \mathbb{R}$ by $L(x) = \ln x$ for all real $x > 0$. We obtain a related function with a larger domain by defining $L_1 : \mathbb{R}_{\neq 0} \to \mathbb{R}$ by $L_1(x) = \ln |x|$ for all real $x \neq 0$. Note $L_1(x) = L(A(x))$ for all $x \neq 0$.

The next examples illustrate the pitfalls that can occur when we use function notation before checking the three defining properties above.

5.39. Example. Suppose we try to define $f : \mathbb{R} \to \mathbb{R}$ by $f(x) = 1/x$ for all real x. This definition is incorrect, because the input $x = 0$ has no associated output in $\mathbb{R}$. We could remedy this by defining $f(x) = 1/x$ for all real nonzero x, and setting $f(0) = 0$ (say). Alternatively, we could shrink the domain, defining $f_0 : \mathbb{R}_{\neq 0} \to \mathbb{R}$ by $f_0(x) = 1/x$ for all real $x \neq 0$.

5.40. Example. Suppose we try to define $g : \mathbb{Z} \to \mathbb{Z}$ by $g(m) = m/2$ for all $m \in \mathbb{Z}$. The graph of g is $\{(m, m/2) : m \in \mathbb{Z}\}$, but this is not a subset of $\mathbb{Z} \times \mathbb{Z}$. In other words, the proposed function does not map the given domain $\mathbb{Z}$ *into the given codomain* $\mathbb{Z}$. We could fix this by enlarging the codomain to $\mathbb{Q}$ (say), defining $g : \mathbb{Z} \to \mathbb{Q}$ by $g(m) = m/2$ for all $m \in \mathbb{Z}$.

5.41. Example. Suppose we try to define $h : \mathbb{Q} \to \mathbb{Z}$ by $h(m/n) = m$ for all rational numbers m/n. More precisely, the graph of h is $H = \{(m/n, m) : m \in \mathbb{Z}, n \in \mathbb{Z}_{\neq 0}\}$. In fact, h is not a function because its graph fails the Uniqueness of Outputs condition. For instance, choosing $x = 1/2 = 2/4$, we find that $(1/2, 1) \in H$ and $(2/4, 2) = (1/2, 2) \in H$, yet $1 \neq 2$. Rephrasing this example using the (technically forbidden) function notation, we have $1/2 = 2/4$ and yet $h(1/2) = 1 \neq 2 = h(2/4)$. We often say that "$h$ is not a *well-defined* function" to indicate that the relation H assigns multiple outputs to a given input. In this situation, it is clearer and safer to discuss the graph H instead of using the function notation $h(m/n)$ for the non-function h.

5.42. Example. Finally, let us try to define $f : \mathbb{Q}_{>0} \to \mathbb{Q}_{>0}$ by $f(m/n) = n/m$ for all positive rationals m/n. More precisely, we are setting $f = (\mathbb{Q}_{>0}, \mathbb{Q}_{>0}, F)$, where the graph F is the set $\{(m/n, n/m) : m, n \in \mathbb{Z}_{>0}\}$. We might at first think that f is not single-valued, as in the previous example. But in fact, the f considered here *is* a legitimate, well-defined function! We verify Uniqueness of Outputs as follows. Suppose x, y, z satisfy $(x, y) \in F$ and $(x, z) \in F$; we must prove $y = z$. By definition of F, there exist $m, n, p, q \in \mathbb{Z}_{>0}$ with $(x, y) = (m/n, n/m)$ and $(x, z) = (p/q, q/p)$. Comparing first coordinates, $m/n = x = p/q$, so $mq = pn$. Using this and comparing second coordinates, we see that $y = n/m = q/p = z$, as needed. In function notation, we have just shown: for all m/n and p/q in $\mathbb{Q}_{>0}$, if $m/n = p/q$ then $f(m/n) = f(p/q)$, i.e., $n/m = q/p$. (To finish the proof that f is a function, we must also check Existence of Outputs and that $F \subseteq \mathbb{Q}_{>0} \times \mathbb{Q}_{>0}$.)

5.43. Remark. Some texts adopt a different definition of functions that essentially identifies a function $g = (A, B, G)$ with its graph G. On one hand, the domain A of g can be reconstructed from G, since it is the set of all first coordinates of the ordered pairs in G. On the other hand, the codomain B of g cannot be inferred from G alone. For example, for the three functions shown in the top row of Figure 5.1, knowing where all the arrows go does not give us enough information to deduce what the codomain B is. Similarly, we cannot determine the codomain of a real-valued function from inspection of its graph, although we often assume an intended codomain of $\mathbb{R}$. In many areas of mathematics (including abstract algebra, topology, and differential geometry), it is crucial to distinguish functions with the same domain and the same graph but different codomains. Some of the reasons for this appear below when we discuss surjections, bijections, and invertible functions. Thus we must use the more elaborate definition of functions that explicitly includes the codomain.

Section Summary

1. *Definition of a Function.* A function is an ordered triple $g = (A, B, G)$ where A and B are sets, $G \subseteq A \times B$, and $\forall x \in A, \exists! y, (x, y) \in G$. A is the domain of g, B is the codomain of g, and G is the graph of g. The notation $g : A \to B$ means g is a function with domain A and codomain B. For all x and y, $y = g(x)$ means $(x, y) \in G$. We have $G = \{(x, g(x)) : x \in A\}$.

2. *Arrow Diagrams of Functions.* In an arrow diagram for a function $g : A \to B$, every input $x \in A$ has exactly one arrow leaving it, and every arrow arrives at some output $y \in B$. It is allowed for a value in the codomain to have no arrows hitting it, or multiple arrows hitting it.

3. *Graphs of Real-Valued Functions.* A relation $G \subseteq \mathbb{R}^2$ drawn in the xy-plane is the graph of a function $g : A \to \mathbb{R}$ iff for each $x_0 \in A$, the vertical line $x = x_0$ intersects G in exactly one point, and for each $x_0 \notin A$, the vertical line $x = x_0$ does not intersect G. Each horizontal line $y = y_0$ may intersect G once, never, or multiple times.

4. *Three Conditions to be a Function.* We often define functions $g : A \to B$ by giving a formula specifying $g(x)$ for each $x \in A$. To be sure we do have a well-defined function, we must check three things: $g(x)$ is in the codomain B for all $x \in A$; for each $x \in A$ there exists at least one associated output $g(x)$; and for all $x_1, x_2 \in A$, if $x_1 = x_2$, then $g(x_1) = g(x_2)$. The last condition can be stated more precisely using the graph G of g: we must check for all x, y, z, if $(x, y) \in G$ and $(x, z) \in G$ then $y = z$.

Exercises

1. Formally describe the functions g_1 and g_3 pictured in Figure 5.1 as ordered triples. Also give informal definitions of these functions using function notation.

2. Suppose we reverse all the arrows in Figure 5.1. Which of the resulting diagrams represents functions from B into A? Which of these diagrams represents functions from some domain C into A?

3. Which of the arrow diagrams in Figure 5.3 represent functions from A to B? Explain.

4. Let $A = \{1, 2, 3, 4, 5, 6, 7\}$. Which relations are graphs of functions from A to A? Explain. (a) $R_1 = \{(i, i + 1) : i = 1, 2, 3, \ldots, 6\}$. (b) $R_2 = R_1 \cup \{(7, 1)\}$. (c) $R_3 = A \times \{5\}$. (d) $R_4 = \{5\} \times A$. (e) $R_5 = \{(a, 8 - a) : a \in A\}$. (f) $R_6 = I_A$.

5. For each relation G_i shown in Figure 5.2, find $G_i[\mathbb{R}]$ and $G_i^{-1}[\mathbb{R}]$.

6. (a) Which of the graphs in Figure 5.4 are graphs of functions from $\mathbb{R}$ to $\mathbb{R}$? Explain. (b) Repeat (a), replacing each graph in the figure by its inverse.

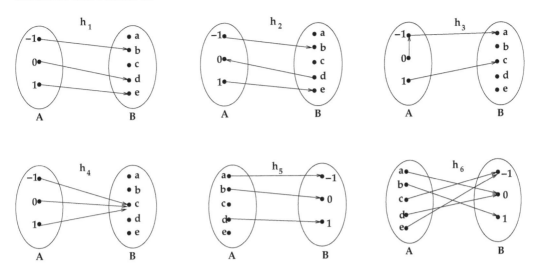

FIGURE 5.3
More arrow diagrams.

7. Let $A = \{1, 2, 3\}$ and $B = \{4, 5, 6, 7\}$. For which of the following choices of the graph G is (A, B, G) a function? Explain briefly. (a) $\{(2, 4), (3, 7)\}$. (b) $\{(1, 6), (2, 4), (3, 4)\}$. (c) $\{(x, x + 4) : x \in A\}$. (d) $\{(x, 2x) : x \in A\}$. (e) $A \times B$.

8. (a) Draw arrow diagrams for all functions $f : \{1, 2, 3\} \to \{a, b\}$.
   (b) For fixed $n \in \mathbb{Z}_{>0}$, how many functions $f : \{1, 2, \ldots, n\} \to \{a, b\}$ are there?

9. (a) Draw graphs for all functions $g : \{1, 4\} \to \{2, 3, 5\}$.
   (b) For fixed $m \in \mathbb{Z}_{>0}$, how many functions $g : \{1, 4\} \to \{1, 2, \ldots, m\}$ are there?

10. Which of the following relations are graphs of functions? For those relations that are graphs of functions, determine the domain and smallest possible codomain for a function with that graph.
    (a) $\{(x, y) \in \mathbb{R} \times \mathbb{R} : x = y\}$.
    (b) $\{(x, y) \in \mathbb{R} \times \mathbb{R} : x^2 = y^2\}$.
    (c) $\{(x, y) \in \mathbb{R} \times \mathbb{R} : x^3 = y^3\}$.
    (d) $\{(x, y) \in \mathbb{R} \times \mathbb{R} : x = y^2\}$.
    (e) $\{(x, y) \in \mathbb{R} \times \mathbb{R} : x^2 = y\}$.
    (f) $\{(x, y) \in \mathbb{R} \times \mathbb{R} : x^2 = y^3\}$.
    (g) $\{(x, y) \in \mathbb{R} \times \mathbb{R} : x^3 = y^2\}$.

11. Which of the following relations are graphs of functions? For those relations that are graphs of functions, determine the domain and smallest possible codomain for a function with that graph.
    (a) $\{(x, y) \in \mathbb{R} \times \mathbb{R} : xy = 5\}$.
    (b) $\{(x, y) \in \mathbb{R} \times \mathbb{R} : y = \frac{x^2 + 5}{x^2 - 6x + 8}\}$.
    (c) $\{(x, y) \in \mathbb{R} \times \mathbb{R} : y = e^{\tan x}\}$.
    (d) $\mathbb{R} \times \{4\}$.
    (e) $\{0\} \times \mathbb{R}$.
    (f) $\mathbb{Q} \times \mathbb{Z}$.
    (g) $(\mathbb{Q} \times \{1\}) \cup ((\mathbb{R} - \mathbb{Q}) \times \{0\})$.
    (h) $\emptyset$.
    (i) $\{((x, y), z) \in (\mathbb{R} \times \mathbb{R}) \times \mathbb{R} : z = xy\}$.
    (j) $\{((x, y), z) \in (\mathbb{R} \times \mathbb{R}) \times \mathbb{R} : zy = x\}$.

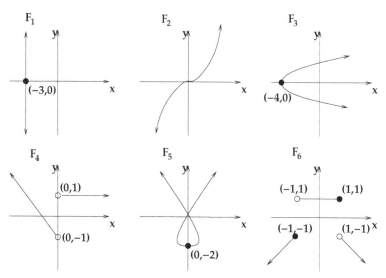

FIGURE 5.4
More graphs in $\mathbb{R}^2$.

12. Let $X = \{1, 2, 3\}$ and define $h : (\mathcal{P}(X) - \{\emptyset\}) \to X$ by letting $h(A)$ be the least element of A for every nonempty $A \subseteq X$. Give a formal description of the function h as an ordered triple.

13. Let $Y = \{2, \{3\}, \{2, \{3\}\}\}$ and define $f : Y \to \mathcal{P}(Y)$ by $f(y) = \{y\}$ for all $y \in Y$. Describe the graph of f as a set of ordered pairs.

14. Which of the following are well-defined functions? Explain. (Assume $m, n, a, b, c, d \in \mathbb{Z}$ with $n, b, d > 0$.)
    (a) $f : \mathbb{Q} \to \mathbb{Z}$ given by $f(m/n) = m + n$.
    (b) $g : \mathbb{Q} \to \mathbb{Q}$ given by $g(m/n) = (m + 2n)/n$.
    (c) $h : \mathbb{R} \to \{0, 1, \ldots, 9\}$ given by $h(x) =$ the second digit after the decimal point in the base-10 representation of x.
    (d) $m : \mathbb{Q} \times \mathbb{Q} \to \mathbb{Q}$ given by $m((a/b, c/d)) = (ac)/(bd)$.
    (e) $p : \mathbb{Q} \times \mathbb{Q} \to \mathbb{Q}$ given by $p((a/b, c/d)) = (a + c)/(b + d)$.
    (f) $s : \mathbb{Q} \times \mathbb{Q} \to \mathbb{Q}$ given by $s((a/b, c/d)) = (ad + bc)/(bd)$.

15. Prove that for all sets A and B, $(A, B, \emptyset)$ is a function iff $A = \emptyset$.

16. *Enlarging the Codomain.* Prove that for all sets A, B, C, if $B \subseteq C$ and (A, B, F) is a function, then (A, C, F) is a function.

17. Prove or disprove: for all A, B, G, if (A, B, G) is a function, then (B, A, G^{-1}) is a function.

5.5 Examples of Functions and Function Equality

This section defines some basic examples of functions, such as constant functions, identity functions, inclusion functions, characteristic functions, and arithmetic functions. We also introduce operations (called pointwise sum, difference, product, and quotient) that can be used to construct new functions from old functions. We then develop a proof template for showing that two functions are equal. We use this template to prove some algebraic properties of the pointwise operations on functions, analogous to corresponding properties for addition and multiplication of real numbers.

Constant Functions, Inclusion Functions, and Identity Functions

5.44. Definition: Constant Functions. For any sets A and B and any fixed object $c \in B$, the function $f : A \to B$ defined by $f(x) = c$ for all $x \in A$ is called a *constant function with value c*.

The graph of the constant function defined above is $G = \{(x, c) : x \in A\} = A \times \{c\}$. Let us prove that $f = (A, B, G)$ really is a function by checking the three conditions from §5.4. First, every member of G is an ordered pair (x, c) with $x \in A$ and $c \in B$, so $G \subseteq A \times B$. Second, for a fixed $x \in A$, there exists y with $(x, y) \in G$, namely $y = c$. Third, fix x, y, z, assume $(x, y) \in G$ and $(x, z) \in G$, and prove $y = z$. By definition of G, we must have $y = c$ and $z = c$, so $y = z$ follows.

5.45. Definition: Inclusion Functions. Let A and B be two sets such that $A \subseteq B$. Define a function $j = j_{A,B} : A \to B$ by setting $j_{A,B}(x) = x$ for all $x \in A$. We call $j_{A,B}$ the *inclusion function mapping A into B*.

The graph of the inclusion function defined above is $G = \{(x, x) : x \in A\} = I_A$, the identity relation on A. Let us prove that $j_{A,B} = (A, B, G)$ really is a function. First, every member of G is an ordered pair (x, x) with $x \in A$ and $x \in B$ (since $A \subseteq B$), so that $G \subseteq A \times B$. Second, for a fixed $x \in A$, there exists y with $(x, y) \in G$, namely $y = x$. Third, fix x, y, z, assume $(x, y) \in G$ and $(x, z) \in G$, and prove $y = z$. By definition of G, we must have $y = x$ and $z = x$, so $y = z$ follows.

5.46. Definition: Identity Functions. Given any set A, define a function $\mathrm{Id}_A : A \to A$ by setting $\mathrm{Id}_A(x) = x$ for all $x \in A$. We call Id_A the *identity function on the set A*.

Note that the identity function Id_A is the same as the inclusion function $j_{A,A}$, so Id_A is a function. On the other hand, if A is a proper subset of B, then Id_A is *not* the same function as the inclusion $j_{A,B}$. To check this carefully, note that Id_A is the ordered triple (A, A, G), where $G = \{(x, x) : x \in A\}$, whereas $j_{A,B}$ is the ordered triple (A, B, G). Since $A \neq B$, these ordered triples are not equal. Less formally, Id_A and $j_{A,B}$ are unequal functions since their codomains are different (even though their domains and graphs agree). We return to this point later when we discuss function equality.

5.47. Example. Figure 5.5 gives arrow diagrams for Id_A and $j_{A,B}$ when $A = \{2, 4, 5\}$ and $B = \{1, 2, 3, 4, 5\}$.

Arithmetic Functions and Pointwise Operations on Functions

We are all familiar with the arithmetic operations on real numbers: addition, subtraction, multiplication, and division. We do not attempt to define these operations, instead viewing

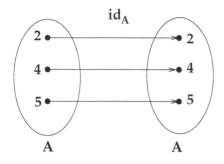
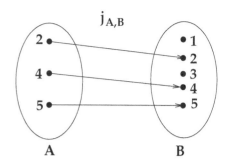

FIGURE 5.5
Contrasting identity functions and inclusion functions.

them as undefined concepts that obey certain axioms (see §2.1 and §8.1). However, now that the language of functions is available, we can at least describe more precisely what kind of objects these undefined operations *are*.

For example, consider the undefined operation of adding two real numbers. For any two real numbers x and y, we can apply the addition operation to these two numbers to obtain a unique output $x + y$, called the *sum* of x and y. More formally, addition transforms the ordered pair of inputs (x, y) into the real output $x + y$. So we could describe addition as a function ADD : $\mathbb{R} \times \mathbb{R} \to \mathbb{R}$, where ADD$((x, y))$ is defined to be the sum of x and y. We would almost always write $x + y$ as a shorthand for ADD$((x, y))$. Similarly, multiplication is a function MULT : $\mathbb{R} \times \mathbb{R} \to \mathbb{R}$, where MULT$((x, y)) = x \cdot y$ is the product of x and y. Subtraction and division could be defined similarly, provided we use $\mathbb{R} \times \mathbb{R}_{\neq 0}$ as the domain of the division function (since we cannot divide by zero). Alternatively, we could first introduce an *additive inverse* function AINV : $\mathbb{R} \to \mathbb{R}$, denoted AINV$(z) = -z$ for $z \in \mathbb{R}$, and then define subtraction via $x - y = x + (-y)$ for $x, y \in \mathbb{R}$. Similarly, we could introduce a *multiplicative inverse* function MINV : $\mathbb{R}_{\neq 0} \to \mathbb{R}$, denoted MINV$(z) = z^{-1}$ for $z \in \mathbb{R}_{\neq 0}$, and then define division via $x/y = x \cdot (y^{-1})$ for $x, y \in \mathbb{R}$ with $y \neq 0$.

We can use the addition operation *on real numbers* to create an addition operation acting *on functions*, as follows.

5.48. Definition: Pointwise Sum of Functions. Suppose A is a set and $f : A \to \mathbb{R}$, $g : A \to \mathbb{R}$ are two real-valued functions with domain A. Define the *pointwise sum*
$\boxed{f + g : A \to \mathbb{R}}$ by letting $\boxed{(f + g)(x) = f(x) + g(x)}$ for all $x \in A$.

The graph of $f + g$ is the set $\{(x, f(x) + g(x) : x \in A)\}$. You can check that $f + g$ is a function, using the fact that f and g are known to be functions.

5.49. Example. If $f : \mathbb{R} \to \mathbb{R}$ is the exponential function and $g : \mathbb{R} \to \mathbb{R}$ is the cosine function, then $f + g : \mathbb{R} \to \mathbb{R}$ is given by $(f + g)(x) = e^x + \cos x$ for all real x. For another example, the function $h : \mathbb{R} \to \mathbb{R}$ defined by $h(x) = x + 3$ is the pointwise sum of the identity function on $\mathbb{R}$ and the constant function with value 3.

Note that the domain A need not be a subset of $\mathbb{R}$ to form pointwise sums. It is required that the functions being combined share the same domain. We can define more pointwise operations as follows.

5.50. Definition: More Pointwise Operations. Given functions $f, g : A \to \mathbb{R}$, we define the following operations (called pointwise product, pointwise additive inverse, pointwise difference, pointwise reciprocal, and pointwise quotient). For parts (d) and (e), we assume $g(x) \neq 0$ for all $x \in A$.

(a) Define $\boxed{f \cdot g : A \to \mathbb{R}}$ by setting $\boxed{(f \cdot g)(x) = f(x) \cdot g(x)}$ for x in A.

(b) Define $\boxed{-g : A \to \mathbb{R}}$ by setting $\boxed{(-g)(x) = -(g(x))}$ for x in A.

(c) Define $\boxed{f - g : A \to \mathbb{R}}$ by setting $\boxed{(f - g)(x) = f(x) - g(x)}$ for x in A.

(d) Define $\boxed{1/g : A \to \mathbb{R}}$ by setting $\boxed{(1/g)(x) = 1/g(x)}$ for x in A.

(e) Define $\boxed{f/g : A \to \mathbb{R}}$ by setting $\boxed{(f/g)(x) = f(x)/g(x)}$ for x in A.

Take care not to confuse the pointwise product $f \cdot g$ (also denoted fg) with the composition of functions $f \circ g$ (to be introduced later). Similarly, we write $1/g$ for the pointwise reciprocal to avoid confusion with the inverse function g^{-1} (to be studied later). Also, be very careful not to confuse a *function* such as $f + g$ with $(f + g)(x)$, which is the *output* of that function at input x. In particular, $(-g)(x)$ is the output of the function $-g$ at input x; this output is defined to be $-(g(x))$, the additive inverse of the real number $g(x)$.

5.51. Example. The function $q : \mathbb{R} \to \mathbb{R}$ given by $q(x) = 5x$ is the pointwise product of the constant function with value 5 and the identity function on $\mathbb{R}$. The squaring function $s : \mathbb{R} \to \mathbb{R}$ given by $s(x) = x^2$ is the pointwise product $s = \mathrm{Id}_\mathbb{R}\, \mathrm{Id}_\mathbb{R}$, since $(\mathrm{Id}_\mathbb{R}\, \mathrm{Id}_\mathbb{R})(x) = (\mathrm{Id}_\mathbb{R}(x)) \cdot (\mathrm{Id}_\mathbb{R}(x)) = x \cdot x = x^2 = s(x)$ for all $x \in \mathbb{R}$.

By starting with the identity function and constant functions on $\mathbb{R}$ and repeatedly iterating the pointwise sum and product constructions, we can build the *polynomial functions* on $\mathbb{R}$. For example, $\mathrm{Id}_\mathbb{R}\, \mathrm{Id}_\mathbb{R}\, \mathrm{Id}_\mathbb{R} + 2\, \mathrm{Id}_\mathbb{R} - 5$ is the polynomial $p : \mathbb{R} \to \mathbb{R}$ given by $p(x) = x^3 + 2x - 5$ for $x \in \mathbb{R}$. The general definition is as follows.

5.52. Definition: Polynomials. A function $p : \mathbb{R} \to \mathbb{R}$ is a *polynomial* (in one real variable) iff there exist an integer $n \geq 0$ and constants $a_0, \ldots, a_n \in \mathbb{R}$ such that for all $x \in \mathbb{R}$,

$$p(x) = a_0 + a_1 x + a_2 x^2 + \cdots + a_n x^n = \sum_{i=0}^{n} a_i x^i.$$

Equality of Functions

Recall that a function $f : A \to B$ was formally defined as an ordered triple $f = (A, B, G)$, where $G = \{(x, f(x)) : x \in A\}$ is the graph of f. Given another function $h = (A', B', G')$, when are the functions f and h *equal*? By the fundamental property of ordered triples, we see that $f = h$ iff $A = A'$ and $B = B'$ and $G = G'$. In other words, *two functions are equal iff they have the same domain, the same codomain, and the same graph*. Suppose we know that $A = A'$ and $B = B'$. You can check that the graph $G = \{(x, f(x)) : x \in A\}$ equals the graph $G' = \{(x, h(x)) : x \in A\}$ iff for all $x \in A$, $f(x) = h(x)$. Thus we can say that *two functions are equal iff they have the same domain, the same codomain, and the same values on their common domain*. This leads to the following proof template for proving the equality of two functions. (Compare this to earlier proof templates for proving equality of sets or equality of numbers.)

5.53. Proof Template for Proving Function Equality. Let functions $f : A \to B$ and $g : A \to B$ have the *same domain* A and the *same codomain* B. To **prove** $f = g$:
Fix an arbitrary object $x \in A$. **Use** the definitions of f and g to **prove** $f(x) = g(x)$.

The last step (proving equality of $f(x)$ and $g(x)$, which are objects in the codomain B) is often achieved by a chain proof, writing down a chain of known equalities linking the left side $f(x)$ to the right side $g(x)$. We now illustrate this proof template by proving some algebraic laws satisfied by pointwise operations on functions.

5.54. Theorem on Pointwise Operations. Suppose A is any set and $f, g, h : A \to \mathbb{R}$ are functions. Let 0_A and 1_A denote the constant functions from A to $\mathbb{R}$ with values 0 and 1 (respectively). The following function equalities hold:

(a) *Commutativity:* $f + g = g + f$ and $f \cdot g = g \cdot f$.

(b) *Associativity:* $(f + g) + h = f + (g + h)$ and $(f \cdot g) \cdot h = f \cdot (g \cdot h)$.

(c) *Identity:* $f + 0_A = f = 0_A + f$ and $f \cdot 1_A = f = 1_A \cdot f$.

(d) *Inverses:* $f + (-f) = 0_A = (-f) + f$.
If $g(x) \neq 0$ for all $x \in A$, then $g \cdot (1/g) = 1_A = (1/g) \cdot g$.

(e) *Distributive Laws:* $f \cdot (g + h) = (f \cdot g) + (f \cdot h)$ and $(f + g) \cdot h = (f \cdot h) + (g \cdot h)$.

Proof. We prove a few identities here, leaving the others as exercises. To prove the first commutative law in (a), first note that $f + g$ and $g + f$ are both functions with domain A and codomain $\mathbb{R}$ (by definition of pointwise sum). Following the proof template for function equality, fix an arbitrary object $x \in A$, and prove $(f + g)(x) = (g + f)(x)$. We compute $(f + g)(x) = f(x) + g(x) = g(x) + f(x) = (g + f)(x)$, where the first step uses the definition of $f + g$, the second step uses the known commutativity of addition of real numbers, and the third step uses the definition of $g + f$.

Next we prove $f \cdot 1_A = f$. Both $f \cdot 1_A$ and f have domain A and codomain $\mathbb{R}$. Fix $x \in A$. We know $(f \cdot 1_A)(x) = f(x) \cdot 1_A(x) = f(x) \cdot 1 = f(x)$, using the definition of pointwise product, then the definition of the constant function 1_A, and then the multiplicative identity axiom for $\mathbb{R}$.

Finally, we prove $f \cdot (g + h) = (f \cdot g) + (f \cdot h)$. Repeated use of the definitions shows that $g + h$, $f \cdot (g + h)$, $f \cdot g$, $f \cdot h$, and $(f \cdot g) + (f \cdot h)$ all have domain A and codomain $\mathbb{R}$. Fix $x \in A$. Compute

$$
\begin{aligned}
[f \cdot (g + h)](x) &= f(x) \cdot (g + h)(x) \quad \text{(by def. of pointwise product)} \\
&= f(x) \cdot [g(x) + h(x)] \quad \text{(by def. of pointwise sum)} \\
&= (f(x) \cdot g(x)) + (f(x) \cdot h(x)) \quad \text{(by the distributive law in } \mathbb{R}) \\
&= (f \cdot g)(x) + (f \cdot h)(x) \quad \text{(by def. of pointwise product)} \\
&= [(f \cdot g) + (f \cdot h)](x) \quad \text{(by def. of pointwise sum)}.
\end{aligned}
$$

So $f \cdot (g + h) = (f \cdot g) + (f \cdot h)$, as needed. $\square$

In the previous computation, be sure you understand why the input x does or does not appear at various places in each expression. For example, the initial expression $[f \cdot (g + h)](x)$ is the output of the function $f \cdot (g + h)$ at input x, while the next expression $f(x) \cdot (g + h)(x)$ is the product of the two real numbers $f(x)$ and $(g + h)(x)$. These two quantities are equal by the definition of the pointwise product of the functions f and $g + h$.

5.55. Example: Changing the Codomain. Suppose $f : A \to B$ is a function with graph $G = \{(x, f(x)) : x \in A\}$. For any set C such that $B \subseteq C$, we can create a new function $f^* : A \to C$ by setting $f^*(x) = f(x)$ for all $x \in A$. Formally, $f = (A, B, G)$ and $f^* = (A, C, G)$. We have constructed f^* from f by *enlarging the codomain*. More generally, the new codomain C can be any set such that $G \subseteq A \times C$. This means that for all $x \in A$, we must have $f(x) \in C$. This condition holds iff $G[A]$ (the image of A under the relation G) is a subset of C. We must remember that when $B \neq C$, f and f^* are *not* the same function, although some texts use the same letter f to denote both functions.

Characteristic Functions (Optional)

We end this section with another class of examples of functions that are often used in probability.

5.56. Definition: Characteristic Functions. Let X be any set, and let A be a fixed subset of X. Define $\chi_A : X \to \mathbb{R}$ by setting $\chi_A(x) = 1$ for all $x \in A$, and $\chi_A(x) = 0$ for all $x \notin A$. We call χ_A the *characteristic function of A in X*.

The graph of χ_A is $(A \times \{1\}) \cup ((X - A) \times \{0\})$. You can prove that χ_A is a function. Intuitively, χ_A is an indicator function that tells us which inputs x belong to A and which do not. For example, taking $X = \mathbb{R}$, we have

$$\chi_\mathbb{Q}(0.73) = \chi_\mathbb{Q}(0) = \chi_\mathbb{Q}(-11/3) = 1; \qquad \chi_\mathbb{Q}(\pi) = \chi_\mathbb{Q}(e) = \chi_\mathbb{Q}(\sqrt{3}) = 0.$$

When using characteristic functions, it is critical to specify what the intended domain X is, since this is not part of the notation χ_A.

We can combine characteristic functions with the same domain using the pointwise operations introduced earlier. We can then verify various identities that connect operations on sets to pointwise operations on functions. The next theorem gives a sample of such identities.

5.57. Theorem on Characteristic Functions. For all sets X and all $A, B \subseteq X$, the following function equalities hold.

(a) $\chi_{A \cap B} = \chi_A \cdot \chi_B$.

(b) $\chi_{X-A} = 1 - \chi_A$.

(c) $\chi_\emptyset = 0_X$ (where $0_X : X \to \mathbb{R}$ is the constant function with value 0).

(d) $\chi_X = 1_X$, where $1_X : X \to \mathbb{R}$ is the constant function with value 1.

(e) $\chi_{A \cup B} = \chi_A + \chi_B - \chi_{A \cap B}$.

(f) If $A \cap B = \emptyset$, then $\chi_{A \cup B} = \chi_A + \chi_B$.

Proof. We prove part (a) and leave the others as exercises. First note that $\chi_{A \cap B}$ and $\chi_A \chi_B$ are both functions with domain X and codomain $\mathbb{R}$. Fix $x \in X$, and consider four cases.
<u>Case 1.</u> Assume $x \in A$ and $x \in B$. Then $\chi_{A \cap B}(x) = 1 = 1 \cdot 1 = \chi_A(x)\chi_B(x) = (\chi_A \chi_B)(x)$.
<u>Case 2.</u> Assume $x \in A$ and $x \notin B$. Then $\chi_{A \cap B}(x) = 0 = 1 \cdot 0 = \chi_A(x)\chi_B(x) = (\chi_A \chi_B)(x)$.
<u>Case 3.</u> Assume $x \notin A$ and $x \in B$. Then $\chi_{A \cap B}(x) = 0 = 0 \cdot 1 = \chi_A(x)\chi_B(x) = (\chi_A \chi_B)(x)$.
<u>Case 4.</u> Assume $x \notin A$ and $x \notin B$. Then $\chi_{A \cap B}(x) = 0 = 0 \cdot 0 = \chi_A(x)\chi_B(x) = (\chi_A \chi_B)(x)$.
Thus, $\chi_{A \cap B}(x) = (\chi_A \chi_B)(x)$ in all cases. So $\chi_{A \cap B} = \chi_A \cdot \chi_B$. $\qquad\square$

Section Summary

1. *Constants, Inclusions, and Identity Functions.* For fixed $c \in B$, the function $f : A \to B$ given by $f(x) = c$ for all $x \in A$ is the constant function on A with value c. For sets $A \subseteq B$, the function $j_{A,B} : A \to B$ given by $j_{A,B}(x) = x$ for all $x \in A$ is the inclusion function mapping A into B. For any set A, the function $\mathrm{Id}_A : A \to A$ given by $\mathrm{Id}_A(x) = x$ for all $x \in A$ is the identity function on A.

2. *Pointwise Operations on Real-Valued Functions.* For functions $f, g : A \to \mathbb{R}$, we can define new functions $f + g$, $f \cdot g$, $-g$, $f - g$, and (when $g(x) \neq 0$ for all $x \in A$) f/g by performing the indicated arithmetic operation one input at a time. For instance, $(f + g)(x) = f(x) + g(x)$ for all $x \in A$, and $(f \cdot g)(x) = f(x)g(x)$ for all $x \in A$. Pointwise operations on functions satisfy algebraic laws such as commutativity, associativity, and the distributive laws.

3. *Function Equality.* Two functions are equal iff they have the same domain, the same codomain, and the same graph (i.e., the same value at each point in the common domain). To prove that $f : A \to B$ and $g : A \to B$ are equal functions, fix $x \in A$ and use known facts to prove $f(x) = g(x)$. Changing the codomain of a function produces a different function with the same domain and graph.

4. *Characteristic Functions.* For $A \subseteq X$, define $\chi_A : X \to \mathbb{R}$ by $\chi_A(x) = 1$ for $x \in A$ and $\chi_A(x) = 0$ for $x \in X - A$. For $A, B \subseteq X$, we have $\chi_{A \cap B} = \chi_A \cdot \chi_B$, $\chi_{X-A} = 1 - \chi_A$, and $\chi_{A \cup B} = \chi_A + \chi_B - \chi_{A \cap B}$.

Exercises

1. Draw arrow diagrams for the following functions. (a) $\mathrm{Id}_{[-2,2] \cap \mathbb{Z}}$ (b) $j_{\{0,3\},\{0,1,2,3,4\}}$ (c) $j_{\mathbb{Z},\mathbb{Q}}$ (d) $\chi_A : X \to \mathbb{R}$, where $X = \{1, 2, \ldots, 10\}$ and A consists of the prime numbers in X.

2. Draw graphs in the xy-plane for the following functions. Mark the domain on the x-axis and the codomain on the y-axis. (a) $\mathrm{Id}_{[-\infty,3] \cup \mathbb{Z}}$ (b) $\mathrm{Id}_{[-\infty,3] \cap \mathbb{Z}}$ (c) $j_{\mathbb{Z},\mathbb{R}}$ (d) $\chi_{(-3,-1] \cup (1,3]} : \mathbb{R} \to \mathbb{R}$ (e) $\chi_{\{0\}} : \mathbb{R} \to \mathbb{R}$ (f) $\chi_{\{x : x^3 + 2x^2 - 13x + 10 = 0\}} : \mathbb{R} \to \mathbb{R}$.

3. Consider the following six functions:

$$\begin{aligned}
&f_1 : \mathbb{R} \to \mathbb{R} &&\text{given by} &&f_1(x) = |x|; \\
&f_2 : \mathbb{R} \to [0, \infty) &&\text{given by} &&f_2(x) = |x|; \\
&f_3 : [0, \infty) \to [0, \infty) &&\text{given by} &&f_3(x) = |x|; \\
&f_4 : \mathbb{R} \to \mathbb{R} &&\text{given by} &&f_4(x) = \sqrt{x^2}; \\
&f_5 : [0, \infty) \to [0, \infty) &&\text{given by} &&f_5(x) = x; \\
&f_6 : \mathbb{R} \to [0, \infty) &&\text{given by} &&f_6(x) = x.
\end{aligned}$$

(a) Which pairs of functions in this list are equal?
(b) Which pairs of functions have the same graph?
(c) Sketch the graph of each function, indicating the domain and codomain of each.

4. Define $f, g, h, k : \mathbb{R} \to \mathbb{R}$ by setting $f(x) = x^3$, $g(x) = x - 4$, $h(x) = 7x$, and $k(x) = 2^x$ for $x \in \mathbb{R}$. Find simplified formulas for the following functions constructed using pointwise operations. (a) $f + g$ (b) fh (c) $-2g$ (d) kk (e) $fk(h - g)$ (f) $ggh - 7f$ (g) f/k.

5. Suppose $f, g : \mathbb{R} \to \mathbb{R}$ satisfy $f(1) = 5$, $g(1) = 2$, $f(3) = -4$, $g(3) = -3$, and $g(x) \neq 0$ for all $x \in \mathbb{R}$. Compute: (a) $(f + g)(1)$ (b) $(f - g)(3)$ (c) $(fg)(1)$ (d) $(f/g)(3)$ (e) $(f \mathrm{Id}_{\mathbb{R}} + gg)(1)$.

6. Let $X = [0, 4]$, $A = [0, 2]$, $B = (1, 3)$, $C = (2, 4]$. Draw the graphs of the following functions from X to $\mathbb{R}$. (a) χ_A (b) $3\chi_B$ (c) $2\chi_A + 5\chi_C$ (d) $3\chi_A - \chi_B + 2\chi_C$.

7. Let $f, g, h : A \to \mathbb{R}$ be functions. Prove the following identities from Theorem 5.54, explaining every step with a definition or an algebraic fact about real numbers.
(a) $(f + g) + h = f + (g + h)$. (b) $f + (-f) = 0_A = (-f) + f$. (c) $f \cdot g = g \cdot f$. (d) $(f \cdot g) \cdot h = f \cdot (g \cdot h)$. (e) $f + 0_A = f = 0_A + f$. (f) $(f + g) \cdot h = (f \cdot h) + (g \cdot h)$.

8. (a) Suppose $f, g : A \to \mathbb{R}$. Prove carefully that $f + g$, $f \cdot g$, and $-g$ are indeed functions. (b) Let $Z = \{x \in A : g(x) = 0\}$. Show that the generalized pointwise quotient $f/g : (A - Z) \to \mathbb{R}$, defined by $(f/g)(x) = f(x)/g(x)$ for all $x \in A - Z$, is a function.

9. Suppose X is a set and $A \subseteq X$. Prove that the characteristic function χ_A is indeed a function.

10. Compute the following quantities. All characteristic functions mentioned have domain and codomain $\mathbb{R}$. (a) $\chi_\mathbb{Q}(1.107)$ (b) $\chi_{\mathbb{R}-\mathbb{Q}}(\sqrt{2})$ (c) $\chi_\mathbb{Z}(111111/37)$ (d) $\chi_{(0,\infty)}(e^\pi - \pi^e)$ (e) $\sum_{i=1}^{10} \chi_{\{1,2,\ldots,i\}}(7)$ (f) $\prod_{i=1}^{5} \chi_{\{1,2,\ldots,i\}}(2)$

11. Let X be a set and let A, B be subsets of X. Prove parts (b) through (f) of Theorem 5.57.

12. Prove: for all sets X and all $A, B, C \subseteq X$,

$$\chi_{A\cup B\cup C} = \chi_A + \chi_B + \chi_C - \chi_{A\cap B} - \chi_{A\cap C} - \chi_{B\cap C} + \chi_{A\cap B\cap C}.$$

13. Suppose $n \in \mathbb{Z}_{>0}$ and $f_1, \ldots, f_n : A \to \mathbb{R}$ are functions. Give recursive definitions of the pointwise sum $\sum_{i=1}^{n} f_i$ and the pointwise product $\prod_{i=1}^{n} f_i$ of these n functions.

14. Suppose $n \in \mathbb{Z}_{>0}$ and $A_1, \ldots, A_n$ are subsets of X.
    (a) Prove by induction that $\chi_{A_1\cap\cdots\cap A_n} = \prod_{i=1}^{n} \chi_{A_i}$.
    (b) Prove: if $A_i \cap A_j = \emptyset$ for all $i \neq j$, then $\chi_{A_1\cup\cdots\cup A_n} = \sum_{i=1}^{n} \chi_{A_i}$.

15. Suppose X is a set, $I \neq \emptyset$ is an index set, and for each $i \in I$, A_i is a subset of X.
    (a) Prove: for all $x \in X$, $\chi_{\bigcap_{i\in I} A_i}(x) = \min\{\chi_{A_i}(x) : i \in I\}$.
    (b) Prove: for all $x \in X$, $\chi_{\bigcup_{i\in I} A_i}(x) = \max\{\chi_{A_i}(x) : i \in I\}$.

16. Suppose G is a relation satisfying the uniqueness condition

$$\forall x, \forall y, \forall z, ((x, y) \in G \wedge (x, z) \in G) \Rightarrow (y = z).$$

Prove that for all sets A and B, (A, B, G) is a function iff $A = \{x : \exists y, (x, y) \in G\}$ and $G[A] \subseteq B$.

17. (a) Prove: for all $f, g : \mathbb{R} \to \mathbb{R}$, if f and g are polynomials, then $f + g$ is a polynomial. (b) Prove: for all $f, g : \mathbb{R} \to \mathbb{R}$, if f and g are polynomials, then $f \cdot g$ is a polynomial.

18. *Even Functions.* A function $f : \mathbb{R} \to \mathbb{R}$ is *even* iff for all $x \in \mathbb{R}$, $f(-x) = f(x)$.
    (a) Prove: for all $f, g : \mathbb{R} \to \mathbb{R}$, if f and g are even, then $f + g$ is even.
    (b) Prove: for all $f, g : \mathbb{R} \to \mathbb{R}$, if f and g are even, then $f \cdot g$ is even.
    (c) Prove: for all $f : \mathbb{R} \to \mathbb{R}$, the function $g : \mathbb{R} \to \mathbb{R}$, given by $g(x) = [f(x) + f(-x)]/2$ for $x \in \mathbb{R}$, is even.

19. *Odd Functions.* A function $f : \mathbb{R} \to \mathbb{R}$ is *odd* iff for all $x \in \mathbb{R}$, $f(-x) = -f(x)$.
    (a) Prove: for all $f, g : \mathbb{R} \to \mathbb{R}$, if f and g are odd, then $f + g$ is odd.
    (b) Prove: for all $f, g : \mathbb{R} \to \mathbb{R}$, if f and g are odd, then $f \cdot g$ is even.
    (c) If f is odd and g is even, what can you say about $f + g$ and $f \cdot g$?

20. Prove: for all $f : \mathbb{R} \to \mathbb{R}$, there exist $g, h : \mathbb{R} \to \mathbb{R}$ such that g is even, h is odd, and $f = g + h$.

21. *Linear Functions.* A function $f : \mathbb{R} \to \mathbb{R}$ is *linear* iff for all $c, x, y \in \mathbb{R}$, $f(x + y) = f(x) + f(y)$ and $f(cx) = cf(x)$.
    (a) Prove: for all $f, g : \mathbb{R} \to \mathbb{R}$, if f and g are linear, then $f + g$ is linear.
    (b) Prove: for all $f : \mathbb{R} \to \mathbb{R}$ and all $b \in \mathbb{R}$, if f is linear, then bf is linear.
    (c) Prove or disprove: for all $f, g : \mathbb{R} \to \mathbb{R}$, if f and g are linear, then $f \cdot g$ is linear.

5.6 Composition, Restriction, and Gluing

This section investigates three more operations for building new functions from given functions: composition, restriction, and gluing. The composition operation combines functions $f : A \to B$ and $g : B \to C$ to get a new function $g \circ f : A \to C$ that acts by doing f and then doing g. The restriction operation takes one function $f : A \to B$ and builds a new function by shrinking the domain of f to some subset of A. The gluing operation allows us to combine two or more functions (under certain conditions) to build a new function with a larger domain. For instance, we need gluing to define the absolute value function and other functions defined by cases.

Composition of Functions

5.58. Definition: Function Composition. Given functions $f : A \to B$ and $g : B \to C$, define the *composition* $g \circ f : A \to C$ by setting $(g \circ f)(x) = g(f(x))$ for all $x \in A$.

Note that the composition $g \circ f$ is only defined when the codomain of f equals the domain of g. In this case, the output of the composite function on an input x in A is found by first performing f on x to obtain an intermediate output $f(x)$ in B, and then performing g on $f(x)$ to obtain the final output $g(f(x))$ in C. The graph of $g \circ f$ is $\{(x, g(f(x))) : x \in A\}$. It is routine to check that $g \circ f$ really is a function, using the assumption that f and g are functions. You can also verify that if F is the graph of f and G is the graph of g, then the composition of relations $G \circ F$ is the graph of $g \circ f$.

5.59. Example: Composition via Arrow Diagrams. Let $A = \{1, 2, 3, 4, 5\}$ and $B = \{a, b, c, d, e\}$. Figure 5.6 shows the arrow diagrams for functions $f : A \to B$ and $g : B \to B$. For each $x \in A$, we obtain $(g \circ f)(x) = g(f(x))$ by following an f-arrow and then a g-arrow in succession. The resulting arrow diagram is shown at the bottom of the figure. The graph of $g \circ f : A \to B$ is $\{(1, c), (2, e), (3, d), (4, d), (5, a)\}$. In this example, the composite function $f \circ g$ is not defined, since the codomain of g (the set B) does not equal the domain of f (the set A).

5.60. Example: Composition via Formulas. Let $f : \mathbb{R} \to \mathbb{R}$ be given by $f(x) = x^2$, and let $g : \mathbb{R} \to \mathbb{R}$ be given by $g(x) = x + 2$. In this case, both $f \circ g$ and $g \circ f$ are functions with domain $\mathbb{R}$ and codomain $\mathbb{R}$. For each real x, we compute $(g \circ f)(x) = g(f(x)) = g(x^2) = x^2 + 2$, whereas $(f \circ g)(x) = f(g(x)) = f(x + 2) = (x + 2)^2 = x^2 + 4x + 4$. In particular, $(g \circ f)(1) = 3 \neq 9 = (f \circ g)(1)$, so we conclude that $g \circ f \neq f \circ g$. This shows that *function composition is not commutative in general*.

We can also combine function composition with pointwise operations. For instance, $g \circ f + fg$ is the function sending $x \in \mathbb{R}$ to $g(f(x)) + f(x)g(x) = x^2 + 2 + x^2(x + 2)$. We have $f \circ f = f \cdot f$, since for each $x \in \mathbb{R}$, $(f \circ f)(x) = f(f(x)) = (x^2)^2 = x^4$ and $(f \cdot f)(x) = f(x)f(x) = x^2 \cdot x^2 = x^4$. On the other hand, $g \circ g \neq g \cdot g$, since $(g \circ g)(x) = g(g(x)) = (x + 2) + 2 = x + 4$ and $(g \cdot g)(x) = g(x)g(x) = (x + 2)^2$, and these functions disagree at $x = 1$.

5.61. Example. Let $A = \{1, 2, 3, 4\}$, and let functions $f, g, h : A \to A$ have graphs

$$G_f = \{(1, 3), (2, 4), (3, 1), (4, 2)\},$$

$$G_g = \{(1, 3), (2, 2), (3, 1), (4, 4)\},$$

$$G_h = \{(1, 2), (2, 3), (3, 4), (4, 1)\}.$$

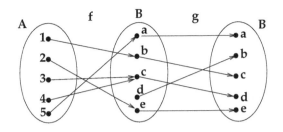

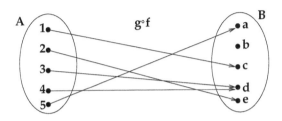

FIGURE 5.6
Arrow diagrams illustrating composition.

You can check that $g \circ f$ and $f \circ g$ both have graph $\{(1,1),(2,4),(3,3),(4,2)\}$, so that $g \circ f = f \circ g$. But $g \circ h \neq h \circ g$, since (for instance) $(g \circ h)(1) = g(h(1)) = g(2) = 2$ and $(h \circ g)(1) = h(g(1)) = h(3) = 4$.

Although function composition is not commutative, the following algebraic laws do hold.

5.62. Theorem on Function Composition.

(a) *Associativity:* For all $f : A \to B$, $g : B \to C$, and $h : C \to D$, $h \circ (g \circ f) = (h \circ g) \circ f$.

(b) *Identity:* For all $f : A \to B$, $f \circ \mathrm{Id}_A = f = \mathrm{Id}_B \circ f$.

(c) *Right Distributive Laws:* For all $f : A \to \mathbb{R}$ and all $g, h : \mathbb{R} \to \mathbb{R}$, $(g+h) \circ f = (g \circ f)+(h \circ f)$ and $(g \cdot h) \circ f = (g \circ f) \cdot (h \circ f)$.

Proof. We use the proof template for proving equality of functions (see page 241). For part (a), fix $f : A \to B$, $g : B \to C$, and $h : C \to D$. We have functions $(g \circ f) : A \to C$ and $(h \circ g) : B \to D$, so that the functions $h \circ (g \circ f)$ and $(h \circ g) \circ f$ are both defined with domain A and codomain D. Now, fix $x \in A$. Repeatedly using the definition of function composition, we find that

$$[h \circ (g \circ f)](x) = h((g \circ f)(x)) = h(g(f(x)));$$

$$[(h \circ g) \circ f](x) = (h \circ g)(f(x)) = h(g(f(x))).$$

Thus, $h \circ (g \circ f) = (h \circ g) \circ f$, as needed.

For part (b), fix $f : A \to B$. Recall $\mathrm{Id}_A : A \to A$ is defined by $\mathrm{Id}_A(x) = x$ for all $x \in A$. We now see that $f \circ \mathrm{Id}_A : A \to B$ is defined and has the same domain and codomain as $f : A \to B$. Fix $x \in A$, and compute $(f \circ \mathrm{Id}_A)(x) = f(\mathrm{Id}_A(x)) = f(x)$. So $f \circ \mathrm{Id}_A = f$. The equality $f = \mathrm{Id}_B \circ f$ is proved similarly.

Finally, we prove the first equality in part (c). Fix $f : A \to \mathbb{R}$ and $g, h : \mathbb{R} \to \mathbb{R}$. Both $(g + h) \circ f$ and $(g \circ f) + (h \circ f)$ are functions from A to $\mathbb{R}$. Fix $x \in A$, and compute

$$((g + h) \circ f)(x) = (g + h)(f(x)) = g(f(x)) + h(f(x)) \tag{5.1}$$

$$= (g \circ f)(x) + (h \circ f)(x) = [(g \circ f) + (h \circ f)](x). \tag{5.2}$$

So the functions $(g + h) \circ f$ and $(g \circ f) + (h \circ f)$ are equal. □

In the exercises, we see that the *left distributive law* $f \circ (g + h) = (f \circ g) + (f \circ h)$ is *not* always true, though it does hold when f has a special property called *linearity*.

Restriction of Functions

Our next operation on functions, called restriction, shrinks the domain of a given function to a smaller set of inputs.

5.63. Definition: Restriction. For any function $f : A \to B$ and any set $S \subseteq A$, the *restriction of f to S* is the function $f|_S : S \to B$ defined by $f|_S(x) = f(x)$ for all $x \in S$.

The graph of $f|_S$ is $\{(x, f(x)) : x \in S\}$, which can be obtained by intersecting the graph of f with $S \times B$. You can check that $f|_S$ really is a function. Note that $f|_S$ and f are different functions (except when $S = A$), since these functions have different domains.

5.64. Example: Arrow Diagrams of Restricted Functions. Let $A = \{1, 2, 3, 4, 5\}$, $B = \{a, b, c, d, e\}$, and define $f : A \to B$ by $f(1) = c$, $f(2) = c$, $f(3) = e$, $f(4) = b$, and $f(5) = e$. Restricting f to the set $S = \{1, 3, 5\}$ gives the function $f|_S : S \to B$ with $f|_S(1) = c$, $f|_S(3) = e$, and $f|_S(5) = e$. If we restrict f to $T = \{2, 4\}$, we get a function $f|_T : T \to B$ with graph $\{(2, c), (4, b)\}$. Note $f|_A = f$, whereas $f|_\emptyset$ is the empty function mapping $\emptyset$ into the codomain B. Arrow diagrams for f, $f|_S$, $f|_T$, and $f|_\emptyset$ are shown in Figure 5.7. Note that all of these restricted functions have the same codomain as the original function. We could obtain more new functions from these by enlarging or shrinking the codomain, as discussed in Example 5.55.

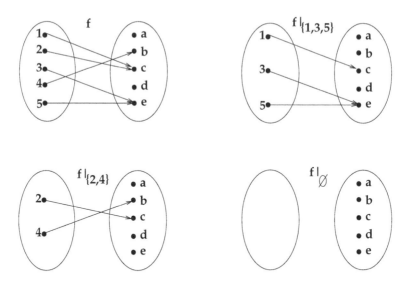

FIGURE 5.7
Arrow diagrams for restricted functions.

5.65. Example: Graphs of Restricted Functions. Suppose $g : \mathbb{R} \to \mathbb{R}$ is the squaring function $g(x) = x^2$ for $x \in \mathbb{R}$. Figure 5.8 displays the graphs of g, $g|_{[0,\infty)}$, $g|_{(-\infty,0]}$, $g|_{(-2,2]}$, $g|_{\{-1,0,1\}}$, and $g|_\emptyset$. For each $S \subseteq \mathbb{R}$, we obtain the graph of $g|_S$ from the graph of g by intersecting the full graph with the product set $S \times \mathbb{R}$. This means that we only keep those points $(x, g(x))$ in the original graph where the first coordinate x is in the set S.

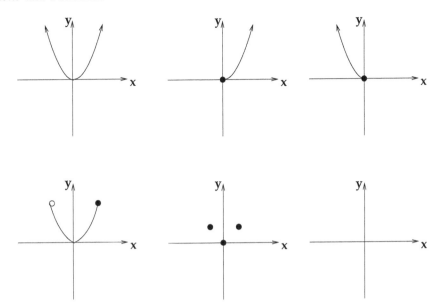

FIGURE 5.8
Graphs for restrictions of the squaring function.

Gluing Functions Together (Optional)

We now discuss the gluing operation, which is used to combine two or more functions to obtain a single new function with a larger domain. Before presenting the general definition, we consider two examples.

5.66. Example. Consider the three functions f, g, h defined as follows:

$$f : \{1,2,3,4\} \to \{a,b,c,d\} \quad \text{with graph} \quad \{(1,a),(2,a),(3,d),(4,b)\};$$
$$g : \{1,3,5,7\} \to \{a,b,c,d\} \quad \text{with graph} \quad \{(1,a),(3,d),(5,b),(7,a)\};$$
$$h : \{0,1,5\} \to \{a,b,c,d\} \quad \text{with graph} \quad \{(0,d),(1,a),(5,d)\}.$$

Arrow diagrams for these functions appear in the top half of Figure 5.9. Consider the effect of merging (gluing together) the diagrams for f and g. The result is shown in the bottom-left panel of the figure. In this case, a new function results, namely

$$\text{glue}(f,g) : \{1,2,3,4,5,7\} \to \{a,b,c,d\} \text{ with graph } \{(1,a),(2,a),(3,d),(4,b),(5,b),(7,a)\}.$$

Similarly, combining the diagrams for f and h produces the new function

$$\text{glue}(f,h) : \{0,1,2,3,4,5\} \to \{a,b,c,d\} \text{ with graph } \{(0,d),(1,a),(2,a),(3,d),(4,b),(5,d)\}.$$

On the other hand, if we try to combine the diagrams for g and h, the new graph is

$$\{(0,d),(1,a),(3,d),(5,b),(5,d),(7,a)\},$$

which is *not* the graph of a function. The trouble is that 5 is in the domain of both g and h, but $g(5) = b \neq d = h(5)$. Thus, when we combine g and h, the input 5 is associated with two distinct outputs.

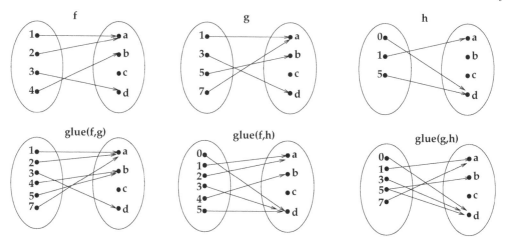

FIGURE 5.9
Arrow diagrams illustrating gluing.

5.67. Example. Consider the function $f : (-\infty, 0] \to \mathbb{R}$ defined by $f(x) = -x$ for $x \leq 0$ and the function $g : [0, \infty) \to \mathbb{R}$ defined by $g(x) = x$ for $x \geq 0$. Combining the graphs of f and g produces a new function $h = \text{glue}(f, g) : \mathbb{R} \to \mathbb{R}$ such that $h(x) = -x$ for $x \leq 0$, and $h(x) = x$ for $x \geq 0$. This function h is the familiar absolute value function from calculus.

On the other hand, suppose we try to glue together the functions $f_1 : \mathbb{R} \to \mathbb{R}$ and $g_1 : \mathbb{R} \to \mathbb{R}$ defined by $f_1(x) = -x$ and $g_1(x) = x$ for all $x \in \mathbb{R}$. In this case, the union of the graphs of f_1 and g_1 is *not* the graph of a function. The trouble is that there exist elements x in the common domain of f_1 and g_1 such that $f_1(x) \neq g_1(x)$. Indeed, this inequality holds for all x except for $x = 0$. Note that this difficulty did not arise when we glued together f and g, because these two functions agree on the common part of their domains (which is the one-point set $\{0\}$).

Now we define the gluing operation for a family of functions.

5.68. Definition: Gluing. Suppose I is an index set, and for each $i \in I$ we have a function $f_i : A_i \to B_i$ with graph G_i. We call $\{f_i : i \in I\}$ a *gluable family* iff for all $i, j \in I$ and all $x \in A_i \cap A_j$, $f_i(x) = f_j(x)$. When this condition holds, we define a new function $f : \bigcup_{i \in I} A_i \to \bigcup_{i \in I} B_i$ with graph $G = \bigcup_{i \in I} G_i$. We use the notation $f = \text{glue}(f_i : i \in I)$.

Given $x \in \bigcup_{i \in I} A_i$, we can compute $f(x)$ as follows. Choose any index $i \in I$ such that $x \in A_i$; then $f(x) = f_i(x)$. Note that the output obtained does not depend on the choice of i; if we choose some other index $j \in I$ with $x \in A_j$, we know $f_j(x) = f_i(x)$ since $\{f_i : i \in I\}$ is assumed to be a gluable family. Thus f is a well-defined function.

Here is a more formal proof that f is a function, based on checking that the graph G satisfies the three conditions in the definition of a function. Write $A = \bigcup_{i \in I} A_i$ and $B = \bigcup_{i \in I} B_i$. First, for each $i \in I$, we know $G_i \subseteq A_i \times B_i \subseteq A \times B$, so that $G = \bigcup_{i \in I} G_i$ is also a subset of $A \times B$. Second, given $x \in A$, we know there exists $i \in I$ with $x \in A_i$. Since f_i is a function, there exists y with $(x, y) \in G_i$, and hence $(x, y) \in G$. Third, fix x, y, z; assume $(x, y) \in G$ and $(x, z) \in G$; prove $y = z$. By definition of G, there exist $i, j \in I$ with $(x, y) \in G_i$ and $(x, z) \in G_j$. The only ordered pair in G_i with first coordinate x is $(x, f_i(x))$, so $y = f_i(x)$. Similarly, the only ordered pair in G_j with first coordinate x is $(x, f_j(x))$, so $z = f_j(x)$. We also know $x \in A_i$ (since $(x, y) \in G_i \subseteq A_i \times B_i$) and $x \in A_j$ (since $(x, z) \in G_j \subseteq A_j \times B_j$). By the gluable family condition, we deduce $y = f_i(x) = f_j(x) = z$, as needed. So $f = (A, B, G)$ is indeed a function. Conversely, it can be shown that if (A, B, G) is a function, then $\{f_i : i \in I\}$ must be a gluable family (Exercise 25(b)).

We remark that if the sets A_i are pairwise disjoint (i.e., for all $i \neq j$ in I, $A_i \cap A_j = \emptyset$), then $\{f_i : i \in I\}$ is automatically a gluable family.

5.69. Example: Piecewise Functions. Define a function $f : \mathbb{R} \to \mathbb{R}$ by setting

$$f(x) = \begin{cases} \sin x & \text{for } x \leq 0; \\ x^2 & \text{for } 0 \leq x \leq 1; \\ 1 & \text{for } 1 \leq x \leq 3; \\ e^{x-3} & \text{for } x \geq 3. \end{cases}$$

Formally, we can build f by gluing together the following four functions:

$$
\begin{array}{lll}
f_1 : (-\infty, 0] \to \mathbb{R} & \text{given by} & f_1(x) = \sin x; \\
f_2 : [0, 1] \to \mathbb{R} & \text{given by} & f_2(x) = x^2; \\
f_3 : [1, 3] \to \mathbb{R} & \text{given by} & f_3(x) = 1; \\
f_4 : [3, \infty) \to \mathbb{R} & \text{given by} & f_4(x) = e^{x-3}.
\end{array}
$$

For all $i \neq j$, f_i and f_j agree on the overlapping part of their domains. Thus, we can construct the overall function $f = \text{glue}(f_1, f_2, f_3, f_4)$ by taking the union of the graphs of the f_i.

Section Summary

1. *Composition of Functions.* Given $f : A \to B$ and $g : B \to C$, the composition $g \circ f : A \to C$ is defined by $(g \circ f)(x) = g(f(x))$ for all $x \in A$. To form $g \circ f$, the codomain of f must equal the domain of g. Function composition is associative (meaning $h \circ (g \circ f) = (h \circ g) \circ f$, when both sides are defined) and satisfies the identity property $f \circ \text{Id}_A = f = \text{Id}_B \circ f$. But function composition is *not* commutative: $f \circ g \neq g \circ f$ for most choices of f and g.

2. *Restriction of a Function.* Given $f : A \to B$ and any subset $S \subseteq A$, the restriction $f|_S : S \to B$ is defined by $f|_S(x) = f(x)$ for all $x \in S$. The graph of $f|_S$ is the intersection of the graph of f with $S \times B$.

3. *Gluing Functions.* Suppose for each i in an index set I, we are given a function $f_i : A_i \to B_i$ with graph G_i. These functions form a gluable family iff for all $i, j \in I$ and all $x \in A_i \cap A_j$, $f_i(x) = f_j(x)$. In this case, we get a new function $f = \text{glue}(f_i : i \in I)$ with domain $\bigcup_{i \in I} A_i$, codomain $\bigcup_{i \in I} B_i$, and graph $\bigcup_{i \in I} G_i$. For x in the domain of f, we have $f(x) = f_i(x)$ for any choice of $i \in I$ such that $x \in A_i$. When the sets A_i are pairwise disjoint, $\{f_i : i \in I\}$ is always a gluable family.

Exercises

1. In Example 5.61, compute $f \circ h$ and $h \circ f$. Are these functions equal?

2. Let $A = \{1, 2, 3, 4\}$ and $B = \{v, w, x, y, z\}$.
   Define $f : A \to B$ by $f(1) = w$, $f(2) = z$, $f(3) = y$, $f(4) = w$.
   Define $g : B \to A$ by $g(v) = 4$, $g(w) = 3$, $g(y) = 1$, $g(x) = 2$, $g(z) = 4$.
   Let $h : A \to A$ have graph $\{(1, 4), (2, 4), (3, 1), (4, 3)\}$.
   Let $k : B \to B$ have graph $\{(v, z), (z, y), (y, v), (w, x), (x, w)\}$.
   Draw arrow diagrams for the following functions (when defined): (a) f, g, h, k

(b) $f \circ f$ (c) $f \circ g$ (d) $f \circ h$ (e) $f \circ k$ (f) $g \circ f$ (g) $g \circ g$ (h) $g \circ h$ (i) $g \circ k$
(j) $h \circ f$ (k) $h \circ g$ (l) $h \circ h$ (m) $h \circ k$ (n) $k \circ f$ (o) $k \circ g$ (p) $k \circ h$ (q) $k \circ k$.

3. Define functions $f, g, h, k : \mathbb{R} \to \mathbb{R}$ by setting $f(x) = x^2$, $g(x) = x - 3$, $h(x) = 2^x$, and $k(x) = 5x$ for all $x \in \mathbb{R}$. Find simplified formulas for the following functions. Take care not to confuse compositions like $f \circ g$ with pointwise products like fg.
   (a) $f \circ g$ (b) $g \circ f$ (c) fg (d) gf (e) $f \circ h$ (f) $h \circ f$ (g) $g \circ k$ (h) $k \circ g$
   (i) $k \circ (f + g)$ (j) $(k \circ f) + (k \circ g)$ (k) $h \circ (f + g)$ (l) $(h \circ f) + (h \circ g)$
   (m) $(h \circ f)(h \circ g)$ (n) $(f + g) \circ h$ (o) $(f \circ h) + (g \circ h)$ (p) $(f \circ h)(g \circ h)$.

4. Define functions $f : \mathbb{R}_{>0} \to \mathbb{R}_{>0}$ by $f(x) = \sqrt{x}$, $g : \mathbb{R}_{>0} \to \mathbb{R}$ by $g(x) = \ln x$, $h : \mathbb{R} \to \mathbb{R}_{>0}$ by $h(x) = e^x$, $k : \mathbb{R} \to \mathbb{R}$ by $k(x) = x^2$, and $j : \mathbb{R}_{>0} \to \mathbb{R}$ by $j(x) = x$.
   (a) For which $p, q \in \{f, g, h, k, j\}$ is the composition $p \circ q$ defined?
   (b) Check that $h \circ g = \mathrm{Id}_{\mathbb{R}_{>0}}$ and $g \circ h = \mathrm{Id}_{\mathbb{R}}$.
   (c) Show that $k|_{\mathbb{R}_{>0}} \circ f = j$.

5. (a) Prove: for all sets A, B, C, if $A \subseteq B \subseteq C$, then $j_{B,C} \circ j_{A,B} = j_{A,C}$.
   (b) Prove: for all sets A and B, if $A \subseteq B$, then $j_{A,B} = \mathrm{Id}_B|_A$.

6. Let A be a set and $g : A \to A$ a function. For $n \geq 0$, we recursively define the powers $g^n : A \to A$ as follows: $g^0 = \mathrm{Id}_A$, and $g^n = g^{n-1} \circ g$ for $n \geq 1$. Let $A = \{1, 2, 3, 4, 5, 6, 7, 8\}$, and let $g : A \to A$ have graph $\{(1, 4), (4, 7), (7, 2), (2, 5), (5, 1), (3, 8), (8, 6), (6, 3)\}$.
   (a) Compute the functions g^n for $0 \leq n \leq 5$. (b) Does there exist $n > 0$ such that $g^n = \mathrm{Id}_A$? If so, what is the smallest such n? If not, why not?

7. Repeat Exercise 6 for $g : A \to A$ with graph

$$\{(2, 6), (6, 4), (4, 8), (8, 2), (1, 3), (3, 2), (5, 8), (7, 7)\}.$$

8. Repeat Exercise 6 for $g : A \to A$ defined by $g(x) = x + 1$ for $1 \leq x \leq 7$, and $g(8) = 1$.

9. Let $A = \{1, 2, \ldots, 10\}$. Find functions $g : A \to A$ such that: (a) $g \neq \mathrm{Id}_A$ but $g^2 = \mathrm{Id}_A$; (b) $g^{10} = \mathrm{Id}_A$ but $g^j \neq \mathrm{Id}_A$ for $0 < j < 10$; (c) $g^{21} = \mathrm{Id}_A$ but $g^j \neq \mathrm{Id}_A$ for $0 < j < 21$; (d) $g^j \neq \mathrm{Id}_A$ for all $j > 0$.

10. Supply a reason for each equality in (5.1) and (5.2).

11. (a) Prove: for all functions $f : A \to B$, $\mathrm{Id}_B \circ f = f$.
    (b) Prove: for all $f : A \to \mathbb{R}$ and all $g, h : \mathbb{R} \to \mathbb{R}$, $(g \cdot h) \circ f = (g \circ f) \cdot (h \circ f)$.

12. (a) Give an example of $f, g, h : \mathbb{R} \to \mathbb{R}$ where $h \circ (f + g) \neq (h \circ f) + (h \circ g)$.
    (b) Give an example of $f, g, h : \mathbb{R} \to \mathbb{R}$ where $h \circ (f \cdot g) \neq (h \circ f) \cdot (h \circ g)$.

13. Let $f, g, h : \mathbb{R} \to \mathbb{R}$ be functions. Suppose h has the following *linearity* property: $h(x + y) = h(x) + h(y)$ for all $x, y \in \mathbb{R}$. Prove that $h \circ (f + g) = (h \circ f) + (h \circ g)$.

14. Let $f, g, h : \mathbb{R} \to \mathbb{R}$ be functions. Find a condition on h that ensures that $h \circ (f \cdot g) = (h \circ f) \cdot (h \circ g)$ for all $f, g : \mathbb{R} \to \mathbb{R}$. Prove your answer.

15. Let $f : \mathbb{R} \to \mathbb{R}$ be defined by $f(x) = \sin x$. Draw graphs of the following restrictions of f: (a) $f|_{[-\pi/2, \pi/2]}$ (b) $f|_{[0, 2\pi]}$ (c) $f|_{(-\infty, \pi/4)}$ (d) $f|_{\{0, \pi/6, \pi/4, \pi/3, \pi/2, 3\pi/4, \pi\}}$ (e) $f|_{\emptyset}$ (f) $f|_{\{k\pi/2 : k \in \mathbb{Z}\}}$.

16. Let $f : \mathbb{R} \to \mathbb{R}$ be defined by $f(x) = |x|$. Draw graphs of the following restrictions of f: (a) $f|_{[0, 3]}$ (b) $f|_{(-3, -1) \cup [1, 2)}$ (c) $f|_{\{0\}}$ (d) $f|_{\mathbb{Z}}$ (e) $f|_{\mathbb{R} - \mathbb{Z}}$ (f) $f|_{\mathbb{Q}}$.

17. Consider the function $g : \{1, 2, 3, \ldots, 8\} \to \{v, w, x, y, z\}$ with graph

$$\{(1, w), (2, z), (3, y), (4, w), (5, x), (6, x), (7, y), (8, x)\}.$$

Draw arrow diagrams for: (a) g (b) $g|_{\{1,3,5,7\}}$ (c) $g|_{\{3,4,5\}}$ (d) $g|_{\{8\}}$ (e) $g|_\emptyset$ (f) $g|_{\{a \in A : g(a) = x\}}$ (g) $g|_{\{a \in A : g(a) = v\}}$.

18. Let $A = [0, 2]$ and $B = [2, \infty)$. Which of the following pairs of functions $f : A \to \mathbb{R}$ and $g : B \to \mathbb{R}$ are gluable? (a) $f(x) = 2x$, $g(x) = 4$. (b) $f(x) = x^2$, $g(x) = 2x$. (c) $f(x) = x^3$, $g(x) = x^2$. (d) $f(x) = e^{x-2}$, $g(x) = x/2$. (e) $f(x) = \sqrt{x}$, $g(x) = x/\sqrt{2}$. (f) $f(x) = |x|$, $g(x) = 3 - x$.

19. Consider the following four functions:
    $f : \{1, 2, 3\} \to \mathbb{Z}$ given by $f(1) = 4$, $f(2) = 0$, $f(3) = 4$;
    $g : \{0, 2, 4\} \to \mathbb{Z}$ given by $g(0) = 2$, $g(2) = 2$, $g(4) = 2$;
    $h : \{0, 1, 2\} \to \mathbb{Z}$ given by $h(0) = 2$, $h(1) = 4$, $h(2) = 2$;
    $k : \{0, 1, 3\} \to \mathbb{Z}$ given by $k(0) = 2$, $k(1) = 4$, $k(3) = 3$.
    Which of the following pairs of functions are gluable? For those that are, draw an arrow diagram illustrating the glued function. (a) f and g; (b) f and h; (c) f and k; (d) g and h; (e) g and k; (f) h and k; (g) f and f.

20. Assume $f : X \to Y$ and $g : Y \to Z$ are functions and $S \subseteq X$. Prove that $(g \circ f)|_S = g \circ (f|_S)$.

21. Assume $f : A \to Y$ and $g : B \to Y$ are gluable functions and $S \subseteq A \cup B$. Prove that $f_1 = f|_{A \cap S}$ and $g_1 = g|_{B \cap S}$ are gluable functions, and that $\text{glue}(f_1, g_1) = \text{glue}(f, g)|_S$.

22. Let X be a set. For fixed $A \subseteq X$, show how the characteristic function χ_A can be built from the constant functions 0_X and 1_X on X by using restriction and gluing operations.

23. Suppose $f : X \to Y$ is a function and $S \subseteq T \subseteq X$. (a) Prove that $(f|_T)|_S = f|_S$. (b) Prove that $f|_X = f$. (c) Prove that $f|_\emptyset = (\emptyset, Y, \emptyset)$.

24. Suppose $f, g : A \to \mathbb{R}$ are functions and $S \subseteq A$. Prove the following facts relating restriction to pointwise operations: (a) $(f + g)|_S = (f|_S) + (g|_S)$; (b) $(fg)|_S = (f|_S)(g|_S)$; (c) the restriction of a constant function on A with value c_0 is the constant function on S with value c_0.

25. Suppose that for all $i \in I$, $f_i : A_i \to B_i$ is a function with graph G_i. (a) Show that $\{f_i : i \in I\}$ is a gluable family iff for all $i, j \in I$, $f_i|_{A_i \cap A_j} = f_j|_{A_i \cap A_j}$. (b) Show that if $\bigcup_{i \in I} G_i$ is the graph of a function, then $\{f_i : i \in I\}$ is a gluable family.

26. Suppose $\{f_i : i \in I\}$ is a gluable family, where $f_i : A_i \to B_i$ for each $i \in I$, and let $h = \text{glue}(f_i : i \in I)$. (a) Prove: for all $i \in I$, $h|_{A_i} = f_i$. (b) Prove: for any $g : \bigcup_{i \in I} A_i \to \bigcup_{i \in I} B_i$, if $g|_{A_i} = f_i$ for all $i \in I$, then $g = h$.

27. Suppose that $h : A \to B$ is any function, and $\{A_i : i \in I\}$ is a family of sets with $A = \bigcup_{i \in I} A_i$. Let $h_i = h|_{A_i}$ for $i \in I$. Prove $\{h_i : i \in I\}$ is a gluable family such that $\text{glue}(h_i : i \in I) = h$.

5.7   Direct Images and Preimages

In earlier sections, we discussed the image of a set A under a relation R or the inverse relation R^{-1}. Intuitively, $R[A]$ is the set of all outputs that are related under R to at least one input in the set A. On the other hand, $R^{-1}[B]$ is the set of all inputs that are related under R to at least one output in the set B. Here we introduce similar constructions where the relation R is replaced by a function f.

Direct Images and Preimages

5.70. Definition: Direct Image of a Set. Given a function $f : X \to Y$ and a set $A \subseteq X$, the *direct image of A under f* is $f[A] = \{f(x) : x \in A\}$. Equivalently, for all objects y, $\boxed{y \in f[A]}$ iff $\boxed{\exists x \in A, y = f(x).}$

Intuitively, $f[A]$ is the set of all outputs we get by applying f to every input in the set A. If G is the graph of f, we could rephrase this definition by writing $f[A] = G[A] = \{y : \exists x \in A, (x, y) \in G\}$. Thus, the direct image of a set under a function is a special case of the image of a set under a relation. In fact, this formulation is a little more general since we could apply it to any set A (not just subsets of the domain X). However, when dealing with functions, it is sometimes easier to use the version of the definition phrased in terms of function notation.

5.71. Definition: Preimage of a Set. Given a function $f : X \to Y$ and a set $B \subseteq Y$, the *preimage of B under f* is $f^{-1}[B] = \{x \in X : f(x) \in B\}$. Equivalently, for all objects x, $\boxed{x \in f^{-1}[B]}$ iff $\boxed{x \in X \text{ and } f(x) \in B}$. The set $f^{-1}[B]$ is also called the *inverse image of B under f*.

Intuitively, the preimage $f^{-1}[B]$ is the set of all inputs in the domain of f that map to an output in the set B when we apply f. As before, if f has graph G, we could have also defined $f^{-1}[B] = G^{-1}[B]$, which is the image of B under the inverse of the relation G.

5.72. Warning. The notation $f^{-1}[B]$ does *not* mean that the function f has an inverse function f^{-1}. (We discuss inverse functions in §5.9.) The preimage construction defined here can be applied to any function, not just those functions that have inverses. We also call your attention to our use of *square brackets* for images of sets under relations, direct images of sets under functions, and preimages of sets under functions. We use brackets rather than round parentheses to avoid confusion with function value notation, as shown in Example 5.74 below.

5.73. Example. Let f and g be the functions shown earlier in Figure 5.6. Recall f has graph $\{(1, b), (2, e), (3, c), (4, c), (5, a)\}$, and g has graph $\{(a, a), (b, c), (c, d), (d, b), (e, e)\}$. We compute $f[\{3, 5\}] = \{c, a\}$, $f^{-1}[\{c, a\}] = \{3, 4, 5\}$, $f^{-1}[\{b, d, e\}] = \{1, 2\}$, $f[\{1, 2\}] = \{b, e\}$. Moreover, $g[\{b, c, d\}] = \{b, c, d\} = g^{-1}[\{b, c, d\}]$, $g[\{c\}] = \{d\}$, $g^{-1}[\{a, d, e\}] = \{a, c, e\}$, and $g[\emptyset] = \emptyset = g^{-1}[\emptyset]$. Finally, $(g \circ f)[\{3, 4, 5\}] = \{d, a\}$, $(g \circ f)^{-1}[\{b\}] = \emptyset$, and

$$(g \circ f)[f^{-1}[\{a, d\}]] = (g \circ f)[\{5\}] = \{a\}.$$

5.74. Example. Let $X = \{4, 5, 6, \{4\}, \{4, 5, 6\}, \emptyset\}$, and define a function $f : X \to X$ by setting $f(4) = \{4\}$, $f(5) = 4$, $f(6) = \emptyset$, $f(\{4\}) = 5$, $f(\{4, 5, 6\}) = \emptyset$, and $f(\emptyset) = \{4\}$. An arrow diagram for this function appears below.

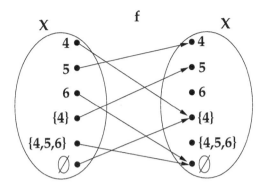

The set X is interesting because some of the members of X (namely $\{4\}$ and $\{4,5,6\}$ and $\emptyset$) are also subsets of X. Let us compute some direct images, function values, and preimages for f. First, $f[\{5,6\}] = \{4,\emptyset\}$ since the arrows leading away from the inputs 5 and 6 lead to 4 and $\emptyset$. Similarly, $f[\{4,\{4\},\emptyset\}] = \{\{4\},5\}$. Computing some function values now, note that $f(5) = 4$, $f(4) = \{4\}$, $f(\{4\}) = 5$, $f(\{4,5,6\}) = \emptyset$, and $f(\emptyset) = \{4\}$. In contrast, if we switch to square brackets, we compute the direct images $f[\{4\}] = \{\{4\}\}$, $f[\{4,5,6\}] = \{\{4\},4,\emptyset\}$, and $f[\emptyset] = \emptyset$. Turning to preimages, we find that $f^{-1}[\{4,6\}] = \{5\}$, $f^{-1}[\{4,5,6\}] = \{5,\{4\}\}$, $f^{-1}[\{\{4,5,6\}\}] = \emptyset$, $f^{-1}[\emptyset] = \emptyset$, and $f^{-1}[\{\emptyset\}] = \{6,\{4,5,6\}\}$. Finally, $f^{-1}[5]$ is not defined since 5 is not a subset of X; on the other hand, $f^{-1}[\{5\}] = \{\{4\}\}$.

5.75. Example. Define $f : [-2\pi, 4\pi] \to \mathbb{R}$ by letting $f(x) = \cos x$ for all $x \in [-2\pi, 4\pi]$. The graph of f is shown below.

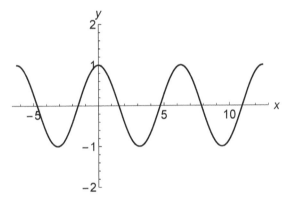

We have, for example, $f[[0, \pi/2]] = [0, 1]$, $f[\mathbb{R}_{>0}] = [-1, 1]$, $f^{-1}[\{1\}] = \{-2\pi, 0, 2\pi, 4\pi\}$, $f^{-1}[\mathbb{R}_{<0}] = (-3\pi/2, -\pi/2) \cup (\pi/2, 3\pi/2) \cup (5\pi/2, 7\pi/2)$, $f^{-1}[f[\{\pi\}]] = f^{-1}[\{-1\}] = \{-\pi, \pi, 3\pi\}$, and $f^{-1}[(1,2]] = \emptyset$. Looking at some restrictions of f, we have $(f|_{[0,\pi)})^{-1}[\mathbb{R}_{<0}] = (\pi/2, \pi]$, $(f|_{\mathbb{R}_{>0}})^{-1}[[1, \infty)] = \{2\pi, 4\pi\}$, and $(f|_{\mathbb{Z}})^{-1}[\mathbb{Z}] = \{0\}$.

Properties of Direct Images and Preimages

The Theorem on Images (for relations) specializes to give theorems on direct images and preimages of sets under functions. As we see below, the preimage construction has some especially nice additional properties.

5.76. Theorem on Direct Images. Let $f : X \to Y$ be a fixed function, let I be a fixed nonempty index set, and let A, C, A_i (for each $i \in I$) be subsets of X. Then:

(a) *Empty Image:* $f[\emptyset] = \emptyset$.

(b) *Codomain Property:* $f[A] \subseteq Y$.

(c) *Monotonicity:* If $A \subseteq C$, then $f[A] \subseteq f[C]$.

(d) *Image of Union:* $f[A \cup C] = f[A] \cup f[C]$ and $f[\bigcup_{i \in I} A_i] = \bigcup_{i \in I} f[A_i]$.

(e) *Image of Intersection:* $f[A \cap C] \subseteq f[A] \cap f[C]$ and $f[\bigcap_{i \in I} A_i] \subseteq \bigcap_{i \in I} f[A_i]$.
    Equality is not always true.

(f) *Image of Set Difference:* $f[A] - f[C] \subseteq f[A - C]$. Equality is not always true.

(g) *Image Under a Composition:* For all $g : Y \to Z$, $(g \circ f)[A] = g[f[A]]$.

Proof. This theorem follows immediately from the corresponding theorem for relations, and the fact that $f[A] = G[A]$ where G is the graph of f. To illustrate the definition of direct image using function notation, we prove part (f). Fix $f : X \to Y$ and subsets $A, C \subseteq X$. Fix an object $y \in f[A] - f[C]$; prove $y \in f[A - C]$. We know $y \in f[A]$ and $y \notin f[C]$. On one hand, $y \in f[A]$ means there is an $x \in A$ with $y = f(x)$. This x cannot belong to C, since otherwise $y \in f[C]$, which is not true. So $x \in A - C$. Since $y = f(x)$ with $x \in A - C$, we have $y \in f[A - C]$.

   To see that equality is not always true in (f), we give an example. Define $f : \{1, 2\} \to \{3, 4\}$ by $f(1) = 3 = f(2)$. Take $A = \{1\}$ and $C = \{2\}$. Then $f[A] = f[C] = \{3\}$, so $f[A] - f[C] = \emptyset$. On the other hand, $A - C = \{1\}$, so $f[A - C] = \{3\}$. We see that $\emptyset \subseteq \{3\}$, but equality does not hold. $\square$

   Now we consider preimages. Some parts of the next theorem follow by applying the Theorem on Images to the relation G^{-1} (the inverse of the graph of f), but here some key additional properties turn out to be true.

5.77. Theorem on Preimages. Let $f : X \to Y$ be a fixed function, let I be a fixed nonempty index set, and let B, D, B_i (for each $i \in I$) be subsets of Y. Then:

(a) *Empty Preimage:* $f^{-1}[\emptyset] = \emptyset$.

(b) *Domain Property:* $f^{-1}[B] \subseteq X$.

(c) *Preimage of Codomain:* $f^{-1}[Y] = X$.

(d) *Monotonicity:* If $B \subseteq D$, then $f^{-1}[B] \subseteq f^{-1}[D]$.

(e) *Preimage of Union:* $f^{-1}[B \cup D] = f^{-1}[B] \cup f^{-1}[D]$ and $f^{-1}\left[\bigcup_{i \in I} B_i\right] = \bigcup_{i \in I} f^{-1}[B_i]$.

(f) *Preimage of Intersection:* $f^{-1}[B \cap D] = f^{-1}[B] \cap f^{-1}[D]$ and $f^{-1}\left[\bigcap_{i \in I} B_i\right] = \bigcap_{i \in I} f^{-1}[B_i]$.

(g) *Preimage of Set Difference:* $f^{-1}[B] - f^{-1}[D] = f^{-1}[B - D]$.

(h) *Image of Preimage:* $f[f^{-1}[B]] \subseteq B$. Equality is not always true.

(i) *Preimage of Image:* For all $A \subseteq X$, $A \subseteq f^{-1}[f[A]]$. Equality is not always true.

(j) *Preimage Under a Composition:* For all $h : W \to X$, $(f \circ h)^{-1}[B] = h^{-1}[f^{-1}[B]]$.

   We prove some parts of this theorem to illustrate the definition of preimages in terms of function notation.

5.78. Proof of Preimage Property for Intersections. With the notation of the theorem, let us prove that $f^{-1}\left[\bigcap_{i \in I} B_i\right] = \bigcap_{i \in I} f^{-1}[B_i]$ via a chain proof. The sets appearing on the left and right sides of this identity are both subsets of the domain X. So, to prove they are equal, we can fix an arbitrary object x in X and show x is in the left set iff x is in the right set. For fixed $x \in X$, note $x \in f^{-1}[\bigcap_{i \in I} B_i]$ iff $f(x) \in \bigcap_{i \in I} B_i$ iff for all $i \in I$, $f(x) \in B_i$ iff for all $i \in I$, $x \in f^{-1}[B_i]$ iff $x \in \bigcap_{i \in I} f^{-1}[B_i]$. The property

$f^{-1}[B \cap D] = f^{-1}[B] \cap f^{-1}[D]$ is a special case of this result (let the index set I have two elements).

5.79. Proof of Image of Preimage Property. Fix $B \subseteq Y$; we must prove $f[f^{-1}[B]] \subseteq B$. Fix y; assume $y \in f[f^{-1}[B]]$; prove $y \in B$. By definition of direct image, there exists $x \in f^{-1}[B]$ with $y = f(x)$. By definition of preimage, $x \in X$ and $f(x) \in B$. Since $y = f(x)$, we conclude that $y \in B$.

Here is an example to show that $f[f^{-1}[B]] = B$ is not always true. Define $f : \{1, 2, 3\} \to \{4, 5, 6\}$ by letting $f(1) = f(2) = f(3) = 5$. Choose $B = \{4, 5\}$. Then $f^{-1}[B] = \{1, 2, 3\}$, so $f[f^{-1}[B]] = \{5\}$. Note that $\{5\}$ is a subset of B, as promised by the theorem, but this subset is unequal to B.

5.80. Proof of Preimage of Image Property. Fix $A \subseteq X$; we must prove $A \subseteq f^{-1}[f[A]]$. Fix x; assume $x \in A$; prove $x \in f^{-1}[f[A]]$. We must prove $x \in X$ and $f(x) \in f[A]$. Since $x \in A$ and $A \subseteq X$, we see that $x \in X$ is true. To prove $f(x) \in f[A]$, we must prove $\exists a \in A, f(x) = f(a)$. Choose $a = x$, which is in A by assumption. Then $f(x) = f(a)$ certainly holds.

Here is an example to show that $A = f^{-1}[f[A]]$ is not always true. Define $f : \{1, 2, 3\} \to \{4, 5, 6\}$ by letting $f(1) = f(2) = f(3) = 5$. Choose $A = \{3\}$. Then $f[A] = \{5\}$, so $f^{-1}[f[A]] = \{1, 2, 3\}$, which contains A but is unequal to A.

5.81. Example. Define a function $f : \mathbb{R} \to \mathbb{R}$ by $f(x) = x^2$ for all $x \in \mathbb{R}$. The graph of f is shown here.

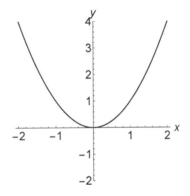

Let $A = [1, 2]$, considered as a subset of the domain of f. We compute $f[A] = [1, 4]$, and $f^{-1}[f[A]] = f^{-1}[[1, 4]] = [-2, -1] \cup [1, 2]$. Thus A is a proper subset of $f^{-1}[f[A]]$. Next let $B = [-3, 4]$, considered as a subset of the codomain of f. We compute $f^{-1}[B] = [-2, 2]$, and $f[f^{-1}[B]] = f[[-2, 2]] = [0, 4]$. Thus $f[f^{-1}[B]]$ is a proper subset of B.

Section Summary

1. *Direct Images and Preimages.* Let $f : X \to Y$ be a function. For all subsets A of X and all objects y, $y \in f[A]$ iff $\exists x \in A, y = f(x)$. For all subsets B of Y and all objects x, $x \in f^{-1}[B]$ iff $x \in X$ and $f(x) \in B$. If G is the graph of f, we have $f[A] = G[A]$ and $f^{-1}[B] = G^{-1}[B]$. The preimage notation $f^{-1}[B]$ can be used for all functions, not just functions that have inverses.

2. *Properties of Preimages.* The preimage construction preserves set concepts such as union, intersection, set difference, set inclusion, and the empty set. For instance, the preimage of an arbitrary union of sets is the union of the preimages of these sets.

3. *Properties of Direct Images.* The direct image construction preserves set inclusions, arbitrary unions, and the empty set. But the set inclusions $f[A \cap C] \subseteq f[A] \cap f[C]$, $f[A] - f[C] \subseteq f[A - C]$, $f[f^{-1}[B]] \subseteq B$, and $A \subseteq f^{-1}[f[A]]$ are not equalities in general.

Exercises

1. Let $X = \{1, 2, 3, 4, 5, 6\}$, and let $f : X \to X$ have graph $\{(1, 4), (2, 3), (3, 4), (4, 1),$ $(5, 6), (6, 4)\}$. Draw an arrow diagram for f. Use this to compute
   $f[\{2, 3, 4\}]$, $f^{-1}[\{2, 3, 4\}]$, $f[\{2, 5\}]$, $f^{-1}[\{2, 5\}]$, $f[f^{-1}[\{4, 5, 6\}]]$, $f^{-1}[f[\{4, 5, 6\}]]$,
   $(f \circ f)[\{1, 3, 5\}]$ and $(f \circ f)^{-1}[\{1, 3, 5\}]$.

2. For each $f : \mathbb{R} \to \mathbb{R}$, sketch the graph of f. Use this to compute the direct image $f[(-1, 1)]$, the preimage $f^{-1}[(-1, 1)]$, and $f^{-1}[\{-4\}]$.
   (a) $f(x) = 3x + 5$.
   (b) $f(x) = 4x^2$.
   (c) $f(x) = |x|$ for $x \neq 0$, $f(0) = -4$.
   (d) $f = \chi_{\mathbb{R}-\mathbb{Z}}$ (a characteristic function).
   (e) $f(x) = \sin x$.
   (f) $f(x) = \ln|x|$ for $x \neq 0$, $f(0) = 0$.

3. For each real-valued function below, describe the preimage sets $f^{-1}[\{c\}]$ for each $c \in \mathbb{R}$ (these sets are called *level sets* of the function). Sketch the level sets for $c = -1, 0, 1, 4$ in each case.
   (a) $f : \mathbb{R}^2 \to \mathbb{R}$ given by $f(x, y) = y$.
   (b) $f : \mathbb{R}^2 \to \mathbb{R}$ given by $f(x, y) = x + y$.
   (c) $f : \mathbb{R}^2 \to \mathbb{R}$ given by $f(x, y) = 4$.
   (d) $f : \mathbb{R}^2 \to \mathbb{R}$ given by $f(x, y) = x^2 + y^2$.
   (e) $f : \mathbb{R}^2 \to \mathbb{R}$ given by $f(x, y) = \sqrt{x^2 + y^2}$.
   (f) $f : \mathbb{R}^2 \to \mathbb{R}$ given by $f(x, y) = \sqrt{1 + x^2 + y^2}$.

4. Let $X = \{1, 2, 3\} \cup \mathcal{P}(\{1, 2\})$. Define $f : X \to X$ by letting $f(1) = \{1\}$, $f(2) = \emptyset$, $f(3) = 2$, and $f(A) = \{1, 2\} - A$ for all $A \subseteq \{1, 2\}$.
   (a) Draw an arrow diagram for f.
   (b) Compute (if defined) $f(1)$, $f(\{1\})$, $f(\{\{1\}\})$, $f[1]$, $f[\{1\}]$, and $f[\{\{1\}\}]$.
   (c) Compute (if defined) $f(\{1, 2\})$, $f(\{\{1, 2\}\})$, $f[\{1, 2\}]$, $f[\{\{1, 2\}\}]$, $f[\{1, \{2\}\}]$, $f[\{\{1\}, 2\}]$, and $f[\{\{1\}, \{2\}\}]$.
   (d) Compute $f^{-1}[\{2\}]$, $f^{-1}[\{\{2\}\}]$, $f^{-1}[\emptyset]$, $f^{-1}[\{\emptyset\}]$, $f^{-1}[\{1, \{2\}\}]$, and $f^{-1}[\{\{1\}, 2\}]$.

5. Use the definition of direct image based on function notation to prove parts (c) and (e) of the Theorem on Direct Images.

6. (a) Prove part (g) of the Theorem on Direct Images.
   (b) Prove part (j) of the Theorem on Preimages.

7. Given sets $A \subseteq X$, let $\chi_A : X \to \mathbb{R}$ be the characteristic function for A.
   (a) Find all possible direct images $\chi_A[B]$, as B ranges over subsets of X. Which choices of B produce each possible answer?
   (b) Find all possible preimages $\chi_A^{-1}[C]$, as C ranges over subsets of $\mathbb{R}$. Which choices of C produce each possible answer?

8. Suppose $f : X \to Y$ and $S \subseteq X$. Prove: for all $B \subseteq Y$, $f|_S^{-1}[B] = f^{-1}[B] \cap S$.

9. Suppose $f : X \to Y$, $A \subseteq X$, and $B \subseteq Y$.
   (a) Prove: $B \cap f[A] = f[f^{-1}[B] \cap A]$.
   (b) Prove: $f[f^{-1}[B] \cup A] \subseteq B \cup f[A]$.

(c) Prove: $f^{-1}[B] \cap A \subseteq f^{-1}[B \cap f[A]]$.

(d) Prove: $f^{-1}[B] \cup A \subseteq f^{-1}[B \cup f[A]]$.

10. Give examples to show that the set inclusions in parts (b), (c), and (d) of the previous exercise are not always equalities.

11. (a) Disprove: for all relations R and all sets A, $A \subseteq R^{-1}[R[A]]$.

(b) Prove or disprove: for all relations R and all sets B, $R[R^{-1}[B]] \subseteq B$.

12. Prove: for all functions $f : X \to Y$ and all $A \subseteq X$, $f[f^{-1}[f[A]]] = f[A]$.

13. Prove: for all functions $f : X \to Y$ and all $B \subseteq Y$, $f^{-1}[f[f^{-1}[B]]] = f^{-1}[B]$.

14. Do the statements in the previous two exercises remain true if we only assume that f is a relation from X to Y?

15. (a) Find (with proof) a property of a function $f : X \to Y$ that guarantees $f[A \cap C] = f[A] \cap f[C]$ for all $A, C \subseteq X$.

(b) Find (with proof) a property of a function $f : X \to Y$ that guarantees $f[A] - f[C] = f[A - C]$ for all $A, C \subseteq X$.

16. (a) Find (with proof) a property of a function $f : X \to Y$ that guarantees $f^{-1}[f[A]] = A$ for all $A \subseteq X$.

(b) Find (with proof) a property of a function $f : X \to Y$ that guarantees $f[f^{-1}[B]] = B$ for all $B \subseteq Y$.

5.8   Injective, Surjective, and Bijective Functions

This section studies three special kinds of functions, called injections, surjections, and bijections. Intuitively, a function is injective (one-to-one) iff distinct inputs to the function always map to distinct outputs. A function is surjective (onto) iff every potential output in the codomain of the function is actually attained as the value of the function at some input. A function is bijective iff it is both one-to-one and onto.

Definition of Injections, Surjections, and Bijections

We introduce the concepts of injective, surjective, and bijective functions using arrow diagrams. Informally, a function $f : X \to Y$ is *injective* (or *one-to-one*) iff every point in Y has *at most one* arrow entering it. This means that every potential output occurs as the value of the function for either zero or one inputs. The function f is *surjective* (or *onto*) iff every point in Y has *at least one* arrow entering it. This means that every *potential* output in the codomain really does get used as an *actual* output assigned to one of the inputs in the domain. Finally, the function f is *bijective* (or *a one-to-one correspondence between X and Y*) iff every point in Y has *exactly one* arrow entering it. This means that for every output in Y, there exists exactly one input in X mapping to that output. Compare this to the requirement in the definition of a function that every input in X has exactly one arrow leaving it. The relation between these conditions is explained in the next section when we discuss inverse functions.

5.82. Example. Consider the following four arrow diagrams.

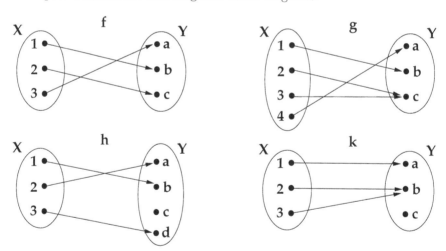

First of all, each of the arrow diagrams does represent a function, since each input x in the domain X has exactly one arrow leaving it, and all arrows go from X to Y. Second, f and h are injective functions, because every output in Y has at most one arrow coming in. The function g is not injective because the output c has two arrows coming in: $g(2) = c = g(3)$. Similarly, k is not one-to-one because $k(2) = b = k(3)$. Third, f and g are surjective functions, because every output in Y has at least one arrow coming in. The function h is not surjective because c is not hit by an arrow; similarly, k is not onto. Fourth, f is a bijection because each output in Y has exactly one arrow coming in. The other three functions are not bijective.

Next we give the formal definitions of injections, surjections, and bijections. At this time, it might be helpful to review §1.6 to recall how the phrases "at most one" and "exactly one" can be encoded using logical symbols.

5.83. Definition: Injective (One-to-One) Function. Given a function $f : X \to Y$,

$\boxed{f \text{ is injective}}$ iff $\boxed{\forall w \in X, \forall x \in X, f(w) = f(x) \Rightarrow w = x}$.

When this holds, we also say $\boxed{f \text{ is one-to-one}}$ or $\boxed{f \text{ is an injection}}$.

The definition of injective function does not explicitly mention members of Y, but these outputs appear implicitly in the expression $f(w) = f(x)$. The implication in the definition says that if two inputs w and x ever map to the same output under f, then the two inputs must in fact be the same. The contrapositive of the definition may also have more intuitive appeal, though it may be more difficult to use in proofs: f is one-to-one iff for all $w, x \in X$, $w \neq x$ implies $f(w) \neq f(x)$.

5.84. Definition: Surjective (Onto) Function. Given a function $f : X \to Y$,

$\boxed{f \text{ is surjective}}$ iff $\boxed{\forall y \in Y, \exists x \in X, y = f(x)}$.

When this holds, we also say $\boxed{f \text{ is onto}}$ or $\boxed{f \text{ is a surjection}}$.

5.85. Definition: Bijective Function. Given a function $f : X \to Y$,

$\boxed{f \text{ is bijective}}$ iff $\boxed{f \text{ is injective and } f \text{ is surjective}}$.

Equivalently, $\boxed{f \text{ is a bijection}}$ iff $\boxed{\forall y \in Y, \exists ! x \in X, y = f(x)}$.

5.86. Example: Denials of the Definitions. What does it mean to say that a function $f : X \to Y$ is: (a) not injective; (b) not surjective; (c) not bijective?
Solution. Using the denial rules, we find:

(a) $\boxed{f \text{ is not injective}}$ iff $\boxed{\exists w \in X, \exists x \in X, f(w) = f(x) \wedge w \neq x}$.

(b) $\boxed{f \text{ is not surjective}}$ iff $\boxed{\exists y \in Y, \forall x \in X, y \neq f(x)}$.

(c) $\boxed{f \text{ is not bijective}}$ iff $\boxed{f \text{ is not injective or } f \text{ is not surjective}}$.

Examples of Injections, Surjections, and Bijections

5.87. Example: Exponential Function. Define $f : \mathbb{R} \to \mathbb{R}$ by $f(x) = e^x$ for all $x \in \mathbb{R}$. The graph of f is shown here.

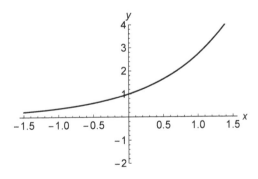

Each vertical line $x = x_0$ (for $x_0 \in \mathbb{R}$) intersects the graph of f in exactly one point; for this reason, f is a function. On the other hand, each horizontal line $y = y_0$ (for $y_0 \in \mathbb{R}$) intersects the graph of f in at most one point. This means that every output y_0 has the form $f(x_0)$ for at most one input x_0. So, f is injective. (Testing injectivity in this way is called the *Horizontal Line Test*.) But, for certain choices of $y_0 \in \mathbb{R}$, the horizontal line $y = y_0$ does not intersect the graph of f anywhere. For instance, this holds for $y_0 = 0$, $y_0 = -1$, or any negative value of y_0. This means that f is not surjective, and hence not bijective. On the other hand, consider the function $f_1 : \mathbb{R} \to \mathbb{R}_{>0}$, given by $f_1(x) = e^x$ for all $x \in \mathbb{R}$. The function f_1 is *not* the same as f, because the two functions have different codomains. We see that f_1 is injective, surjective, and bijective, since every y_0 in the new codomain $\mathbb{R}_{>0}$ has the form $y_0 = f(x_0)$ for exactly one input $x_0 \in \mathbb{R}$, namely $x_0 = \ln y_0$.

5.88. Example: Functions from Calculus. Define relations F, G, H, K, by setting $F = \{(x, x^3) : x \in \mathbb{R}\}$, $G = \{(x, x^2) : x \in \mathbb{R}\}$, $H = \{(x, \sin x) : x \in \mathbb{R}\}$, and

$$K = \{(x, y) : x \in \mathbb{R}, \tan x \text{ is defined, and } y = \tan x\}.$$

The graphs of these relations are shown here:

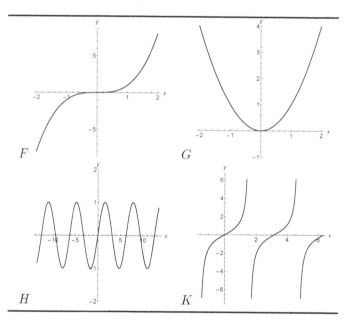

We cannot apply the adjectives injective, surjective, or bijective to any of the graphs F, G, H, or K, because these graphs are relations, not functions. Each graph can be used to produce many different functions, depending on which domain and codomain are chosen. Consider these examples.

(a) $f_1 : \mathbb{R} \to \mathbb{R}$ with graph F is injective and surjective and bijective.

(b) $f_2 : \mathbb{R} \to \mathbb{R}$ with graph G is not injective and not surjective and not bijective.

(c) $f_3 : \mathbb{R} \to \mathbb{R}_{\geq 0}$ with graph G is surjective but not injective.

(d) $f_4 : \mathbb{R}_{\geq 0} \to \mathbb{R}_{\geq 0}$ with graph G is not a function, because the graph G is not a subset of $\mathbb{R}_{\geq 0} \times \mathbb{R}_{\geq 0}$.

(e) $f_5 : R_{>0} \to \mathbb{R}_{>0}$ with graph $G \cap (\mathbb{R}_{>0} \times \mathbb{R}_{>0})$ is injective and surjective and bijective. In this case, the domain is the positive x-axis, the codomain is the positive y-axis, and the graph of f_5 is the part of the parabola G in the first quadrant.

(f) $f_6 : \mathbb{R} \to \mathbb{R}$ with graph H is not injective (e.g., because $f_6(\pi) = 0 = f_6(-\pi)$. f_6 is not surjective (e.g., because there does not exist $x_0 \in \mathbb{R}$ with $f_6(x_0) = 2$). f_6 is not bijective (e.g., because it is not injective).

(g) $f_7 : \mathbb{R} \to [-1, 1]$ with graph H is surjective but not injective. Note that we obtain f_7 from f_6 by restricting the codomain.

(h) $f_8 : [-\pi/2, \pi/2] \to \mathbb{R}$ with graph $H \cap ([-\pi/2, \pi/2] \times \mathbb{R})$ is injective but not surjective. Note $f_8 = f_6|_{[-\pi/2, \pi/2]}$ is a restriction of f_6.

(i) $f_9 : [-\pi/2, \pi/2] \to [-1, 1]$ with graph $H \cap ([-\pi/2, \pi/2] \times \mathbb{R})$ is injective and surjective and bijective. Note that f_9 is obtained from f_6 by restricting both the domain and the codomain; these modifications have converted a non-bijection into a bijection.

(j) $f_{10} : [0, \pi/2] \to [0, 1]$ with graph $H \cap ([0, \pi/2] \times \mathbb{R})$ is also injective and surjective and bijective. This is a different restriction of f_6 that produces a bijection.

(k) $f_{11} : \mathbb{R} \to \mathbb{R}$ with graph K is not a function, because there exist inputs x_0 in the domain $\mathbb{R}$ with no associated outputs. For example, for all y, $(\pi/2, y) \notin K$, so the input $\pi/2$ has no corresponding output.

(l) To fix the previous example, let $A = \{\pi/2 + k\pi : k \in \mathbb{Z}\}$ be the set of real numbers x where $\tan x$ is undefined. Then $f_{12} : \mathbb{R} - A \to \mathbb{R}$ with graph K is a function, and f_{12} is surjective but not injective.

(m) $f_{13} : (-\pi/2, \pi/2) \to \mathbb{R}$ with graph $K \cap ((-\pi/2, \pi/2) \times \mathbb{R})$ is a function that is injective, surjective, and bijective. The same is true of $f_{14} : (\pi/2, 3\pi/2) \to \mathbb{R}$ with graph $K \cap ((\pi/2, 3\pi/2) \times \mathbb{R})$.

These examples show that a non-surjective function $f : X \to Y$ can be changed into a new, surjective function by shrinking the codomain from Y to the image $f[X]$. Similarly, we can change a non-injective function $g : X \to Y$ into a new function that is injective if we are willing to shrink the domain X sufficiently. We might need to shrink the domain by a lot, however. For instance, let $g : \mathbb{R} \to \mathbb{R}$ be the constant function given by $g(x) = 3$ for all $x \in \mathbb{R}$. In order to manufacture an injective function from g, we need to shrink the domain all the way down to a single point! The new function is not surjective either; to obtain surjectivity, we must also shrink the codomain to the set $\{3\}$. The resulting function (mapping a one-point set to a one-point set) is not very interesting, but it is a bijection.

Proofs of Injectivity, Surjectivity, and Bijectivity

Here we consider some examples to see how to prove that a function is or is not injective, surjective, or bijective.

5.89. Example. For any set X, the identity function $\mathrm{Id}_X : X \to X$ is a bijection. To prove injectivity, fix $w, x \in X$; assume $\mathrm{Id}_X(w) = \mathrm{Id}_X(x)$; prove $w = x$. Using the definition of Id_X, the assumption becomes $w = x$, as needed. To prove surjectivity, fix $y \in X$; we must prove $\exists x \in X, \mathrm{Id}_X(x) = y$. Choose $x = y$; then $x \in X$ and $\mathrm{Id}_X(x) = x = y$, as needed.

5.90. Example. Define $f : (\mathbb{R} - \mathbb{Z}) \to \mathbb{Z}$ by $f(x) = \lfloor x \rfloor$, which is x rounded down to the nearest integer. Let us show that f is surjective but not injective. For surjectivity, fix $y_0 \in \mathbb{Z}$. Choose $x_0 = y_0 + 1/2$, which does lie in the domain $\mathbb{R} - \mathbb{Z}$ of f. Rounding x_0 down to the nearest integer produces y_0, so $f(x_0) = y_0$. Next, to prove f is not injective, choose $x = 1/3$ and $w = 2/3$, which lie in $\mathbb{R} - \mathbb{Z}$. Note $x \neq w$, but $f(x) = 0 = f(w)$.

5.91. Example. Define $g : \mathbb{Z}_{\geq 0} \to \mathbb{Z}_{\geq 0}$ by $g(n) = 6^n$. Let us show that g is injective but not surjective. For injectivity, fix $n, m \in \mathbb{Z}_{\geq 0}$ and assume $g(n) = g(m)$. We must prove $n = m$.

The assumption means $6^n = 6^m$, which is equivalent to $2^n 3^n = 2^m 3^m$. By uniqueness of prime factorizations, we conclude that $n = m$. So g is injective. To show that g is not surjective, choose $y_0 = 5$. We must prove $g(x) \neq 5$ for all $x \in \mathbb{Z}_{\geq 0}$. Consider two cases. If $x = 0$, then $g(x) = 1 \neq 5$. If $x > 0$, then $g(x) = 6^x \geq 6^1 = 6 > 5$, so again $g(x) \neq 5$. So g is not surjective. (Alternatively, we can see that $g(x) \neq 5$ by noting that $g(x) = 2^x 3^x = 5 = 5^1$ would violate the uniqueness of the prime factorization of 5.)

5.92. Example. Define $h : \mathbb{R} - \{3\} \to \mathbb{R} - \{2\}$ by $h(x) = \frac{2x+1}{x-3}$ for all $x \neq 3$. Let us prove that h is a bijection. First we show h is injective. Fix $x, z \in \mathbb{R} - \{3\}$; assume $h(x) = h(z)$; prove $x = z$. Our assumption means $\frac{2x+1}{x-3} = \frac{2z+1}{z-3}$. Using algebra to rewrite these fractions, we deduce

$$2 + \frac{7}{x-3} = 2 + \frac{7}{z-3}.$$

Subtracting 2 from both sides gives $\frac{7}{x-3} = \frac{7}{z-3}$. Dividing by 7 gives $\frac{1}{x-3} = \frac{1}{z-3}$. Multiplying by the nonzero numbers $z - 3$ and $x - 3$ yields $z - 3 = x - 3$. Finally, adding 3 to both sides gives $z = x$, so $x = z$ is indeed true.

Next, we show that h is surjective. Fix $y_0 \in \mathbb{R} - \{2\}$; we must find $x \in \mathbb{R} - \{3\}$ such that $h(x) = y_0$, or equivalently $2 + \frac{7}{x-3} = y_0$. To see which x to choose, we do a scratch calculation in which we solve for x in terms of y_0. We get $\frac{7}{x-3} = y_0 - 2$, then $\frac{x-3}{7} = \frac{1}{y_0-2}$, then $x - 3 = \frac{7}{y_0-2}$, then $x = 3 + \frac{7}{y_0-2}$. Returning to the official proof, we now choose $x = 3 + \frac{7}{y_0-2}$. Observe that this number is defined, since $y_0 \neq 2$. Furthermore, $\frac{7}{y_0-2}$ is nonzero, and so x is unequal to 3. This proves that x does belong to the domain $\mathbb{R} - \{3\}$ of h. Finally, we conclude the proof by computing

$$h(x) = \frac{2(3 + 7/(y_0 - 2)) + 1}{(3 + 7/(y_0 - 2)) - 3} = \frac{7 + 14/(y_0 - 2)}{7/(y_0 - 2)} = \left(\frac{7(y_0 - 2) + 14}{y_0 - 2} \right) \cdot \left(\frac{y_0 - 2}{7} \right) = y_0.$$

Properties of Injections, Surjections, and Bijections

We conclude this section with a fundamental theorem showing how composition of functions is related to the properties of being injective, surjective, or bijective.

5.93. Theorem on Injections, Surjections, and Bijections.
For all functions $f : X \to Y$ and $g : Y \to Z$:
(a) If f is injective and g is injective, then $g \circ f : X \to Z$ is injective.
(b) If f is surjective and g is surjective, then $g \circ f : X \to Z$ is surjective.
(c) If f is bijective and g is bijective, then $g \circ f : X \to Z$ is bijective.
(d) If $g \circ f$ is injective, then f is injective (but g may or may not be injective).
(e) If $g \circ f$ is surjective, then g is surjective (but f may or may not be surjective).

We prove parts (a), (b), and (e) here, asking you to prove (c) and (d) in the exercises. For all proofs, fix arbitrary functions $f : X \to Y$ and $g : Y \to Z$; we know $g \circ f : X \to Z$ is also a function, and $(g \circ f)(x) = g(f(x))$ for all $x \in X$.

5.94. Proof of (a). Assume f is one-to-one and g is one-to-one; prove $g \circ f$ is one-to-one. We have assumed

$$\forall x_1 \in X, \forall x_2 \in X, f(x_1) = f(x_2) \Rightarrow x_1 = x_2 \tag{5.3}$$

and

$$\forall y_1 \in Y, \forall y_2 \in Y, g(y_1) = g(y_2) \Rightarrow y_1 = y_2. \tag{5.4}$$

Keeping in mind that $g \circ f$ has domain X, we must prove

$$\forall w_1 \in X, \forall w_2 \in X, (g \circ f)(w_1) = (g \circ f)(w_2) \Rightarrow w_1 = w_2. \tag{5.5}$$

Fix arbitrary $w_1 \in X$ and $w_2 \in X$. Assume $(g \circ f)(w_1) = (g \circ f)(w_2)$. Prove $w_1 = w_2$. We have assumed $g(f(w_1)) = g(f(w_2))$. We intend to use (5.4) and the Inference Rule for ALL, taking $y_1 = f(w_1)$ and $y_2 = f(w_2)$. These choices are legitimate, because $f(w_1) \in Y$ and $f(w_2) \in Y$ (recall f maps X into Y). Now our assumption $g(f(w_1)) = g(f(w_2))$ becomes $g(y_1) = g(y_2)$. By the Inference Rule for IF and (5.4), $y_1 = y_2$ follows. This means $f(w_1) = f(w_2)$. Now we use the rules for ALL and IF again, taking $x_1 = w_1 \in X$ and $x_2 = w_2 \in X$ in (5.3). We deduce $w_1 = w_2$, which is the goal.

Here is a less formal, but potentially more intuitive, contrapositive proof of the same result. Start with two arbitrary unequal inputs $w_1 \neq w_2$ in X. Since f is injective, $y_1 = f(w_1)$ and $y_2 = f(w_2)$ are unequal intermediate outputs in Y. Since g is injective, $g(f(w_1))$ and $g(f(w_2))$ are unequal final outputs in Z. So, $w_1 \neq w_2$ implies $(g \circ f)(w_1) \neq (g \circ f)(w_2)$, which means that $g \circ f$ is one-to-one.

5.95. Proof of (b). Assume $f : X \to Y$ is onto and $g : Y \to Z$ is onto; prove $g \circ f : X \to Z$ is onto. We have assumed $\forall y \in Y, \exists x \in X, y = f(x)$ and $\forall z \in Z, \exists y \in Y, z = g(y)$. We must prove $\forall z \in Z, \exists x \in X, z = (g \circ f)(x)$. To do so, fix an arbitrary $z_0 \in Z$. By our assumption on g, there is a fixed object $y_0 \in Y$ with $z_0 = g(y_0)$. Now by our assumption on f, there is a fixed object $x_0 \in X$ with $y_0 = f(x_0)$. We see that $z_0 = g(y_0) = g(f(x_0)) = (g \circ f)(x_0)$. Thus to prove $\exists x \in X, z_0 = (g \circ f)(x)$, we may choose $x = x_0$.

5.96. Proof of (e). Assume $g \circ f$ is surjective. Prove g is surjective. We have assumed $\forall z \in Z, \exists x \in X, z = (g \circ f)(x)$. We must prove $\forall z \in Z, \exists y \in Y, z = g(y)$. Fix arbitrary $z_0 \in Z$. Prove $\exists y \in Y, z_0 = g(y)$. Using our assumption, we know there is some object $x_0 \in X$ with $z_0 = (g \circ f)(x_0)$. This means $z_0 = g(f(x_0))$. Comparing to the goal, we may choose $y_0 = f(x_0)$ to make $z_0 = g(y_0)$ be true. Observe that y_0 does lie in the required set Y, since f is a function with codomain Y.

To justify the parenthetical remark in (e), we give an example where $g \circ f$ is surjective but f is not surjective. Let $X = \{1, 2\}$, $Y = \{1, 2, 3\}$, and $Z = \{4, 5\}$. Define $f : X \to Y$ by $f(1) = 1$ and $f(2) = 2$. Define $g : Y \to Z$ by $g(1) = 4$ and $g(2) = 5$ and $g(3) = 5$. Now $g \circ f : X \to Z$ satisfies $(g \circ f)(1) = 4$ and $(g \circ f)(2) = 5$. It is immediate that $g \circ f$ is surjective (in fact, bijective), yet f is not surjective since $3 \in Y$ does not have the form $f(x)$ for any $x \in X$.

Section Summary

1. *Injective Functions.* A function $f : X \to Y$ is injective (one-to-one) iff for all $w, x \in X$, if $f(w) = f(x)$, then $w = x$. In an arrow diagram for an injective function, every output gets hit by at most one arrow. A real-valued function is injective iff every horizontal line $y = y_0$ intersects the graph in at most one point.

2. *Surjective Functions.* A function $f : X \to Y$ is surjective (onto) iff $\forall y \in Y, \exists x \in X, y = f(x)$. In an arrow diagram for a surjective function, every output gets hit by at least one arrow. A real-valued function is surjective iff every horizontal line $y = y_0$ intersects the graph in at least one point.

3. *Bijective Functions.* A function $f : X \to Y$ is bijective iff f is injective and surjective iff $\forall y \in Y, \exists! x \in X, y = f(x)$. In an arrow diagram for a bijective function, every output gets hit by exactly one arrow. A real-valued function is bijective iff every horizontal line $y = y_0$ intersects the graph in exactly one point.

4. *Properties of Injections, Surjections, and Bijections.* The composition of two injective functions is injective. The composition of two surjective functions is surjective. The composition of two bijective functions is bijective. If $g \circ f$ is injective, then f is injective. If $g \circ f$ is surjective, then g is surjective.

Exercises

1. Let $A = \{1, 2, 3, 4\}$, $B = \{v, w, x, y, z\}$, and consider the following four functions:
   $f : A \rightarrow B$ with graph $\{(1, w), (2, z), (3, y), (4, w)\}$;
   $g : B \rightarrow A$ with graph $\{(v, 4), (w, 3), (y, 1), (x, 2), (z, 4)\}$;
   $h : A \rightarrow A$ with graph $\{(1, 4), (2, 4), (3, 1), (4, 3)\}$;
   $k : B \rightarrow B$ with graph $\{(v, z), (z, y), (y, v), (w, x), (x, w)\}$.
   Decide whether each function is injective, surjective, or bijective.

2. For each of the following functions $f : \mathbb{R} \rightarrow \mathbb{R}$, is f injective? surjective? bijective?
   (a) $f(x) = \cos x$. (b) $f(x) = e^{-x}$. (c) $f(x) = x^4$. (d) $f(x) = \sqrt[5]{x}$. (e) $f(x) = 1/x$
   for $x \neq 0$, and $f(0) = 0$. (f) $f(x) = \lceil x \rceil$ (x rounded up to the nearest integer).

3. Let $a \neq 0$ and b be real constants. Prove that $f : \mathbb{R} \rightarrow \mathbb{R}$, given by $f(x) = ax + b$
   for $x \in \mathbb{R}$, is a bijection.

4. Define $g : \mathbb{R}_{\neq 0} \rightarrow \mathbb{R}_{\neq 0}$ by $g(x) = 1/x$ for nonzero real x. Prove that g is a
   bijection.

5. Define $f : \mathbb{R} \rightarrow \mathbb{R}$ by $f(x) = \cos x$ and define $g : \mathbb{R} \rightarrow [-1, 1]$ by $g(x) = \cos x$.
   Is each of the following functions injective? surjective? bijective? (a) f (b) g
   (c) $f|_{[0,\pi]}$ (d) $g|_{[0,\pi]}$ (e) $g|_{[0,2\pi]}$ (f) $g|_{[3\pi/2, 5\pi/2]}$ (g) $f \circ f$ (h) $g \circ (f|_{[0,1]})$.

6. Give an example of a function $f : \mathbb{Z}_{\geq 0} \rightarrow \mathbb{Z}_{\geq 0}$ with the stated properties. Prove
   that your examples work. (a) f is injective but not surjective. (b) f is surjective
   but not injective. (c) f is not injective, not surjective, and not constant. (d) f is
   injective and surjective, but f is not the identity function.

7. Determine whether each of the following functions is: (i) injective; (ii) surjective;
   (iii) bijective. Prove your answers.
   (a) $f : \mathbb{R} \rightarrow \mathbb{R}$ defined by $f(x) = x^3 - x$.
   (b) $g : \mathbb{R} \rightarrow [1, \infty)$ defined by $g(x) = e^{x^2}$.
   (c) $h : \mathbb{R} - \{3/7\} \rightarrow \mathbb{R} - \{1/7\}$ defined by $h(x) = \frac{x+5}{7x-3}$.
   (d) $k : \mathbb{R}_{\neq 0} \rightarrow \mathbb{R}$ defined by $k(x) = \ln|x|$.
   (e) $p : (0, \infty) \rightarrow (0, \infty)$ defined by $p(x) = x^x$.

8. Determine whether each of the following functions is: (i) injective; (ii) surjective;
   (iii) bijective. Explain informally.
   (a) $p : \mathbb{Z}_{\geq 0} \rightarrow \mathbb{Z}_{\geq 0}$ defined by $p(n) = n^2$ for $n \in \mathbb{Z}_{\geq 0}$.
   (b) $q : \mathbb{Z} \rightarrow \mathbb{Z}$ defined by $q(n) = n^2$ for $n \in \mathbb{Z}$.
   (c) $r : \mathbb{R} \rightarrow \mathbb{R}$ defined by $r(x) = x^2$ for $x \in \mathbb{R}$.
   (d) $s : \mathbb{Z} \rightarrow \mathbb{Z}$ defined by $s(m) = m^3$ for $m \in \mathbb{Z}$.
   (e) $t : \mathbb{R} \rightarrow \mathbb{R}$ defined by $t(x) = x^3$ for $x \in \mathbb{R}$.
   (f) $u : \mathbb{C} \rightarrow \mathbb{C}$ defined by $u(z) = z^3$ for $z \in \mathbb{C}$.

9. Let $f : X \rightarrow Y$ be a function with graph F. Prove carefully that f is injective iff
   for all x, y, z, $(x, y) \in F$ and $(z, y) \in F$ implies $x = z$.

10. (a) Is the function $(\emptyset, \emptyset, \emptyset)$ injective? surjective? bijective? (b) Repeat part (a) for
    the function $(\emptyset, B, \emptyset)$, where B is a nonempty set.

11. For each part, find functions $f, g : \mathbb{R} \rightarrow \mathbb{R}$ satisfying the stated conditions.
    (a) f and g are injective, but $f + g$ is not injective.
    (b) f and g are not injective, but $f + g$ is injective.
    (c) f and g are surjective, but $f + g$ is not surjective.
    (d) f and g are not surjective, but $f + g$ is surjective.
    (e) f and g are bijective, but $f \cdot g$ is not bijective.
    (f) f and g are not bijective, but $f \cdot g$ is bijective.

12. Prove parts (c) and (d) of Theorem 5.93.

13. Prove: for all $f : X \to Y$, f is injective iff for all $A, C \subseteq X$, $f[A \cap C] = f[A] \cap f[C]$.

14. Prove: for all $f : X \to Y$, f is injective iff for all $A, C \subseteq X$, $f[A - C] = f[A] - f[C]$.

15. Prove: for all $f : X \to Y$, f is injective iff for all $A \subseteq X$, $f^{-1}[f[A]] = A$.

16. Prove: for all $f : X \to Y$, f is surjective iff for all $B \subseteq Y$, $f[f^{-1}[B]] = B$.

17. Suppose $h : X \to X$ is a bijection. (a) Prove: for all functions $f : X \to Y$, f is injective iff $f \circ h$ is injective; f is surjective iff $f \circ h$ is surjective; and f is bijective iff $f \circ h$ is bijective. (b) Prove: for all functions $g : W \to X$, g is injective iff $h \circ g$ is injective; g is surjective iff $h \circ g$ is surjective; and g is bijective iff $h \circ g$ is bijective. (c) Which implications in (a) and (b) remain true if h is only assumed to be an injection? (d) Which implications in (a) and (b) remain true if h is only assumed to be a surjection?

18. (a) Use uniqueness of prime factorization to show that $g : \mathbb{Z}_{\geq 0} \times \mathbb{Z}_{\geq 0} \to \mathbb{Z}_{\geq 0}$ defined by $g(a, b) = 2^a 3^b$ is injective. (b) Generalize this idea to define an injection $h_k : \mathbb{Z}_{\geq 0}^k \to \mathbb{Z}_{\geq 0}$ for any fixed integer $k \geq 1$. (c) Define $f : \mathbb{Z}_{\geq 0} \times \mathbb{Z}_{\geq 0} \to \mathbb{Z}_{\geq 0}$ by $f(a, b) = 2^a(2b + 1) - 1$. Prove that f is a bijection. (d) Use the function f in part (c) to recursively construct bijections $p_k : \mathbb{Z}_{\geq 0}^k \to \mathbb{Z}_{\geq 0}$ for all $k \geq 2$.

19. Suppose $f : X \to Y$ is a function and $S \subseteq X$.
    (a) Prove: if f is injective, then $f|_S$ is injective.
    (b) Prove: if $f|_S$ is surjective, then f is surjective.
    (c) Give an example to show that the converse of (a) can be false.
    (d) Give an example to show that the converse of (b) can be false.

20. Consider a function given by the formula $f(x) = \frac{ax+b}{cx+d}$ for some real constants a, b, c, d such that $ad - bc \neq 0$. (a) Find the largest possible domain of f contained in $\mathbb{R}$, and the smallest possible codomain corresponding to this domain. (Consider three cases: (i) $a = 0$; (ii) $c = 0$; (iii) $a \neq 0 \neq c$.) (b) Prove that the function f with domain and codomain found in part (a) is a bijection. (c) Where in your proofs did you use the assumption $ad - bc \neq 0$? What happens if $ad - bc = 0$?

21. Suppose $f : A \to C$ and $g : B \to C$ are gluable functions. Let $h = \text{glue}(f, g) : A \cup B \to C$. (a) Prove: If f or g is surjective, then h is surjective. (b) Prove: If h is injective, then f and g are injective. (c) Show that the converse of part (a) can be false. (d) Show that the converse of part (b) can be false.

22. Suppose $f : A \to B$, $g : C \to D$, and $A \cap C = B \cap D = \emptyset$. Let $h = \text{glue}(f, g) : A \cup C \to B \cup D$. (a) Prove: h is injective iff f and g are injective. (b) Prove: h is surjective iff f and g are surjective. (c) Prove: h is bijective iff f and g are bijective. (d) What can be said if $B \cap D \neq \emptyset$? (e) What happens if $A \cap C \neq \emptyset$?

23. Suppose $f : X \to Y$ is a function. Prove: f is injective iff for all sets W and all functions $g, h : W \to X$, $f \circ g = f \circ h$ implies $g = h$.

24. Suppose $f : X \to Y$ is a function. Prove: f is surjective iff for all sets Z and all functions $g, h : Y \to Z$, $g \circ f = h \circ f$ implies $g = h$.

5.9   Inverse Functions

In calculus, it is often helpful to pass from a function f to the inverse function f^{-1}. The idea is that the inverse function interchanges the roles of inputs and outputs, so that for all inputs x and outputs y for f, $y = f(x)$ iff $x = f^{-1}(y)$. A key subtlety is that many functions do not have inverse functions (in contrast to relations, where the inverse of a relation R is always another relation). It turns out that the inverse of a function f exists iff f is a bijection. The optional second half of this section introduces left inverses, right inverses, and their connections to injective and surjective functions.

The Inverse of a Function

The idea of interchanging inputs and outputs leads to the following definition of inverse functions.

5.97. Definition: Inverse Function. Given a function $f : X \to Y$, an *inverse function* for f is a function $g : Y \to X$ satisfying $\boxed{\forall x \in X, \forall y \in Y, y = f(x) \Leftrightarrow x = g(y).}$

When such a function g exists, we say that f is *invertible* and write $g = f^{-1}$.

The next example shows that *inverse functions do not always exist*. But if a function does have an inverse, it is unique, as we show later. This justifies the introduction of the notation f^{-1} for the inverse function.

5.98. Example. Define $f : \{1, 2, 3\} \to \{a, b, c\}$ by $f(1) = b$, $f(2) = c$, and $f(3) = a$. An inverse function for f is the function $g : \{a, b, c\} \to \{1, 2, 3\}$ given by $g(a) = 3$, $g(b) = 1$, and $g(c) = 2$. On the other hand, consider $h : \{1, 2, 3\} \to \{a, b\}$ given by $h(1) = b$, $h(2) = a$, and $h(3) = b$. We claim h has no inverse function. To get a contradiction, assume $k : \{a, b\} \to \{1, 2, 3\}$ is an inverse for h. Since $h(1) = b$, we must have $k(b) = 1$. Since $h(3) = b$, we must also have $k(b) = 3$. Now k is not a function, since the input b is associated with two different outputs 1 and 3. Similarly, consider $p : \{1, 2\} \to \{a, b, c\}$ given by $p(1) = a$ and $p(2) = c$. If $q : \{a, b, c\} \to \{1, 2\}$ were an inverse function for p, we must have $q(b) = 1$ or $q(b) = 2$. On one hand, if $q(b) = 1$, we deduce that $p(1) = b$, which is false. On the other hand, if $q(b) = 2$, we see that $p(2) = b$, which is false. So p has no inverse function.

The next theorem shows how the concept of an inverse function is related to composition of functions and identity functions. The condition in this theorem is sometimes used as the definition of inverse functions.

5.99. Theorem on Inverses and Composition. For all functions $f : X \to Y$ and $g : Y \to X$, g is an inverse function for f iff $f \circ g = \mathrm{Id}_Y$ and $g \circ f = \mathrm{Id}_X$.

Proof. Fix functions $f : X \to Y$ and $g : Y \to X$. <u>Part 1.</u> Assume g is an inverse function for f; prove $f \circ g = \mathrm{Id}_Y$ and $g \circ f = \mathrm{Id}_X$. Both $f \circ g$ and Id_Y are functions from Y to Y. To prove they are equal, we fix $y \in Y$ and show $(f \circ g)(y) = \mathrm{Id}_Y(y)$. Write $x = g(y) \in X$. By definition of inverse functions, we know $y = f(x)$. Now compute $(f \circ g)(y) = f(g(y)) = f(x) = y = \mathrm{Id}_Y(y)$. A similar calculation shows that $g \circ f = \mathrm{Id}_X$.

<u>Part 2.</u> Assume $f \circ g = \mathrm{Id}_Y$ and $g \circ f = \mathrm{Id}_X$; prove $\forall x \in X, \forall y \in Y, y = f(x) \Leftrightarrow x = g(y)$. Fix $x \in X$ and $y \in Y$. *Part 2a.* Assume $y = f(x)$; prove $x = g(y)$. We compute $x = \mathrm{Id}_X(x) = (g \circ f)(x) = g(f(x)) = g(y)$, as needed. *Part 2b.* Assume $x = g(y)$; prove $y = f(x)$. Here, we compute $y = \mathrm{Id}_Y(y) = (f \circ g)(y) = f(g(y)) = f(x)$, as needed. $\square$

5.100. Corollary: Uniqueness of Inverse Functions. Every $f : X \to Y$ has at most one inverse function.

Proof. Fix $f : X \to Y$; assume $g : Y \to X$ and $h : Y \to X$ are both inverse functions for f; prove $g = h$. [The direct approach is to fix $y \in Y$ and prove $g(y) = h(y)$. But here is a faster way using Theorem 5.99 and the identity and associativity properties from Theorem 5.62.] We compute $g = g \circ \mathrm{Id}_Y = g \circ (f \circ h) = (g \circ f) \circ h = \mathrm{Id}_X \circ h = h$. $\qquad\square$

Properties of Inverse Functions

We are about to show that a function f is invertible iff f is bijective. First we give an intuitive version of the proof based on arrow diagrams. Given a function $f : X \to Y$, we know each $x \in X$ has exactly one arrow leaving it. We attempt to form the inverse function $g : Y \to X$ by reversing all the arrows. In order for g to be a function, it must be true that each $y \in Y$ has exactly one g-arrow leaving it. Since g-arrows are the reversals of f-arrows, this means that each $y \in Y$ should have exactly one f-arrow entering it. This is precisely the criterion for f to be a bijection. The next proof formalizes this argument.

5.101. Theorem on Inverses and Bijections. For all $f : X \to Y$, f has an inverse function iff f is a bijection.

Proof. Fix $f : X \to Y$. <u>Part 1.</u> Assume f has an inverse function $g : Y \to X$. Prove f is a bijection, which means f is surjective and injective. By Theorem 5.99, we know $g \circ f = \mathrm{Id}_X$ and $f \circ g = \mathrm{Id}_Y$. Now Id_X is injective, so $g \circ f$ is injective, so f is injective by Theorem 5.93(d). Similarly, Id_Y is surjective, so $f \circ g$ is surjective, so f is surjective by Theorem 5.93(e).

    <u>Part 2.</u> Assume f is a bijection (hence surjective and injective); prove f has an inverse function. Let $F = \{(x, f(x)) : x \in X\}$ be the graph of f. Define $g = (Y, X, F^{-1})$, where $F^{-1} = \{(y, x) : (x, y) \in F\}$ is the inverse of the relation F. We prove below that g really is a function (with domain Y and codomain X). Granting this fact, we can then conclude that for all $x \in X$ and $y \in Y$, $x = g(y)$ iff $(y, x) \in F^{-1}$ iff $(x, y) \in F$ iff $y = f(x)$, so that g is the required inverse function for f. To prove that g is a function, we check the three conditions appearing after Definition 5.35. First, F is a relation from X to Y (i.e., a subset of $X \times Y$), so we know F^{-1} is a relation from Y to X (i.e., a subset of $Y \times X$).

    Second, to prove Existence of Outputs for g, we must prove $\forall y \in Y, \exists x \in X, (y, x) \in F^{-1}$. Note $(y, x) \in F^{-1}$ iff $(x, y) \in F$ iff $y = f(x)$. So we must prove $\forall y \in Y, \exists x \in X, y = f(x)$. This is true since our assumption guarantees f is surjective.

    Third, to prove Uniqueness of Outputs for g, we must prove for all y, x, w, if $(y, x) \in F^{-1}$ and $(y, w) \in F^{-1}$, then $x = w$. Fix y, x, w. Assume $(y, x) \in F^{-1}$ and $(y, w) \in F^{-1}$; prove $x = w$. Our assumption can be rewritten $(x, y) \in F$ and $(w, y) \in F$, i.e., $y = f(x)$ and $y = f(w)$. Then $f(x) = f(w)$, so $x = w$ because f is injective. $\qquad\square$

5.102. Example. The function $f : \mathbb{R} \to \mathbb{R}$, given by $f(x) = e^x$ for $x \in \mathbb{R}$, is injective but not surjective, so it has no inverse. By restricting the codomain, we can obtain a bijective function $\exp : \mathbb{R} \to \mathbb{R}_{>0}$ that does have an inverse. The inverse is the natural logarithm function $\ln : \mathbb{R}_{>0} \to \mathbb{R}$. In this setting, Theorem 5.99 states that $\ln(e^x) = x$ for all $x \in \mathbb{R}$, and $e^{\ln y} = y$ for all $y \in \mathbb{R}_{>0}$.

    The function $f : \mathbb{R} \to \mathbb{R}$ given by $f(x) = x^2$ for $x \in \mathbb{R}$ is not a bijection, so it has no inverse. On the other hand, $f_1 : \mathbb{R}_{\geq 0} \to \mathbb{R}_{\geq 0}$ given by $f_1(x) = x^2$ for $x \in \mathbb{R}_{\geq 0}$ is a bijection; the inverse function is $f_1^{-1}(y) = \sqrt{y}$ for $y \in \mathbb{R}_{\geq 0}$. The function $f_2 : \mathbb{R}_{\leq 0} \to \mathbb{R}_{\geq 0}$ given by $f_2(x) = x^2$ for $x \in \mathbb{R}_{\leq 0}$ is also a bijection; the inverse satisfies $f_2^{-1}(y) = -\sqrt{y}$ for $y \in \mathbb{R}_{\geq 0}$.

Next consider $h : \mathbb{R} \to \mathbb{R}$ given by $h(x) = \sin x$. The function h is neither injective nor surjective, so it has no inverse. By restricting the domain and codomain of h, we can create a new bijective function $S : [-\pi/2, \pi/2] \to [-1, 1]$ given by $S(x) = \sin x$ for $x \in [-\pi/2, \pi/2]$. The inverse function $S^{-1} : [-1, 1] \to [-\pi/2, \pi/2]$ is called the *arcsine function* in calculus.

5.103. Theorem on Invertible Functions. For all functions $f : X \to Y$ and $g : Y \to Z$:

(a) $\mathrm{Id}_X : X \to X$ is invertible, and $\mathrm{Id}_X^{-1} = \mathrm{Id}_X$.

(b) If f is invertible, then f^{-1} is invertible and $(f^{-1})^{-1} = f$.

(c) If f and g are invertible, then $g \circ f$ is invertible and $(g \circ f)^{-1} = f^{-1} \circ g^{-1}$.

(d) If f is a bijection, then f^{-1} exists and is a bijection.

Proof. Fix functions $f : X \to Y$ and $g : Y \to Z$. We prove each part using the criterion in Theorem 5.99. For (a), we take $Y = X$ and $f = g = \mathrm{Id}_X$ in that theorem. Since "$\mathrm{Id}_X \circ \mathrm{Id}_X = \mathrm{Id}_X$ and $\mathrm{Id}_X \circ \mathrm{Id}_X = \mathrm{Id}_X$" is true, the theorem tells us that Id_X is the inverse function for Id_X. To prove (b), assume f is invertible. We know "$f^{-1} \circ f = \mathrm{Id}_X$ and $f \circ f^{-1} = \mathrm{Id}_Y$" is true. Now use Theorem 5.99, replacing X by Y, Y by X, f by f^{-1}, and g by f. We deduce that f is an inverse function for f^{-1}, i.e., $(f^{-1})^{-1} = f$. To prove (c), assume f and g are invertible. By Theorem 5.99 (replacing Y by Z, f by $g \circ f$, and g by $f^{-1} \circ g^{-1}$), it suffices to prove that $(g \circ f) \circ (f^{-1} \circ g^{-1}) = \mathrm{Id}_Z$ and $(f^{-1} \circ g^{-1}) \circ (g \circ f) = \mathrm{Id}_X$. To prove the first equation, use associativity of function composition to compute

$$(g \circ f) \circ (f^{-1} \circ g^{-1}) = g \circ (f \circ (f^{-1} \circ g^{-1})) = g \circ ((f \circ f^{-1}) \circ g^{-1}) = g \circ (\mathrm{Id}_Y \circ g^{-1}) = g \circ g^{-1} = \mathrm{Id}_Z .$$

The second equation is proved similarly. Finally, part (d) follows from part (b) and the fact that f is invertible iff f is bijective. $\qquad\square$

Left Inverses (Optional)

Theorem 5.99 tells us that g is an inverse function for f iff $f \circ g$ and $g \circ f$ are both identity functions. We now investigate what happens when only one of these conditions holds.

5.104. Definition: Left Inverses. For all functions $f : X \to Y$ and $g : Y \to X$, g is a *left inverse for f* iff $g \circ f = \mathrm{Id}_X$.

5.105. Example. Figure 5.10 gives arrow diagrams for two functions f and h. A left inverse for f is a function $g : \{a, b, c, d, e\} \to \{1, 2, 3\}$ satisfying $(g \circ f)(x) = \mathrm{Id}_X(x) = x$ for all $x \in \{1, 2, 3\}$. Taking $x = 1, 2, 3$, we see that g must satisfy $g(d) = 1$, $g(b) = 2$, $g(a) = 3$. However, no requirement has been imposed on $g(c)$ and $g(e)$, since c and e are not in the image of f. We obtain one left inverse for f by setting $g(c) = 1$ and $g(e) = 1$. We could build other left inverses for f by choosing other values for $g(c)$ and $g(e)$. In this example, f has nine different left inverses. So, unlike the two-sided inverses studied earlier, *left inverses need not be unique*. Also, no left inverse g can satisfy $f \circ g = \mathrm{Id}_Y$, since f is not a bijection. For instance, the g chosen above has $(f \circ g)(c) = f(g(c)) = f(1) = d \neq c = \mathrm{Id}_Y(c)$.

Furthermore, *left inverses do not always exist*. Suppose we try to find a left inverse $g : \{a, b, c\} \to \{1, 2, 3, 4, 5, 6\}$ for the function h in the figure. Writing out the requirement $g \circ h = \mathrm{Id}_X$, we obtain (among other conditions) $g(h(2)) = 2$ and $g(h(5)) = 5$. Now $h(2) = c = h(5)$, so we must have $g(c) = 2$ and $g(c) = 5$. But then g is not a function, since input c has two associated outputs. This shows that h does not have any left inverses. The trouble arises because h was not injective: distinct inputs 2 and 5 to h both mapped to the same output c.

The previous example illustrates the following characterization of which functions have left inverses.

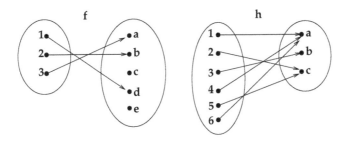

FIGURE 5.10
Arrow diagrams of two functions.

5.106. Theorem on Left Inverses. Let X be a nonempty set. For all functions $f : X \to Y$, f has a left inverse iff f is injective.

Proof. Fix $f : X \to Y$, where X is nonempty. First assume f has a left inverse $g : Y \to X$. Since $g \circ f = \mathrm{Id}_X$ is an injective function, Theorem 5.93(d) tells us that f is injective.

Conversely, assume f is injective; we must construct a left inverse $g : Y \to X$. To do so, let $Z = f[X] \subseteq Y$ by the image of f, and define $f_1 : X \to Z$ by $f_1(x) = f(x)$ for all $x \in X$. The function f_1 (obtained by restricting the codomain of f) is injective and surjective, so it has a two-sided inverse $g_1 : Z \to X$. In particular, $g_1 \circ f_1 = \mathrm{Id}_X$. To build the function $g : Y \to X$, select one fixed element $x_0 \in X$; this is possible since X is nonempty. Now define $g(y) = g_1(y)$ for $y \in Z$, and $g(y) = x_0$ for $y \in Y - Z$. You can check that g really is a function from Y to X. For instance, this is a special case of the gluing construction in Definition 5.68; here we are gluing together g_1 and the constant function with domain $Y - Z$ and constant value x_0.

To finish, we must check that $g \circ f = \mathrm{Id}_X$. Fix $x \in X$, and set $y = f(x)$, which is in the image Z of f. We now compute $(g \circ f)(x) = g(f(x)) = g(y) = g_1(y) = g_1(f_1(x)) = (g_1 \circ f_1)(x) = \mathrm{Id}_X(x)$. $\qquad\square$

Right Inverses (Optional)

5.107. Definition: Right Inverses. For all functions $f : X \to Y$ and $g : Y \to X$, g is a *right inverse for* f iff $f \circ g = \mathrm{Id}_Y$.

5.108. Example. Consider once again the functions f and h in Figure 5.10. A function $g : \{a, b, c\} \to \{1, 2, 3, 4, 5, 6\}$ is a right inverse for h iff $h \circ g = \mathrm{Id}_{\{a,b,c\}}$ iff $h(g(a)) = a$ and $h(g(b)) = b$ and $h(g(c)) = c$. Now, $h(x) = a$ is true when x is 1, 4, or 6. So we could define $g(a)$ to be any one of these three values and thereby satisfy $h(g(a)) = a$. Similarly, $h(x) = b$ is solved by $x = 3$, and $h(x) = c$ is true when x is 2 or 5. Thus, g is a right inverse for h iff $g(a) \in \{1, 4, 6\} = h^{-1}[\{a\}]$ and $g(b) \in \{3\} = h^{-1}[\{b\}]$ and $g(c) \in \{2, 5\} = h^{-1}[\{c\}]$. For example, one right inverse is given by $g(a) = 1$, $g(b) = 3$, and $g(c) = 2$. Another right inverse is given by $g_1(a) = 4$, $g_1(b) = 3$, and $g_1(c) = 5$. In this example, there are $3 \cdot 1 \cdot 2 = 6$ possible right inverses for h. So, *right inverses need not be unique*. Also, no right inverse g can satisfy $g \circ h = \mathrm{Id}_X$, since h is not a bijection. For instance, the right inverse g chosen above has $(g \circ h)(4) = g(h(4)) = g(a) = 1 \neq 4 = \mathrm{Id}_X(4)$.

Furthermore, *right inverses do not always exist*. Suppose we try to find a right inverse $g : \{a, b, c, d, e\} \to \{1, 2, 3\}$ for the function f in the figure. Writing out the requirement $f \circ g = \mathrm{Id}_Y$, we obtain (among other conditions) $f(g(c)) = c$. But there is no x in the domain of f solving $f(x) = c$, since $f^{-1}[\{c\}] = \emptyset$, so it is impossible to define $g(c)$ in a way that makes $f(g(c)) = c$ true. So f does not have any right inverses. This happened because f was not surjective: there exist points (such as c and e) in the codomain of f that do not have the form $f(x)$ for any x in the domain of f.

The previous example illustrates the following characterization of which functions have right inverses.

5.109. Theorem on Right Inverses. For all functions $f : X \to Y$, f has a right inverse iff f is surjective.

Proof. Fix $f : X \to Y$. First assume f has a right inverse $g : Y \to X$. Since $f \circ g = \mathrm{Id}_Y$ is a surjective function, Theorem 5.93(e) tells us that f is surjective.

Conversely, assume f is surjective; we must construct a right inverse $g : Y \to X$. Informally, we can build g as follows. For each $y \in Y$, we know that the preimage $f^{-1}[\{y\}]$ is a *nonempty* subset of X, because f is surjective. Define $g(y)$ to be any particular element of this nonempty set (chosen arbitrarily). To check that $f \circ g = \mathrm{Id}_Y$, fix $y \in Y$, and compute $(f \circ g)(y) = f(g(y)) = y = \mathrm{Id}_Y(y)$. The equality $f(g(y)) = y$ is true because $g(y) \in f^{-1}[\{y\}]$. Thus the proof appears to be complete.

[This paragraph should be considered optional.] There is a technical subtlety, however. To give a completely rigorous definition of g, we must construct the graph of g, which is a certain set of ordered pairs $G = \{(y, g(y)) : y \in Y\}$. To prove that this set exists working within axiomatic set theory, we must use the Axiom of Choice (see §3.7). Specifically, let Z be the collection of all sets $f^{-1}[\{y\}]$, for $y \in Y$. The set Z exists by the Power Set Axiom and the Axiom of Specification, since we can write $Z = \{S \in \mathcal{P}(X) : \exists y \in Y, S = f^{-1}[\{y\}]\}$. Since f is surjective, the empty set is not a member of Z. As discussed in §3.7, the Axiom of Choice provides us with a set $G' \subseteq Z \times X$ such that for all $(S, x) \in G'$, $x \in S$, and for each $S \in Z$, there exists a unique $x \in X$ with $(S, x) \in G'$. (The set G' is the formal mechanism by which we choose one element x from each nonempty set $S = f^{-1}[\{y\}]$.) Now define $G = \{(y, x) \in Y \times X : \exists S \in Z, S = f^{-1}[\{y\}] \wedge (S, x) \in G'\}$. It must now be checked that $g = (Y, X, G)$ is a function; the key point is that for each $y \in Y$, there exists a unique $S \in Z$ with $S = f^{-1}[\{y\}]$, and hence there exists a unique $x \in X$ with $(y, x) \in G$. Finally, by definition of G and Z, $(y, x) \in G$ implies $x \in f^{-1}[\{y\}]$, so that $f(g(y)) = f(x) = y$ for each $y \in Y$. □

Section Summary

1. *Inverse Functions.* Given $f : X \to Y$, a function $g : Y \to X$ is an inverse function for f iff $\forall x \in X, \forall y \in Y, y = f(x) \Leftrightarrow x = g(y)$. Equivalently, g is an inverse function for f iff $f \circ g = \mathrm{Id}_Y$ and $g \circ f = \mathrm{Id}_X$. Inverse functions are unique when they exist; the inverse of f is denoted f^{-1}.

2. *Properties of Invertibility.* A function $f : X \to Y$ is invertible iff f is a bijection. Identity functions are invertible. The inverse of an invertible function is invertible. The composition of invertible functions $f : X \to Y$ and $h : Y \to Z$ is invertible, and $(h \circ f)^{-1} = f^{-1} \circ h^{-1}$.

3. *Left Inverses.* A left inverse of $f : X \to Y$ is a function $g : Y \to X$ such that $g \circ f = \mathrm{Id}_X$. Assuming X is nonempty, f has a left inverse iff f is injective. Left inverses are not unique in general.

4. *Right Inverses.* A right inverse of $f : X \to Y$ is a function $g : Y \to X$ such that $f \circ g = \mathrm{Id}_Y$. A function f has a right inverse iff f is surjective. Right inverses are not unique in general.

Exercises

1. Let $A = \{1, 2, 3, 4\}$, $B = \{v, w, x, y, z\}$, and consider the following four functions:

   $f : A \to B$ defined by $f(1) = w$, $f(2) = z$, $f(3) = y$, $f(4) = w$;
   $g : B \to A$ defined by $g(v) = 4$, $g(w) = 3$, $g(y) = 1$, $g(x) = 2$, $g(z) = 4$;
   $h : A \to A$ defined by $h(1) = 4$, $h(2) = 4$, $h(3) = 1$, $h(4) = 3$;
   $k : B \to B$ defined by $k(v) = z$, $k(z) = y$, $k(y) = v$, $k(w) = x$, $k(x) = w$.

   Which functions have: (a) a left inverse? (b) a right inverse? (c) a two-sided inverse? Give an example of each type of inverse when it exists.

2. For each of the following functions from calculus, say whether the function has a left, right, or two-sided inverse. Give an example of each type of inverse when it exists.
   (a) $f : \mathbb{R} \to \mathbb{R}_{\geq 0}$ given by $f(x) = x^2$.
   (b) $f : \mathbb{R} \to \mathbb{R}$ given by $f(x) = \cos x$.
   (c) $f : \mathbb{R} \to [-1, 1]$ given by $f(x) = \cos x$.
   (d) $f : (0, \pi/2) \to \mathbb{R}$ given by $f(x) = \cos x$.
   (e) $f : [0, \pi] \to [-1, 1]$ given by $f(x) = \cos x$.
   (f) $f : \mathbb{R} \to \mathbb{R}$ given by $f(x) = x^3 - x$.
   (g) $f : \mathbb{R}_{\geq 0} \to \mathbb{R}$ given by $f(x) = 1/(1 + x^2)$.
   (h) $f : \mathbb{R}_{\leq 0} \to (0, 1]$ given by $f(x) = 1/(1 + x^2)$.

3. Let $A = \{1, 2, 3, 4, 5, 6\}$ and $B = \{w, x, y, z\}$. Consider the functions:

   $f : A \to B$ with graph $\{(1, x), (2, z), (3, w), (4, x), (5, z), (6, y)\}$;
   $g : B \to A$ with graph $\{(w, 4), (x, 6), (y, 1), (z, 3)\}$;
   $h : A \to A$ with graph $\{(1, 3), (2, 4), (3, 6), (4, 2), (5, 1), (6, 5)\}$;
   $k : B \to B$ with graph $\{(w, w), (x, w), (y, w), (z, z)\}$.

   (a) For each function, find all possible left inverses for the function, or explain why none exist. (b) Repeat part (a) for right inverses. (c) Repeat part (a) for two-sided inverses.

4. For each of the following functions, say whether the function has a left, right, or two-sided inverse. Give an example of each type of inverse when it exists.
   (a) $f : \mathbb{Z} \to \mathbb{Z}$ given by $f(n) = 2n$.
   (b) $f : \mathbb{Z}_{\geq 0} \to \mathbb{Z}_{\geq 0}$ given by $f(n) = n + 1$.
   (c) $f : \mathbb{Z} \to \mathbb{Z}$ given by $f(n) = n + 1$.
   (d) $f : \mathbb{Z} \to \mathbb{Z}$ given by $f(n) = -n$.
   (e) $f : \mathbb{Z}_{\geq 0} \to \mathbb{Z}_{\geq 0}$ given by $f(n) = n^2$.
   (f) $f : \mathbb{Z}_{\geq 0} \to \mathbb{Z}$ given by $f(n) = n$.
   (g) $f : \mathbb{Z}_{\geq 0} \to \mathbb{Z}_{\geq 0}$ given by $f(0) = 0$ and $f(n) = n - 1$ for $n > 0$.

5. Define $f : \mathbb{Z}_{\geq 0} \to \{0, 1, 2, \ldots, 9\}$ by letting $f(n)$ be the last digit in the base-10 expansion of n. For example, $f(1407) = 7$. Describe three different right inverses for f.

6. (a) Finish Part 1 of the proof of Theorem 5.99 by showing that $g \circ f = \mathrm{Id}_X$.
   (b) Finish the proof of part (c) of Theorem 5.103.

7. Let B be a fixed set, and let $f : \emptyset \to B$ be the empty function. Under what conditions does f have a left, right, or two-sided inverse?

8. Suppose $f : A \to \mathbb{R}$ is a function. Find necessary and sufficient conditions to ensure that there exists a function $g : A \to \mathbb{R}$ such that $f \cdot g = g \cdot f = 1$. (The

dot denotes pointwise product of functions, and 1 denotes the constant function on A with value 1.)

9. Suppose $f : \mathbb{R} \to \mathbb{R}$ is a function. Find necessary and sufficient conditions to ensure that there exists a function $g : \mathbb{R} \to \mathbb{R}$ such that $f \cdot g = g \cdot f = \mathrm{Id}_{\mathbb{R}}$.

10. Suppose $f : \mathbb{R} \to \mathbb{R}$ is a function. (a) Under what conditions does there exist $g : \mathbb{R} \to \mathbb{R}$ such that $f \circ g = 1$ (the constant function 1)? Describe all functions g that satisfy this property (when such g exist). (b) Under what conditions does there exist $g : \mathbb{R} \to \mathbb{R}$ such that $g \circ f = 1$? Describe all functions g that satisfy this property (when such g exist).

11. Suppose $f : X \to Y$ and $g, h : Y \to Z$ are functions. Prove: if f has a right inverse and $g \circ f = h \circ f$, then $g = h$. Give an example to show that this *right cancellation* property can fail if f does not have a right inverse.

12. Suppose $f : X \to Y$ and $g, h : W \to X$ are functions. Prove: if f has a left inverse and $f \circ g = f \circ h$, then $g = h$. Give an example to show that this *left cancellation* property can fail if f does not have a left inverse.

13. (a) Suppose $f : X \to Y$ and $g : Y \to Z$ both possess left inverses. Show that $g \circ f$ also has a left inverse. (b) Prove a similar result for right inverses.

14. Suppose $n \in \mathbb{Z}_{>0}$, $X_0, \ldots, X_n$ are sets, and $f_i : X_{i-1} \to X_i$ are bijections for $1 \le i \le n$. Use induction on n to prove that $(f_n \circ \cdots \circ f_1)^{-1}$ exists, and $(f_n \circ \cdots \circ f_2 \circ f_1)^{-1} = f_1^{-1} \circ f_2^{-1} \cdots \circ f_n^{-1}$.

15. Prove: for all sets X, Y and all $f : X \to Y$, $g : Y \to X$, g is a left inverse of f iff $(\forall x \in X, \forall y \in Y, y = f(x) \Rightarrow x = g(y))$.

16. Prove: for all sets X, Y and all $f : X \to Y$, $g : Y \to X$, g is a right inverse of f iff $(\forall x \in X, \forall y \in Y, x = g(y) \Rightarrow y = f(x))$.

17. Use right inverses to prove: for all nonempty sets X, Y, Z, all $f : X \to Y$, and all $g : Y \to Z$, if $g \circ f$ is one-to-one and f is onto, then g is one-to-one.

18. Use left inverses to prove: for all nonempty sets X, Y, Z, all $f : X \to Y$, and all $g : Y \to Z$, if $g \circ f$ is surjective and g is injective, then f is surjective.

19. Suppose f is an injection mapping an n-element set into an m-element set, where $m > n$. How many left inverses does f have?

20. Let $f : X \to Y$ be a surjection. Suppose $Y = \{y_1, y_2, \ldots, y_m\}$, and suppose that, for $1 \le i \le m$, there are exactly n_i elements $x \in X$ with $f(x) = y_i$. How many right inverses does f have?

21. Let X be an n-element set. How many bijections are there from X onto X?

22. Suppose $f : X \to Y$ is a function with graph F. Prove that the inverse relation F^{-1} is the graph of a function (not necessarily a function *from Y to X*) iff f is injective.

23. Suppose $f : X \to Y$ is a function with graph F. Prove that the inverse relation F^{-1} contains the graph of some function $g : Y \to X$ iff f is surjective.

24. Prove: if $f : X \to Y$ has exactly one right inverse g, then f is invertible and $g = f^{-1}$.

25. If $f : X \to Y$ has exactly one left inverse, must f be invertible? What if X has more than one element?

6

Equivalence Relations and Partial Orders

6.1 Reflexive, Symmetric, and Transitive Relations

In this chapter, we continue our study of relations. So far, we have been thinking of a relation R as a generalization of a function, where $(x, y) \in R$ means that the input x is associated with the output y under R. Now, we change our viewpoint slightly, interpreting $(x, y) \in R$ to mean that x is "related" to y in some way, without insisting on the notion that x is the "input" and y is the "output." For example, $(x, y) \in R$ might mean x is equal to y, or x is greater than y, or x is a subset of y, or x is a sibling of y, and so on. We begin this section by developing a new way of visualizing R, called a *digraph*, that emphasizes this "relational" viewpoint. Later, we will be especially interested in relations that share certain properties with the equality relation, called reflexivity, symmetry, and transitivity. We define these properties here and give many examples.

Digraph of a Relation

Previously, we looked at relations R from X to Y, where X was an "input set" and Y was an "output set." Now, we focus attention on a single set of objects X, thinking of R as specifying a relationship that may or may not hold between any two given objects in X. This leads to the following definitions.

6.1. Definition: Relation on a Set X. For all sets X and R, $\boxed{R \text{ is a relation on } X}$ iff $\boxed{R \subseteq X \times X}$. For all objects x and y, we write $\boxed{xRy}$ iff $\boxed{(x, y) \in R}$. The notation xRy is called *infix notation* and can be read "x is related to y under R."

6.2. Example. Let $X = \{1, 2, 3\}$. The following sets of ordered pairs are relations on X:

$$R_1 = \{(1, 1), (2, 2), (3, 3)\}; \quad R_2 = \{(1, 2), (2, 3), (1, 3)\}; \quad R_3 = R_1 \cup R_2;$$

$$R_4 = \{(1, 3), (2, 2), (3, 1)\}; \quad R_5 = \{(1, 1), (1, 2), (1, 3)\}; \quad R_6 = \{(1, 3), (2, 3)\}.$$

Relation R_1 was formerly called the identity relation I_X on the set X. Now, we can think of R_1 as the *equality relation* on this set, since for all $x, y \in X$, $(x, y) \in R_1$ iff $x = y$. Similarly, R_2 is the *less-than relation* on X, since for all $x, y \in X$, $(x, y) \in R_2$ iff $x < y$. Next, for all $x, y \in X$, xR_3y iff $x \leq y$. We see that the notation xRy for relations imitates the way we write arithmetical relations such as $=$, $<$, and $\leq$.

   Turning to R_4, we have xR_4y iff $x + y = 4$, for all $x, y \in X$. As for R_5, note xR_5y iff $x = 1$, so that this "relation" between x and y does not actually involve y. Relation R_6 does not describe any particularly meaningful relationship between x and y; it illustrates that a relation on X can be *any* set of ordered pairs contained in $X \times X$.

   Earlier, we introduced arrow diagrams and (Cartesian) graphs as ways to visualize relations from X to Y. When R is a relation on a finite set X, we can instead draw a modified

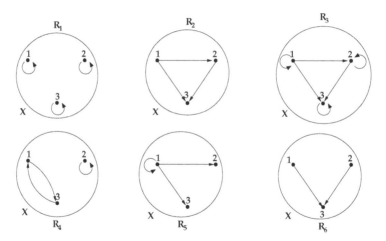

FIGURE 6.1
Digraphs for six relations on $X = \{1, 2, 3\}$.

version of the arrow diagram called the *digraph* (directed graph) of the relation. We make a single copy of the set X, drawing a dot labeled x for each object $x \in X$. For each ordered pair $(x, y) \in R$, we draw an arrow from x to y. If $y = x$, this arrow becomes a little loop that starts and ends at x. Figure 6.1 displays the digraphs for the six relations in Example 6.2.

6.3. Example. Define a relation R on the set $X = \{1, 2, 3, 4, 5, 6\}$ by

$$R = \{(1, 1), (2, 2), (3, 3), (4, 4), (2, 5), (5, 2), (1, 3), (3, 4), (4, 1), (2, 6)\}.$$

Figure 6.2 displays both the arrow diagram and the digraph for this relation. We will see that certain structural features of R can be seen more easily in the digraph picture.

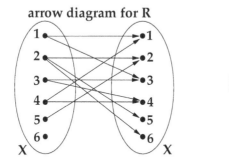
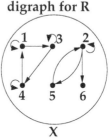

FIGURE 6.2
Arrow diagram and digraph for a relation R.

Reflexive Relations

Now we introduce three properties of abstract relations that are based on corresponding properties of logical equality. The first property, called *reflexivity*, comes from the fact that $x = x$ for all objects x.

6.4. Definition: Reflexive Relations. For all relations R and all sets X:

$\boxed{R \text{ is reflexive on } X}$ iff $\boxed{\forall x \in X, (x, x) \in R}$.

Using infix notation, we can also say that R is reflexive on X iff for all $x \in X$, xRx.

Informally, R is reflexive on X iff every object x in X is related to itself under R. This means that the digraph for R has a loop at every vertex in the set X. Applying the negation rules, we see that $\boxed{R \text{ is } not \text{ reflexive on } X}$ iff $\boxed{\exists x \in X, (x, x) \notin R}$.

6.5. Example. Relations R_1 and R_3 in Figure 6.1 are reflexive on $X = \{1, 2, 3\}$, but the other four relations are not. In particular, even though $(2, 2) \in R_4$, R_4 is not reflexive on X because $(1, 1)$ is not in R_4. Similarly, the relation R in Figure 6.2 is not reflexive on the set $X = \{1, 2, 3, 4, 5, 6\}$ because $(5, 5) \notin R$. However, R *is* reflexive on the smaller set $X' = \{1, 2, 3, 4\}$, although R is not a relation on X'. Thus, reflexivity is a property of both the relation R and the set X. Usually, however, we are dealing with relations on a fixed set X, and we are only interested in reflexivity with respect to the full set X. In such circumstances, we often say "R is reflexive" as an abbreviation for "R is reflexive on X."

6.6. Example. Suppose X is any collection of sets, and R is the relation on X defined by $(A, B) \in R$ iff $A \subseteq B$, for all $A, B \in X$. This "subset relation" R is reflexive on X, because $A \subseteq A$ for every set A.

6.7. Example. For any set X, the equality (or identity) relation $I_X = \{(x, x) : x \in X\}$ is reflexive on X. The product set $X \times X$ is a relation on X that is reflexive on X. The empty set $\emptyset$ is a relation on X that is not reflexive on X, assuming X itself is nonempty.

6.8. Example. Define relations $R = \{(x, y) \in \mathbb{Z}^2 : x + y \text{ is even}\}$, $S = \{(x, y) \in [-1, 1]^2 : x^2 + y^2 = 1\}$, and $T = \{(A, B) \in \mathcal{P}(\mathbb{Z})^2 : 8 \notin A - B\}$. Determine (with proof) which of these relations are reflexive. *Solution.* First, we show R is reflexive on $\mathbb{Z}$. Fix $x \in \mathbb{Z}$; we must prove $(x, x) \in R$, which means $x + x$ is even. Since $x + x = 2x$ where x is an integer, $x + x$ is even. Second, we show S is not reflexive on $[-1, 1]$. Pick $x = 1 \in [-1, 1]$, and note $(x, x) \notin S$ because $x^2 + x^2 = 2 \neq 1$. Third, we show T is reflexive on $\mathcal{P}(\mathbb{Z})$. Fix $A \in \mathcal{P}(\mathbb{Z})$; we must prove $(A, A) \in T$, which means $8 \notin A - A$. Since $A - A = \emptyset$ by the Theorem on Set Equality, 8 is not in this set.

Symmetric Relations

We next generalize the *symmetry* property of equality, which says that for all objects x, y, if $x = y$ then $y = x$.

6.9. Definition: Symmetric Relations. For all relations R:

$\boxed{R \text{ is symmetric}}$ iff $\boxed{\forall x, y, (x, y) \in R \Rightarrow (y, x) \in R}$.

Using infix notation, we can also say R is symmetric iff for all x and y, xRy implies yRx.

Thus, symmetry of R means that whenever x is related to y, y is also related to x. Pictorially, this means that for any arrow from x to y in the digraph for R, the reversed arrow from y back to x must also appear in the digraph. Applying the negation rules, we see that $\boxed{R \text{ is } not \text{ symmetric}}$ iff $\boxed{\exists x, \exists y, (x, y) \in R \land (y, x) \notin R}$.

6.10. Example. In Figure 6.1, relations R_1 and R_4 are symmetric, but the other relations are not. For instance, R_2 is not symmetric because $(1, 2) \in R_2$ but $(2, 1) \notin R_2$. The relation R in Figure 6.2 is not symmetric because, for instance, $(2, 6) \in R$ but $(6, 2) \notin R$.

6.11. Example. Let X be the set of all subsets of $\{1, 2, 3\}$, and let R be the subset relation given by aRb iff $a \subseteq b$ (for $a, b \in X$). R is not symmetric, since $\{1\} \subseteq \{1, 2\}$ but $\{1, 2\} \nsubseteq \{1\}$.

6.12. Example. For any set X, the equality relation I_X is symmetric. $X \times X$ is also symmetric. The empty relation $\emptyset$ *is* symmetric, since the implication $(x, y) \in \emptyset \Rightarrow (y, x) \in \emptyset$ has a false hypothesis and is therefore true. The relation $<$ on $\mathbb{R}$ is not symmetric since $3 < 5$ is true but $5 < 3$ is false.

6.13. Example. Decide (with proof) whether the relations R, S, and T from Example 6.8 are symmetric. *Solution.* First, we prove R is symmetric. Fix arbitrary (x, y); assume $(x, y) \in R$; prove $(y, x) \in R$. We have assumed $x, y \in \mathbb{Z}$ and $x + y$ is even; we must prove $x, y \in \mathbb{Z}$ and $y + x$ is even. Since $x + y = y + x$, the conclusion holds. Second, we prove S is symmetric. Fix arbitrary $(x, y) \in \mathbb{R}^2$; assume $(x, y) \in S$; prove $(y, x) \in S$. We have assumed $x^2 + y^2 = 1$; we must prove $y^2 + x^2 = 1$. Since $x^2 + y^2 = y^2 + x^2$, this conclusion holds. Third, we prove T is not symmetric. Choose $A = \{1, 2, 3\}$ and $B = \{6, 7, 8\}$, which are elements of $\mathcal{P}(\mathbb{Z})$. Note $(A, B) \in T$, since $A - B = A$, and $8 \notin A$. But $(B, A) \notin T$, since $B - A = B$, and $8 \in B$.

Transitive Relations

The *transitive* property of equality says that for all x, y, z, if $x = y$ and $y = z$, then $x = z$. This leads to the following definition.

6.14. Definition: Transitive Relations. For all relations R:

$\boxed{R \text{ is transitive}}$ iff $\boxed{\forall x, y, z, \text{ if } (x, y) \in R \text{ and } (y, z) \in R, \text{ then } (x, z) \in R}$.

Using infix notation, we can say R is transitive iff for all x, y, z, if xRy and yRz, then xRz.

Transitivity means that whenever x is related to y and y is related to z, then x must be related to z. To detect transitivity from a digraph, we must check that whenever there are arrows from x to y to z, there is also an arrow directly from x to z. We must be sure to check all cases, including those where $x = z$. By the negation rules, $\boxed{R \text{ is } not \text{ transitive}}$ iff $\boxed{\exists x, y, z, (x, y) \in R \wedge (y, z) \in R \wedge (x, z) \notin R}$.

6.15. Example. All of the relations in Figure 6.1 are transitive except for R_4. We see that R_4 is not transitive because $(1, 3) \in R_4$ and $(3, 1) \in R_4$, but $(1, 1) \notin R_4$. Relation R_6 is transitive because there are no choices for x, y, z that make the hypothesis $(x, y) \in R_6$ and $(y, z) \in R_6$ true; so the IF-statement defining transitivity is true. The relation R in Figure 6.2 is not transitive because, for example, $(5, 2) \in R$ and $(2, 6) \in R$ but $(5, 6) \notin R$.

6.16. Example. If X is any collection of sets, the subset relation $\subseteq$ on X is transitive since $A \subseteq B$ and $B \subseteq C$ imply $A \subseteq C$. For any set X, the equality relation I_X is transitive, as is the relation $X \times X$. The empty set is a transitive relation, by the definition of IF. The relation $<$ on $\mathbb{R}$ is transitive as well.

6.17. Example. Decide (with proof) whether the relations R, S, and T from Example 6.8 are transitive. *Solution.* First, we show R is transitive. Fix $x, y, z \in \mathbb{Z}$, assume $(x, y) \in R$ and $(y, z) \in R$, and prove $(x, z) \in R$. We have assumed $x + y$ is even and $y + z$ is even, which means $x + y = 2a$ and $y + z = 2b$ for some integers a, b. We must prove $x + z$ is even, which means $\exists c \in \mathbb{Z}, x + z = 2c$. Compute $x + z = (2a - y) + (2b - y) = 2(a + b - y)$, where $a + b - y \in \mathbb{Z}$. So we may choose $c = a + b - y$ to see that $x + z$ is even. Second, we show S is not transitive. Choose $x = 0$, $y = 1$, and $z = 0$. Then $(x, y) \in S$ because $0^2 + 1^2 = 1$; $(y, z) \in S$ because $1^2 + 0^2 = 1$; but $(x, z) \notin S$ because $0^2 + 0^2 \neq 1$. Third, we show T is transitive using a proof by contradiction. Assume, to get a contradiction, that there exist

$A, B, C \in \mathcal{P}(\mathbb{Z})$ with $(A, B) \in T$ and $(B, C) \in T$ and $(A, C) \notin T$. By definition, this means $8 \notin A - B$ and $8 \notin B - C$ and $8 \in A - C$. The last condition tells us that $8 \in A$ and $8 \notin C$. Since 8 is in A but not in $A - B$, we must have $8 \in B$. But now 8 is in B and not in C, so $8 \in B - C$. This contradicts the assumption $8 \notin B - C$.

6.18. Remark. A relation R on a fixed set X may be reflexive or not, symmetric or not, and transitive or not. In general, all eight possible combinations can occur (see the exercises).

Properties of Reflexivity, Symmetry, and Transitivity

The next theorem lists some facts about reflexivity, symmetry, and transitivity.

6.19. Theorem on Reflexivity, Symmetry, and Transitivity. For all relations R and all sets X:
(a) R is reflexive on X iff $I_X \subseteq R$.
(b) R is reflexive on X iff R^{-1} is reflexive on X.
(c) Any union or intersection of reflexive relations on X is reflexive on X.
(d) R is symmetric iff R^{-1} is symmetric.
(e) R is symmetric iff $R \subseteq R^{-1}$ iff $R = R^{-1}$.
(f) Any union or intersection of symmetric relations is symmetric.
(g) R is transitive iff R^{-1} is transitive.
(h) R is transitive iff $R \circ R \subseteq R$.
(i) Any intersection of transitive relations is transitive.

Proof. We prove a few parts here and leave the rest as exercises. Fix a relation R and a set X. To prove (b), we first assume R is reflexive on X and prove R^{-1} is reflexive on X. Fix $x \in X$; prove $(x, x) \in R^{-1}$. Because R is reflexive on X, we know $(x, x) \in R$. Switching the order of the components, we conclude $(x, x) \in R^{-1}$ as needed. The converse is proved similarly, or it can be deduced from the IF-statement we just proved by replacing the arbitrary relation R by R^{-1} and recalling $(R^{-1})^{-1} = R$.

Next we prove that the union of any collection of symmetric relations is symmetric (part of (f)). Let $\{R_i : i \in I\}$ be a collection of relations, and define $S = \bigcup_{i \in I} R_i$. Assume R_i is symmetric for all $i \in I$; prove S is symmetric. Fix objects x, y; assume $(x, y) \in S$; prove $(y, x) \in S$. We know there exists $i \in I$ with $(x, y) \in R_i$. Since R_i is symmetric, it follows that $(y, x) \in R_i$. By definition of union, $(y, x) \in S$, as needed.

Finally, we prove that if R is transitive, then $R \circ R \subseteq R$ (part of (h)). We have assumed $\forall x, y, z, (x, y) \in R \wedge (y, z) \in R \Rightarrow (x, z) \in R$. To prove $R \circ R \subseteq R$, fix an ordered pair $(u, v) \in R \circ R$ and prove $(u, v) \in R$. By definition of composition of relations, there exists w with $(u, w) \in R$ and $(w, v) \in R$. By transitivity of R, $(u, v) \in R$ follows. $\square$

6.20. Example. It is false that the union of transitive relations must always be transitive. For example, let $R = \{(1, 2)\}$ and $S = \{(2, 3)\}$, so $R \cup S = \{(1, 2), (2, 3)\}$. Both R and S are transitive relations, but their union is not.

Section Summary

1. *Relations and Digraphs.* R is a relation on a set X iff $R \subseteq X \times X$. In this case, the digraph of R consists of a point for each $x \in X$, and an arrow from x to y for each ordered pair $(x, y) \in R$.

2. *Reflexive Relations.* R is a reflexive relation on X iff for all $x \in X$, $(x, x) \in R$.

3. *Symmetric Relations.* A relation R is symmetric iff for all a, b, if $(a, b) \in R$, then $(b, a) \in R$.

4. *Transitive Relations.* A relation R is transitive iff for all a, b, c, if $(a, b) \in R$ and $(b, c) \in R$, then $(a, c) \in R$.

5. *Infix Notation.* Given a relation R, aRb means $(a, b) \in R$. In this notation, R is reflexive on X iff $\forall x \in X, xRx$; R is symmetric iff $\forall a, b, aRb \Rightarrow bRa$; and R is transitive iff $\forall a, b, c, (aRb \wedge bRc) \Rightarrow aRc$.

6. *Properties of Reflexivity, Symmetry, and Transitivity.* If R is reflexive on X, or symmetric, or transitive, then R^{-1} has the same property. Reflexivity, symmetry, and transitivity are inherited by intersections of relations. Reflexivity and symmetry are inherited by unions of relations, but transitivity is not. R is reflexive on X iff $I_X \subseteq R$; R is symmetric on X iff $R \subseteq R^{-1}$ iff $R = R^{-1}$; R is transitive on X iff $R \circ R \subseteq R$.

Exercises

1. The digraphs of various relations are shown in Figure 6.3. Is each relation reflexive on X? symmetric? transitive?

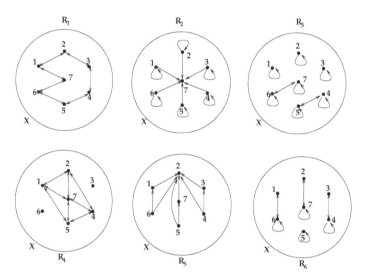

FIGURE 6.3
Digraphs for some relations.

2. Draw an arrow diagram and a digraph for each relation. Is each relation reflexive on $X = \{1, 2, 3, 4\}$? symmetric? transitive?
   (a) $R = \{(1,1), (2,2), (3,3), (4,4), (1,3), (3,4), (1,4), (2,4)\}$.
   (b) $R = \{(1,2), (2,1), (3,4), (4,3)\}$.
   (c) $R = (X \times X) - I_X$.
   (d) $R = \{(1,1), (3,3), (1,3), (3,1), (2,2), (4,4), (2,4), (4,2)\}$.

3. Is each relation reflexive on $\mathbb{Z}$? symmetric? transitive? Explain.
   (a) $R = \{(a, b) \in \mathbb{Z}^2 : a < b\}$.
   (b) $R = \{(a, b) \in \mathbb{Z}^2 : a^2 \leq b^2\}$.
   (c) $R = \{(a, b) \in \mathbb{Z}^2 : ab \text{ is even}\}$.
   (d) $R = \{(a, b) \in \mathbb{Z}^2 : |a - b| \leq 1\}$.
   (e) $R = \{(a, b) \in \mathbb{Z}^2 : a \text{ divides } b\}$.

(f) $R = \{(a, b) \in \mathbb{Z}^2 : 5 \text{ divides } ab\}$.

(g) $R = \{(a, b) \in \mathbb{Z}^2 : 7 \text{ divides } a - b\}$.

4. Let S be a relation on $\mathbb{R}$. (a) Describe how to determine if S is reflexive on $\mathbb{R}$ by visual inspection of the graph of S in the xy-plane. (b) Similarly, describe how to determine if S is symmetric by inspection of its graph.

5. Is each relation reflexive on $\mathbb{R}$? symmetric? transitive? Explain.

(a) $R = \{(x, y) \in \mathbb{R}^2 : \cos x = \cos y\}$.

(b) $R = \{(x, y) \in \mathbb{R}^2 : xy = 1\}$.

(c) $R = \{(x, y) \in \mathbb{R}^2 : y = x \vee y = -x \vee x^2 + y^2 = 1\}$.

(d) $R = \{(x, y) \in \mathbb{R}^2 : xy < 0\}$.

(e) $R = \{(x, y) \in \mathbb{R}^2 : xy \geq 0\}$.

(f) $R = \{(x, y) \in \mathbb{R}^2 : xy > 0\} \cup \{(0, 0)\}$.

(g) $R = \{(x, y) \in \mathbb{R}^2 : y - x \in \mathbb{Z}\}$.

6. Is each relation reflexive on $\mathbb{R}^2$? symmetric? transitive? Explain.

(a) $(a, b)R(c, d)$ means $a^2 + b^2 = c^2 + d^2$.

(b) $(a, b)R(c, d)$ means $ad = bc$.

(c) $(a, b)R(c, d)$ means $a + d = b + c$.

(d) $(a, b)R(c, d)$ means $a < c$ or ($a = c$ and $b \leq d$).

7. How can you decide by looking at the digraph of a relation G on X whether G is the graph of a function $f : X \to X$?

8. Let $X = \{a, b, c\}$. For each part, give two different examples of relations on X with the indicated properties (if possible).

(a) R is reflexive on X and symmetric and transitive.

(b) R is reflexive on X and symmetric and not transitive.

(c) R is reflexive on X and not symmetric and transitive.

(d) R is reflexive on X and not symmetric and not transitive.

(e) R is not reflexive on X and symmetric and transitive.

(f) R is not reflexive on X and symmetric and not transitive.

(g) R is not reflexive on X and not symmetric and transitive.

(h) R is not reflexive on X and not symmetric and not transitive.

9. Which parts of the previous exercise can be solved (with either one or two examples) using relations on the two-element set $X = \{1, 2\}$?

10. Prove (a) and (c) of Theorem 6.19.

11. Prove (d), (e), and the second part of (f) in Theorem 6.19.

12. Prove (g), (i), and the second part of (h) in Theorem 6.19.

13. (a) Suppose R is a reflexive relation on a nonempty set X. What can you say about whether $(X \times X) - R$ is reflexive on X? (b) Repeat (a) replacing "reflexive" by "symmetric." (c) Repeat (a) replacing "reflexive" by "transitive."

14. (a) If R and S are reflexive relations on X, must $R \circ S$ be reflexive on X? (b) If R and S are symmetric relations, must $R \circ S$ be symmetric? (c) If R and S are transitive relations, must $R \circ S$ be transitive?

15. Prove: for any relation R and any set X, $R \cup R^{-1} \cup I_X$ is reflexive on X and symmetric.

16. (a) For a relation R on X, prove that the intersection T of all transitive relations on X containing R is a transitive relation on X containing R. (b) Recursively define $R^1 = R$ and $R^{n+1} = R^n \circ R$ for $n \in \mathbb{Z}_{\geq 1}$. Prove: for all $n, m \in \mathbb{Z}_{\geq 1}$, $R^{n+m} = R^n \circ R^m$. (c) Prove that $T = \bigcup_{n=1}^{\infty} R^n$.

6.2   Equivalence Relations

In mathematics, it often happens that two objects are *equivalent* in a certain respect, even though the objects in question are not equal to each other. For instance, two triangles (thought of as subsets of $\mathbb{R}^2$) can be geometrically congruent even if they are not equal as sets of points. As further examples, two unequal shapes may be geometrically similar; two unequal vectors may have the same length; two unequal lines may have the same slope; two unequal times (differing by a multiple of 24 hours) may represent the same hour of the day; two unequal real numbers (differing by a multiple of 2π) may represent the same geometric angle; and so on. The abstract definition of an *equivalence relation* gives a precise mathematical formulation of various intuitive notions of equivalence. We define equivalence relations in this section and consider many examples, including congruence of integers modulo n and the equivalence relation induced by a function.

Definition and Examples of Equivalence Relations

Any intuitive concept of "equivalence" ought to share some of the basic properties of logical equality. It turns out that reflexivity, symmetry, and transitivity are the key properties we need.

6.21. Definition: Equivalence Relations. Given a relation R on a set X,

$\boxed{R \text{ is an equivalence relation on } X}$ iff $\boxed{R \text{ is reflexive on } X, \text{ symmetric, and transitive}}$.

Expanding this definition, we see that R is an equivalence relation on X iff the following three conditions hold for all $a, b, c \in X$: (i) aRa; (ii) if aRb then bRa; (iii) if aRb and bRc, then aRc.

6.22. Example: Equality Relation. For any set X, the identity relation $I_X = \{(x, x) : x \in X\}$ is an equivalence relation on X. If we write $x = y$ instead of xI_Xy, the three conditions above become: (i) $a = a$; (ii) if $a = b$, then $b = a$; (iii) if $a = b$ and $b = c$, then $a = c$.

6.23. Example. For any set X, $R = X \times X$ is an equivalence relation on X. In this case, xRy is true for *all* pairs of objects $x, y \in X$. So it is immediate that conditions (i), (ii), and (iii) hold for this R.

6.24. Example. Let $X = \mathbb{Z}$ and $R = \{(x, y) \in \mathbb{Z}^2 : x + y \text{ is even}\}$. We saw in the last section that R is reflexive on X, symmetric, and transitive. Thus, R is an equivalence relation on X. On the other hand, let $S = \{(x, y) \in \mathbb{Z}^2 : x + y \text{ is odd}\}$. S is not reflexive on $\mathbb{Z}$, since $(1, 1) \notin S$, so S is not an equivalence relation on $\mathbb{Z}$.

6.25. Example. Define $T = \{(x, y) \in \mathbb{R}^2 : xy \geq 0\}$. You can check that T is reflexive on $\mathbb{R}$ and symmetric. But T is not transitive, because $(1, 0) \in T$ and $(0, -1) \in T$ but $(1, -1) \notin T$. So T is not an equivalence relation on $\mathbb{R}$.

6.26. Example. Figure 6.4 shows the digraphs for six relations $R_1, \ldots, R_6$ on the set $X = \{1, 2, 3, 4, 5, 6\}$. We see from inspection of the figure that R_1, R_2, and R_3 are equivalence relations on X, but R_4, R_5, and R_6 are not. R_4 is not symmetric because $(2, 3) \in R_4$ but $(3, 2) \notin R_4$. R_5 is not reflexive on X because $(5, 5) \notin R_5$. R_6 is not transitive because $(1, 3) \in R_6$ and $(3, 6) \in R_6$, but $(1, 6) \notin R_6$.

In Figure 6.1 from the previous section, only R_1 (the equality relation) is an equivalence relation on $X = \{1, 2, 3\}$. Digraphs of equivalence relations have special structural features that we will explore later, when we discuss equivalence classes and set partitions.

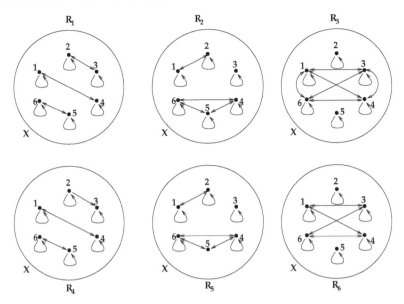

FIGURE 6.4
Digraphs for some equivalence relations and non-equivalence relations.

Equivalence Relation Induced by a Function

Given a function $f : X \to Y$, we can create an equivalence relation on the domain of f by the following construction.

6.27. Definition: Equivalence Relation Induced by a Function. Given $f : X \to Y$, the *equivalence relation induced by* f is $R_f = \{(a,b) \in X \times X : f(a) = f(b)\}$. We often denote R_f by the symbol $\sim_f$; thus, for all $a, b \in X$, $\boxed{a \sim_f b}$ means $\boxed{f(a) = f(b)}$.

Let us check that $\sim_f$ always is an equivalence relation. Fix an arbitrary function $f : X \to Y$, and fix $a, b, c \in X$. We have $a \sim_f a$ because $f(a) = f(a)$, so $\sim_f$ is reflexive on X. To prove symmetry, assume $a \sim_f b$ and prove $b \sim_f a$. We have assumed $f(a) = f(b)$; we must prove $f(b) = f(a)$; this follows since equality is symmetric. To prove transitivity, assume $a \sim_f b$ and $b \sim_f c$; prove $a \sim_f c$. We have assumed $f(a) = f(b)$ and $f(b) = f(c)$; we must prove $f(a) = f(c)$. This follows since equality is transitive. We have now proved $\sim_f$ is an equivalence relation on X.

By choosing different functions f, we can use this construction to build a wide variety of equivalence relations. (In fact, we will see later that *all* equivalence relations arise in this way for an appropriate choice of f.) Here are some typical examples.

6.28. Example. For $f : \mathbb{R} \to \mathbb{R}$ given by $f(x) = |x|$, we get an equivalence relation $\sim_f$ on $\mathbb{R}$ such that $x \sim_f y$ iff $|x| = |y|$. So, for instance, $-3 \sim_f 3$, $4 \sim_f 4$, $0 \sim_f 0$, but $-3 \not\sim_f 5$. Intuitively, this relation regards two real numbers as equivalent iff they have the same magnitude (disregarding sign). The function $g : \mathbb{R} \to \mathbb{R}$ given by $g(x) = x^2$ induces the same equivalence relation as f. For, given any $x, y \in \mathbb{R}$, $x \sim_g y$ iff $x^2 = y^2$ iff $\sqrt{x^2} = \sqrt{y^2}$ iff $|x| = |y|$ iff $x \sim_f y$.

6.29. Example. Define $\text{sgn} : \mathbb{R} \to \{-1, 0, 1\}$ by setting $\text{sgn}(x) = -1$ if $x < 0$, $\text{sgn}(x) = 0$ if $x = 0$, and $\text{sgn}(x) = 1$ if $x > 0$. The equivalence relation $\sim_{\text{sgn}}$ regards two real numbers as equivalent iff they have the same sign. For example, $-3 \sim_{\text{sgn}} -7$, $0 \sim_{\text{sgn}} 0$, $2 \sim_{\text{sgn}} 7$, but $2 \nsim_{\text{sgn}} -2$.

6.30. Example. Define $f : \mathbb{R} \to \mathbb{R}^2$ by $f(t) = (\cos t, \sin t)$ for all $t \in \mathbb{R}$. From trigonometry, we know that for all $s, t \in \mathbb{R}$, $f(s) = f(t)$ iff $s = t + 2\pi k$ for some integer k. Intuitively, $\sim_f$ regards two real numbers as equivalent iff those numbers represent the same angle going around the unit circle. The individual sine and cosine functions induce more complicated equivalence relations on $\mathbb{R}$. For instance, $\pi/3 \sim_{\sin} 2\pi/3$ and $\pi/3 \sim_{\cos} -\pi/3$, although no two of the numbers $-\pi/3$, $\pi/3$, $2\pi/3$ are related under $\sim_f$.

6.31. Example. Define $f : \mathbb{Z}_{>0} \to \{0, 1, \ldots, 9\}$ by letting $f(n)$ be the rightmost digit in the decimal representation of n. For instance, $f(5247) = 7$, $f(10^3 - 1) = 9$, and $f(10k) = 0$ for any $k \in \mathbb{Z}_{>0}$. The relation $\sim_f$ regards two positive integers as equivalent iff they have the same last digit.

Congruence Modulo n

We now introduce a family of equivalence relations that play a fundamental role in number theory.

6.32. Definition: Congruence Modulo n. For each positive integer $n \geq 1$, define a relation $\equiv_n$ on $\mathbb{Z}$ as follows: for all $a, b \in \mathbb{Z}$, $\boxed{a \equiv_n b}$ iff $\boxed{n \text{ divides } a - b}$. When $a \equiv_n b$, we also use the notation $\boxed{a \equiv b \pmod{n}}$, which can be read "$a$ is congruent to b modulo n."

The next theorem says that congruence modulo n is an equivalence relation that is compatible with the operations of integer addition and multiplication.

6.33. Theorem on Congruence Modulo n. For all $n \in \mathbb{Z}_{\geq 1}$ and all $a, b, c, d \in \mathbb{Z}$:

(a) $\equiv_n$ is an equivalence relation on $\mathbb{Z}$.

(b) If $a \equiv_n b$ and $c \equiv_n d$, then $a + c \equiv_n b + d$.

(c) If $a \equiv_n b$ and $c \equiv_n d$, then $ac \equiv_n bd$.

Proof. For (a), we must show $\equiv_n$ is reflexive on $\mathbb{Z}$ and symmetric and transitive. To prove $\equiv_n$ is reflexive on $\mathbb{Z}$, fix $a \in \mathbb{Z}$. We prove $a \equiv_n a$, which means n divides $a - a$. Since $a - a = 0 = 0n$ where 0 is an integer, n does divide $a - a$. To prove symmetry, fix $a, b \in \mathbb{Z}$; assume $a \equiv_n b$; prove $b \equiv_n a$. We have assumed n divides $a - b$, so $a - b = kn$ for some integer k. We must prove n divides $b - a$. Since $b - a = -(a - b) = -(kn) = (-k)n$ where $-k$ is an integer, n does divide $b - a$. To prove transitivity, fix $a, b, c \in \mathbb{Z}$; assume $a \equiv_n b$ and $b \equiv_n c$; prove $a \equiv_n c$. We have assumed n divides $a - b$, so $a - b = dn$ for some integer d; and n divides $b - c$, so $b - c = en$ for some integer e. We must prove n divides $a - c$. Note $a - c = (a + b) + (b - c) = dn + en = (d + e)n$ where $d + e$ is an integer; so n does divide $a - c$.

We prove (c), leaving (b) as an exercise. Assume $a \equiv_n b$ and $c \equiv_n d$, so $a - b = jn$ and $c - d = kn$ for some $j, k \in \mathbb{Z}$. We must prove $ac \equiv_n bd$, i.e., $ac - bd = pn$ for some $p \in \mathbb{Z}$. Compute

$$ac - bd = ac - ad + ad - bd = a(c - d) + (a - b)d = a(kn) + (jn)d = (ak + jd)n.$$

Choosing $p = ak + jd$, which is an integer, we then have $ac - bd = pn$ as needed. $\square$

6.34. Example. When $n = 2$, $a \equiv_2 b$ iff 2 divides $a - b$. By considering cases, we see that for all $a, b \in \mathbb{Z}$, $a \equiv_2 b$ iff a and b are both even, or a and b are both odd. Thus, $\equiv_2$ regards two integers as being equivalent iff these integers have the same parity (odd or even).

6.35. Example. Taking $n = 10$, we have $5247 \equiv_{10} 7$, $115 \equiv_{10} 9015$, but $4441 \not\equiv_{10} 4444$. For all *positive* integers a, b, we have $a \equiv_{10} b$ iff a and b have the same last digit written in decimal iff $a \sim_f b$, where f is the function from Example 6.31. However, we must be careful when negative integers appear. For instance, $17 \equiv_{10} -3$ since 10 divides $17 - (-3) = 20$, but $17 \not\equiv_{10} -7$ since 10 does not divide $17 - (-7) = 24$.

Further Examples of Equivalence Relations

When considering an equivalence relation (or any relation) on a set X, the objects in X can be anything. In the next example, the objects are ordered pairs (points or vectors in the xy-plane), so that the relation itself is a set of ordered pairs where each component is an ordered pair. In this situation, infix notation greatly improves readability.

6.36. Example. Let $X = \mathbb{R} \times \mathbb{R}$. Define a relation R on X by setting $(a, b)R(c, d)$ iff there exists $r \in \mathbb{R}_{>0}$ with $c = ra$ and $d = rb$. We now prove that R is an equivalence relation on X. For reflexivity, fix $(a, b) \in X$. To prove $(a, b)R(a, b)$, we must prove $\exists r \in \mathbb{R}_{>0}$, $a = ra$ and $b = rb$. Choosing $r = 1$, these equations do hold. For symmetry, fix $(a, b), (c, d) \in X$, assume $(a, b)R(c, d)$, and prove $(c, d)R(a, b)$. By assumption, there is $r \in \mathbb{R}_{>0}$ with $c = ra$ and $d = rb$. We must prove $\exists s \in \mathbb{R}_{>0}$, $a = sc$ and $b = sd$. Choose $s = r^{-1}$, which exists and is a positive real number since r is positive. The assumptions $c = ra$ and $d = rb$ become $a = r^{-1}c = sc$ and $b = r^{-1}d = sd$, as needed. For transitivity, fix $(a, b), (c, d), (e, f) \in X$, assume $(a, b)R(c, d)$ and $(c, d)R(e, f)$, and prove $(a, b)R(e, f)$. By assumption, there exist $t, u \in \mathbb{R}_{>0}$ with $c = ta$, $d = tb$, $e = uc$, and $f = ud$. Combining these equations, we see that $e = uc = u(ta) = (ut)a$ and $f = ud = u(tb) = (ut)b$, where $ut \in \mathbb{R}_{>0}$ since ut is a product of two positive real numbers. We now see that $(a, b)R(e, f)$, as needed.

In the next example, the elements of X are functions.

6.37. Example. Let X be the set of all functions $f : \mathbb{R} \to \mathbb{R}$. Define a relation R on X by setting fRg iff $f(3) = g(3)$ for all $f, g \in X$. It is routine to prove that R is an equivalence relation by verifying the definition. Alternatively, we can deduce this using a previous construction, as follows. Define an "evaluation function" $E : X \to \mathbb{R}$ by setting $E(f) = f(3)$ for all $f \in X$. Note that E is a function that takes another function as its input and produces a real number as its output. For all $f, g \in X$, fRg iff $f(3) = g(3)$ iff $E(f) = E(g)$ iff $f \sim_E g$. Thus R is the equivalence relation induced by the function E.

In the next example, the elements of X are sets.

6.38. Example. Let $X = \mathcal{P}(\mathbb{Z})$, and for $A, B \in X$, define $A \approx B$ iff $A \cap \mathbb{Z}_{>0} = B \cap \mathbb{Z}_{>0}$. The relation $\approx$ is an equivalence relation on X. In fact, $\approx$ is the equivalence relation $\sim_f$ induced by the function $f : X \to \mathcal{P}(\mathbb{Z})$ such that $f(A) = A \cap \mathbb{Z}_{>0}$ for all $A \in X$.

We now give a method for obtaining a new equivalence relation from existing equivalence relations.

6.39. Lemma on Intersection of Equivalence Relations. For all sets X, the intersection of any nonempty set of equivalence relations on X is an equivalence relation on X.

Proof. Fix a set X, and let $\{R_i : i \in I\}$ be a nonempty set of equivalence relations on X. We must prove $R = \bigcap_{i \in I} R_i$ is also an equivalence relation on X. Since every R_i is reflexive on X, the intersection R is also reflexive on X by Theorem 6.19. Similarly, by the same theorem, R is symmetric since all R_i are symmetric, and R is transitive since all R_i are transitive. So R is an equivalence relation on X. $\square$

6.40. Example. Given $X = \mathbb{Z}$, we claim that $\equiv_3 \cap \equiv_5$ equals $\equiv_{15}$. Writing R_n for the relation $\equiv_n$, we are claiming that the two sets of ordered pairs $R_3 \cap R_5$ and R_{15} are equal. [We prove the set equality using a chain proof, observing that all members of both sets are ordered pairs of integers.] Fix an arbitrary pair $(a, b) \in \mathbb{Z} \times \mathbb{Z}$. Note that $(a, b) \in R_3 \cap R_5$ iff $(a, b) \in R_3$ and $(a, b) \in R_5$ iff 3 divides $a - b$ and 5 divides $a - b$ iff 15 divides $a - b$ iff $(a, b) \in R_{15}$. The next-to-last equivalence can be proved by considering the prime factorization of $a - b$. More generally, you can check that for all $m, n \in \mathbb{Z}_{>0}$, $\equiv_m \cap \equiv_n$ equals $\equiv_{\mathrm{lcm}(m,n)}$.

Section Summary

1. *Equivalence Relations.* R is an equivalence relation on a set X iff R is a relation on X that is reflexive on X, symmetric, and transitive. The intersection of any nonempty set of equivalence relations on X is also an equivalence relation on X.

2. *Equivalence Relation Induced by a Function.* Given any function $f : X \to Y$, there is an induced equivalence relation $\sim_f$ on X defined by $a \sim_f b$ iff $f(a) = f(b)$ (where $a, b \in X$).

3. *Congruence mod n.* For each $n \geq 1$, the relation $\equiv_n$ defined by $a \equiv_n b$ iff n divides $a - b$ (for $a, b \in \mathbb{Z}$) is an equivalence relation on $\mathbb{Z}$. This relation is compatible with addition and multiplication: for all $a, a', b, b' \in \mathbb{Z}$, if $a \equiv_n a'$ and $b \equiv_n b'$, then $a + b \equiv_n a' + b'$ and $ab \equiv_n a'b'$. When $a \equiv_n b$, we also write $a \equiv b \pmod{n}$.

Exercises

1. (a) Which relations shown in Figure 6.3 are equivalence relations on X? (b) Repeat (a) replacing each relation R_i by $R_i \cup I_X$.

2. Let $X = \{1, 2, 3, 4, 5, 6\}$. For each function $f : X \to \{a, b, c, d, e\}$, describe the equivalence relation $\sim_f$ as a digraph and as a set of ordered pairs.
   (a) $f(x) = c$ for all $x \in X$.
   (b) $f(x) = a$ for odd $x \in X$, $f(x) = e$ for even $x \in X$.
   (c) $f(1) = b$, $f(2) = e$, $f(3) = a$, $f(4) = b$, $f(5) = e$, $f(6) = a$.
   (d) $f(1) = a$, $f(2) = b$, $f(3) = c$, $f(4) = d$, $f(5) = e$, $f(6) = b$.

3. Let $X = \{1, 2, 3, 4, 5, 6, 7, 8\}$. It is known that R is an equivalence relation on X containing these ordered pairs: $(2, 4)$, $(5, 6)$, $(1, 3)$, $(4, 6)$, $(8, 3)$. Find the relation R (satisfying these conditions) such that R has the smallest possible number of elements. Display R as a set of ordered pairs and as a digraph. [*Hint:* Since $(2, 4) \in R$ and R is symmetric, R must also contain $(4, 2)$. Make further deductions of this kind to see what other ordered pairs must be in R.]

4. Draw the digraph of the smallest equivalence relation R on the set $X = \{1, 2, 3, 4, 5, 6, 7, 8\}$ such that R contains the ordered pairs $(1, 3)$, $(1, 5)$, $(1, 7)$, $(2, 2)$, and $(8, 6)$.

5. Show that each relation R is an equivalence relation on X by finding a function $f : X \to Y$ such that R is $\sim_f$.
   (a) $X = \mathbb{Z}$, $R = I_{\mathbb{Z}}$.
   (b) $X = \mathbb{Z}$, $R = \mathbb{Z} \times \mathbb{Z}$.
   (c) $X = \mathbb{Z}$, R is $\equiv_2$.
   (d) $X = \mathbb{R}^2$, $(a, b)R(c, d)$ iff (a, b) and (c, d) lie on the same vertical line.
   (e) $X = \mathbb{R}^2$, $(a, b)R(c, d)$ iff (a, b) and (c, d) lie on the same circle centered at the

origin.

(f) $X = \mathbb{R}^2_{>0}$, $(a, b)R(c, d)$ iff (a, b) and (c, d) lie on the same line through the origin.

(g) $X = \mathbb{R}$, xRy iff $x, y \in \mathbb{R}$ belong to the same interval $(n, n + 1]$ with $n \in \mathbb{Z}$.

6. (a) Find all $b \in \mathbb{Z}$ such that $7 \equiv_2 b$. (b) Find all $b \in \mathbb{Z}$ such that $7 \equiv_5 b$. (c) Find all $b \in \mathbb{Z}$ such that $7 \equiv_{10} b$.

7. Find all n such that $111 \equiv_n 471$.

8. Prove part (b) of the Theorem on Congruence Modulo n.

9. Fix $n \geq 1$. (a) Prove: for all $a, b \in \mathbb{Z}$, if $a \equiv_n b$, then $-a \equiv_n -b$. (b) Prove: for all $a, b, c, d \in \mathbb{Z}$, if $a \equiv_n b$ and $c \equiv_n d$, then $a - c \equiv_n b - d$.

10. Prove by induction on k: for all $n \in \mathbb{Z}_{\geq 1}$ and all $k \in \mathbb{Z}_{\geq 0}$ and all $a, b \in \mathbb{Z}$, if $a \equiv b$ (mod n), then $a^k \equiv b^k$ (mod n).

11. Draw the digraphs of all equivalence relations on the set $X = \{1, 2, 3\}$.

12. Give a specific example to show that the union of two equivalence relations on the set $\{1, 2, 3, 4\}$ need not be an equivalence relation on this set.

13. Let $X = \mathbb{Z} \times \mathbb{Z}$. Define a relation R on X by setting $(a, b)R(c, d)$ iff $a + d = b + c$. Prove that R is an equivalence relation on X.

14. Let $X = \mathbb{Z} \times \mathbb{Z}_{\neq 0}$. Define a relation R on X by setting $(a, b)R(c, d)$ iff $ad = bc$. Prove that R is an equivalence relation on X. Would this still be true if we had used $X = \mathbb{Z} \times \mathbb{Z}$?

15. Define a relation $\sim$ on $\mathbb{R}^n$ by setting $\mathbf{v} \sim \mathbf{w}$ iff there exists $r \in \mathbb{R}_{\neq 0}$ with $\mathbf{w} = r\mathbf{v}$ (i.e., $w_i = rv_i$ for $i = 1, 2, \ldots, n$). Prove that $\sim$ is an equivalence relation on $\mathbb{R}^n$. Would this still be true if we had used $r \in \mathbb{R}$ instead of $r \in \mathbb{R}_{\neq 0}$?

16. Given sets A and X with $A \subseteq X$, let $R = (A \times A) \cup ((X - A) \times (X - A))$. Prove R is an equivalence relation on X.

17. Let S be any relation from X to Y. Define a relation $\sim_S$ on X by setting (for $a, b \in X$) $a \sim_S b$ iff $S[\{a\}] = S[\{b\}]$. Is $\sim_S$ an equivalence relation on X?

18. Fix $n \geq 1$, and define $f : \mathbb{Z} \to \{0, 1, \ldots, n - 1\}$ by letting $f(a)$ be the unique remainder in $\{0, 1, \ldots, n - 1\}$ when a is divided by n. Prove that $\equiv_n$ is the equivalence relation $\sim_f$.

19. Let X be any collection of functions with a common domain D. Let S be a fixed subset of D. For $f, g \in X$, define $f \sim g$ iff $f(s) = g(s)$ for all $s \in S$. Prove $\sim$ is an equivalence relation on X.

20. Let X be any collection of sets. For $A, B \in X$, define $A \sim B$ iff there exists a bijection $f : A \to B$. Prove $\sim$ is an equivalence relation on X.

21. (a) Given functions $f : X \to Y$ and $g : Y \to Z$, prove that $\sim_f$ is a subset of $\sim_{g \circ f}$. (b) Prove that if g is injective, then $\sim_f$ equals $\sim_{g \circ f}$.

22. Write R_m for the relation $\equiv_m$. Given positive integers $n_1, \ldots, n_k$, prove that $\bigcap_{i=1}^k R_{n_i} = R_N$, where $N = \text{lcm}(n_1, \ldots, n_k)$.

23. Given any relation S on a set X, let R be the intersection of all equivalence relations on X containing S. (a) Show that R is an equivalence relation on X containing S, such that $R \subseteq R'$ for all equivalence relations R' on X containing S. (b) Let $T = \bigcup_{n=1}^\infty (S \cup S^{-1} \cup I_X)^n$ (this notation is defined in Exercise 16 of §6.1). Prove $R = T$.

6.3   Equivalence Classes

Intuitively, an equivalence relation R on a set X lets us "clump together" all the objects in X that are the same in some way, creating a collection of subsets of X. For instance, congruence modulo 2 decomposes the set of integers into the subset of odd integers and the subset of even integers. The subset of X consisting of all objects in X related under R to a given object b is called the *equivalence class* of b determined by R. We can gain a lot of insight into an equivalence relation by studying its equivalence classes. We consider many examples of equivalence classes in this section. We then show that any two equivalence classes of R are either equal or do not overlap at all, so that every element of X belongs to exactly one equivalence class.

Definition and Examples of Equivalence Classes

Consider again the digraphs of the three equivalence relations shown in the top row of Figure 6.4, which are redrawn in Figure 6.5 below. We notice that the arrows in the digraph divide the points of X into different clumps, where all pairs of points in a given clump are connected by relation arrows, but no arrow goes between points in two different clumps. The next definition helps us understand this situation better by giving names to the clumps — they are the equivalence classes of the given equivalence relation.

6.41. Definition: Equivalence Classes. Let R be an equivalence relation on a set X. For all $b \in X$, the *equivalence class of b determined by R*, denoted $[b]_R$, is the set of all objects related to b by R. In symbols,

$$[b]_R = \{x \in X : (x, b) \in R\}.$$

Since R is symmetric, we can restate this definition in infix notation as follows:

$$\text{for all } b, x \in X, \quad \boxed{x \in [b]_R} \Leftrightarrow \boxed{(x, b) \in R} \Leftrightarrow \boxed{xRb} \Leftrightarrow \boxed{bRx}.$$

If R is understood from context, we may write $[b]$ to abbreviate $[b]_R$. Given any set S, $\boxed{S \text{ is an equivalence class of } R}$ iff $\boxed{\exists b \in X, S = [b]_R}$.

6.42. Example. Let us compute the equivalence classes for the equivalence relations shown in Figure 6.5. For R_1, we find that

$$[1]_{R_1} = \{1, 4\}, \ [2]_{R_1} = \{2, 3\}, \ [3]_{R_1} = \{2, 3\}, \ [4]_{R_1} = \{1, 4\}, \ [5]_{R_1} = \{5, 6\}, \ [6]_{R_1} = \{5, 6\}.$$

Notice that $[1]_{R_1} = [4]_{R_1}$, $[2]_{R_1} = [3]_{R_1}$, and $[5]_{R_1} = [6]_{R_1}$. In general, for an equivalence relation R, *it often happens that $a \neq b$ and yet $[a]_R = [b]_R$.* This means that a given equivalence class S may have several different *names* $[a]_R$ for various choices of $a \in X$. In fact, we will prove shortly that for all $a, b \in X$, $[a]_R = [b]_R$ iff aRb. In the case of R_1, there are only three distinct equivalence classes, specifically $\{1, 4\}$ and $\{2, 3\}$ and $\{5, 6\}$, but each equivalence class has two different names. The fact that a given equivalence class may have multiple names is a confusing but fundamental feature of this concept.

Turning to R_2, inspection of the digraph shows that

$$[1]_{R_2} = \{1, 2\} = [2]_{R_2}, \ [3]_{R_2} = \{3\}, \ [4]_{R_2} = \{4, 5, 6\} = [5]_{R_2} = [6]_{R_2}.$$

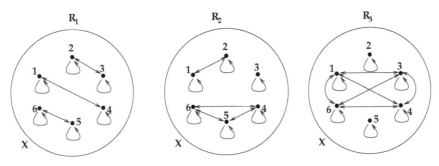

FIGURE 6.5
Digraphs for some equivalence relations.

We see that there are three distinct equivalence classes of R_2, namely $\{1, 2\}$ and $\{3\}$ and $\{4, 5, 6\}$. The class $\{1, 2\}$ has two members and two names; the class $\{3\}$ has one member and one name; the class $\{4, 5, 6\}$ has three members and three names. In general, if S is an equivalence class of an equivalence relation R, each member $a \in S$ gives us a possible name $[a]_R$ for the equivalence class S. For this reason, each $a \in S$ is called a *representative* of the equivalence class S.

Finally, R_3 has three equivalence classes: $[2]_{R_3} = \{2\}$, $[5]_{R_3} = \{5\}$, and $[1]_{R_3} = \{1, 3, 4, 6\} = [3]_{R_3} = [4]_{R_3} = [6]_{R_3}$.

6.43. Example. For any set X, consider the equality relation $I_X = \{(x, x) : x \in X\}$. For all $a, x \in X$, $x I_X a$ holds iff $x = a$, so that $[a]_{I_X} = \{a\}$. In this case, each equivalence class consists of a single element. Intuitively, the relation of logical equality never clumps together two unequal objects. (See the first digraph in Figure 6.1.) At the other extreme, consider the equivalence relation $R = X \times X$. For all $a, x \in X$, $x R a$ is true, so $[a]_R = X$ for all $a \in X$. In this case, there is a single equivalence class consisting of every element of X. This equivalence relation clumps together all objects in X.

Equivalence Classes of $\sim_f$

We saw in §6.2 that for any function $f : X \to Y$, there is an associated equivalence relation $\sim_f$ on X such that for all $a, b \in X$, $a \sim_f b$ iff $f(a) = f(b)$. So for each $a \in X$, the equivalence class of a is

$$[a]_{\sim_f} = \{x \in X : x \sim_f a\} = \{x \in X : f(x) = f(a)\} = f^{-1}[\{f(a)\}].$$

This equivalence class is sometimes called the *fiber* of f over $f(a)$. The figure in the next example explains this terminology.

6.44. Example. Let $X = \{1, 2, 3, 4, 5, 6\}$, $Y = \{r, s, t, u\}$, and define $f : X \to Y$ by $f(1) = f(2) = f(3) = s$, $f(4) = f(6) = t$, and $f(5) = u$. Figure 6.6 shows the arrow diagram for f and the digraph for $\sim_f$ in a single picture. The arrows leading into a given point in the image of f resemble threads in a fiber starting at that point and leading back into the domain of f. The distinct equivalence classes of $\sim_f$ are $[1]_{\sim_f} = \{1, 2, 3\} = [2]_{\sim_f} = [3]_{\sim_f} = f^{-1}[\{s\}]$, $[4]_{\sim_f} = \{4, 6\} = [6]_{\sim_f} = f^{-1}[\{t\}]$, and $[5]_{\sim_f} = \{5\} = f^{-1}[\{u\}]$. As before, we can specify an equivalence class by listing its elements or by picking a representative b and using the name $[b]_{\sim_f}$. In this situation, we can also use fiber notation for an equivalence class, e.g., denoting $\{1, 2, 3\}$ as $f^{-1}[\{s\}]$. The fiber $f^{-1}[\{r\}]$ is the empty set, which is not an equivalence class.

arrow diagram for f

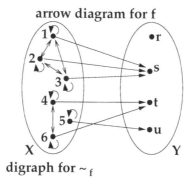

digraph for ∼ f

FIGURE 6.6
A function f and its associated equivalence relation $\sim_f$.

6.45. Example. For $f : \mathbb{R} \to \mathbb{R}$ given by $f(x) = |x|$, the equivalence classes (fibers) are $[x]_{\sim_f} = \{x, -x\}$ for each $x \in \mathbb{R}$. For example, $[3]_{\sim_f} = \{3, -3\} = [-3]_{\sim_f}$, $[-7]_{\sim_f} = \{-7, 7\} = [7]_{\sim_f}$, and $[0]_{\sim_f} = \{0, -0\} = \{0\}$. The function $g : \mathbb{R} \to \mathbb{R}$ given by $g(x) = x^2$ has the same induced equivalence relation and hence the same equivalence classes. The function sgn $: \mathbb{R} \to \mathbb{R}$ has three distinct equivalence classes: the set $\mathbb{R}_{<0}$ of negative numbers, the set $\{0\}$, and the set $\mathbb{R}_{>0}$ of positive numbers. For instance, $[-8]_{\sim_{\mathrm{sgn}}} = [-0.001]_{\sim_{\mathrm{sgn}}} = [-2\pi]_{\sim_{\mathrm{sgn}}} = \mathbb{R}_{<0}$.

6.46. Example. Define $f : \mathbb{R} \to \mathbb{R}^2$ by $f(t) = (\cos t, \sin t)$ for $t \in \mathbb{R}$. By trigonometry, $f(s) = f(t)$ iff $s = t + 2\pi k$ for some integer k. Therefore,

$$[t]_{\sim_f} = \{s \in \mathbb{R} : f(s) = f(t)\} = \{s \in \mathbb{R} : \exists k \in \mathbb{Z}, s = t + 2\pi k\}.$$

For instance, $[0]_{\sim_f} = \{0, \pm 2\pi, \pm 4\pi, \ldots\} = [2\pi]_{\sim_f}$, and $[\pi/3]_{\sim_f} = \{\pi/3, 7\pi/3, 13\pi/3, \ldots, -5\pi/3, -11\pi/3, \ldots\}$. In this example, each equivalence class has infinitely many members (representatives) and hence infinitely many names.

More Examples of Equivalence Classes

6.47. Example. Fix $n \geq 1$, and consider congruence modulo n on the set $\mathbb{Z}$. For brevity, let us write $[a]_n$ instead of $[a]_{\equiv_n}$. For each $a \in \mathbb{Z}$, we have

$$\begin{aligned}
[a]_n &= \{x \in \mathbb{Z} : x \equiv_n a\} = \{x \in \mathbb{Z} : n \text{ divides } x - a\} = \{x \in \mathbb{Z} : \exists k \in \mathbb{Z}, x - a = kn\} \\
&= \{x \in \mathbb{Z} : \exists k \in \mathbb{Z}, x = a + kn\} = \{a + kn : k \in \mathbb{Z}\}.
\end{aligned}$$

So, for example, $[0]_5 = \{0, \pm 5, \pm 10, \pm 15, \ldots\} = [5]_5 = [10]_5 = [-15]_5 = [5k]_5$ for any k in $\mathbb{Z}$; $[1]_4 = \{1, 5, 9, 13, \ldots, -3, -7, -11, \ldots\}$; $[777]_{10} = \{7, 17, 27, 37, \ldots, -3, -13, -23, \ldots\} = [7]_{10}$, and so on. When $n = 2$, the relation $\equiv_2$ has two distinct equivalence classes, the set of even integers and the set of odd integers. We can describe these sets as $[0]_2$ and $[1]_2$, respectively. More generally, we will prove later that $\equiv_n$ has exactly n distinct equivalence classes, which can be named as $[0]_n, [1]_n, \ldots, [n-1]_n$. The set of these n equivalence classes is called *the set of integers modulo n*. We discuss the algebraic structure of the integers modulo n in §6.6.

6.48. Example. Let $X = \mathbb{R} \times \mathbb{R}$. For all $(a, b), (c, d) \in X$, define $(a, b)R(c, d)$ iff $a^2 + b^2 = c^2 + d^2$. R is the equivalence relation induced by the function $f : X \to \mathbb{R}$ given by $f(a, b) = a^2 + b^2$. The equivalence class of $(3, 4) \in X$ is $[(3, 4)]_R = \{(x, y) \in X : x^2 + y^2 = 3^2 + 4^2 = 25\}$, which is the circle of radius 5 centered at the origin. More generally, for any $a_0, b_0 \in \mathbb{R}$, $[(a_0, b_0)] = \{(x, y) \in X : x^2 + y^2 = a_0^2 + b_0^2\}$ is the circle of radius $r_0 = \sqrt{a_0^2 + b_0^2}$ centered at the origin. A degenerate case occurs at the origin: $[(0, 0)]_R = \{(0, 0)\}$. Note that every point in the plane belongs to exactly one circle centered at the origin. So the full set X has been decomposed into a union of equivalence classes (the circles), where any two unequal equivalence classes are disjoint (have empty intersection).

Now consider another relation S on X, defined by $(a, b)S(c, d)$ iff there exists $r \in \mathbb{R}_{>0}$ with $c = ra$ and $d = rb$. We saw in the last section that S is an equivalence relation on X. The equivalence class of $(3, 4) \in X$ determined by S is $[(3, 4)]_S = \{(3r, 4r) : r \in \mathbb{R}_{>0}\}$, which is the ray from the origin through the point $(3, 4)$ [excluding the origin itself]. More generally, each equivalence class $[(a_0, b_0)]_S$ is a ray starting at (but excluding) the origin and passing through (a_0, b_0). As before, the origin is a special case: $[(0, 0)]_S = \{(0, 0)\}$. Once again, X has been decomposed into a union of equivalence classes (the rays, along with the origin), where any two equivalence classes are either disjoint or equal.

6.49. Example. Let X be the set of subsets of $\{1, 2, 3\}$, and define a relation $\sim$ on X by letting $A \sim B$ iff A and B have the same number of elements. Then $\sim$ is an equivalence relation on X with the following equivalence classes: $[\emptyset]_\sim = \{\emptyset\}$;

$$[\{1\}]_\sim = \{\{1\}, \{2\}, \{3\}\} = [\{2\}]_\sim = [\{3\}]_\sim;$$

$$[\{1, 2\}]_\sim = \{\{1, 2\}, \{1, 3\}, \{2, 3\}\} = [\{1, 3\}]_\sim = [\{2, 3\}]_\sim;$$

and $[\{1, 2, 3\}]_\sim = \{\{1, 2, 3\}\}$. Here too, every member of X belongs to exactly one equivalence class, and two distinct equivalence classes do not share any common members.

6.50. Example. Let X be the set of all functions $f : \mathbb{R} \to \mathbb{R}$, and for all $f, g \in X$, define $f \sim g$ iff $f(3) = g(3)$. The equivalence class of the squaring function s (given by $s(x) = x^2$ for $x \in \mathbb{R}$) is $[s]_\sim = \{f \in X : f(3) = s(3)\} = \{f \in X : f(3) = 9\}$. The equivalence class of the identity function is $[\text{Id}_\mathbb{R}]_\sim = \{f \in X : f(3) = 3\}$. The equivalence class of a constant function f_c with constant value $c \in \mathbb{R}$ is $[f_c]_\sim = \{f \in X : f(3) = c\}$. In this example, each equivalence class contains exactly one constant function, which could be singled out as a particularly natural choice of a representative for that equivalence class. The constant tells us the value of f at input 3 for all functions f in the equivalence class.

Properties of Equivalence Classes

The following theorem lists some key features of equivalence classes illustrated by the preceding examples.

6.51. Theorem on Equivalence Classes. Let R be an equivalence relation on a set X.

(a) *When Equivalence Classes are Equal:* For all $a, b \in X$, the following five conditions are equivalent: $[a]_R = [b]_R$; $b \in [a]_R$; bRa; aRb; $a \in [b]_R$.

(b) *Disjointness of Equivalence Classes:* For all equivalence classes S, T of R, if $S \neq T$, then $S \cap T = \emptyset$. So two equivalence classes are either *disjoint* or *equal*.

(c) *Partitioning Property:* For all $a \in X$, there exists a unique equivalence class S of R with $a \in S$, namely $S = [a]_R$. Every equivalence class S is nonempty.

Proof. (a) Fix $a, b \in X$. We give a circle proof that the indicated conditions are equivalent. First assume $[a]_R = [b]_R$. We know bRb since R is reflexive on X, so $b \in [b]_R$ by definition. As the sets $[a]_R$ and $[b]_R$ are equal, we see that $b \in [a]_R$. Next, assume $b \in [a]_R$. By definition of equivalence class, our assumption means that bRa. Next, assume bRa. Then aRb because

R is symmetric. Next, assume aRb. Then $a \in [b]_R$ by definition. Finally, assume $a \in [b]_R$, and prove $[a]_R = [b]_R$. Our assumption means aRb, and hence bRa by symmetry. Let us prove $[a]_R \subseteq [b]_R$. Fix c; assume $c \in [a]_R$; prove $c \in [b]_R$. We have assumed cRa and must prove cRb. Using cRa, aRb, and transitivity, we deduce cRb, as needed. Now we prove $[b]_R \subseteq [a]_R$. Fix d; assume $d \in [b]_R$; prove $d \in [a]_R$. We have assumed dRb and must prove dRa. Since dRb and bRa are known, transitivity gives dRa, as needed. We have now shown that $[a]_R = [b]_R$.

(b) Fix two equivalence classes S, T of R. Using a contrapositive proof, we assume $S \cap T \neq \emptyset$ and prove $S = T$. By definition of equivalence class, there exist $a, b \in X$ with $S = [a]_R$ and $T = [b]_R$. Our assumption means that there exists $c \in S \cap T$. Thus, $c \in [a]_R$ and $c \in [b]_R$, so that cRa and cRb by definition. Now symmetry gives aRc, and transitivity gives aRb. By part (a), we conclude that $[a]_R = [b]_R$, which means that $S = T$.

(c) Fix $a \in X$. By reflexivity, aRa, and hence $a \in [a]_R$. Thus, $S = [a]_R$ is an equivalence class of R containing a. To prove uniqueness, suppose T and U are any two equivalence classes of R with $a \in T$ and $a \in U$; we must prove $T = U$. This follows from part (b), since $a \in T \cap U$ and hence $T \cap U \neq \emptyset$. Finally, any equivalence class V has the form $V = [b]_R$ for some $b \in X$. Since $b \in [b]_R$ as remarked earlier, V must be nonempty. □

Regarding part (c) of the theorem, note that each $a \in X$ belongs to a unique equivalence class of R; but the *name* of this equivalence class is usually not unique. Part (a) of the theorem tells us all possible names of a given equivalence class: $[b]_R$ is another name for the set $[a]_R$ iff $[b]_R = [a]_R$ iff aRb iff $b \in [a]_R$. This confirms our earlier remark that each representative (member) of a given equivalence class provides one of the possible names for this equivalence class.

Section Summary

1. *Equivalence Classes.* Given an equivalence relation R on a set X and given $b \in X$, the equivalence class of b determined by R is $[b]_R = \{x \in X : xRb\}$. So $x \in [b]_R$ iff xRb iff bRx. A set S is an equivalence class of R iff $\exists b \in X, S = [b]_R$.

2. *Fibers of a Function.* Given $f : X \to Y$ and $y \in Y$, the fiber of f over y is $f^{-1}[\{y\}]$. For any $a \in X$, $[a]_{\sim_f} = \{x \in X : f(a) = f(x)\} = f^{-1}[\{f(a)\}]$, which is the fiber of f over $f(a)$.

3. *Equivalence Classes of Congruence mod n.* For all $n \geq 1$ and $a \in \mathbb{Z}$, $[a]_n = \{x \in \mathbb{Z} : x \equiv a \pmod{n}\} = \{a + kn : k \in \mathbb{Z}\}$.

4. *Theorem on Equivalence Classes.* Given an equivalence relation R on a set X and $a, b \in X$, $[a]_R = [b]_R$ iff aRb iff bRa iff $b \in [a]_R$ iff $a \in [b]_R$. For all equivalence classes S, T of R, if $S \neq T$, then $S \cap T = \emptyset$. Equivalence classes are nonempty, and every $a \in X$ belongs to exactly one equivalence class of R, namely $[a]_R$.

Exercises

1. For the equivalence relation R_3 shown in Figure 6.3, explicitly list all elements in $[k]_{R_3}$ for $1 \leq k \leq 7$.

2. Let $X = \{1, 2, 3, 4, 5\}$ and $R = I_X \cup \{(1,3), (3,1), (3,5), (5,3), (1,5), (5,1)\}$. Find all equivalence classes of R.

3. Let $X = \{1, 2, 3, 4, 5, 6\}$. For each function $f : X \to \{a, b, c, d, e\}$, find all equivalence classes of $\sim_f$.
   (a) $f(x) = c$ for all $x \in X$.
   (b) $f(x) = a$ for odd $x \in X$, $f(x) = e$ for even $x \in X$.

(c) $f(1) = b$, $f(2) = e$, $f(3) = a$, $f(4) = b$, $f(5) = e$, $f(6) = a$.

(d) $f(1) = a$, $f(2) = b$, $f(3) = c$, $f(4) = d$, $f(5) = e$, $f(6) = b$.

4. For each function $f : \mathbb{R} \to \mathbb{R}$, describe the requested equivalence classes of $\sim_f$.
   (a) $f(x) = x^4$; find $[-3]$, $[2]$, and $[0]$.
   (b) $f(x) = \sin x$; find $[0]$, $[\pi/2]$, and $[4\pi/3]$.
   (c) $f(x) = 3x - 7$; find $[2]$, $[5]$, and $[-3]$.
   (d) $f(x) = 1$ if x is rational, and 0 otherwise; find $[3]$, $[9/4]$, $[\pi]$, and $[e]$.
   (e) $f(x) = |x|$ if x is an integer, and 0 otherwise; find $[4]$, $[4.5]$, and $[0]$.

5. For each set X, equivalence relation R, and object $b \in X$, describe the equivalence class $[b]_R$ as explicitly as possible. Informally explain each answer.
   (a) $X = \{1, 2, 3, 4, 5, 6\}$, $R = I_X$, $b = 3$.
   (b) $X = \{1, 2, 3, 4, 5, 6\}$, $R = X \times X$, $b = 3$.
   (c) $X = \{1, 2, 3, 4, 5, 6\}$, $R = (\{1, 3, 5\} \times \{1, 3, 5\}) \cup (\{2, 4, 6\} \times \{2, 4, 6\})$, $b = 3$.
   (d) $X = \mathbb{Z}$, R is congruence mod 2, $b = 7$.
   (e) $X = \mathbb{Z}$, R is congruence mod 7, $b = 7$.

6. For $x \in \mathbb{R}$, let $\lceil x \rceil$ be the least integer $\geq x$. For instance, $\lceil 5 \rceil = 5$, $\lceil \pi \rceil = 4$, and $\lceil -1/2 \rceil = 0$. For $x, y \in \mathbb{R}$, define $(x, y) \in R$ iff $\lceil x \rceil = \lceil y \rceil$. (a) State why R is an equivalence relation on $\mathbb{R}$. (b) Describe the equivalence class $[\pi]_R$. (c) Find three different real numbers x, y, z such that $[\pi]_R = [x]_R = [y]_R = [z]_R$. (d) Draw a picture of the real line showing all the distinct equivalence classes of R.

7. For each equivalence relation R on $X = \mathbb{R}^2$, draw the equivalence classes $[(1, 2)]_R$, $[(3, 4)]_R$, $[(-4, -3)]_R$, $[(1, 1)]_R$, and $[(0, 0)]_R$. Are any of these classes equal?
   (a) $(a, b)R(c, d)$ means $a = c$.
   (b) $(a, b)R(c, d)$ means $b = d$.
   (c) $(a, b)R(c, d)$ means $a - b = c - d$.
   (d) $(a, b)R(c, d)$ means $a^2 + b^2 = c^2 + d^2$.
   (e) $(a, b)R(c, d)$ means $(c, d) = (ra, rb)$ for some $r \in \mathbb{R}_{\neq 0}$.

8. Define an equivalence relation $\sim$ on the set of nonempty subsets of $\{1, 2, 3, 4\}$ by letting $A \sim B$ mean that A and B have the same least element. Find all equivalence classes of $\sim$.

9. Describe an equivalence relation on $\mathbb{R}^2$ whose equivalence classes are squares centered at the origin.

10. Describe an equivalence relation R on $\mathbb{Z}$ such that for each $k \in \mathbb{Z}_{>0}$, R has an equivalence class of size k.

11. Find all positive integers n such that $[-13]_n = [37]_n$.

12. Find all equivalence relations R on $X = \{1, 2, 3, 4, 5\}$ such that $[1]_R = [4]_R$ and $[2]_R = [3]_R$.

13. Prove: for all sets X, all equivalence relations R on X, and all $b \in X$, $[b]_R = R[\{b\}]$.

14. Prove: for all sets X and all equivalence relations R on X, $R = \bigcup_{a \in X}[a]_R \times [a]_R$.

15. Prove: for all sets X and all equivalence relations R, S on X, $R \subseteq S$ iff for all $a \in X$, $[a]_R \subseteq [a]_S$.

16. Prove: for all sets X, all equivalence relations R, S on X, and all $a \in X$, $[a]_{R \cap S} = [a]_R \cap [a]_S$.

6.4   Set Partitions

This section introduces *set partitions*, which are ways of dividing a given set into collections of non-overlapping subsets. For example, the set of equivalence classes of a given equivalence relation on a set X is always a set partition of X. It turns out that this set partition contains enough information to reconstruct the equivalence relation that produced it. This observation leads us to the main theorem of this section, which states (intuitively) that equivalence relations and set partitions are two different ways of capturing the same underlying concept. More formally, for any set X, there is a natural one-to-one correspondence (bijection) between the set of all equivalence relations on X and the set of all set partitions of X.

Definition and Examples of Set Partitions

To motivate the introduction of set partitions, consider Figure 6.7. The top row shows the digraphs of three equivalence relations on the set $X = \{1, 2, 3, 4, 5, 6\}$. The digraphs are rather cluttered because of all the arrows joining pairs of objects in the same equivalence class. The second row of the figure reduces the clutter by omitting all the arrows and circling elements that belong to the same equivalence class. Intuitively, the new diagrams contain exactly the same information as the old diagrams, since we can reconstruct the arrows by drawing all possible arrows between two objects in the same little oval.

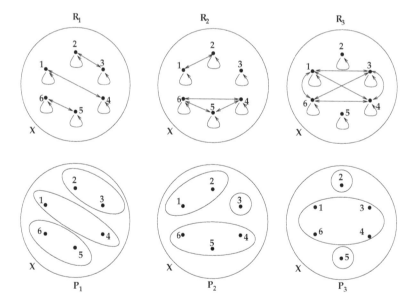

FIGURE 6.7
Digraphs for equivalence relations and the associated set partitions.

We can mathematically model each picture in the second row of Figure 6.7 as a *set of subsets* of X. For example, the first picture is encoded by $P_1 = \{\{2, 3\}, \{1, 4\}, \{6, 5\}\}$. The next two pictures correspond to the sets $P_2 = \{\{1, 2\}, \{3\}, \{4, 5, 6\}\}$ and $P_3 = \{\{2\}, \{1, 3, 4, 6\}, \{5\}\}$. Note that the order in which we list the subsets, as well as the order in which the members of each subset are listed, does not matter. For instance, we could also write $P_2 = \{\{3\}, \{5, 6, 4\}, \{2, 1\}\}$. Also note that the subsets in each P_i do not overlap each other, and every element of X belongs to one of the subsets. Each P_i is an example of a set partition, which we formally define as follows.

6.52. Definition: Set Partitions. For all sets X and P, $\boxed{P \text{ is a set partition of } X}$ iff:

(a) $\forall S \in P, S \neq \emptyset$ and $S \subseteq X$;

(b) $\forall x \in X, \exists S \in P, x \in S$; and

(c) $\forall S, T \in P, S \neq T \Rightarrow S \cap T = \emptyset$.

Each set $S \in P$ is called a *block* of the set partition P.

Condition (a) says that a set partition is a set of nonempty subsets of X. Condition (b) says that every element of X belongs to some subset in the set partition. Condition (c) says that any two blocks of the set partition are either equal or disjoint. When proving part (c), we often check the contrapositive: for all $S, T \in P$, if $S \cap T \neq \emptyset$, then $S = T$. Conditions (a) and (b) together show that $X = \bigcup_{S \in P} S$, so that the whole set X is the union of all the blocks of the set partition. Conditions (b) and (c) are equivalent to the requirement that for all $x \in X$, there *exists* a *unique* block $S \in P$ with $x \in S$ (see Exercise 9).

6.53. Example. Which of the following are set partitions of $X = \{1, 2, 3, 4, 5\}$?

$P_1 = \{\{1, 3, 4\}, \{2, 5\}\}$; $P_2 = \{1, 2, 3, 4, 5\}$; $P_3 = \{\{1, 2, 3, 4, 5\}\}$;
$P_4 = \{\{1, 5\}, \{2, 4\}\}$; $P_5 = \{\{1, 3, 5\}, \{2, 4\}, \emptyset\}$; $P_6 = \{\{1, 2\}, \{3, 4, 5\}, \{2, 1\}\}$;
$P_7 = \{\{1, 2, 3\}, \{3, 4, 5\}\}$; $P_8 = \{\{1\}, \{2\}, \{3\}, \{4\}, \{5\}\}$; $P_9 = \{\{1, 2, 3\}, \{4, 5, 6\}\}$.

Solution. P_1 is a set partition of X with blocks $\{1, 3, 4\}$ and $\{2, 5\}$. P_2 is not a set partition of X because the elements of P_2 are members of X, rather than subsets of X. P_3 is a set partition of X whose only block is the full set $X = \{1, 2, 3, 4, 5\}$. At the other extreme, P_8 is a set partition with five blocks each containing a single element. P_4 is not a set partition of X because $3 \in X$ does not belong to any block of P_4; note P_4 is a set partition of the smaller set $\{1, 2, 4, 5\}$. P_5 is not a set partition because one of its members is the empty set, and all blocks are required to be nonempty. P_6 is a set partition with two distinct blocks $\{1, 2\}$ and $\{3, 4, 5\}$; it does not matter that the block $\{1, 2\}$ was listed twice. P_7 is not a set partition because the two unequal subsets $\{1, 2, 3\}$ and $\{3, 4, 5\}$ have nonempty intersection. Finally, P_9 is not a set partition of X since one of the members of P_9, namely $\{4, 5, 6\}$, is not a subset of X. However, P_9 is a set partition of the larger set $\{1, 2, 3, 4, 5, 6\}$.

The Set Partition Induced by an Equivalence Relation

The major point of this section is that every equivalence relation on X determines a set partition of X, and vice versa. Intuitively, given an equivalence relation R on X, we obtain the associated set partition P by forming the set of all equivalence classes of R. The next definition gives a name to this set partition.

6.54. Definition: The Quotient Set X/R. Given a set X and an equivalence relation R on X, the *quotient set* X/R is the set of all equivalence classes of R. Formally, $X/R = \{[a]_R : a \in X\}$. So for all sets S, $\boxed{S \in X/R}$ iff $\boxed{\exists a \in X, S = [a]_R}$. X/R is also called the *set partition induced by* R.

The fact that X/R really is a set partition of X follows from the Theorem on Equivalence Classes, parts (b) and (c).

6.55. Example. Let $X = \{1, 2, 3, 4, 5, 6\}$. For any equivalence relation R on X, we always have $X/R = \{[1]_R, [2]_R, [3]_R, [4]_R, [5]_R, [6]_R\}$. But this presentation is misleading because a given equivalence class may be listed more than once (compare to the set partition P_6 in Example 6.53). We get a better understanding of the quotient set by choosing a single name for each distinct equivalence class. For example, consider the equivalence relations R_1, R_2, and R_3 from Figure 6.7. We find that

$$X/R_1 = \{\{1, 4\}, \{2, 3\}, \{5, 6\}\} = \{[1]_{R_1}, [2]_{R_1}, [5]_{R_1}\} = \{[4]_{R_1}, [3]_{R_1}, [6]_{R_1}\};$$

$$X/R_2 = \{\{1, 2\}, \{3\}, \{4, 5, 6\}\} = \{[2]_{R_2}, [3]_{R_2}, [5]_{R_2}\};$$

$$X/R_3 = \{\{1, 3, 4, 6\}, \{2\}, \{5\}\} = \{[1]_{R_3}, [2]_{R_3}, [5]_{R_3}\}.$$

We have $X/I_X = \{\{1\}, \{2\}, \{3\}, \{4\}, \{5\}, \{6\}\}$. For $R = X \times X$, $X/R = \{\{1, 2, 3, 4, 5, 6\}\} = \{[k]_R\}$ for any $k \in X$. For the function f from Example 6.44,

$$X/\sim_f = \{\{1, 2, 3\}, \{4, 6\}, \{5\}\} = \{[1]_{\sim_f}, [4]_{\sim_f}, [5]_{\sim_f}\} = \{f^{-1}[\{s\}], f^{-1}[\{t\}], f^{-1}[\{u\}]\}.$$

6.56. Example. For $f : \mathbb{R} \to \mathbb{R}$ given by $f(x) = |x|$,

$$\mathbb{R}/\sim_f = \{[x]_{\sim_f} : x \in \mathbb{R}\} = \{\{x, -x\} : x \in \mathbb{R}_{\geq 0}\}.$$

For the sign function, $\mathbb{R}/\sim_{\text{sgn}} = \{\mathbb{R}_{<0}, \{0\}, \mathbb{R}_{>0}\}$. For $g : \mathbb{R}^2 \to \mathbb{R}$ given by $g(x, y) = x^2 + y^2$, $\mathbb{R}^2/\sim_g$ is the set of all circles centered at the origin (including the degenerate circle consisting of just the origin). For the relation S on $\mathbb{R}^2$ from Example 6.48, $\mathbb{R}^2/S$ is the set of all rays starting at the origin, together with $\{0\}$.

6.57. Example: Integers Modulo n. For fixed $n \geq 1$, the quotient set $\mathbb{Z}/\equiv_n$ is called the *set of integers modulo n*. We often write $\mathbb{Z}/n$ instead of $\mathbb{Z}/\equiv_n$. By definition, $\mathbb{Z}/n = \{[a]_n : a \in \mathbb{Z}\}$. However, this presentation of the elements of $\mathbb{Z}/n$ contains a lot of redundancy. For instance, consider the case $n = 5$. By our calculation of equivalence classes in the previous section, we know that $[0]_5 = \{5k+0 : k \in \mathbb{Z}\}$, $[1]_5 = \{5k+1 : k \in \mathbb{Z}\}$, $[2]_5 = \{5k+2 : k \in \mathbb{Z}\}$, $[3]_5 = \{5k+3 : k \in \mathbb{Z}\}$, and $[4]_5 = \{5k+4 : k \in \mathbb{Z}\}$. These five subsets of $\mathbb{Z}$ already exhaust the entire set $\mathbb{Z}$; see Figure 6.8. Note, for example, that $[5]_5 = [0]_5$ since $5 \in [0]_5$; similarly $[10]_5 = [15]_5 = [-5]_5 = [0]_5$, $[7]_5 = [12]_5 = [17]_5 = [-3]_5 = [2]_5$, and so on. To summarize, $\mathbb{Z}/5$ is the five-element set $\mathbb{Z}/5 = \{[0]_5, [1]_5, [2]_5, [3]_5, [4]_5\}$, where each element of $\mathbb{Z}/5$ is an infinite subset of $\mathbb{Z}$. In §6.6, we prove the general fact that $\mathbb{Z}/n$ consists of the n distinct equivalence classes $[0]_n, [1]_n, \ldots, [n-1]_n$.

The Equivalence Relation Induced by a Set Partition

So far, we have seen how an equivalence relation R on a set X generates an associated set partition X/R. Next, let us start with an arbitrary set partition P on a set X and see how to use P to build an equivalence relation $\sim_P$ on X. The key idea is that two objects in X are related under $\sim_P$ iff those objects belong to the same block of P; compare to Figure 6.7.

6.58. Definition: Equivalence Relation Induced by a Set Partition. Given a set partition P of a set X, define a relation $\sim_P$ on X as follows: for all $a, b \in X$, $\boxed{a \sim_P b}$ iff $\boxed{\exists S \in P, a \in S \text{ and } b \in S}$. We call $\sim_P$ the *equivalence relation on X induced by P*.

We check in a moment that $\sim_P$ really is an equivalence relation on X. First, however, let us consider some examples.

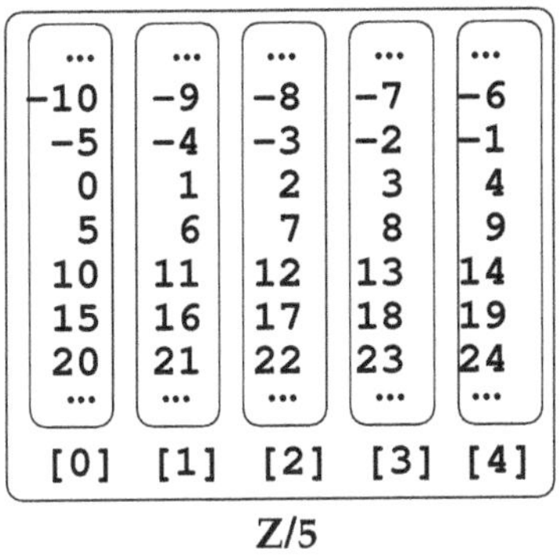

...	...	...	...	...
-10	-9	-8	-7	-6
-5	-4	-3	-2	-1
0	1	2	3	4
5	6	7	8	9
10	11	12	13	14
15	16	17	18	19
20	21	22	23	24
...	...	...	...	...
[0]	[1]	[2]	[3]	[4]

Z/5

FIGURE 6.8
The quotient set $\mathbb{Z}/5$, consisting of five equivalence classes.

6.59. Example. Suppose $X = \{1, 2, 3, 4, 5\}$ and consider the set partitions P_1, P_3, P_6, and P_8 from Example 6.53. Applying the definition to $P_1 = \{\{1, 3, 4\}, \{2, 5\}\}$, we find that

$$\sim_{P_1} = \{(1, 1), (3, 3), (4, 4), (1, 3), (3, 1), (1, 4), (4, 1), (3, 4), (4, 3), (2, 2), (5, 5), (2, 5), (5, 2)\}.$$

Since every $x \in X$ belongs to the same block of P_3, $\sim_{P_3} = \{(a, b) : a, b \in X\} = X \times X$. At the other extreme, $\sim_{P_8} = \{(a, a) : a \in X\} = I_X$. We can write $\sim_{P_6}$ as $(\{1, 2\} \times \{1, 2\}) \cup (\{3, 4, 5\} \times \{3, 4, 5\})$. If we try this construction on $P_4 = \{\{1, 5\}, \{2, 4\}\}$ (which is not a set partition of X), we get a relation

$$\sim_{P_4} = \{(1, 1), (5, 5), (1, 5), (5, 1), (2, 2), (4, 4), (2, 4), (4, 2)\},$$

which is not reflexive on X since $3 \not\sim_{P_4} 3$. However, $\sim_{P_4}$ is an equivalence relation on the smaller set $\{1, 2, 4, 5\}$. Similarly, $\sim_{P_9}$ is not an equivalence relation on X since it is not a relation on X, but it is an equivalence relation on the larger set $\{1, 2, 3, 4, 5, 6\}$. Finally, if we apply the definition above to P_7 (which is not a set partition of any set), we get a non-transitive relation $\sim_{P_7}$, since $1 \sim_{P_7} 3$ and $3 \sim_{P_7} 5$ but $1 \not\sim_{P_7} 5$.

6.60. Lemma. For any set partition P of any set X, $\sim_P$ is an equivalence relation on X.

Proof. Fix a set partition P of a set X. By definition, $\sim_P$ is a relation *on* X; we check $\sim_P$ is reflexive on X, symmetric, and transitive. Fix $a, b, c \in X$. By (b) in the definition of set partition, there exists $S \in P$ such that $a \in S$. So "$\exists S \in P, a \in S \wedge a \in S$" is true, which means that $a \sim_P a$. Thus, $\sim_P$ is reflexive on X. To prove symmetry, assume $a \sim_P b$ and prove $b \sim_P a$. We have assumed $\exists S \in P, a \in S \wedge b \in S$. By commutativity of AND, we see that $\exists S \in P, b \in S \wedge a \in S$, so $b \sim_P a$ holds. To prove transitivity, assume $a \sim_P b$ and $b \sim_P c$; prove $a \sim_P c$. We have assumed there exist blocks $S, T \in P$ with $a \in S$, $b \in S$, $b \in T$, and $c \in T$. The critical observation is that $b \in S \cap T$, so $S \cap T$ is nonempty. By (c) in the definition of set partition, $S = T$ follows. So now $a \in S$ and $c \in S$ with $S \in P$, which means $a \sim_P c$. $\square$

6.61. Example. Let $X = \mathbb{R}$ and $P = \{[n, n+1) : n \in \mathbb{Z}\}$. By drawing a picture of the real number line, it is intuitively evident that P is a set partition of X (see Exercise 11 in §8.5 for a formal proof). The associated equivalence relation $\sim_P$ can be described as follows: for $x, y \in \mathbb{R}$, $x \sim_P y$ iff we obtain the same answer when we round x and y *down* to the next closest integer. Letting $f(x) = \lfloor x \rfloor$ denote x rounded down to the next integer, we see that $\sim_P$ is the equivalence relation $\sim_f$ induced by the *floor function* $f : \mathbb{R} \to \mathbb{Z}$.

The Bijection Between Equivalence Relations and Set Partitions

Fix a set X. We have now discussed a construction that maps any equivalence relation R on X to an associated set partition X/R on X, along with another construction that maps any set partition P on X to an associated equivalence relation $\sim_P$ on X. We now prove the key point that these constructions are inverses of each other. Intuitively, this means that equivalence relations and set partitions are really just two different formalizations of the same intuitive concept of dividing the objects in X into non-overlapping clumps, where all objects in a given clump share some common property.

6.62. Theorem on Equivalence Relations and Set Partitions. Fix a set X, let $\mathcal{EQ}$ be the set of all equivalence relations on X, and let $\mathcal{SP}$ be the set of all set partitions of X. The map $f : \mathcal{EQ} \to \mathcal{SP}$ given by $f(R) = X/R$ for $R \in \mathcal{EQ}$ is a bijection with inverse $g : \mathcal{SP} \to \mathcal{EQ}$ given by $g(P) = \sim_P$ for $P \in \mathcal{SP}$.

Proof. [This proof may be considered optional.] We have already seen that f maps $\mathcal{EQ}$ into the codomain $\mathcal{SP}$ (by the Theorem on Equivalence Classes), and g maps $\mathcal{SP}$ into the codomain $\mathcal{EQ}$ (by Lemma 6.60). It suffices to check that $g \circ f = \mathrm{Id}_{\mathcal{EQ}}$ and $f \circ g = \mathrm{Id}_{\mathcal{SP}}$. For $g \circ f$, fix an equivalence relation $R \in \mathcal{EQ}$, let $P = f(R) = X/R$, and let $R' = g(f(R)) = g(P) = \sim_P$. We must prove $R = R'$. To do so, we fix $a, b \in X$, and we prove aRb iff $aR'b$. First assume aRb, and prove $aR'b$. Since aRb and bRb, we know that $a \in [b]_R$ and $b \in [b]_R$. Now, $[b]_R$ is one of the blocks in the set partition P. Since a and b both belong to this block of P, we have $a \sim_P b$ by definition of $\sim_P$, and hence $aR'b$. Conversely, assume $aR'b$, and prove aRb. We have assumed $a \sim_P b$, which means $\exists S \in P, a \in S \land b \in S$. Now, every block S in $P = X/R$ is an equivalence class of R. So S must have the form $[c]_R$ for some $c \in X$. Since $a, b \in [c]_R$, we know aRc and bRc. Using symmetry and transitivity of R, we conclude aRb.

Next we check $f \circ g$ is the identity function on $\mathcal{SP}$. Fix a set partition $P \in \mathcal{SP}$, let $R = g(P) = \sim_P \in \mathcal{EQ}$, and let $P' = f(g(P)) = f(R) = X/R$. We must prove $P = P'$. First, fix S, assume $S \in P$, and prove $S \in P'$. Since P is a set partition of X, there exists $a \in X$ with $a \in S$. The equivalence class $[a]_R$ is one of the blocks in P', so it suffices to show $S = [a]_R$. On one hand, fix $b \in S$, and prove $b \in [a]_R$. Since b and a both belong to the block S of P, we have $b \sim_P a$, hence bRa, hence $b \in [a]_R$. On the other hand, fix $d \in [a]_R$, and prove $d \in S$. Here dRa, so $d \sim_P a$, so $\exists T \in P, d \in T \land a \in T$. Now, S is the *unique* block of P containing a, so we must have $T = S$. Then $d \in S$, as needed.

Conversely, fix a new S, assume $S \in P'$, and prove $S \in P$. Here we know that S is an equivalence class of $\sim_P$, say $S = [a]_{\sim_P}$ for some $a \in X$. Let T be the unique block in P containing a; it suffices to show $S = T$. On one hand, fix $b \in S$, and prove $b \in T$. Here $b \in [a]_{\sim_P}$, so $b \sim_P a$, so there exists a block U of P containing b and a. By uniqueness, the only block containing a is T; so $U = T$, and $b \in T$ as needed. On the other hand, fix $c \in T$, and prove $c \in S$. Here c and a are in the same block T of the set partition P, so $c \sim_P a$, so $c \in [a]_{\sim_P} = S$, as needed. $\square$

Section Summary

1. *Set Partitions.* For all sets X and P, P is a set partition of X iff every $S \in P$ is a nonempty subset of X; every $x \in X$ belongs to some $S \in P$; and for all $S, T \in P$, if $S \neq T$, then $S \cap T = \emptyset$. Given a set partition P of X, every $a \in X$ belongs to a unique block of P.

2. *Quotient Sets.* Given an equivalence relation R on a set X, the quotient set X/R is the set of all equivalence classes of R, namely $X/R = \{[a]_R : a \in X\}$. So

$S \in X/R$ iff $\exists a \in X, S = [a]_R$. X/R is always a set partition of X, called the set partition induced by the equivalence relation R.

3. *Equivalence Relation Induced by a Set Partition.* Given a set partition P of a set X, the relation $\sim_P$, defined by $a \sim_P b$ iff $\exists S \in P, a \in S \wedge b \in S$ (where $a, b \in X$), is an equivalence relation on X. Informally, this equivalence relation relates each pair of objects that belong to the same block of P.

4. *Correspondence between Equivalence Relations and Set Partitions.* For any set X, there is a bijection from the set of equivalence relations on X to the set of set partitions of X. This bijection sends an equivalence relation R to the quotient set X/R. The inverse bijection sends a set partition P to the equivalence relation $\sim_P$. Figure 6.7 gives visual intuition for how this correspondence works.

Exercises

1. Which of the following are set partitions of $X = \{v, w, x, y, z\}$? Explain.
   (a) $\{\{v, x, z\}, \{w, y\}\}$
   (b) $\{\{v\}, \{w, z\}, \{x, y\}, \{z, w\}, \{v\}\}$
   (c) $\{\{v, w\}, \{w, x\}, \{y, z\}\}$
   (d) $\{\{v, x, y\}, \{z\}\}$
   (e) $\{v, w, x, y, z\}$
   (f) $\{\{v, w, x, y, z\}\}$
   (g) $\{\{\{v, w, x, y, z\}\}\}$

2. Which of the following are set partitions of $\mathbb{R}$? Explain.
   (a) $\{\mathbb{Z}, \mathbb{R} - \mathbb{Z}\}$ (b) $\{\mathbb{R}\}$ (c) $\{\{x\} : x \in \mathbb{R}\}$ (d) $\{(n - 1, n) : n \in \mathbb{Z}\}$
   (e) $\{(n - 1, n] : n \in \mathbb{Z}\}$ (f) $\{[n - 1, n] : n \in \mathbb{Z}\}$ (g) $\{(-n, n) : n \in \mathbb{Z}\}$.

3. Which of the following are set partitions of some set X? For those that are, say what X is. (a) $\{1, 2, \{3, 4\}\}$ (b) $\{\{1, 2\}, \{3, 4\}\}$ (c) $\{\{\{1, 2\}\}, \{3, 4\}\}$ (d) $\{\{1, 3\}, \{4, 6\}, \{2, 8\}\}$ (e) $\{\emptyset, \{1\}, \{2, 3\}\}$ (f) $\{\{\emptyset\}, \{\{1\}\}, \{\{1, 2\}\}\}$.

4. (a) List all possible set partitions of $X = \{1, 2, 3, 4\}$.
   (b) Use (a) to count the number of equivalence relations on X.

5. For each set X and equivalence relation R on X, explicitly describe the associated set partition X/R.
   (a) $X = \{a, b, c, d, e\}$, $R = I_X \cup \{(a, c), (c, a), (c, e), (e, c), (a, e), (e, a)\}$.
   (b) $X = \{w, x, y, z\}$, $R = X \times X$.
   (c) $X = \mathbb{Z}$, R is $\equiv_7$ (congruence mod 7).
   (d) $X = \mathbb{R}$, xRy means $y - x \in \mathbb{Z}$.
   (e) $X = \mathcal{P}(\{1, 2, 3\})$, $(S, T) \in R$ iff S and T have the same size.
   (f) $X = \mathbb{R}^2$, $(a, b)R(c, d)$ means $b - a^2 = d - c^2$.

6. For each function f in Exercise 2 of §6.2, explicitly describe the set partition associated to the equivalence relation $\sim_f$.

7. For each function $f : X \to Y$, explicitly describe the set partition associated to the equivalence relation $\sim_f$.
   (a) $f : [-1, 1] \to \mathbb{R}$ given by $f(x) = \sqrt{1 - x^2}$ for $x \in [-1, 1]$.
   (b) $f : \mathbb{Z} \to \mathbb{Z}$ given by $f(n) = -7$ for all $n \in \mathbb{Z}$.
   (c) $f : \mathbb{Z}_{\geq 0} \to \mathbb{Z}_{\geq 0}$ given by $f(0) = 0$ and $f(n) = n - 1$ for all $n > 0$.
   (d) $f : \mathbb{Z} \to \mathbb{Z}$ given by $f(n) = n/2$ for n even, and $f(n) = (n - 1)/2$ for n odd.
   (e) $f : \mathbb{R} \to \mathbb{R}$ given by $f(x) = x - \lfloor x \rfloor$ for $x \in \mathbb{R}$.

8. For each set partition P of each set X, describe the induced equivalence relation $\sim_P$ on X.
   (a) $X = \{a, b, c, d\}$, $P = \{\{a, c\}, \{b, d\}\}$.
   (b) $X = \{1, 2, 3, 4, 5, 6\}$, $P = \{\{1, 3, 6\}, \{2, 5\}, \{4\}\}$.
   (c) $X = \mathbb{Z}$, $P = \{\mathbb{Z}_{\leq 0}, \mathbb{Z}_{>3}, \{1, 2, 3\}\}$.
   (d) $X = \mathbb{R}^2$, $P = \{\{a\} \times \mathbb{R} : a \in \mathbb{R}\}$.
   (e) $X = \mathbb{R}^2$, $P = \{\{(a, a + c) : a \in \mathbb{R}\} : c \in \mathbb{R}\}$.

9. Let P be a collection of nonempty subsets of a set X. Prove: P is a set partition of X iff $\forall x \in X, \exists! S \in P, x \in S$.

10. Let A, B, C, D be pairwise disjoint nonempty sets.
    (a) Show that $R = (A \times A) \cup (B \times B) \cup (C \times C) \cup (D \times D)$ is an equivalence relation on $X = A \cup B \cup C \cup D$.
    (b) What is the set partition X/R?

11. (a) Prove: for all sets X, Y, P, Q, if P is a set partition of X and Q is a set partition of Y and $X \cap Y = \emptyset$, then $P \cup Q$ is a set partition of $X \cup Y$. (b) Show that (a) can be false without the hypothesis $X \cap Y = \emptyset$.

12. Prove or disprove: for all sets X, Y, P, Q, if P is a set partition of X and Q is a set partition of Y, then $P \cap Q$ is a set partition of $X \cap Y$.

13. Give an example of a set X and set partitions P, Q of X such that P is a four-element set, Q is an eight-element set, and every member of Q is a subset of some member of P. Is $Q \subseteq P$ here?

14. Prove or disprove: for all sets X, P, Q, if P and Q are set partitions of X and $P \subseteq Q$, then $P = Q$.

15. Let R be an equivalence relation on a set X, and define $p : X \to X/R$ by $p(a) = [a]_R$ for all $a \in X$. (a) Prove p is surjective. (b) Prove that R equals $\sim_p$. [This means that *every* equivalence relation has the form $\sim_f$ for an appropriate choice of f.] (c) Prove p is injective iff $R = I_X$.

6.5   Partially Ordered Sets

Recall that the abstract idea of an equivalence relation came from the concrete concept of equality by isolating three key properties of equality: reflexivity, symmetry, and transitivity. In this section, we formulate the abstract concept of ordering by singling out appropriate properties of the concrete ordering relation $\leq$ on real numbers. This relation has the following properties: for all $x, y, z \in \mathbb{R}$, (1) $x \leq x$ (reflexivity); (2) if $x \leq y$ and $y \leq z$, then $x \leq z$ (transitivity); (3) if $x \leq y$ and $y \leq x$, then $x = y$ (antisymmetry); and (4) $x \leq y$ or $y \leq x$ (comparability). It turns out that many mathematical relations, like subset inclusion and divisibility, satisfy the first three of these properties. Thus we use these properties to define *partial order* relations. Partial orderings that also possess property (4) are called *total* orders. We also discuss *well-ordered* sets, in which every nonempty subset of the set has a least element.

Antisymmetric Relations and Partial Orders

For all real numbers x and y, if $x \leq y$ and $y \leq x$, then $x = y$. The generalization of this property to arbitrary relations is given in the next definition.

6.63. Definition: Antisymmetric Relations. For all relations R:

$\boxed{R \text{ is antisymmetric}}$ iff $\boxed{\forall a, b, \text{ if } (a, b) \in R \text{ and } (b, a) \in R, \text{ then } a = b.}$

Using infix notation, R is antisymmetric iff for all a, b, if aRb and bRa, then $a = b$.

In terms of the digraph for R, antisymmetry means that for any two distinct vertices a and b, the arrows from a to b and from b to a cannot both appear in the digraph. In Figure 6.1 of §6.1, every relation shown is antisymmetric except for R_4. Relation R_1 is both symmetric and antisymmetric.

6.64. Definition: Partial Orders and Posets. For all sets X and all relations R on X:

$\boxed{R \text{ is a partial order on } X}$ iff $\boxed{R \text{ is reflexive on } X, \text{ antisymmetric, and transitive}}$.

The ordered pair (X, R) is called a *partially ordered set* or *poset*.

We often use the $\leq$ symbol, or related symbols such as $\preceq$, to denote a partial order relation.

6.65. Example: Subset Ordering Relation. Let X be any collection of sets. We can define a relation R on X as follows: for all $A, B \in X$, $(A, B) \in R$ iff $A \subseteq B$. We know that for all $A, B, C \in X$, $A \subseteq A$ (reflexivity); if $A \subseteq B$ and $B \subseteq A$, then $A = B$ (antisymmetry); and if $A \subseteq B$ and $B \subseteq C$, then $A \subseteq C$ (transitivity). Thus, R is a partial ordering relation on X, and $(X, \subseteq)$ is a partially ordered set.

6.66. Example: Divisibility Ordering Relation. Let X be any set of positive integers. Define a relation R on X as follows: for all $a, b \in X$, $(a, b) \in R$ iff a divides b. The notation $a|b$ means that a divides b. We know that for all $a, b, c \in \mathbb{Z}_{>0}$, $a|a$ (reflexivity); if $a|b$ and $b|a$, then $a = b$ (antisymmetry); and if $a|b$ and $b|c$, then $a|c$ (transitivity). Thus, $(X, |)$ is a partially ordered set. In contrast, $(\mathbb{Z}, |)$ is *not* a poset, because antisymmetry fails: $2|(-2)$, and $(-2)|2$, yet $2 \neq -2$. To see why antisymmetry works in $\mathbb{Z}_{>0}$, fix positive integers a, b, and assume $a|b$ and $b|a$. Because a and b are positive, we can deduce $a \leq b$ and $b \leq a$ (see Example 2.36). Then $a = b$ because the standard ordering on integers is antisymmetric.

When X is finite, we can visualize a poset $(X, \leq)$ using a picture called a *Haase diagram*. This picture is obtained from the digraph of the relation $\leq$ by omitting all arrows that can be deduced from reflexivity or transitivity. It is customary to position the elements so that each relation arrow goes from a lower element to a higher element; this is possible by antisymmetry. For example, Figure 6.9 shows the Haase diagrams of four posets: the set $\mathcal{P}(\{1, 2, 3\})$ ordered by set inclusion, the set of positive divisors of 30 ordered by divisibility, the set of positive divisors of 12 ordered by divisibility, and the set of positive divisors of 12 ordered by the standard ordering of integers.

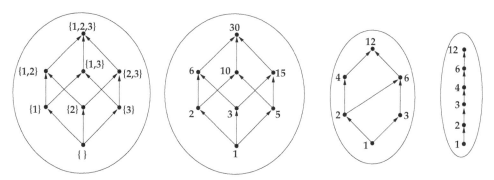

FIGURE 6.9
Haase diagrams for four posets.

The last two examples show that there can be more than one partial order relation on a given set. It also frequently happens in a poset $(X, \preceq)$ that there are *incomparable* pairs of elements a, b where neither $a \preceq b$ nor $b \preceq a$ is true. For instance, in the poset $(\mathbb{Z}_{>0}, |)$, $3|4$ and $4|3$ are both false. In contrast, for any two real numbers x and y, $x \leq y$ or $y \leq x$ must be true. Generalizing this latter property, we arrive at the definition of *total* ordering relations.

6.67. Definition: Total Orders. A relation $\preceq$ on a set X is a $\boxed{\text{total order on } X}$ iff $\preceq$ is a partial order on X such that $\boxed{\forall a, b \in X, \ a \preceq b \text{ or } b \preceq a.}$ The pair $(X, \preceq)$ is called a *totally ordered set*.

We have seen above that $(\mathbb{R}, \leq)$ is a totally ordered set, but $(\mathbb{Z}_{>0}, |)$ is not totally ordered. Similarly, a poset $(X, \subseteq)$ is not totally ordered if we can find subsets $A, B \in X$ such that $A \not\subseteq B$ and $B \not\subseteq A$. For example, if $X = \mathcal{P}(\{1, 2, 3, 4\})$, we may take $A = \{1, 2\}$ and $B = \{2, 3\}$ to see that $(X, \subseteq)$ is not totally ordered. On the other hand, if $S = \{\emptyset, \{1\}, \{1, 3\}, \{1, 3, 4\}, \{1, 2, 3, 4\}\}$, then $(S, \subseteq)$ is a totally ordered set.

Constructing New Posets

We now give more examples of posets by introducing methods of constructing new posets from old ones. The first construction lets us turn a subset of a poset into a partially ordered set.

6.68. Definition: Subposets. Given a poset $(X, \leq)$ and a subset S of X, define a relation $\leq'$ on S as follows: for all $a, b \in S$, let $a \leq' b$ iff $a \leq b$. The relation $\leq'$ is called the *restriction* of $\leq$ to S, and $(S, \leq')$ is called a *subposet* of $(X, \leq)$.

The poset axioms for $(S, \leq')$ follow immediately from the poset axioms for $(X, \leq)$. For example, to check antisymmetry of $\leq'$, suppose $a, b \in S$ satisfy $a \leq' b$ and $b \leq' a$. Then $a \leq b$ and $b \leq a$, so $a = b$ because $\leq$ is antisymmetric. Every subposet of a totally ordered set is also totally ordered. On the other hand, it is quite possible for $(S, \leq')$ to be totally ordered even when $(X, \leq)$ is not totally ordered. We saw an example of this above when $(X, \leq)$ is $(\mathcal{P}(\{1, 2, 3, 4\}), \subseteq)$. For another example, taking $S = \{2^k : k \in \mathbb{Z}_{\geq 0}\}$ gives a totally ordered subposet of the divisibility poset $(\mathbb{Z}_{>0}, |)$. In general, given a poset $(X, \leq)$ and a subset S of X, S is called a *chain* in X iff the subposet $(S, \leq')$ is totally ordered.

We can concatenate a list of non-overlapping posets to obtain a new poset, as follows.

6.69. Definition: Concatenation Posets. Let $(X_1, \leq_1), (X_2, \leq_2), \ldots, (X_n, \leq_n)$ be a list of posets such that the sets $X_1, X_2, \ldots, X_n$ are pairwise disjoint (i.e., $X_i \cap X_j = \emptyset$ for all $i \neq j$). Define a relation $\leq$ on $X = X_1 \cup X_2 \cup \cdots \cup X_n$ as follows: for $a, b \in X$ with $a \in X_i$ and $b \in X_j$, let $a \leq b$ iff $i < j$, or $i = j$ and $a \leq_i b$. The pair $(X, \leq)$ is a poset called the *concatenation* of $X_1, \ldots, X_n$.

Let us check that $\leq$ is transitive. Fix $a, b, c \in X$, assume $a \leq b$ and $b \leq c$, and prove $a \leq c$. We know there are unique $i, j, k \in \{1, 2, \ldots, n\}$ with $a \in X_i$, $b \in X_j$, and $c \in X_k$. Consider various cases. Case 1. Assume $i < j < k$. Then $i < k$ (by transitivity of $<$ on $\mathbb{Z}$), so $a \leq c$. Case 2. Assume $i < j = k$. Here $i < k$, so $a \leq c$. Case 3. Assume $i = j < k$. Here $i < k$, so $a \leq c$. Case 4. Assume $i = j = k$. Here we know $a \leq_i b$ and $b \leq_i c$, so $a \leq_i c$ by transitivity of $\leq_i$. Thus, $a \leq c$ follows by definition of $\leq$. You can check that $\leq$ is reflexive on X and antisymmetric. Moreover, if every $(X_i, \leq_i)$ is totally ordered, then $(X, \leq)$ is also totally ordered.

Our next construction gives a way to turn the product of posets into a poset.

6.70. Definition: Product Posets. Given posets $(X_1, \leq_1), (X_2, \leq_2), \ldots, (X_n, \leq_n)$, define a relation $\preceq$ on $X = X_1 \times X_2 \times \cdots \times X_n$ as follows: for all $a = (a_1, \ldots, a_n)$ and $b = (b_1, \ldots, b_n)$ in X, define $a \preceq b$ iff $a_i \leq b_i$ for all $i \in \{1, 2, \ldots, n\}$. The pair $(X, \preceq)$ is called the *product poset* with factors $(X_i, \leq_i)$.

We check that $(X, \preceq)$ is transitive, letting the reader verify reflexivity and antisymmetry. Fix $a, b, c \in X$, and write $a = (a_1, \ldots, a_n)$, $b = (b_1, \ldots, b_n)$, $c = (c_1, \ldots, c_n)$ for some $a_i, b_i, c_i \in X_i$. Assume $a \preceq b$ and $b \preceq c$, which means $a_i \leq b_i$ and $b_i \leq c_i$ for all $i \in \{1, 2, \ldots, n\}$. Since each relation $\leq_i$ is transitive, we deduce $a_i \leq c_i$ for all i, so $a \preceq c$ holds.

The product of totally ordered posets is usually *not* totally ordered. For example, let $X = \{1, 2, 3\}$ and $Y = \{1, 2, 3, 4\}$ considered as subposets of $(\mathbb{Z}, \leq)$. In the product poset $(X \times Y, \preceq)$, we have $(1, 4) \not\preceq (2, 3)$ and $(2, 3) \not\preceq (1, 4)$, so this poset is not totally ordered. In the exercises, we describe another ordering on product sets called *lexicographic* ordering, which is a total ordering if all the factors are totally ordered.

Greatest and Least Elements

Given a subset S of a partially ordered set, it is often useful to single out the largest and smallest elements in S (when such elements exist).

6.71. Definition: Greatest and Least Elements. Let $(X, \leq)$ be a poset and $S \subseteq X$.

For all z, $\boxed{z \text{ is the greatest element of } S}$ iff $\boxed{z \in S \text{ and } \forall y \in S, y \leq z.}$

For all x, $\boxed{x \text{ is the least element of } S}$ iff $\boxed{x \in S \text{ and } \forall y \in S, x \leq y.}$

We sometimes write $\boxed{z = \max S}$ to mean that z is the greatest (maximum) element of S, and $\boxed{x = \min S}$ to mean that x is the least (minimum) element of S.

The next examples show that not all subsets *have* greatest and least elements. But when these elements exist, they are unique. For suppose z_1 and z_2 are both greatest elements of S. Since z_1 is a greatest element and $z_2 \in S$, $z_2 \leq z_1$. Since z_2 is also a greatest element and $z_1 \in S$, $z_1 \leq z_2$. Then $z_1 = z_2$ follows from antisymmetry. Similarly, least elements are unique when they exist.

6.72. Example. In the poset $(\mathbb{Z}, \leq)$, the subset $\mathbb{Z}_{>0}$ has least element 1 but no greatest element. The subset $\mathbb{Z}_{\leq 5}$ has greatest element 5 and no least element. The subset $\emptyset$ has no least element and no greatest element. Similarly, $\mathbb{Z}$ itself has no least element and no greatest element. Let us prove carefully that the set $\mathbb{Z}_{odd}$ of all odd integers has no greatest element. Assume, to get a contradiction, that $z \in \mathbb{Z}_{odd}$ is the greatest element of this subset. Then $z + 2$ is also odd, and $z + 2 \leq z$ is not true. This contradiction proves that no greatest element exists.

6.73. Example. In the poset $(\mathbb{R}, \leq)$, the half-open interval $(0, 1] = \{x \in \mathbb{R} : 0 < x \leq 1\}$ has greatest element 1 but no least element. Note that 0 is not the least element of $(0, 1]$ because it does not belong to this subset. For any $x_0 \in (0, 1]$, x_0 cannot be the smallest element of $(0, 1]$ since $x_0/2$ is in this subset and $x_0/2 < x_0$. Similarly, for all real $a < b$, we see that $[a, b)$ has least element a but no greatest element, and (a, b) has no least element and no greatest element.

6.74. Example. For any set X, the poset $(\mathcal{P}(X), \subseteq)$ consists of all subsets of X ordered by set inclusion. This poset has least element $\emptyset$ and greatest element X. However, you can show that the subset S of $\mathcal{P}(X)$ consisting of all *nonempty proper* subsets of X has no least element and no greatest element. For example, consider $X = \{1, 2, 3\}$, so $S = \{\{1\}, \{2\}, \{3\}, \{1, 2\}, \{1, 3\}, \{2, 3\}\}$. No one of these six sets is a subset of all other members of S, so S has no least element. Similarly, no one of these six sets contains all the other sets as subsets, so S has no greatest element.

6.75. Example. Let X be the set of positive divisors of 36, ordered by divisibility. X has least element 1 and greatest element 36. Removing these two elements, we get the subset $S = \{2, 3, 4, 6, 9, 12, 18\}$ ordered by divisibility. You may check that S has no least element and no greatest element *with respect to this partial order*. Of course, with respect to the standard ordering $\leq$, S has least element 2 and greatest element 18. On the other hand, now take X to be the set of positive divisors of 32, ordered by divisibility. Here $X = \{1, 2, 4, 8, 16, 32\}$ is totally ordered by $|$. You may verify that every nonempty subset of X has a least element and a greatest element with respect to $|$.

6.76. Example. Suppose $(X_1, \leq_1), \ldots, (X_n, \leq_n)$ are nonempty posets. The concatenation poset $(X, \leq)$ has a least element iff $(X_1, \leq_1)$ has a least element, and these elements are the same. Similarly, the greatest element of $(X, \leq)$ is the greatest element of $(X_n, \leq_n)$ if the latter exists. On the other hand, $a = (a_1, \ldots, a_n)$ is the least element of the product poset $(X_1 \times \cdots \times X_n, \preceq)$ iff a_i is the least element of $(X_i, \leq_i)$ for all i; likewise for greatest elements.

Well-Ordered Posets

In the examples above, we have seen many posets $(X, \leq)$ with no least element. We have also seen that X itself may have a least element, but there could exist nonempty subsets S of X where S has no least element. For example, this happened when $X = \mathbb{R}$ and S is an open interval (a, b). We now give a special name to partially ordered sets where every subset (other than $\emptyset$) has a least element.

6.77. Definition: Well-Ordered Posets. A poset $(X, \leq)$ is $\boxed{\text{well-ordered}}$ iff $\boxed{\text{every nonempty subset of } X \text{ has a least element.}}$

A well-ordered poset must be totally ordered. For if $(X, \leq)$ is well-ordered and $a, b \in X$, then $S = \{a, b\}$ is a nonempty subset of X. So this subset has a least element. If a is the least element of S, then $a \leq b$. If b is the least element of S, then $b \leq a$. Thus, $a \leq b$ or $b \leq a$ holds.

It is intuitively plausible that *every totally ordered finite set is well-ordered.* We delay the formal proof until §7.1, where finite sets are officially defined. On the other hand, $(\mathbb{Z}, \leq)$ and $(\mathbb{R}, \leq)$ are not well-ordered sets, since they have no least element. Although $(\mathbb{R}_{\geq 0}, \leq)$ has a least element (zero), this poset is not well-ordered either, since its subset $\mathbb{R}_{>0}$ has no least element. Intuitively, $(\mathbb{Z}_{\geq 0}, \leq)$ and $(\mathbb{Z}_{>0}, \leq)$ are well-ordered sets, although the proof of this fact is not obvious. In some formal developments of the integers, the well-ordering of $\mathbb{Z}_{>0}$ is postulated as an axiom.

6.78. Well-Ordering Property of $\mathbb{Z}_{>0}$. The poset $(\mathbb{Z}_{>0}, \leq)$ is well-ordered.

We now show that the Well-Ordering Property is logically equivalent to the Induction Axiom from §4.1. [This proof may be considered optional.] First, we assume the Induction Axiom holds and prove the Well-Ordering Property. Fix a subset S of $\mathbb{Z}_{>0}$ that has no least element; we must prove that $S = \emptyset$. To do so, we use strong induction to prove the statement $\forall n \in \mathbb{Z}_{>0}, n \notin S$. Fix $n_0 \in \mathbb{Z}_{>0}$. By strong induction, we may assume that for all integers m in the range $0 < m < n_0$, $m \notin S$. Using this, we must prove that $n_0 \notin S$. Assume, to get a contradiction, that $n_0 \in S$. We derive a contradiction to our original assumption on S by showing that n_0 is the least element of S. Fix $k \in S$, so k is a positive integer. We cannot have $k < n_0$, since otherwise the induction hypothesis would give $k \notin S$. Because $(\mathbb{Z}_{>0}, \leq)$ is totally ordered, we conclude that $n_0 \leq k$. We have now shown that n_0 is the least element of S. This contradiction forces $n_0 \notin S$, completing the proof by induction.

Next, we assume the Well-Ordering Property holds and prove the Induction Axiom. Fix an open sentence $P(n)$. As discussed in §4.1, it is enough to prove that the statement Q: "$P(1) \wedge \forall n \in \mathbb{Z}_{>0}, (P(n) \Rightarrow P(n+1))$" implies the statement R: "$\forall n \in \mathbb{Z}_{>0}, P(n)$." We use proof by contradiction, assuming that Q is true and R is false. To use the Well-Ordering Property, we need a nonempty subset of the positive integers. We define this subset to be $S = \{n \in \mathbb{Z}_{>0} : \sim P(n)\}$, the set of positive integers for which $P(n)$ is false. A useful denial of R is $\exists n \in \mathbb{Z}_{>0}, \sim P(n)$, which means that S is nonempty. The Well-Ordering Property states that there is a least element of S, say n_0. Either $n_0 = 1$ or $n_0 > 1$. In the case $n_0 = 1$, we see that $P(1)$ is false because n_0 belongs to S, but also $P(1)$ is true by statement Q. This is a contradiction. In the case $n_0 > 1$, then $n_0 - 1$ is a positive integer less than n_0. On one hand, n_0 is the *least* element of S, so $n_0 - 1 \notin S$ and $n_0 \in S$. This means that $P(n_0 - 1)$ is true and $P(n_0)$ is false. On the other hand, taking $n = n_0 - 1$ in statement Q, we see that $P(n_0 - 1) \Rightarrow P(n_0)$ is true. This contradicts the truth table for IF, and the proof is complete. [The preceding proof tacitly uses certain additional facts about integers. For example, we needed to know that for any integer $n_0 > 1$, $n_0 - 1$ is also a positive integer. For a careful development of the integers where all necessary properties are derived from axioms, see Chapter 8. In particular, compare the discussion here to Lemma 8.41 and the proof of Theorem 8.51.]

Section Summary

1. *Partial Orders and Total Orders.* A relation $\leq$ on a set X is a partial order iff $\leq$ is reflexive on X, transitive, and antisymmetric. This means that for all $x, y, z \in X$,

$x \leq x$; if $x \leq y$ and $y \leq z$, then $x \leq z$; and if $x \leq y$ and $y \leq x$, then $x = y$. A total order is a partial order such that for all $x, y \in X$, $x \leq y$ or $y \leq x$.

2. *Examples of Posets.* Any set of sets is partially ordered by set inclusion. Any set of positive integers is partially ordered by divisibility. Any subset of a poset is a poset with the restricted ordering. The concatenation of a list of pairwise disjoint posets is a poset. The product of posets is a poset if we let $(a_1, \ldots, a_n) \leq (b_1, \ldots, b_n)$ mean $a_i \leq_i b_i$ for all i.

3. *Greatest and Least Elements.* If S is any subset of a poset $(X, \leq)$, z is the greatest element of S iff $z \in S$ and $\forall y \in S, y \leq z$; x is the least element of S iff $x \in S$ and $\forall y \in S, x \leq y$. Greatest and least elements do not always exist, but they are unique when they do exist.

4. *Well-Ordered Sets.* A poset $(X, \leq)$ is well-ordered iff every nonempty subset of X has a least element. The Well-Ordering Property states that $\mathbb{Z}_{>0}$ with the standard ordering is well-ordered. This property is equivalent to the Induction Axiom.

Exercises

1. Which relations shown in Figure 6.3 are antisymmetric?

2. Give a specific example of a relation R that is not symmetric and not antisymmetric.

3. Find all total order relations on the set $X = \{1, 2, 3\}$.

4. Draw Haase diagrams for each poset.
   (a) the set of positive divisors of 6 ordered by divisibility.
   (b) the set of positive divisors of 8 ordered by divisibility.
   (c) the set of positive divisors of 100 ordered by divisibility.
   (d) the set of all subsets of $\{a, b, c, d\}$ ordered by set inclusion.

5. Let $X = \{1, 2, 3, 4, 5\}$. A Haase diagram for the poset $(X, \leq)$ has arrows from 1 to 2 to 4 to 5 and arrows from 1 to 3 to 5. (a) Draw this diagram. (b) Describe the relation $\leq$ as a set of ordered pairs. (c) Confirm that $\leq$ really is a partial order on X.

6. (a) Find two different chains of maximum size in the poset $(\mathcal{P}(\{1, 2, 3, 4\}), \subseteq)$. (b) Find all chains of maximum size in the poset of positive divisors of 75 ordered by divisibility.

7. Suppose $(X, \leq)$ is a poset. Define a new relation $\geq$ on X as follows: for all $a, b \in X$, define $a \geq b$ iff $b \leq a$. Prove that $\geq$ is a partial order on X, and $\geq$ is a total ordering iff $\leq$ is a total ordering.

8. (a) Suppose X is a fixed set, and for each i in an index set I, R_i is a partial order on X. Show that $R = \bigcap_{i \in I} R_i$ is also a partial order on X. (b) If every R_i is a total order, must R be a total ordering on X? Prove your answer. (c) Is $S = \bigcup_{i \in I} R_i$ a partial order on X? Prove your answer.

9. (a) Let $(X, \leq)$ be the concatenation of n posets $(X_i, \leq_i)$. Prove $\leq$ satisfies reflexivity and antisymmetry. (b) Prove $\leq$ is a total order on X iff every $\leq_i$ is a total order on X_i.

10. Let $(X, \preceq)$ be the product of n posets $(X_i, \leq_i)$. Prove $\preceq$ satisfies reflexivity and antisymmetry.

11. Suppose $(I, \preceq)$ is a partially ordered index set, and for each $i \in I$, $(X_i, \leq_i)$ is a poset. Assume the sets X_i are pairwise disjoint. Define a relation $\leq$ on $X = \bigcup_{i \in I} X_i$ as in Definition 6.69. Prove that $(X, \leq)$ is a poset.

12. Suppose $(X_i, \leq_i)$ is a poset for $1 \leq i \leq n$. Let $X = X_1 \times X_2 \times \cdots \times X_n$. Define the *lexicographic order* $\leq_{\text{lex}}$ on X as follows: for $a = (a_1, \ldots, a_n)$ and $b = (b_1, \ldots, b_n)$ in X, let $a \leq_{\text{lex}} b$ iff $a = b$ or there exists $i \in \{1, 2, \ldots, n\}$ with $a_j = b_j$ for all $j < i$, $a_i \neq b_i$, and $a_i \leq_i b_i$. (a) Prove the poset axioms for $\leq_{\text{lex}}$. (b) Prove: if every $\leq_i$ is a total ordering, then $\leq_{\text{lex}}$ is a total ordering.

13. (a) Show that the concatenation of n well-ordered sets is well-ordered. (b) Show that the product of n well-ordered sets, ordered lexicographically (see the previous exercise), is well-ordered.

14. Let $Y = \mathcal{P}(\{1, 2\})$, $Z = \{3, 4, 5, 6\}$, and $R = I_Z \cup \{(3, 4), (3, 5), (3, 6)\}$. (a) Draw Haase diagrams for $(Y, \subseteq)$ and (Z, R). (b) Draw a Haase diagram for the concatenation of Y and Z (in this order). (c) Draw a Haase diagram for the concatenation of Z and Y (in this order).

15. Let $X = \{1, 2, 3\}$ and $Y = \{0, 2, 4, 6\}$. Draw a Haase diagram for the product poset $(X, \leq) \times (Y, \leq)$, where $\leq$ is the standard ordering on $\mathbb{Z}$.

16. Determine (with explanation) whether each poset below is well-ordered.
    (a) $\mathbb{Z}_{>0}$ ordered by divisibility.
    (b) $\mathbb{Z}_{\leq a}$ ordered by $\leq$, for fixed $a \in \mathbb{Z}$.
    (c) the set of positive divisors of 80 ordered by divisibility.
    (d) the set of positive divisors of 81 ordered by divisibility.
    (e) $[1, 5]$ ordered by $\leq$.
    (f) $\{1, 2, 3, 4, 5\}$ ordered by $\leq$.

17. Use the Well-Ordering Property for $\mathbb{Z}_{>0}$ to prove that for each $a \in \mathbb{Z}$, $(\mathbb{Z}_{\geq a}, \leq)$ is a well-ordered set.

18. Prove that every nonempty subset of $(\mathbb{Z}_{<0}, \leq)$ has a greatest element.

19. Define a relation R on $\mathbb{Z}$ by setting, for all $a, b \in \mathbb{Z}$,

$$aRb \Leftrightarrow ((0 \leq a \leq b) \vee (b \leq a < 0) \vee (a \geq 0 \wedge b < 0)).$$

Prove that $(\mathbb{Z}, R)$ is a poset, and this poset is well-ordered.

20. Carefully prove the assertions in the last sentence of Example 6.73.

21. Prove the assertions made in Example 6.76.

22. Show that a relation R on a set X is both symmetric and antisymmetric iff $R \subseteq I_X$.

23. A relation $<$ on a set X is called a *strict partial order* on X iff $<$ is transitive and for all $a \in X$, $a < a$ is false. (The last condition is called *irreflexivity*.)
    (a) Given a poset $(X, \leq)$, define $a < b$ to mean $a \leq b$ and $a \neq b$ for $a, b \in X$. Show that $<$ is a strict partial order on X.
    (b) Conversely, given a strict partial order $<$ on a set X, define $a \leq b$ to mean $a < b$ or $a = b$ for $a, b \in X$. Show that $\leq$ is a partial order relation on X.

24. Count the number of partial order relations on $X = \{1, 2, 3\}$.

25. For a fixed set X, let $\mathcal{SP}$ be the set of all set partitions of X. We introduce a partial ordering on $\mathcal{SP}$ as follows: given $P, Q \in \mathcal{SP}$, let $P \preceq Q$ mean that every block of P is contained in a block of Q; i.e., $\forall S \in P, \exists T \in Q, S \subseteq T$. (a) Prove that $\preceq$ is a partial order on $\mathcal{SP}$. (b) Draw the Haase diagram for $(\mathcal{SP}, \preceq)$ when $X = \{1, 2, 3\}$. (c) Draw the Haase diagram for $(\mathcal{SP}, \preceq)$ when $X = \{a, b, c, d\}$.

6.6   Equivalence Relations and Algebraic Structures (Optional)

In this chapter, we have defined equivalence relations, equivalence classes, and quotient sets. These ideas are very abstract, and you may be wondering how these concepts could be at all useful. It turns out that quotient sets can be used to define new algebraic structures generalizing familiar number systems such as the integers. In particular, starting with $\mathbb{Z}$ and the arithmetic operations of integer addition and multiplication, we use quotient sets to build new number systems: the integers modulo n (denoted $\mathbb{Z}/n$) and the rational numbers (denoted $\mathbb{Q}$). In this section, we investigate how the known arithmetic operations on $\mathbb{Z}$ let us introduce new arithmetic operations on $\mathbb{Z}/n$ and $\mathbb{Q}$. The notion of a *well-defined* (or *single-valued*) operation plays a key role here. We will see that facts in $\mathbb{Z}$, such as the commutative law or the distributive law, can be used to deduce corresponding facts about these new number systems.

The Integers Modulo n

We have already introduced the integers modulo n in some earlier examples. For convenience, we review the relevant definitions here. Fix an integer $n \geq 1$. For all $a, b \in \mathbb{Z}$, recall $a \equiv_n b$ means n divides $a - b$. The relation $\equiv_n$ is an equivalence relation on $\mathbb{Z}$, and the equivalence class of $a \in \mathbb{Z}$ is $[a]_{\equiv_n} = \{a + kn : k \in \mathbb{Z}\}$ by Example 6.47. We abbreviate the equivalence class notation to $[a]_n$ or $[a]$, when n is understood.

6.79. Definition: Integers Modulo n. For each fixed integer $n \geq 1$, $\mathbb{Z}/n$ is the quotient set $\mathbb{Z}/\equiv_n$. Thus, $\mathbb{Z}/n = \{[a]_n : a \in \mathbb{Z}\}$ is the set of all equivalence classes of congruence modulo n.

The next theorem shows that $\mathbb{Z}/n$ consists of n distinct equivalence classes.

6.80. Theorem on the Elements of $\mathbb{Z}/n$. For all $n \geq 1$, $\mathbb{Z}/n = \{[0]_n, [1]_n, \ldots, [n-1]_n\}$. Furthermore, the n equivalence classes $[0]_n, [1]_n, \ldots, [n-1]_n$ are all distinct.

Proof. Fix $n \geq 1$. Since $\mathbb{Z}/n = \{[a]_n : a \in \mathbb{Z}\}$, the set $\{[0]_n, [1]_n, \ldots, [n-1]_n\}$ is a subset of $\mathbb{Z}/n$. To establish the reverse set inclusion, fix $[a]_n \in \mathbb{Z}/n$, where $a \in \mathbb{Z}$. We must prove there exists $r \in \{0, 1, \ldots, n-1\}$ with $[a]_n = [r]_n$. By the Division Theorem, we know $a = qn + r$ for some $q, r \in \mathbb{Z}$ with $0 \leq r < n$. Then n divides $a - r = qn$, so $a \equiv_n r$, so $[a]_n = [r]_n$ by part (a) of the Theorem on Equivalence Classes (§6.3). To see that $[0]_n, \ldots, [n-1]_n$ are all distinct, we assume $s, t \in \{0, 1, \ldots, n-1\}$ satisfy $[s]_n = [t]_n$, and we prove $s = t$. Using the Theorem on Equivalence Classes again, we see that $s \equiv_n t$, so n divides $s - t$. Since $0 \leq t < n$, we see that $-n < -t \leq 0$. Adding this inequality to the inequality $0 \leq s < n$, we conclude $-n < s - t < n$. The only integer multiple of n strictly between $-n$ and n is zero, so $s - t = 0$, giving $s = t$. $\square$

Addition and Multiplication Modulo n

Now we introduce versions of addition and multiplication defined on integers modulo n.

6.81. Definition: Addition and Multiplication on $\mathbb{Z}/n$. Fix $n \geq 1$. For all $a, b \in \mathbb{Z}$, we define *addition mod n* by $[a]_n \oplus_n [b]_n = [a+b]_n$ and *multiplication mod n* by $[a]_n \otimes_n [b]_n = [a \cdot b]_n$.

6.82. Example. Take $n = 5$, so $\mathbb{Z}/5 = \{[0], [1], [2], [3], [4]\}$. We can make an *addition table* and a *multiplication table* for arithmetic modulo 5. In these tables, the entry in the row labeled $[a]_n$ and the column labeled $[b]_n$ is $[a]_n \oplus_n [b]_n$ (left table) or $[a]_n \otimes_n [b]_n$ (right table).

$\oplus_5$	[0]	[1]	[2]	[3]	[4]
[0]	[0]	[1]	[2]	[3]	[4]
[1]	[1]	[2]	[3]	[4]	[5]
[2]	[2]	[3]	[4]	[5]	[6]
[3]	[3]	[4]	[5]	[6]	[7]
[4]	[4]	[5]	[6]	[7]	[8]

$\otimes_5$	[0]	[1]	[2]	[3]	[4]
[0]	[0]	[0]	[0]	[0]	[0]
[1]	[0]	[1]	[2]	[3]	[4]
[2]	[0]	[2]	[4]	[6]	[8]
[3]	[0]	[3]	[6]	[9]	[12]
[4]	[0]	[4]	[8]	[12]	[16]

We can get more informative tables by changing the *names* of the outputs so that every entry in the table is one of the standard names [0], [1], [2], [3], or [4]. For example, $[3] \oplus_5 [4] = [7] = [2]$ and $[3] \otimes_5 [4] = [12] = [2]$. Making these changes produces the following tables:

$\oplus_5$	[0]	[1]	[2]	[3]	[4]
[0]	[0]	[1]	[2]	[3]	[4]
[1]	[1]	[2]	[3]	[4]	[0]
[2]	[2]	[3]	[4]	[0]	[1]
[3]	[3]	[4]	[0]	[1]	[2]
[4]	[4]	[0]	[1]	[2]	[3]

$\otimes_5$	[0]	[1]	[2]	[3]	[4]
[0]	[0]	[0]	[0]	[0]	[0]
[1]	[0]	[1]	[2]	[3]	[4]
[2]	[0]	[2]	[4]	[1]	[3]
[3]	[0]	[3]	[1]	[4]	[2]
[4]	[0]	[4]	[3]	[2]	[1]

We must now discuss a crucial technical point about the definitions of $\oplus_n$ and $\otimes_n$. Just as the *outputs* of these operations have many names, so also do the *inputs*. For instance, when we computed $[3] \otimes_5 [4] = [12]$, we chose the particular *names* [3] and [4] for the equivalence classes $\{3 + 5k : k \in \mathbb{Z}\}$ and $\{4 + 5k : k \in \mathbb{Z}\}$ that are being combined by the new addition operation. What happens if we use different names for these two inputs? For instance, we know $[3] = [8]$ and $[4] = [-6] \pmod 5$ by the Theorem on Equivalence Classes. Using the new names, we would compute $[8] \otimes_5 [-6] = [-48]$. This may at first seem to disagree with our previous answer of [12], until we realize that $[-48] = [2] = [12]$. However, it is not yet clear that we must *always* get the same output (possibly with a new name) whenever we change the names of the two inputs to the operations $\oplus_n$ and $\otimes_n$.

To resolve this issue in general, we need to address the following question. Suppose $n \geq 1$ is fixed, and a, a', b, b' are integers such that $[a]_n = [a']_n$ and $[b]_n = [b']_n$. Using the original names $[a]_n$ and $[b]_n$, we would compute $[a]_n \oplus_n [b]_n = [a + b]_n$. Using the new names $[a']_n$ and $[b']_n$, we would instead compute $[a']_n \oplus_n [b']_n = [a' + b']_n$. Is it always true that $[a + b]_n = [a' + b']_n$? The answer is yes, thanks to some previous theorems. By the Theorem on Equivalence Classes (§6.3), our assumptions on a, a', b, b' mean that $a \equiv_n a'$ and $b \equiv_n b'$. Next, by part (b) of the Theorem on Congruence Modulo n (§6.2), we see that $a + b \equiv_n a' + b'$. So the Theorem on Equivalence Classes tells us that $[a + b]_n = [a' + b']_n$. We have now proved that changing the names of the inputs to $\oplus_n$ does not change the output of the operation (although the name of the output may very well change). More briefly, we say that $\oplus_n$ is a *well-defined* operation. Similarly, to say that $\otimes_n$ is well-defined means that

$$\forall a, b, a', b' \in \mathbb{Z}, ([a]_n = [a']_n \wedge [b]_n = [b']_n) \Rightarrow [ab]_n = [a'b']_n.$$

The proof we gave above works here, this time using part (c) of the Theorem on Congruence Modulo n.

We can rephrase our discussion in a more formal way that makes contact with our definition of a function. Addition mod n is really a *function* $S : \mathbb{Z}/n \times \mathbb{Z}/n \to \mathbb{Z}/n$ such that $S([a]_n, [b]_n) = [a]_n \oplus_n [b]_n = [a+b]_n$ for all $a, b \in \mathbb{Z}$. The graph of this addition function is, by definition, the set of ordered pairs $G = \{(([a]_n, [b]_n), [a + b]_n) : a, b \in \mathbb{Z}\}$. To confirm that S really is a function, we must check three things (see §5.4): (a) $G \subseteq (\mathbb{Z}/n \times \mathbb{Z}/n) \times \mathbb{Z}/n$; (b) for all $x \in \mathbb{Z}/n \times \mathbb{Z}/n$, there exists $y \in \mathbb{Z}/n$ with $(x, y) \in G$; and (c) for all $x, x' \in \mathbb{Z}/n \times \mathbb{Z}/n$ and all $y, y' \in \mathbb{Z}/n$, if $(x, y) \in G$ and $(x', y') \in G$ and $x = x'$, then $y = y'$. Part (c) tells us that the proposed function S is *single-valued*: putting equal inputs x and x' into the function must produce equal outputs y and y'. This is exactly what we checked above, taking x to be the ordered pair $([a]_n, [b]_n)$, x' to be the ordered pair $([a']_n, [b']_n)$, $y = [a + b]_n$, and $y' = [a' + b']_n$. Conditions (a) and (b) are immediate in this example, though in other situations these conditions might need to be checked explicitly.

Algebraic Properties of Operations Modulo n

In §2.1, we listed some algebraic properties of the addition and multiplication operations on the set of integers. Almost all of these properties extend to the new operations of addition and multiplication modulo n on $\mathbb{Z}/n$. Furthermore, now that we know the latter operations are well-defined, we can quickly deduce properties for $\mathbb{Z}/n$ from the corresponding properties for $\mathbb{Z}$.

6.83. Theorem on Addition and Multiplication Modulo n. For all $n \geq 1$, Axioms 2.4(a) through (j) in §2.1 (closure, commutativity, associativity, identity, additive inverses, and the distributive law) hold with $\mathbb{Z}$ replaced by $\mathbb{Z}/n$, $+$ replaced by $\oplus_n$, $\cdot$ replaced by $\otimes_n$, 0 replaced by $[0]_n$, and 1 replaced by $[1]_n$.

Proof. We prove a few axioms to illustrate, leaving the others as exercises. To show that addition modulo n is commutative, fix two elements of $\mathbb{Z}/n$, say $[a]_n$ and $[b]_n$ for some $a, b \in \mathbb{Z}$. We compute

$$
\begin{aligned}
[a]_n \oplus_n [b]_n &= [a + b]_n \quad \text{(by definition of } \oplus_n) \\
&= [b + a]_n \quad \text{(since integer addition is commutative)} \\
&= [b]_n \oplus_n [a]_n \quad \text{(by definition of } \oplus_n).
\end{aligned}
$$

To prove the distributive law, fix three elements of $\mathbb{Z}/n$, say $[a]_n, [b]_n, [c]_n$ where $a, b, c \in \mathbb{Z}$. We compute

$$
\begin{aligned}
[a]_n \otimes_n ([b]_n \oplus_n [c]_n) &= [a]_n \otimes_n [b + c]_n \quad \text{(by definition of } \oplus_n) \\
&= [a \cdot (b + c)]_n \quad \text{(by definition of } \otimes_n) \\
&= [(a \cdot b) + (a \cdot c)]_n \quad \text{(by the distributive law in } \mathbb{Z}) \\
&= [a \cdot b]_n \oplus_n [a \cdot c]_n \quad \text{(by definition of } \oplus_n) \\
&= ([a]_n \otimes_n [b]_n) \oplus_n ([a]_n \otimes_n [c]_n) \quad \text{(by definition of } \otimes_n).
\end{aligned}
$$

Assume we have already proved that $[0]_n$ is the identity element for $\oplus_n$. To prove that additive inverses exist, fix $[a]_n \in \mathbb{Z}/n$, where $a \in \mathbb{Z}$. Note that $-a \in \mathbb{Z}$, so $[-a]_n$ is an element of $\mathbb{Z}_n$, and

$$[a]_n \oplus_n [-a]_n = [a + (-a)]_n = [0]_n.$$

Thus, $[-a]_n$ is an additive inverse for $[a]_n$ in $\mathbb{Z}/n$. □

The question of whether $\mathbb{Z}/n$ has zero divisors or multiplicative inverses for nonzero elements is more subtle. Some of the exercises ask you to show that when n is prime, every nonzero element of $\mathbb{Z}/n$ has a multiplicative inverse modulo n, and there are no zero divisors. On the other hand, when $n > 1$ is not prime, $\mathbb{Z}/n$ has zero divisors and some nonzero elements of $\mathbb{Z}/n$ do not have multiplicative inverses.

Constructing Rational Numbers from Integers

In most of this book, we have taken the set $\mathbb{R}$ of real numbers as our starting point, and we have defined the set $\mathbb{Q}$ of rational numbers as a particular subset of $\mathbb{R}$ (namely, the subset of real numbers of the form a/b, where $a \in \mathbb{Z}$ and $b \in \mathbb{Z}_{\neq 0}$). However, it is also possible to *build* the set of rational numbers from the more basic set of integers. We can then introduce addition, multiplication, and an ordering relation on $\mathbb{Q}$ using the corresponding concepts for $\mathbb{Z}$, and prove that these new entities obey the expected axioms. These constructions provide an excellent example of an algebraic application of equivalence relations and quotient sets.

Intuitively, we are going to build $\mathbb{Q}$ from $\mathbb{Z}$ using the notion of fractions, which we have already mentioned above. We want a rational number to be a ratio a/b of two integers a and b, where b cannot be zero. The main difficulty is that the division operation for real numbers is no longer available, so we cannot interpret a/b as "a divided by b." We could try to get around this by introducing the ordered pair (a, b) as a formal model for the intuitive fraction a/b.

But now a new difficulty arises: there are many ways to represent a particular rational number in the form a/b. For example, $3/5 = 6/10 = 9/15 = -3/-5 = \cdots$. However, the ordered pairs $(3, 5)$, $(6, 10)$, etc., are *not* equal! Now the key idea presents itself: we would like to regard all of these ordered pairs as *equivalent* for the purposes of representing fractions. This suggests introducing an equivalence relation on the set of ordered pairs, and using the equivalence class of (a, b) as the formal model for the fraction a/b. We want (a, b) to be equivalent to (c, d) iff $a/b = c/d$, but the latter condition involves the forbidden notion of division. So we rewrite the condition as $ad = bc$, which is a statement that only involves integers and multiplication of integers. At last, we are ready for our new definition of rational numbers.

6.84. Definition: Rational Numbers. Let $X = \mathbb{Z} \times \mathbb{Z}_{\neq 0}$ be the set of ordered pairs (a, b) where a, b are integers and $b \neq 0$. Define an equivalence relation $\sim$ on X by setting $(a, b) \sim (c, d)$ iff $ad = bc$. The equivalence class $[(a, b)]_\sim$ is denoted a/b or $\frac{a}{b}$. The set $X/\sim$ of all equivalence classes is denoted $\mathbb{Q}$. Elements of $\mathbb{Q}$ are called *rational numbers*.

You checked that $\sim$ is an equivalence relation on X in Exercise 14 of §6.2. Look at your solution to that exercise and make sure your proof did not use division! For instance, here is the verification that $\sim$ is transitive. Fix $a, b, c, d, e, f \in \mathbb{Z}$ with $b, d, f \neq 0$; assume $(a, b) \sim (c, d)$ and $(c, d) \sim (e, f)$; prove $(a, b) \sim (e, f)$. We have assumed $ad = bc$ and $cf = de$; we must prove $af = be$. Multiplying the assumptions, we get $(ad)(cf) = (bc)(de)$, or $(afc)d = (bec)d$ since integer multiplication is commutative and associative. We can rewrite this as $(afc - bec)d = 0$ by the distributive law. Since $d \neq 0$ and $\mathbb{Z}$ has no zero divisors, we conclude $afc - bec = 0$, or $(af)c = (be)c$. If c is nonzero, we can similarly deduce that $af = be$, as needed. If $c = 0$, we have $ad = bc = 0 = cf = de$. Since d is nonzero, we see that $a = e = 0$, so $af = 0 = be$, as needed.

Let $a, b, c, d \in \mathbb{Z}$ with $b, d \neq 0$. By the Theorem on Equivalence Classes and the definition of $\sim$, $[(a, b)]_\sim = [(c, d)]_\sim$ iff $(a, b) \sim (c, d)$ iff $ad = bc$. We can now officially say that $\boxed{a/b = c/d}$ iff $\boxed{ad = bc}$, provided we remember that the expressions a/b and c/d are abbreviations for the equivalence classes $[(a, b)]_\sim$ and $[(c, d)]_\sim$.

Addition and Multiplication of Rational Numbers

The next step in the formal development of $\mathbb{Q}$ is to introduce the operations of addition and multiplication for rational numbers. As in the case of $\mathbb{Z}/n$, the key initial point is to verify that these operations are *well-defined*, not depending on which names we use for the inputs to the operations.

6.85. Definition: Addition and Multiplication in $\mathbb{Q}$. Given rational numbers a/b and c/d, define $\boxed{\dfrac{a}{b} + \dfrac{c}{d} = \dfrac{ad + bc}{bd}}$ and $\boxed{\dfrac{a}{b} \cdot \dfrac{c}{d} = \dfrac{ac}{bd}}$.

Using the more cumbersome equivalence class notation for fractions, this definition says that $[(a,b)] + [(c,d)] = [(ad + bc, bd)]$ and $[(a,b)] \cdot [(c,d)] = [(ac, bd)]$.

6.86. Lemma. The operations $+$ and $\cdot$ on $\mathbb{Q}$ are well-defined.

Proof. Fix $a, b, c, d, a', b', c', d' \in \mathbb{Z}$ with $b, d, b', d' \neq 0$. Assume $a/b = a'/b'$ and $c/d = c'/d'$, which means $ab' = ba'$ and $cd' = dc'$. To show that addition is well-defined, we prove that $a/b + c/d = a'/b' + c'/d'$. Expanding the definition, we must show that $(ad + bc)/(bd) = (a'd' + b'c')/(b'd')$, which means $(ad + bc)b'd' = bd(a'd' + b'c')$. Working in $\mathbb{Z}$, we compute

$$(ad + bc)b'd' = adb'd' + bcb'd' = ab'dd' + cd'bb' = ba'dd' + dc'bb' = bd(a'd' + b'c'),$$

as needed. To show that multiplication is well-defined, we prove that $(a/b) \cdot (c/d) = (a'/b') \cdot (c'/d')$, which means $(ac)/(bd) = (a'c')/(b'd')$. Here we must prove $acb'd' = bda'c'$. We compute $acb'd' = (ab')(cd') = (ba')(dc') = bda'c'$, as needed. $\qquad\square$

Now we can prove that addition and multiplication of rational numbers satisfies axioms similar to those in Item 2.4, and moreover every nonzero rational number has a multiplicative inverse.

6.87. Theorem on Algebraic Operations in $\mathbb{Q}$.

(a) *Closure:* For all $x, y \in \mathbb{Q}$, $x + y \in \mathbb{Q}$ and $x \cdot y \in \mathbb{Q}$.

(b) *Commutativity:* For all $x, y \in \mathbb{Q}$, $x + y = y + x$ and $x \cdot y = y \cdot x$.

(c) *Associativity:* For all $x, y, z \in \mathbb{Q}$, $(x + y) + z = x + (y + z)$ and $(x \cdot y) \cdot z = x \cdot (y \cdot z)$.

(d) *Identity:* For all $x \in \mathbb{Q}$, $x + (0/1) = x$ and $x \cdot (1/1) = x$.

(e) *Additive Inverses:* $\forall x \in \mathbb{Q}, \exists y \in \mathbb{Q}, x + y = 0/1$.

(f) *Multiplicative Inverses:* $\forall x \in \mathbb{Q}$, if $x \neq 0/1$, then $\exists y \in \mathbb{Q}, x \cdot y = 1/1$.

(g) *No Zero Divisors:* For all $x, y \in \mathbb{Q}$, if $x \cdot y = 0/1$, then $x = 0/1$ or $y = 0/1$.

Proof. We prove a few parts as illustrations, leaving the others as exercises. First we prove that addition is commutative. Fix $x, y \in \mathbb{Q}$, and write $x = a/b$ and $y = c/d$ for some $a, b, c, d \in \mathbb{Z}$ with $b, d \neq 0$. Note that $x + y = (a/b) + (c/d) = (ad + bc)/(bd)$, whereas $y + x = (c/d) + (a/b) = (cb + da)/(db)$. Since addition and multiplication in $\mathbb{Z}$ are commutative, $ad + bc = cb + da$ and $bd = db$, so $x + y = y + x$. Next we prove that multiplication is associative. Fix $x, y, z \in \mathbb{Q}$, and write $x = a/b$, $y = c/d$, $z = e/f$ for some $a, b, c, d, e, f \in \mathbb{Z}$ with $b, d, f \neq 0$. On one hand, $(x \cdot y) \cdot z = \frac{ac}{bd} \cdot \frac{e}{f} = \frac{(ac)e}{(bd)f}$. On the other hand, $x \cdot (y \cdot z) = \frac{a}{b} \cdot \frac{ce}{df} = \frac{a(ce)}{b(df)}$. Since multiplication in $\mathbb{Z}$ is associative, $(x \cdot y) \cdot z = x \cdot (y \cdot z)$ follows. To check the additive identity axiom, fix $x \in \mathbb{Q}$, and write $x = a/b$ for some $a, b \in \mathbb{Z}$ with $b \neq 0$. Compute $x + 0/1 = a/b + 0/1 = (a \cdot 1 + b \cdot 0)/(b \cdot 1) = a/b = x$. To check the multiplicative inverse axiom, fix $x \in \mathbb{Q}$ with $x \neq 0/1$. Write $x = a/b$ with $a, b \in \mathbb{Z}$ and $b \neq 0$. Since $a/b \neq 0/1$, $a \cdot 1 \neq b \cdot 0$, and hence $a \neq 0$. Thus, $y = b/a$ is a member of $\mathbb{Q}$, and we compute $x \cdot y = (a/b) \cdot (b/a) = (ab)/(ba)$. The fraction $(ab)/(ba)$ equals $1/1$ because $(ab)1 = (ba)1$. Thus, $xy = 1/1$, as needed. $\qquad\square$

We usually write 0 instead of $0/1$ and 1 instead of $1/1$. More generally, for any integer n, we often identify n with the rational number $n/1$. You can check that for all $n, m \in \mathbb{Z}$, $n/1 + m/1 = (n + m)/1$; $(n/1) \cdot (m/1) = (nm)/1$; and if $n \neq m$, then $n/1 \neq m/1$. So the identification of n with $n/1$ preserves equality, addition, and multiplication of integers. This identification also allows us to view $\mathbb{Z}$ as a subset of $\mathbb{Q}$.

Section Summary

1. *Integers Modulo n.* For $n \in \mathbb{Z}_{>0}$, $\mathbb{Z}/n$ consists of the n distinct equivalence classes $[0]_n, [1]_n, \ldots, [n-1]_n$ of the equivalence relation $\equiv_n$. For all $a, b \in \mathbb{Z}$, $[a]_n = [b]_n$ iff $a \equiv_n b$ iff n divides $a - b$. The operations $[a]_n \oplus_n [b]_n = [a+b]_n$ and $[a]_n \otimes_n [b]_n = [ab]_n$ are well-defined and satisfy closure, commutativity, associativity, identity, distributive laws, and existence of additive inverses.

2. *Well-Defined Operations.* To see that an operation defined on equivalence classes is well-defined, one must check that changing the names of the inputs to the operations does not change the value of the output (although the name of the output may change). For example, saying that $\oplus_n$ is well-defined means $\forall a, b, a', b' \in \mathbb{Z}$, if $[a]_n = [a']_n$ and $[b]_n = [b']_n$, then $[a+b]_n = [a'+b']_n$.

3. *Rational Numbers.* Formally, $\mathbb{Q} = X/\sim$, where $X = \mathbb{Z} \times \mathbb{Z}_{\neq 0}$ and $(a, b) \sim (c, d)$ iff $ad = bc$ (where $a, b, c, d \in \mathbb{Z}$ and $b, d \neq 0$). The equivalence class $[(a, b)]_\sim$ is denoted a/b or $\frac{a}{b}$. The operations $a/b + c/d + (ad + bc)/(bd)$ and $(a/b) \cdot (c/d) = (ac)/(bd)$ are well-defined and satisfy closure, commutativity, associativity, identity, inverse, and distributive laws.

Exercises

1. Make an addition table and multiplication table for $\mathbb{Z}/6$. Does $\mathbb{Z}/6$ have zero divisors?

2. Make an addition table and multiplication table for $\mathbb{Z}/7$. Does $\mathbb{Z}/7$ have zero divisors?

3. Fix $n \in \mathbb{Z}_{>0}$, and define $f : \mathbb{Z}/n \to \mathbb{Z}/n$ by $f([a]_n) = [-a]_n$ for all $a \in \mathbb{Z}$. Prove that f is well-defined.

4. Prove the remaining axioms for addition in Theorem 6.83.

5. Prove the remaining axioms for multiplication in Theorem 6.83.

6. (a) Find all $x \in \mathbb{Z}/12$ for which there exist $y \in \mathbb{Z}/12$ with $x \otimes_{12} y = 1$. (b) Repeat (a) for $\mathbb{Z}/13$. (c) Repeat (a) for $\mathbb{Z}/16$.

7. Suppose p is prime. (a) Prove: for all $x \in \mathbb{Z}/p$, if $x \neq [0]_p$ then there exists $y \in \mathbb{Z}/p$ with $x \otimes_p y = [1]_p$. [*Hint:* What is $\gcd(x, p)$?] (b) Prove $\mathbb{Z}/p$ has no zero divisors, i.e., for all $x, y \in \mathbb{Z}/p$, if $x \otimes_p y = [0]_p$, then $x = [0]_p$ or $y = [0]_p$.

8. Suppose $n > 1$ is not prime. (a) Prove there exist nonzero $x, y \in \mathbb{Z}/n$ with $x \otimes_n y = [0]_n$. (b) Prove that some nonzero elements of $\mathbb{Z}/n$ do not have multiplicative inverses.

9. Prove the remaining parts of Theorem 6.87.

10. Define $i : \mathbb{Z} \to \mathbb{Q}$ by $i(m) = m/1$ for $m \in \mathbb{Z}$. Prove that i is one-to-one and for all $m, n \in \mathbb{Z}$, $i(m + n) = i(m) + i(n)$ and $i(mn) = i(m)i(n)$. [The function i lets us identify $\mathbb{Z}$ with a subset of $\mathbb{Q}$.]

11. Which of the following functions are *well-defined*? Prove your answers.
    (a) $f : \mathbb{Q} \to \mathbb{Z}$ given by $f(m/n) = m$ for $m, n \in \mathbb{Z}$ with $n \neq 0$.
    (b) $g : \mathbb{Q}_{\neq 0} \to \mathbb{Q}$ given by $g(m/n) = n/m$ for $m, n \in \mathbb{Z}_{\neq 0}$.
    (c) $h : \mathbb{Q} \to \mathbb{Q}$ given by $h(m/n) = 3m/n$ for $m, n \in \mathbb{Z}$ with $n \neq 0$.
    (d) $k : \mathbb{Q} \to \mathbb{Q}$ given by $k(m/n) = (m + 5)/n$ for $m, n \in \mathbb{Z}$ with $n \neq 0$.

12. (a) Prove: for all $a, b, c \in \mathbb{Z}$ with $b, c \neq 0$, $a/b = (ac)/(bc)$. (b) Show that for every $x \in \mathbb{Q}$, there exists a *unique* pair $(a, b) \in \mathbb{Z}^2$ with $b > 0$, $\gcd(a, b) = 1$, and

$x = a/b$. (Informally, the particular name a/b for x is called the fraction *in lowest terms* representing x.)

13. *Ordering on* $\mathbb{Q}$. Given $x, y \in \mathbb{Q}$, we know $x = a/b$ and $y = c/d$ for some $a, b, c, d \in \mathbb{Z}$ with $b, d \in \mathbb{Z}_{>0}$ (see the previous exercise). Define a relation $\leq$ on $\mathbb{Q}$ by setting $x \leq y$ iff $ad \leq bc$. (a) Prove that this relation is well-defined. (b) Prove that $\leq$ is a total ordering of $\mathbb{Q}$. Also prove that for all $m, n \in \mathbb{Z}$, $m \leq n$ (in $\mathbb{Z}$) iff $m/1 \leq n/1$ (in $\mathbb{Q}$).

14. Fix $n \in \mathbb{Z}_{>1}$. Suppose we try to define an ordering relation $\leq$ on $\mathbb{Z}/n$ as follows: for all $a, b \in \mathbb{Z}$, let $[a]_n \leq [b]_n$ iff $a \leq b$. Give a precise explanation of why this construction fails.

15. Let X be a fixed set. Define a relation $\sim$ on $X \times X$ as follows: for $a, b, c, d \in X$, let $(a, b) \sim (c, d)$ iff $(a = c \wedge b = d) \vee (a = d \wedge b = c)$.
    (a) Prove $\sim$ is an equivalence relation on $X \times X$.
    (b) Let Y be the set of unordered pairs $\{a, b\}$ with $a, b \in X$. Define $f : X/\sim \rightarrow Y$ by $f([(a, b)]_\sim) = \{a, b\}$. Prove that f is a well-defined bijection.

16. Let R be a reflexive, transitive relation on a set X. (a) Define a relation $\sim$ on X as follows: for all $a, b \in X$, let $a \sim b$ iff aRb and bRa. Prove that $\sim$ is an equivalence relation. (b) Define a relation $\leq$ on the quotient set $X/\sim$ as follows: for all $a, b \in X$, $[a]_\sim \leq [b]_\sim$ iff aRb. Prove that $\leq$ is well-defined and $(X/\sim, \leq)$ is a poset. (c) Discuss the effect of the preceding construction when $X = \mathbb{Z}$ and $R = \{(a, b) \in \mathbb{Z}^2 : a|b\}$.

17. This problem outlines a formal construction of $\mathbb{Z}$ (the set of all integers) from $\mathbb{Z}_{\geq 0}$ (the set of nonnegative integers). When solving this problem, take care not to use subtraction or additive inverses. Assume we know: $\forall x, y \in \mathbb{Z}_{\geq 0}, x + 1 = y + 1 \Rightarrow x + y$. (a) Prove by induction on n: $\forall x, y, n \in \mathbb{Z}_{\geq 0}, x + n = y + n \Rightarrow x = y$. (b) Let $X = \mathbb{Z}_{\geq 0} \times \mathbb{Z}_{\geq 0}$, and define a relation $\sim$ on X by $(a, b) \sim (c, d)$ iff $a + d = b + c$ for $a, b, c, d \in \mathbb{Z}_{\geq 0}$. Prove that $\sim$ is an equivalence relation on X. Denote the equivalence class $[(a, b)]$ by $[a - b]$, and let $\mathbb{Z}$ be the set of all such equivalence classes. (c) For $a, b, c, d \in \mathbb{Z}_{\geq 0}$, define $[a - b] + [c - d] = [(a + c) - (b + d)]$ and $[a - b] \cdot [c - d] = [(ac + bd) - (ad + bc)]$. Prove that these operations are well-defined. (d) Check the axioms in 2.4, assuming the corresponding facts for $\mathbb{Z}_{\geq 0}$ are known. (e) Show: for every $x \in \mathbb{Z}$, there exists a unique $a \in \mathbb{Z}_{\geq 0}$ such that $x = [a - 0]$ or $x = [0 - a]$.

18. Let $f : X \rightarrow Y$ be a function, let $Z = f[X]$ be the image of f, and let R be the equivalence relation $\sim_f$. Show that $g : X/R \rightarrow Z$, given by $g([a]_R) = f(a)$ for $a \in X$, is well-defined and bijective.

7

Cardinality

7.1 Finite Sets

The theory of *cardinality* gives us a rigorous way to measure and compare the size of various sets (finite or infinite). The key idea is that two sets X and Y have the same size (cardinality) iff there exists a bijection from X to Y. This definition has many surprising consequences for infinite sets that may challenge your intuition. For example, not all infinite sets have the same size. We will see that, in a sense to be made precise later, $\mathbb{R}$ is strictly larger infinite set than $\mathbb{Q}$, and $\mathcal{P}(\mathbb{R})$ (the set of all subsets of $\mathbb{R}$) is larger than $\mathbb{R}$. On the other hand, $\mathbb{Z}_{>0}$ and $\mathbb{Z}$ and $\mathbb{Q}$ all have the same cardinality, even though we have strict set inclusions $\mathbb{Z}_{>0} \subsetneq \mathbb{Z} \subsetneq \mathbb{Q}$. We begin our study of cardinality by defining finite sets and proving their basic properties. Our theorems about finite sets should agree with your intuition and may even seem to be obvious, but some of these results are surprisingly tricky to prove carefully.

Definition of Finite Sets

In ordinary life, we count a set of objects by pointing to each object in the set and saying "one, two, three, ..." This verbal procedure is essentially setting up a one-to-one correspondence (bijection) from some set of integers $\{1, 2, \ldots, n\}$ onto the given set, as shown in Figure 7.1. We formalize this idea in the next definition. Here and below, we use the notation $\{1, 2, \ldots, n\}$ to denote the set $\{i \in \mathbb{Z} : 1 \leq i \leq n\}$.

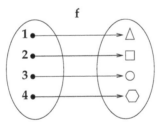

FIGURE 7.1
Counting a set of objects.

7.1. Definition: n-Element Sets. Let X be a set. For all $n \in \mathbb{Z}_{>0}$, $\boxed{X \text{ has } n \text{ elements}}$ iff $\boxed{\text{there exists a bijection } f : \{1, 2, \ldots, n\} \to X}$. We say $\boxed{X \text{ has } 0 \text{ elements}}$ iff $\boxed{X = \emptyset}$. For $n \in \mathbb{Z}_{\geq 0}$, $\boxed{|X| = n}$ means that X has n elements.

Since a function f is a bijection iff f^{-1} exists and is a bijection (see §5.9), we can also say that X has n elements iff there is a bijection $g : X \to \{1, 2, \ldots, n\}$.

7.2. Definition: Finite and Infinite Sets. For all sets X,
$\boxed{X \text{ is finite}}$ iff $\boxed{\exists n \in \mathbb{Z}_{\geq 0}, |X| = n}$. $\boxed{X \text{ is infinite}}$ iff $\boxed{X \text{ is not finite}}$.

Properties of Finite Sets

The next theorem lists the main properties of finite sets.

7.3. Theorem on Finite Sets. For all sets $X, Y, X_1, \ldots, X_s$ (where $s \in \mathbb{Z}_{>0}$):

(a) *Cardinality is Well-Defined:* For all $n, m \in \mathbb{Z}_{\geq 0}$, if $|X| = n$ and $|X| = m$, then $n = m$.

(b) *Subset Property:* If $|X| = n$ and $Y \subseteq X$, then $\exists m \in \mathbb{Z}_{\geq 0}, |Y| = m$ and $m \leq n$.

(c) *Sum Rule:* If $|X| = n$ and $|Y| = k$ and $X \cap Y = \emptyset$, then $|X \cup Y| = n + k$.

(d) *Sum Rule for s Sets:* If $|X_i| = n_i$ for $1 \leq i \leq s$ and $X_i \cap X_j = \emptyset$ for all $i \neq j$, then $|X_1 \cup X_2 \cup \cdots \cup X_s| = n_1 + n_2 + \cdots + n_s$.

(e) *General Sum Rule:* If $|X| = n$, $|Y| = k$, and $|X \cap Y| = i$, then $|X \cup Y| = n + k - i$.

(f) *Product Rule:* If $|X| = n$ and $|Y| = k$, then $|X \times Y| = nk$.

(g) *Product Rule for s Sets:* If $|X_i| = n_i$ for $1 \leq i \leq s$, then $|X_1 \times X_2 \times \cdots \times X_s| = n_1 n_2 \cdots n_s$.

(h) *Power Set Rule:* If $|X| = n$, then $|\mathcal{P}(X)| = 2^n$.

(i) *Finite Set Operations Preserve Finiteness:* If $X_1, \ldots, X_s$ are all finite, then $\bigcup_{i=1}^{s} X_i$, $\bigcap_{i=1}^{s} X_i$, $X_1 \times \cdots \times X_s$, and $\mathcal{P}(X_1)$ are also finite. Every subset of a finite set is finite.

The first two parts are examples of intuitively plausible properties that are not so straightforward to prove rigorously. We give proofs of these results in an optional subsection later. The remaining items are proved below. Throughout this chapter, we frequently use the following terminology. Sets X and Y are called *disjoint* iff $X \cap Y = \emptyset$; and a list of sets $X_1, \ldots, X_s$ is *pairwise disjoint* iff $X_i \cap X_j = \emptyset$ for all $i \neq j$ with $1 \leq i, j \leq s$.

The Sum Rule

Let us now prove the Sum Rule (part (c) of Theorem 7.3). Assume X and Y are disjoint sets with $|X| = n$ and $|Y| = k$; we must prove $|X \cup Y| = n + k$. Our assumption means there exist bijections $f : \{1, 2, \ldots, n\} \to X$ and $g : \{1, 2, \ldots, k\} \to Y$; we need to build a bijection $h : \{1, 2, \ldots, n + k\} \to X \cup Y$. To do this, define $h(i) = f(i)$ for $1 \leq i \leq n$, and define $h(i) = g(i - n)$ for $n + 1 \leq i \leq n + k$ (where i denotes an integer). Figure 7.2 illustrates the construction of h when $n = 3$ and $k = 2$. We prove h is a bijection by exhibiting a two-sided inverse $h' : X \cup Y \to \{1, 2, \ldots, n + k\}$. Each $z \in X \cup Y$ is in exactly one of the sets X or Y, since $X \cap Y = \emptyset$. If $z \in X$, define $h'(z) = f^{-1}(z)$. If $z \in Y$, define $h'(z) = n + g^{-1}(z)$. You can now prove that h and h' are well-defined functions such that $h \circ h' = \mathrm{Id}_{X \cup Y}$ and $h' \circ h = \mathrm{Id}_{\{1,2,\ldots,n+k\}}$. As an example of what needs to be checked, let us see that $h(h'(z)) = z$ when $z \in Y$. By definition of h', we know $h'(z) = n + g^{-1}(z)$. Since g^{-1} maps Y into the set $\{1, 2, \ldots, k\}$, we see that $n + 1 \leq h'(z) \leq n + k$. Hence, applying the definition of h, we get $h(h'(z)) = g(h'(z) - n) = g(g^{-1}(z)) = z = \mathrm{Id}_{X \cup Y}(z) = z$, as needed.

We can now prove the Sum Rule for s Sets (part (d) of the theorem) by induction on s. The result is immediate if $s = 1$, and it holds by what we just proved if $s = 2$. For the induction step, fix $s \in \mathbb{Z}_{\geq 2}$, assume the Sum Rule is known for any union of s pairwise disjoint sets, and prove this rule for a union of $s + 1$ pairwise disjoint sets. Fix arbitrary sets $X_1, \ldots, X_s, X_{s+1}$ with $|X_i| = n_i$ for all i and $X_i \cap X_j = \emptyset$ for all $i \neq j$. The idea is to apply the Sum Rule already proved (part (c) of the theorem) to the sets $X = X_1 \cup \cdots \cup X_s$

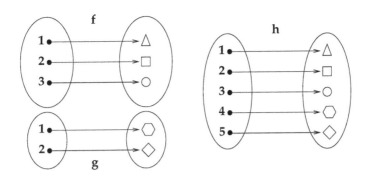

FIGURE 7.2
Building h from f and g.

and $Y = X_{s+1}$. The key observation is that

$$X \cap Y = \left(\bigcup_{i=1}^{s} X_i \right) \cap X_{s+1} = \bigcup_{i=1}^{s} (X_i \cap X_{s+1}) = \bigcup_{i=1}^{s} \emptyset = \emptyset.$$

So we can invoke the Sum Rule to conclude that $|X \cup Y| = |X| + |Y|$. Now, since the sets $X_1, \ldots, X_s$ are pairwise disjoint, the induction hypothesis applies to show that $|X| = n_1 + \cdots + n_s$. Since $|Y| = |X_{s+1}| = n_{s+1}$, we conclude that $|X \cup Y| = (n_1 + \cdots + n_s) + n_{s+1}$, completing the induction step.

Finally, we prove the General Sum Rule (part (e) of the theorem) Fix sets X and Y with $|X| = n$, $|Y| = k$, and $|X \cap Y| = i$. On one hand, $X = (X - Y) \cup (X \cap Y)$ where the two sets $X - Y$ and $X \cap Y$ have empty intersection. By the Sum Rule already proved, $|X| = |X - Y| + |X \cap Y|$. On the other hand, $X \cup Y = (X - Y) \cup Y$ where the two sets $X - Y$ and Y have empty intersection. Using the Sum Rule again, we get $|X \cup Y| = |X - Y| + |Y|$. Inserting the previous equation into this one, we find that $|X \cup Y| = |X| + |Y| - |X \cap Y| = n + k - i$, as needed. This proof tacitly used the fact that $X - Y$ is a *finite* set (being a subset of the finite set X; see part (b) of the theorem), so that $|X - Y| = j$ for some integer j. Similarly, all of our counting formulas make implicit use of the fact that the size of a finite set is well-defined (part (a) in Theorem 7.3).

The Product Rule and Power Set Rule

Before proving the Product Rule (parts (f) and (g) of Theorem 7.3), we need a preliminary lemma. We prove that for any n-element set X and any object b, $|X \times \{b\}| = n$. Fix X and b with $|X| = n$. We know there exists a bijection $f : \{1, 2, \ldots, n\} \to X$, and we must show there is a bijection $g : \{1, 2, \ldots, n\} \to X \times \{b\}$. To do this, define a function $h : X \to X \times \{b\}$ by $h(x) = (x, b)$ for all $x \in X$. We see at once that h is a bijection with two-sided inverse given by $h^{-1}(x, b) = x$ for $(x, b) \in X \times \{b\}$. We may now choose $g = h \circ f$, which is a bijection since it is the composition of two bijections.

Now we prove the Product Rule for two sets. Fix sets X and Y with $|X| = n$ and $|Y| = k$. We know there is a bijection $y : \{1, 2, \ldots, k\} \to Y$. The idea is to slice up the product set $X \times Y$ into k pairwise disjoint sets $X_i = X \times \{y(i)\}$ for $1 \le i \le k$; see Figure 7.3 for an example where $k = 3$, $X = \{1, 2, 3, 4\}$, and $Y = \{1, 2, 4\}$. To see that the sets X_i are pairwise disjoint, fix $i \ne j$ between 1 and k. To get a contradiction, assume there is an ordered pair $(a, b) \in X_i \cap X_j$. Then $y(i) = b = y(j)$, forcing $i = j$ since y is one-to-one. This contradicts $i \ne j$. Next, we claim $X \times Y = X_1 \cup X_2 \cup \cdots \cup X_k$. On one hand, fix

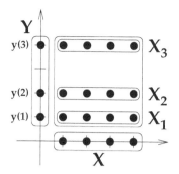

FIGURE 7.3
Decomposing $X \times Y$ into disjoint sets $X_1, \ldots, X_k$.

$(a, b) \in X \times Y$. Then $a \in X$ and $b = y(i)$ for some i, because y is onto. So $(a, b) \in X_i$ for some i, hence $(a, b) \in \bigcup_{i=1}^{k} X_i$. Thus $X \times Y \subseteq \bigcup_{i=1}^{k} X_i$; we let the reader check the reverse set inclusion. By the lemma, $|X_i| = |X| = n$ for $1 \leq i \leq k$. Since $X \times Y$ is the union of the pairwise disjoint sets $X_1, \ldots, X_k$, the Sum Rule gives

$$|X \times Y| = |X_1| + \cdots + |X_k| = n + \cdots + n \ (k \text{ summands}) = nk.$$

We can now prove the Product Rule for s Sets by induction on s, using the case just proved ($s = 2$) to complete the induction step. This proof is very similar to what we did in the proof of the Sum Rule for s Sets, so we leave this as an exercise.

7.4. Example. For $n \in \mathbb{Z}_{>0}$, how many strings of zeroes and ones of length n are there? For example, when $n = 3$, there are eight such strings: 000, 001, 010, 011, 100, 101, 110, and 111. The map sending a binary string $a_1 a_2 \cdots a_n$ to the ordered n-tuple $(a_1, a_2, \ldots, a_n)$ is a bijection from the set of such strings to the product set $\{0, 1\}^n$. You can check that $\{0, 1\}$ is a two-element set. By the Product Rule, $|\{0, 1\}^n| = 2^n$. Thus, there exists a bijection $f : \{1, 2, \ldots, 2^n\} \to \{0, 1\}^n$.

Aided by the preceding example, we can now prove the Power Set Rule. Fix an n-element set X. We can choose a bijection $x : \{1, 2, \ldots, n\} \to X$ so that $X = \{x(1), x(2), \ldots, x(n)\}$. Define a function $g : \{0, 1\}^n \to \mathcal{P}(X)$ as follows. Given $(a_1, \ldots, a_n) \in \{0, 1\}^n$, let $S = g(a_1, \ldots, a_n)$ be the set of all $x(i)$ such that $a_i = 1$. In other words, if $a_i = 0$, then $x(i)$ is not a member of S, and if $a_i = 1$, then $x(i)$ is a member of S. Next define $g' : \mathcal{P}(X) \to \{0, 1\}^n$ as follows. Given $T \subseteq X$, let $g'(T) = (b_1, \ldots, b_n)$, where $b_i = 1$ if $x(i) \in T$, and $b_i = 0$ if $x(i) \notin T$. We ask the reader to check that g and g' are well-defined functions such that $g \circ g'$ and $g' \circ g$ are identity maps. So g is a bijection. Composing g with the bijection f from the example, we get a bijection from $\{1, 2, \ldots, 2^n\}$ onto $\mathcal{P}(X)$. Thus, $|\mathcal{P}(X)| = 2^n$ as needed.

To illustrate this proof, suppose $n = 3$, $X = \{a, b, c\}$, and $x(1) = a$, $x(2) = b$, $x(3) = c$. We find that $g(1, 0, 1) = \{x(1), x(3)\} = \{a, c\}$, $g(0, 1, 0) = \{x(2)\} = \{b\}$, and $g(0, 0, 0) = \{\} = \emptyset$. Going the other way, $g'(\{b, c\}) = g'(\{x(2), x(3)\}) = (0, 1, 1)$, $g'(\{a\}) = g'(\{x(1)\}) = (1, 0, 0)$, and so on.

Preservation of Finiteness

Finally, we use the preceding results to prove part (i) of Theorem 7.3. Informally, this item says that doing *finitely* many set operations on *finite* sets always produces another *finite* set. Fix $s \in \mathbb{Z}_{>0}$ and finite sets $X_1, \ldots, X_s$ with $|X_i| = n_i$ for each i. By the Product Rule,

$|X_1 \times \cdots \times X_s| = n_1 n_2 \cdots n_s$, so $X_1 \times \cdots \times X_s$ is finite. By the Power Set Rule, $|\mathcal{P}(X_1)| = 2^{n_1}$, so $\mathcal{P}(X_1)$ is finite. Item 2 of the theorem immediately implies that any subset of any finite set is finite. Since $\bigcap_{i=1}^{s} X_i$ is a subset of the finite set X_1, this intersection is a finite set.

Proving finiteness of the union of the X_i is a bit more complicated, since the Sum Rule for s Sets required that the given sets be pairwise disjoint. Here we use induction on s. If $s = 1$, $\bigcup_{i=1}^{s} X_i = X_1$, and X_1 was assumed to be finite. Now fix $s \geq 1$, assume that the union of any s finite sets is finite, and prove that the union of any $s + 1$ finite sets is finite. Fix finite sets $X_1, \ldots, X_s, X_{s+1}$. By the induction hypothesis, $X = X_1 \cup \cdots \cup X_s$ is finite, say $|X| = n$. By assumption, X_{s+1} is also finite, say $|X_{s+1}| = k$. The intersection $X \cap X_{s+1}$ is finite, being a subset of the finite set X_{s+1}; say $|X \cap X_{s+1}| = i$. Now the General Sum Rule (part (e) in the theorem) tells us that $|X \cup X_{s+1}| = n + k - i$, so $\bigcup_{i=1}^{s+1} X_i = X \cup X_{s+1}$ is finite.

Two Technical Proofs (Optional)

We now prove part (a) of the Theorem on Finite Sets, which says that the size of a finite set is unique. We use induction on n to prove this statement:

$$\forall n, m \in \mathbb{Z}_{\geq 0}, \forall X, (|X| = n \wedge |X| = m) \Rightarrow n = m.$$

We prove the base case ($n = 0$) by contradiction. Assume, to get a contradiction, that there is a set X and there exists $m \in \mathbb{Z}_{\geq 0}$ with $|X| = 0$ and $|X| = m$ and $0 \neq m$. On one hand, $|X| = 0$ means $X = \emptyset$ by definition. On the other hand, $|X| = m > 0$ means there is a bijection $f : \{1, 2, \ldots, m\} \to X$. Then $f(1) \in X$ contradicts $X = \emptyset$.

For the induction step, fix $n \in \mathbb{Z}_{\geq 0}$, and assume $\forall m \in \mathbb{Z}_{\geq 0}, \forall X, (|X| = n \wedge |X| = m) \Rightarrow n = m$. We must prove $\forall p \in \mathbb{Z}_{\geq 0}, \forall Y, (|Y| = n + 1 \wedge |Y| = p \Rightarrow n + 1 = p$. Fix $p \in \mathbb{Z}_{\geq 0}$, fix a set Y, assume $|Y| = n + 1$ and $|Y| = p$, and prove $n + 1 = p$. Since $n + 1 > 0$, we cannot have $p = 0$; otherwise, the argument in the previous paragraph leads to a contradiction. Thus, we have $p = m + 1$ for some $m \in \mathbb{Z}_{\geq 0}$. By assumption on Y, there are bijections $f : \{1, 2, \ldots, n+1\} \to Y$ and $g : \{1, 2, \ldots, m+1\} \to Y$. Let $y_0 = f(n+1) \in Y$ and $z_0 = g(m + 1) \in Y$. Suppose first that $z_0 = y_0$. You can check that the restrictions $f' = f|_{\{1,2,\ldots,n\}}$ and $g' = g|_{\{1,2,\ldots,m\}}$ are both bijections onto the set $X = Y - \{y_0\}$. Then $|X| = n$ and $|X| = m$, so the induction hypothesis applies to show that $n = m$, and hence $n + 1 = m + 1 = p$ as needed.

If $z_0 \neq y_0$, we reduce to the case just considered as follows. Define $h : Y \to Y$ by setting $h(z_0) = y_0$, $h(y_0) = z_0$, and $h(y) = y$ for all $y \in Y - \{y_0, z_0\}$. You can verify that $h \circ h = \text{Id}_Y$, so h is a bijection with $h^{-1} = h$. The composition $h \circ g$ is a bijection from $\{1, 2, \ldots, m+1\}$ to Y such that $(h \circ g)(m + 1) = h(g(m+1)) = h(z_0) = y_0$. Now we finish as before: the restriction of f to $\{1, 2, \ldots, n\}$ and the restriction of $h \circ g$ to $\{1, 2, \ldots, m\}$ are bijections from their respective domains onto $X = Y - \{y_0\}$, so $|X| = n$ and $|X| = m$, so $n = m$ by induction hypothesis, so $n + 1 = m + 1 = p$. This completes the proof of part (a).

We now use induction on n to prove part (b) of the Theorem on Finite Sets. We must show: $\forall n \in \mathbb{Z}_{\geq 0}, \forall X, \forall Y, (|X| = n \wedge Y \subseteq X) \Rightarrow (\exists m \in \mathbb{Z}_{\geq 0}, |Y| = m \wedge m \leq n)$. For the base case, take $n = 0$, fix sets X and Y, and assume $|X| = 0$ and $Y \subseteq X$. By definition, $|X| = 0$ means $X = \emptyset$. We assumed $Y \subseteq \emptyset$, and we know $\emptyset \subseteq Y$, so $Y = \emptyset$. Thus $Y = \emptyset$ and $|Y| = 0$. So we may choose $m = 0$ in the statement to be proved.

For the induction step, fix $n \in \mathbb{Z}_{\geq 0}$, and assume

$$\forall X', \forall Y', (|X'| = n \wedge Y' \subseteq X') \Rightarrow (\exists m \in \mathbb{Z}_{\geq 0}, |Y'| = m \wedge m \leq n).$$

We must prove

$$\forall X, \forall Y, (|X| = n + 1 \wedge Y \subseteq X) \Rightarrow (\exists p \in \mathbb{Z}_{\geq 0}, |Y| = p \wedge p \leq n + 1).$$

Fix sets X and Y, assume $|X| = n + 1$ and $Y \subseteq X$, and prove $\exists p \in \mathbb{Z}_{\geq 0}, |Y| = p \leq n + 1$. Choose a bijection $f : \{1, 2, \ldots, n + 1\} \to X$, define $x_0 = f(n + 1) \in X$, and note $f' = f|_{\{1,2,\ldots,n\}}$ is a bijection from $\{1, 2, \ldots, n\}$ onto $X' = X - \{x_0\}$. Thus, $|X'| = n$. We know $x_0 \notin Y$ or $x_0 \in Y$, so consider two cases. <u>Case 1.</u> Assume $x_0 \notin Y$. Since we also assumed $Y \subseteq X$, we see that $Y \subseteq X'$ in this case. We can apply the induction hypothesis taking $Y' = Y$ to see that for some nonnegative integer $m \leq n$, $|Y| = m$. Choosing $p = m$, we have $|Y| = p \leq n + 1$ as needed. <u>Case 2.</u> Assume $x_0 \in Y$. In this case, we take $Y' = Y - \{x_0\}$ in the induction hypothesis, which is allowed since Y' is a subset of X'. This leads to the existence of $m \leq n$ with $|Y'| = m$, which means there is a bijection $g : \{1, 2, \ldots, m\} \to Y'$. (If $m = 0$, $g : \emptyset \to \emptyset$ is the empty function.) Define a function $h : \{1, 2, \ldots, m + 1\} \to Y$ by setting $h(i) = g(i)$ for $1 \leq i \leq m$, and $h(m + 1) = x_0$. We ask the reader to check that h is a bijection. Thus, $|Y| = m + 1 \leq n + 1$, so that the required existence statement holds with $p = m + 1$. This completes the proof of part (b).

Section Summary

1. *Finite Sets.* A set X has n elements iff there is a bijection $f : \{1, 2, \ldots, n\} \to X$, in which case we write $|X| = n$. A set X is finite iff $\exists n \in \mathbb{Z}_{\geq 0}, |X| = n$. The size n is uniquely determined by X. A set X is infinite iff X is not finite.

2. *Sum Rule.* Given a finite list of pairwise disjoint finite sets $X_1, \ldots, X_s$,

$$|X_1 \cup X_2 \cup \cdots \cup X_s| = |X_1| + |X_2| + \cdots + |X_s|.$$

   Given any two finite sets X and Y, $|X \cup Y| = |X| + |Y| - |X \cap Y|$.

3. *Product Rule.* Given a finite list of finite sets $X_1, \ldots, X_s$,

$$|X_1 \times X_2 \times \cdots \times X_s| = \prod_{i=1}^{s} |X_i|.$$

4. *Power Set Rule.* Given an n-element set X, $|\mathcal{P}(X)| = 2^n$.

5. *Preservation of Finiteness.* The union, intersection, and product of finitely many finite sets are finite sets. Each subset of a finite set is finite. The power set of a finite set is finite.

Exercises

1. (a) Prove: for all $n \in \mathbb{Z}_{>0}$, $|\{1, 2, \ldots, n\}| = n$.
   (b) Prove: for all $n \in \mathbb{Z}_{>0}$, $\{0, 1, \ldots, n - 1\}$ has n elements.
   (c) Prove: for all $n \in \mathbb{Z}_{>0}$, $\{-1, -2, \ldots, -n\}$ has size n.

2. (a) Prove that $X = \{1, 3, 5, 7, 9\}$ has 5 elements. (b) Prove that $Y = \{2, 4, 6, 8\}$ has 4 elements. (c) Use (a), (b), and the proof of the Sum Rule to build a bijection $h : \{1, 2, \ldots, 9\} \to X \cup Y$.

3. (a) Prove: for all sets X and Y, if X is finite, then $X - Y$ is finite. (b) Prove: for all sets X and Y, if $X - Y$ and Y are finite, then X is finite.

4. Prove: for all sets X and Y, if $Y \subseteq X$ and Y is infinite, then X is infinite.

5. Verify these details in the proof of the Sum Rule: (a) h is a well-defined function mapping $\{1, 2, \ldots, n + k\}$ into $X \cup Y$. (b) h' is a well-defined function mapping $X \cup Y$ into $\{1, 2, \ldots, n + k\}$. (c) $h \circ h' = \text{Id}_{X \cup Y}$. (d) $h' \circ h = \text{Id}_{\{1,2,\ldots,n+k\}}$.

6. Prove the set inclusion $\bigcup_{i=1}^{k} X_i \subseteq X \times Y$ in the proof of the Product Rule.

7. Prove the general Product Rule (part (g) in the Theorem on Finite Sets) by induction on s.

8. Finish the proof of the Power Set Rule by checking that g and g' are well-defined functions such that $g \circ g'$ and $g' \circ g$ are identity maps.

9. (a) Draw the arrow diagram for a bijection $f : \{1, 2, \ldots, 8\} \to \{0, 1\}^3$. (b) Use (a) and the proof of the Power Set Rule to draw an arrow diagram for a bijection from $\{1, 2, \ldots, 8\}$ onto $\mathcal{P}(\{a, b, c\})$.

10. Prove carefully that the restriction of a bijection $f : \{1, 2, \ldots, n+1\} \to Y$ to $\{1, 2, \ldots, n\}$ is a bijection from $\{1, 2, \ldots, n\}$ onto $Y - \{f(n+1)\}$.

11. Prove: for all finite sets X, Y, Z,

$$|X \cup Y \cup Z| = |X| + |Y| + |Z| - |X \cap Y| - |X \cap Z| - |Y \cap Z| + |X \cap Y \cap Z|.$$

12. Given $c, d \in \mathbb{Z}_{>0}$, define $f : \{0, 1, \ldots, c-1\} \times \{0, 1, \ldots, d-1\} \to \{0, 1, \ldots, cd-1\}$ by $f(i, j) = di + j$. Aided by the Division Theorem, prove that f is injective and surjective.

13. Use the previous problem to give a new proof of part (f) in the Theorem on Finite Sets.

14. Prove carefully that $\mathbb{Z}_{>0}$ is infinite. (*Suggestion:* One approach is to show by induction on n that for all $n \in \mathbb{Z}_{\geq 0}$, $\mathbb{Z}_{>0}$ does not have size n.)

15. Prove: for all sets X and Y and all $n \in \mathbb{Z}_{\geq 0}$, if $|Y| = n$ and there exists an injection $g : X \to Y$, then $|X| = m$ for some $m \leq n$.

16. Prove: for all sets X and Y and all $n \in \mathbb{Z}_{\geq 0}$, if $|X| = n$ and there exists a surjection $f : X \to Y$, then $|Y| = m$ for some $m \leq n$.

17. Prove: for all sets X and Y, if X is finite and $Y \subseteq X$ and $|X| = |Y|$, then $X = Y$.

18. Prove: for all finite sets X and all $f : X \to X$, f is injective iff f is surjective.

19. (a) Prove that every nonempty finite totally ordered set has a least element. (b) Prove that every finite totally ordered set is well-ordered.

7.2   Countably Infinite Sets

Now that we understand finite sets, the next step is to examine *countably infinite* sets, which are the smallest infinite sets. Informally, a set X is countably infinite iff it is possible to list all the elements of X (with no repetitions) in a sequence $a_1, a_2, \ldots, a_n, \ldots$, where the terms of the sequence are indexed by positive integers. Formally, this means that there exists a bijection from $\mathbb{Z}_{>0}$ onto X. We will see that the following sets are all countably infinite: $\mathbb{Z}_{>0}$, $\mathbb{Z}_{\geq 0}$, $\mathbb{Z}$, $\mathbb{Q}$, $\mathbb{Z}^k$, and $\mathbb{Q}^k$ (where k is any positive integer). More generally, the product of finitely many countably infinite sets is still countably infinite.

Cardinality Definitions

The following definition allows us to compare the size (cardinality) of two arbitrary sets.

7.5. Definition: Cardinality. For two sets A and B:

(a) $\boxed{|A| = |B|}$ means $\boxed{\text{there exists a bijection } f : A \to B}$.

(b) $\boxed{|A| \leq |B|}$ means $\boxed{\text{there exists an injective function } g : A \to B}$.

When A is finite with n elements, the symbol $|A|$ denotes the nonnegative integer n. For an infinite set A, we do not attempt to assign a meaning to the symbol $|A|$ by itself; only the longer expressions "$|A| = |B|$" and "$|A| \leq |B|$" have been defined here. We now show that these concepts obey some expected rules of equality and order.

7.6. Theorem on Cardinal Equality and Order. For all sets A, B, C, A', and B':

(a) *Reflexivity:* $|A| = |A|$.

(b) *Symmetry:* If $|A| = |B|$, then $|B| = |A|$.

(c) *Transitivity:* If $|A| = |B|$ and $|B| = |C|$, then $|A| = |C|$.

(d) *Reflexivity of Order:* $|A| \leq |A|$.

(e) *Antisymmetry of Order:* If $|A| \leq |B|$ and $|B| \leq |A|$, then $|A| = |B|$.

(f) *Transitivity of Order:* If $|A| \leq |B|$ and $|B| \leq |C|$, then $|A| \leq |C|$.

(g) *Disjoint Union Property:* If $|A| = |A'|$ and $|B| = |B'|$ and $A \cap B = \emptyset = A' \cap B'$, then $|A \cup B| = |A' \cup B'|$.

(h) *Product Property:* If $|A| = |A'|$ and $|B| = |B'|$, then $|A \times B| = |A' \times B'|$.

(i) *Power Set Property:* If $|A| = |A'|$, then $|\mathcal{P}(A)| = |\mathcal{P}(A')|$.

Proof. Fix sets A, B, C, A', and B'. To prove $|A| = |A|$, we must find a bijection from A to A. The identity map $\mathrm{Id}_A : A \to A$ is such a bijection. For part (b), assume $|A| = |B|$ and prove $|B| = |A|$. We assumed there is a bijection $f : A \to B$. Then $f^{-1} : B \to A$ is a bijection, so $|B| = |A|$. For part (c), assume $|A| = |B|$ and $|B| = |C|$, so there exist bijections $f : A \to B$ and $g : B \to C$. The composition $g \circ f$ is a bijection from A to C, so $|A| = |C|$. Parts (d) and (f) are proved similarly, replacing bijections with injections. But part (e) is much more difficult to prove. This is a famous result called the *Schröder–Bernstein Theorem*, which we prove in §7.3.

For the remaining properties, assume $|A| = |A'|$ and $|B| = |B'|$, which means there are bijections $f : A \to A'$ and $g : B \to B'$. To prove part (h), define $h : A \times B \to A' \times B'$ by $h(a, b) = (f(a), g(b))$ for $(a, b) \in A \times B$, and define $h' : A' \times B' \to A \times B$ by $h'(a', b') =$

$(f^{-1}(a'), g^{-1}(b'))$ for $(a', b') \in A' \times B'$. It is routine to check that $h' \circ h = \mathrm{Id}_{A \times B}$ and $h \circ h' = \mathrm{Id}_{A' \times B'}$, so h and h' are bijections. Thus, $|A \times B| = |A' \times B'|$ as needed. To prove part (g), assume $A \cap B = \emptyset = A' \cap B'$. Define $p : A \cup B \to A' \cup B'$ by letting $p(x) = f(x)$ if $x \in A$, and $p(x) = g(x)$ if $x \in B$. Then p is a bijection because it has a two-sided inverse $p' : A' \cup B' \to A \cup B$ defined by letting $p'(y) = f^{-1}(y)$ if $y \in A'$, and $p'(y) = g^{-1}(y)$ if $y \in B'$. You are asked to prove part (i) as an exercise. $\qquad\square$

Countably Infinite Sets

In Exercise 14 of §7.1, you were asked to prove that $\mathbb{Z}_{>0}$ is an infinite set. The next definition introduces countably infinite sets, which are sets with the same cardinality as $\mathbb{Z}_{>0}$.

7.7. Definition: Countably Infinite Sets. For all sets X:

$\boxed{X \text{ is countably infinite}}$ iff $\boxed{\text{there exists a bijection } f : \mathbb{Z}_{>0} \to X}$.

Note that X is countably infinite iff $|\mathbb{Z}_{>0}| = |X|$. For all sets X and Y, if $|X| = |Y|$ then X is countably infinite iff Y is countably infinite. This follows from symmetry and transitivity of cardinal equality. The next few propositions provide specific examples of countably infinite sets.

7.8. Proposition. For all $a \in \mathbb{Z}$, $\mathbb{Z}_{>a}$ and $\mathbb{Z}_{\geq a}$ and $\mathbb{Z}_{<a}$ and $\mathbb{Z}_{\leq a}$ are countably infinite sets.

Proof. We prove the statement about $\mathbb{Z}_{\geq a}$ as an example. Fix $a \in \mathbb{Z}$. Writing $\mathbb{Z}_{>0} = \mathbb{Z}_{\geq 1}$, we need to construct a bijection $f : \mathbb{Z}_{\geq 1} \to \mathbb{Z}_{\geq a}$. Define $f(n) = n + a - 1$ for $n \in \mathbb{Z}_{\geq 1}$. Also define $g : \mathbb{Z}_{\geq a} \to \mathbb{Z}_{\geq 1}$ by $g(m) = m - (a - 1)$ for $m \in \mathbb{Z}_{\geq a}$. You can check that f and g map into their claimed codomains, and g is a two-sided inverse of f. Hence, f is the required bijection. $\qquad\square$

7.9. Proposition. $\mathbb{Z}$ is countably infinite.

Proof. We build a bijection $f : \mathbb{Z}_{>0} \to \mathbb{Z}$ as follows. Each $n \in \mathbb{Z}_{>0}$ is either even or odd, but not both. If n is even, define $f(n) = n/2$, which is in the codomain $\mathbb{Z}$. If n is odd, define $f(n) = -(n-1)/2$, which is in the codomain $\mathbb{Z}$. To see that f is a bijection, we describe its inverse $g : \mathbb{Z} \to \mathbb{Z}_{>0}$. For $m \in \mathbb{Z}$, let $g(m) = 2m$ if $m > 0$, and let $g(m) = -2m + 1$ if $m \leq 0$. The arrow diagram in Figure 7.4 illustrates the action of g. You can prove that g maps into $\mathbb{Z}_{>0}$ and is the two-sided inverse for f. We check one case as an example. Given $m \in \mathbb{Z}$ with $m \leq 0$, we must show $f(g(m)) = m$. First, $g(m) = -2m + 1$, which is odd. So $f(g(m)) = -(-2m + 1 - 1)/2 = 2m/2 = m$, as needed. $\qquad\square$

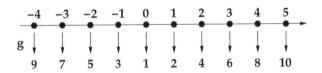

FIGURE 7.4
A bijection g from $\mathbb{Z}_{>0}$ to $\mathbb{Z}_{\geq 0}^2$.

Countability of $\mathbb{Z} \times \mathbb{Z}$

For any set A, recall that A^2 denotes the set $A \times A$, which consists of all ordered pairs (x, y) with $x \in A$ and $y \in A$. We now show that the product set $\mathbb{Z}^2$ and certain related sets are countably infinite. Our proof gives surprising applications of material from Chapter 4.

7.10. Proposition. $\mathbb{Z}_{\geq 0}^2$ and $\mathbb{Z}_{> 0}^2$ and $\mathbb{Z}^2$ are countably infinite sets.

Proof. We first define a bijection $h : \mathbb{Z}_{\geq 0}^2 \to \mathbb{Z}_{>0}$. Let $h(a, b) = 2^a(2b + 1)$ for all $a, b \in \mathbb{Z}_{\geq 0}$. We prove that h is one-to-one and onto. To see that h is injective, fix $a, b, c, d \in \mathbb{Z}_{\geq 0}$ with $h(a, b) = h(c, d)$, and prove $(a, b) = (c, d)$. We have assumed $2^a(2b + 1) = 2^c(2d + 1)$. We know $a > c$ or $a < c$ or $a = c$, so consider three cases.
<u>Case 1.</u> Assume $a > c$. Rewrite the assumption as $2^{a-c}(2b + 1) = 2d + 1$. The left side of this equation is an even integer since $a - c > 0$, but the right side is an odd integer. This contradicts the earlier theorem that no integer is both even and odd.
<u>Case 2.</u> Assume $a < c$. Here we can write $2b + 1 = 2^{c-a}(2d + 1)$ with $c - a > 0$. Again there is a contradiction because the left side is odd and the right side is even.
<u>Case 3.</u> Assume $a = c$. Dividing the assumption by $2^a = 2^c$, we get $2b + 1 = 2d + 1$, and hence $b = d$. Thus, $(a, b) = (c, d)$ as needed.

To see that h is surjective, fix $n \in \mathbb{Z}_{>0}$; we must find $(a, b) \in \mathbb{Z}_{\geq 0}^2$ with $h(a, b) = n$. Using the formula for h, we need to solve $2^a(2b + 1) = n$ for a and b. To do so, use the Fundamental Theorem of Arithmetic to write $n = 2^{e_1} p_2^{e_2} \cdots p_k^{e_k}$, where $k \geq 1$, $e_1, \ldots, e_k \geq 0$, and $p_2, \ldots, p_k$ are distinct odd primes. Comparing to the previous expression, we choose $a = e_1$ and $b = (p_2^{e_2} \cdots p_k^{e_k} - 1)/2$ (which is in $\mathbb{Z}_{\geq 0}$) to achieve $h(a, b) = n$. Note that $b = 0$ when $k = 1$.

So far, we have proved that $|\mathbb{Z}_{\geq 0}^2| = |\mathbb{Z}_{>0}|$. We also know $|\mathbb{Z}| = |\mathbb{Z}_{>0}| = |\mathbb{Z}_{\geq 0}|$, so it follows from part (h) of Theorem 7.6 that $|\mathbb{Z} \times \mathbb{Z}| = |\mathbb{Z}_{>0} \times \mathbb{Z}_{>0}| = |\mathbb{Z}_{\geq 0} \times \mathbb{Z}_{\geq 0}| = |\mathbb{Z}_{>0}|$. By transitivity and symmetry, we conclude that $\mathbb{Z}^2$, $\mathbb{Z}_{>0}^2$, and $\mathbb{Z}_{\geq 0}^2$ are all countably infinite sets. $\square$

We may visualize the map h in the previous proof by labeling each point (a, b) in a picture of $\mathbb{Z}_{\geq 0}^2$ with the corresponding positive integer $h(a, b)$, as shown on the left in Figure 7.5. Pictures like this one provide powerful intuition for defining bijections on infinite sets. For example, the picture on the right in Figure 7.5 suggests another possible bijection $k : \mathbb{Z}_{\geq 0}^2 \to \mathbb{Z}_{>0}$, obtained by traversing integer points in the first quadrant one diagonal at a time.

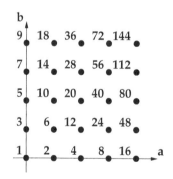
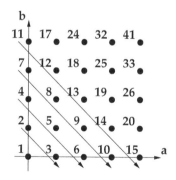

FIGURE 7.5
Pictures of two bijections from $\mathbb{Z}_{>0}$ to $\mathbb{Z}_{\geq 0}^2$.

The new figure gives compelling visual insight into *why* there is a bijection from $\mathbb{Z}_{\geq 0}^2$ to $\mathbb{Z}_{>0}$, but how do we translate the figure into a rigorous proof? One approach is to give a recursive definition of the inverse $f : \mathbb{Z}_{>0} \to \mathbb{Z}_{\geq 0}^2$ for k. For the base case, define $f(1) = (0,0)$. For the recursive case, fix $n \in \mathbb{Z}_{>0}$ and suppose $f(n) = (a,b)$ has already been defined. Motivated by the figure, we set $f(n+1) = (a+1, b-1)$ if $b > 0$, and we set $f(n+1) = (0, a+1)$ if $b = 0$. Intuitively, the first case moves one step down and right along the current diagonal. The second case occurs when n appears at the lower-right end of one diagonal, so that $n+1$ must be placed at the upper-left end of the next diagonal. It can now be proved by induction that f is one-to-one and onto. Remarkably, the map $k = f^{-1}$ is given by a simple non-recursive formula: $k(a,b) = ((a+b)^2 + 3a + b)/2 + 1$ for $(a,b) \in \mathbb{Z}_{\geq 0}^2$. You can check that $k(f(n)) = n$ for all $n \in \mathbb{Z}_{>0}$ by induction on n, and that $f(k(a,b)) = (a,b)$ for all $a, b \in \mathbb{Z}_{\geq 0}^2$ by induction on $a+b$, with an inner induction on a.

Countability of $\mathbb{Q}$ and Finite Products

Now that we know $\mathbb{Z} \times \mathbb{Z}$ is countably infinite, we can show that the set of rational numbers is countably infinite.

7.11. Proposition. $\mathbb{Q}$ is countably infinite.

Proof. By the Schröder-Bernstein Theorem (part (e) of Theorem 7.6), it suffices to show $|\mathbb{Z}_{>0}| \leq |\mathbb{Q}|$ and $|\mathbb{Q}| \leq |\mathbb{Z}_{>0}|$. [In the next section, we give another proof of the countability of $\mathbb{Q}$ not relying on this hard result.] On one hand, the inclusion map $i : \mathbb{Z}_{>0} \to \mathbb{Q}$ given by $i(n) = n/1$ for all $n \in \mathbb{Z}_{>0}$ is one-to-one, so $|\mathbb{Z}_{>0}| \leq |\mathbb{Q}|$ holds by definition. On the other hand, define $f : \mathbb{Q} \to \mathbb{Z} \times \mathbb{Z}$ as follows. Given $q \in \mathbb{Q}$, there exist unique integers m and n with $n > 0$, $\gcd(m,n) = 1$, and $q = m/n$; we define $f(q) = (m,n)$. Now f is injective, because if $q, r \in \mathbb{Q}$ satisfy $f(q) = (m,n) = f(r)$, then $q = m/n = r$. Since $\mathbb{Z} \times \mathbb{Z}$ is known to be countably infinite, there is a bijection $g : \mathbb{Z} \times \mathbb{Z} \to \mathbb{Z}_{>0}$. Then $g \circ f : \mathbb{Q} \to \mathbb{Z}_{>0}$ is injective, being a composition of two injective functions. Therefore $|\mathbb{Q}| \leq |\mathbb{Z}_{>0}|$, as needed. $\square$

Next we prove that a product of finitely many countably infinite sets is also countably infinite.

7.12. Theorem on Countability of Products. For all $k \in \mathbb{Z}_{>0}$ and all sets $X_1, \ldots, X_k$, if every X_i is countably infinite, then $X_1 \times X_2 \times \cdots \times X_k$ is countably infinite.

Proof. We use induction on k. The base case $k = 1$ is immediate. Fix $k \in \mathbb{Z}_{>0}$, and assume the product of any k countably infinite sets is countably infinite. Next, fix $k+1$ countably infinite sets $X_1, \ldots, X_k, X_{k+1}$, and prove $X = X_1 \times \cdots \times X_k \times X_{k+1}$ is countably infinite. Let $Y = X_1 \times \cdots \times X_k$, which is countably infinite by the induction hypothesis. Define $f : X \to Y \times X_{k+1}$ by letting f send $(x_1, \ldots, x_k, x_{k+1})$ to $((x_1, \ldots, x_k), x_{k+1})$ for $(x_1, \ldots, x_{k+1}) \in X$. You can check that f is a bijection, so $|X| = |Y \times X_{k+1}|$. We know $|Y| = |\mathbb{Z}_{>0}|$ and $|X_{k+1}| = |\mathbb{Z}_{>0}|$ since Y and X_{k+1} are countably infinite. By part (h) of Theorem 7.6, we conclude that $|Y \times X_{k+1}| = |\mathbb{Z}_{>0} \times \mathbb{Z}_{>0}|$. Now $\mathbb{Z}_{>0}^2$ is countably infinite, so $|\mathbb{Z}_{>0} \times \mathbb{Z}_{>0}| = |\mathbb{Z}_{>0}|$. Combining all these cardinal equalities using transitivity, we conclude that $|X| = |\mathbb{Z}_{>0}|$, as needed. $\square$

7.13. Corollary. For any countably infinite set X and any $k \in \mathbb{Z}_{>0}$, X^k is countably infinite. In particular, $\mathbb{Z}_{>0}^k$ and $\mathbb{Z}^k$ and $\mathbb{Q}^k$ are countably infinite.

Section Summary

1. *Cardinal Equality.* For all sets A and B, $|A| = |B|$ means there exists a bijection $f : A \to B$. Cardinal equality is reflexive, symmetric, and transitive. If $|A| = |A'|$ and $|B| = |B'|$, then $|A \times B| = |A' \times B'|$ and $|\mathcal{P}(A)| = |\mathcal{P}(A')|$; if also $A \cap B = \emptyset = A' \cap B'$, then $|A \cup B| = |A' \cup B'|$.

2. *Cardinal Ordering.* For all sets A and B, $|A| \leq |B|$ means there exists an injection $g : A \to B$. Cardinal ordering is reflexive, antisymmetric, and transitive. Antisymmetry (for all A and B, if $|A| \leq |B|$ and $|B| \leq |A|$, then $|A| = |B|$) is a hard result called the Schröder–Bernstein Theorem.

3. *Countably Infinite Sets.* A set X is countably infinite iff $|\mathbb{Z}_{>0}| = |X|$. The following sets are countably infinite: $\mathbb{Z}$, $\mathbb{Z}_{\geq a}$ for any $a \in \mathbb{Z}$, $\mathbb{Z} \times \mathbb{Z}$, $\mathbb{Q}$, products of finitely many countably infinite sets, $\mathbb{Z}^k$, and $\mathbb{Q}^k$ (where $k \in \mathbb{Z}_{>0}$).

Exercises

1. Prove: for all sets X and Y, if $|X| = |Y|$, then $|X| \leq |Y|$.

2. (a) Prove: for all sets X, Y, and Z, if $|X| = |Y|$ and $|Y| \leq |Z|$, then $|X| \leq |Z|$.
   (b) Prove: for all sets X, Y, and Z, if $|X| \leq |Y|$ and $|Y| = |Z|$, then $|X| \leq |Z|$.

3. (a) Prove: for all sets X and Y, if $X \subseteq Y$, then $|X| \leq |Y|$. (b) Prove or disprove the converse of part (a).

4. (a) Complete the proof of part (h) of Theorem 7.6 by showing $h' \circ h$ and $h \circ h'$ are identity maps. (b) Finish the proof of part (h) without using h', by proving that h is one-to-one and onto.

5. Complete the proof of part (g) of Theorem 7.6 by showing p and p' are well-defined functions such that $p' \circ p$ and $p \circ p'$ are identity maps. Where is the assumption $A \cap B = \emptyset = A' \cap B'$ needed?

6. Prove part (i) of Theorem 7.6.

7. Prove from the definition that each set below is countably infinite: (a) the set of odd positive integers; (b) the set of positive multiples of a fixed $k \in \mathbb{Z}$; (c) the set of all even integers.

8. Explain why every countably infinite set actually is infinite (not finite).

9. Complete the proof of Proposition 7.8.

10. Complete the proof of Proposition 7.9.

11. (a) Show that $f : \mathbb{Z}_{\geq 0}^3 \to \mathbb{Z}_{>0}$ defined by $f(a, b, c) = 2^a 3^b 5^c$ for $a, b, c \in \mathbb{Z}_{\geq 0}$ is injective. (b) Trace through proofs from this section to find an explicit formula for a bijection $g : \mathbb{Z}_{\geq 0}^3 \to \mathbb{Z}_{>0}$.

12. (a) Prove that the map f defined recursively below Figure 7.5 is one-to-one and onto. (b) Prove that $k(a, b) = ((a + b)^2 + 3a + b)/2 + 1$ is the two-sided inverse of f.

13. The diagram below depicts yet another bijection $p : \mathbb{Z}_{\geq 0}^2 \to \mathbb{Z}$. Formally define the function p (perhaps recursively), and prove p is a bijection.

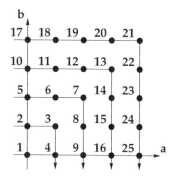

14. Draw a diagram (similar to the one in the previous problem) depicting a bijection $q : \mathbb{Z} \times \mathbb{Z} \to \mathbb{Z}$. Define q formally, and prove q is a bijection.

15. Prove: for all nonempty sets A and B, $|A| \leq |B|$ iff there exists a surjection $h : B \to A$.

16. Suppose I is any index set, and for each $i \in I$, A_i and A_i' are sets such that $|A_i| = |A_i'|$. Prove: if for all $i \neq j$ in I, $A_i \cap A_j = \emptyset = A_i' \cap A_j'$, then $|\bigcup_{i \in I} A_i| = |\bigcup_{i \in I} A_i'|$.

17. Suppose $k \in \mathbb{Z}_{>0}$, and for $1 \leq i \leq k$, A_i and A_i' are sets such that $|A_i| = |A_i'|$. (a) Prove $|A_1 \times \cdots \times A_k| = |A_1' \times \cdots \times A_k'|$ by constructing a bijection. (b) Prove the result in (a) by induction on k, using part (h) of Theorem 7.6.

18. (a) Prove that the union of two countably infinite sets is countably infinite. (b) Prove: for all $k \in \mathbb{Z}_{>0}$ and all sets $X_1, \ldots, X_k$, if every X_i is countably infinite, then $\bigcup_{i=1}^{k} X_i$ is countably infinite.

19. Suppose that for every $n \in \mathbb{Z}_{>0}$, X_n is a given countably infinite set. Prove that $X = \bigcup_{n \in \mathbb{Z}_{>0}} X_n$ is countably infinite. [*Hint:* Use the fact that $\mathbb{Z}_{>0}^2$ is countably infinite.]

20. Let X be any countably infinite set. (a) Prove there exists $f : X \to X$ that is injective but not surjective. (b) Prove there exists $g : X \to X$ that is surjective but not injective.

7.3   Countable Sets

This section studies countable sets, which are sets that are either finite or countably infinite. Our main results are: any subset of a countable set is countable; products of finitely many countable sets are countable; and unions of countably many countable sets are countable.

Countable Sets and Their Subsets

7.14. Definition: Countable Sets. For all sets X,

$$\boxed{X \text{ is countable}} \text{ iff } \boxed{X \text{ is finite or } X \text{ is countably infinite.}}$$

We intend to show that any subset of any countable set is countable. First we consider the special case of subsets of $\mathbb{Z}_{>0}$.

7.15. Lemma on Subsets of $\mathbb{Z}_{>0}$. Every subset of $\mathbb{Z}_{>0}$ is countable.

Proof. Fix an arbitrary subset T of $\mathbb{Z}_{>0}$. To prove T is countable, assume T is not finite and prove T is countably infinite. We construct a bijective function $f : \mathbb{Z}_{>0} \to T$ recursively. Intuitively, $f(n) = t$ will mean that t is the nth smallest element of T. To make this precise, recall from §6.5 the Well-Ordering Property of $\mathbb{Z}_{>0}$, which tells us that every nonempty subset of $\mathbb{Z}_{>0}$ has a least element. Since T is certainly nonempty, we may define $f(1)$ to be the least element of T. Now fix $n \in \mathbb{Z}_{>0}$, and assume that $f(1), f(2), \ldots, f(n)$ have already been defined. We also assume $f(1), \ldots, f(n)$ are pairwise distinct members of T. Note that $T_n = T - \{f(1), f(2), \ldots, f(n)\}$ is a subset of $\mathbb{Z}_{>0}$ that is nonempty. For if T_n were empty, then $T \subseteq \{f(1), f(2), \ldots, f(n)\}$, so that T would be finite, contrary to our assumption. We define $f(n+1)$ to be the least element of T_n, which is a member of T different from $f(i)$ for all i between 1 and n. This completes the recursive definition of f. Figure 7.6 illustrates the definition of f for the set $T = \{3, 6, 10, 11, 13, \ldots\}$.

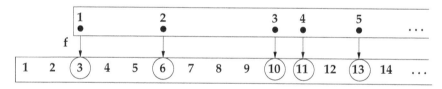

FIGURE 7.6
Example of the bijection $f : \mathbb{Z}_{>0} \to T$.

To show that f is injective, fix $m, p \in \mathbb{Z}_{>0}$, assume $m \neq p$, and prove $f(m) \neq f(p)$. We know $m < p$ or $p < m$. In the case where $m < p$, $f(p) \neq f(m)$ since $f(m)$ is in the set $\{f(1), \ldots, f(p-1)\}$, but $f(p)$ is not in this set by definition. In the case where $p < m$, $f(p) \neq f(m)$ since $f(p)$ is in the set $\{f(1), \ldots, f(m-1)\}$, but $f(m)$ is not in this set.

To show that f is surjective, we first prove that $f(n) \geq n$ for all $n \in \mathbb{Z}_{>0}$. We use induction on n. For the base case, $f(1)$ is a member of T, so $f(1) \in \mathbb{Z}_{>0}$, so $f(1) \geq 1$. Now fix $n \in \mathbb{Z}_{>0}$, assume $f(n) \geq n$, and prove $f(n+1) \geq n+1$. Assume, to get a contradiction, that $f(n+1) < n+1$. Since $f(n+1)$ is an integer, it follows that $f(n+1) \leq n \leq f(n)$. Now, $f(n+1)$ is the least element of $T - \{f(1), f(2), \ldots, f(n)\}$, so $f(n+1) \neq f(n)$. We deduce that $f(n+1) < f(n)$. But now $f(n+1)$ is an element of the set $T - \{f(1), \ldots, f(n-1)\}$ strictly smaller than $f(n)$, which contradicts the definition of $f(n)$ as the least element of this set. So $f(n+1) \geq n+1$, completing the induction.

Now we prove f is surjective. Fix $t \in T$; we must show there exists $n \in \mathbb{Z}_{>0}$ with $f(n) = t$. We have shown that $f(t) \geq t$. If $f(t) = t$, we may choose $n = t$. If $f(t) > t$, then the least element of the set $S = T - \{f(1), f(2), \ldots, f(t-1)\}$ (which is $f(t)$ by definition) is larger than t. It follows that t cannot belong to S. But t does belong to T, which forces $t = f(n)$ for some $n \in \{1, 2, \ldots, t-1\}$. Thus, f is surjective. $\qquad \square$

7.16. Theorem on Countable Sets. For all sets X and Y:

(a) If $|X| = |Y|$, then X is countable iff Y is countable.

(b) $|X| \leq |Y|$ iff $\exists W, W \subseteq Y$ and $|X| = |W|$.

(c) X is countable iff $|X| \leq |\mathbb{Z}_{>0}|$.

(d) If $|X| \leq |Y|$ and Y is countable, then X is countable.

(e) If $X \subseteq Y$ and Y is countable, then X is countable.

Proof. We prove parts (b) through (e), leaving part (a) as an exercise. For part (b), first assume $|X| \leq |Y|$, which means there is an injection $f : X \to Y$. Choose $W = f[X]$ to be the image of f, which is a subset of Y. Let $f' : X \to W$ be obtained from f by restricting the codomain from Y to W. Then f' is injective since f is, and f' is surjective by construction. So $|X| = |W|$. Conversely, assume $|X| = |W|$ for some $W \subseteq Y$. This means there is a bijection $g : X \to W$. Composing g with the injective inclusion map $i : W \to Y$, we obtain an injection $i \circ g : X \to Y$, which proves that $|X| \leq |Y|$.

For part (c), first assume X is countable. In the case where X is finite, we know there exists $n \in \mathbb{Z}_{\geq 0}$ with $|X| = |\{1, 2, \ldots, n\}|$. Since $\{1, 2, \ldots, n\}$ is a subset of $\mathbb{Z}_{>0}$, part (b) applies to show that $|X| \leq |\mathbb{Z}_{>0}|$. Similarly, in the case where X is countably infinite, we know $|X| = |\mathbb{Z}_{>0}|$. Since $\mathbb{Z}_{>0}$ is a subset of itself, part (b) applies to show that $|X| \leq |\mathbb{Z}_{>0}|$. Conversely, assume $|X| \leq |\mathbb{Z}_{>0}|$. By part (b), there exists $W \subseteq \mathbb{Z}_{>0}$ with $|X| = |W|$. By Lemma 7.15, W is countable. By part (a), X is also countable.

For part (d), assume $|X| \leq |Y|$ and Y is countable. Then $|Y| \leq |\mathbb{Z}_{>0}|$ by part (c), so $|X| \leq |\mathbb{Z}_{>0}|$ by transitivity, so X is countable by part (c). For part (e), assume $X \subseteq Y$ and Y is countable. The inclusion $i : X \to Y$ is injective, so $|X| \leq |Y|$. Thus, X is countable by part (d). $\qquad \square$

Using this theorem, we can see that $\mathbb{Q}$ is countably infinite without invoking the Schröder–Bernstein Theorem. Recall that we showed $|\mathbb{Q}| \leq |\mathbb{Z} \times \mathbb{Z}|$ in the proof of Proposition 7.11. Since $|\mathbb{Z} \times \mathbb{Z}|$ is known to be countably infinite and hence countable, we conclude that $\mathbb{Q}$ is countable by part (d) of Theorem 7.16. Now $\mathbb{Q}$ is not finite (since it contains the infinite subset $\mathbb{Z}_{>0}$), so $\mathbb{Q}$ is countably infinite.

Products of Countable Sets

We have seen that the product of finitely many finite sets is finite, and the product of finitely many countably infinite sets is countably infinite. We now establish the corresponding result for countable sets.

7.17. Theorem on Products of Countable Sets. For all positive integers k and all sets $X_1, \ldots, X_k, Y_1, \ldots, Y_k$:

(a) If $|X_i| \leq |Y_i|$ for $1 \leq i \leq k$, then $|X_1 \times X_2 \times \cdots \times X_k| \leq |Y_1 \times Y_2 \times \cdots \times Y_k|$.

(b) If every X_i is countable, then $X_1 \times X_2 \times \cdots \times X_k$ is countable.

Proof. Fix $k \in \mathbb{Z}_{>0}$ and sets X_i, Y_i for $1 \leq i \leq k$. Let $X = X_1 \times \cdots \times X_k$ and $Y = Y_1 \times \cdots \times Y_k$. For part (a), assume $|X_i| \leq |Y_i|$ for all i, which means there exist injections $f_i : X_i \to Y_i$ for all i. We must build an injection $f : X \to Y$. Define $f(x_1, \ldots, x_k) = (f_1(x_1), \ldots, f_k(x_k))$ for $(x_1, \ldots, x_k) \in X$. To see that f is injective, fix $x = (x_1, \ldots, x_k) \in X$, $x' = (x_1', \ldots, x_k') \in X$, assume $f(x) = f(x')$, and prove $x = x'$. We have assumed $(f_1(x_1), \ldots, f_k(x_k)) = (f_1(x_1'), \ldots, f_k(x_k'))$, so $f_i(x_i) = f_i(x_i')$ for $1 \leq i \leq k$. Since each f_i is injective, we conclude that $x_i = x_i'$ for $1 \leq i \leq k$, which means $x = x'$.

For part (b), assume every X_i is countable. By part (c) of Theorem 7.16, $|X_i| \leq |\mathbb{Z}_{>0}|$ for $1 \leq i \leq k$. By what we proved in the last paragraph, $|X| = |X_1 \times \cdots \times X_k| \leq |\mathbb{Z}_{>0} \times \cdots \times \mathbb{Z}_{>0}| = |\mathbb{Z}_{>0}^k| = |\mathbb{Z}_{>0}|$, so $|X| \leq |\mathbb{Z}_{>0}|$. By part (c) of Theorem 7.16, X is countable. $\square$

Unions of Countable Sets

We have seen that the union of finitely many finite sets is finite. Now we can show that the union of countably many countable sets is countable.

7.18. Theorem on Unions of Countable Sets.

(a) Suppose X_k is a countable set for each $k \in \mathbb{Z}_{>0}$. Then $\bigcup_{k=1}^{\infty} X_k$ is countable.

(b) For all $n \in \mathbb{Z}_{>0}$ and all sets $Y_1, \ldots, Y_n$, if every Y_i is countable, then $\bigcup_{k=1}^{n} Y_k$ is countable.

Proof. We first prove part (a) under the additional assumption that the sets X_k are *pairwise disjoint* and countable. We know $|X_k| \leq |\mathbb{Z}_{>0}|$ for each k, so there exist injections $f_k : X_k \to \mathbb{Z}_{>0}$ for each $k \in \mathbb{Z}_{>0}$. Let $X = \bigcup_{k=1}^{\infty} X_k$, and define $f : X \to \mathbb{Z}_{>0}^2$ as follows. Given $x \in X$, there exists a *unique* $k \in \mathbb{Z}_{>0}$ with $x \in X_k$, because the sets X_k are pairwise disjoint. Define $f(x) = (k, f_k(x)) \in \mathbb{Z}_{>0}^2$. Figure 7.7 illustrates how f sends each set X_k into the kth vertical slice of $\mathbb{Z}_{>0}^2$. In this figure, $|X_1| = 4$, $|X_3| = 3$, and X_2 and X_4 are countably infinite.

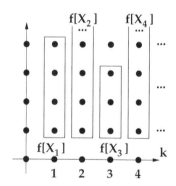

FIGURE 7.7
Illustration of the map $f : \bigcup_{k=1}^{\infty} X_k \to \mathbb{Z}_{>0}^2$.

We now show that f is injective. Fix $x, y \in X$, assume $f(x) = f(y)$, and prove $x = y$. We have $f(x) = (k, f_k(x))$ and $f(y) = (j, f_j(y))$ for some $j, k \in \mathbb{Z}_{>0}$. Since $f(x) = f(y)$, we have $k = j$ and $f_k(x) = f_j(y) = f_k(y)$. Because f_k is injective, $x = y$ follows. Because f is one-to-one and $\mathbb{Z}_{>0}^2$ is countably infinite, we conclude that $|X| \leq |\mathbb{Z}_{>0}^2| = |\mathbb{Z}_{>0}|$. So $|X| \leq |\mathbb{Z}_{>0}|$, hence X is countable by part (c) of Theorem 7.16.

Now we prove the general version of part (a). Assume each X_k (for $k \in \mathbb{Z}_{>0}$) is a countable set. Define $X_1' = X_1$ and, for each $k > 1$, $X_k' = X_k - \bigcup_{i=1}^{k-1} X_i$. For each $k > 0$, $X_k' \subseteq X_k$, so X_k' is countable. You can check that $\bigcup_{k=1}^{\infty} X_k = \bigcup_{k=1}^{\infty} X_k'$ and that the sets X_k' are pairwise disjoint. Since the union of the sets X_k' is countable by the special case of (a) already proved, we conclude that the union of the sets X_k is countable, as needed.

Part (b) follows from part (a) by taking $X_k = Y_k$ for $1 \le k \le n$, and $X_k = \emptyset$ for all $k > n$. $\qquad\qquad\qquad\qquad\qquad\qquad\qquad\qquad\qquad\qquad\qquad\qquad\qquad\qquad\qquad$ $\square$

Section Summary

1. *Countable Sets.* A set X is countable iff X is finite or countably infinite. X is countable iff $|X| \le |\mathbb{Z}_{>0}|$ iff there is a bijection from X onto a subset of $\mathbb{Z}_{>0}$. If $|X| \le |Y|$ and Y is countable, then X is countable.

2. *Operations on Countable Sets.* Any subset of a countable set is countable. A product of finitely many countable sets is countable. A union of countably many countable sets is countable.

Exercises

1. Prove part (a) of Theorem 7.16.

2. Complete the proof of Theorem 7.18 by showing that the sets X'_k are pairwise disjoint, and $\bigcup_{k=1}^{\infty} X_k = \bigcup_{k=1}^{\infty} X'_k$.

3. (a) Prove: for all sets X and Y, if X is countable, then $X - Y$ is countable.
   (b) Prove: for all sets X and Y, if X is countably infinite and $X \cap Y$ is finite, then $X - Y$ is countably infinite.

4. Give a specific example of countably infinite sets X and Y such that $Y \subseteq X$ and $X - Y$ is countably infinite.

5. (a) Give a specific example of pairwise disjoint countably infinite sets X_k (for each $k \in \mathbb{Z}_{>0}$) such that $\mathbb{Z} = \bigcup_{k=1}^{\infty} X_k$. (b) Give a specific example of pairwise disjoint countably infinite sets Y_k (for each $k \in \mathbb{Z}_{>0}$) such that $\mathbb{Q} = \bigcup_{k=1}^{\infty} Y_k$.

6. A *finite sequence* with values in a set X is a function $f : \{1, 2, \ldots, k\} \to X$ for some $k \in \mathbb{Z}_{\ge 0}$. Show: for all countable sets X, the set of all finite sequences with values in X is countably infinite.

7. Show: for all countable sets X, the set of all finite subsets of X is countable.

8. Let T be an infinite subset of $\mathbb{Z}_{>0}$. In the proof of Lemma 7.15, we constructed a bijection $f : \mathbb{Z}_{>0} \to T$. Prove that the function $g : T \to \mathbb{Z}_{>0}$, given by $g(t) = |T \cap \{1, 2, \ldots, t\}|$ for $t \in T$, is the two-sided inverse of f.

9. (a) Suppose we try to prove that every subset of $\mathbb{Z}$ is countable by replacing $\mathbb{Z}_{>0}$ by $\mathbb{Z}$ everywhere in the proof of Lemma 7.15. Explain precisely why the proof does not work.
   (b) Repeat part (a), but now replace $\mathbb{Z}_{>0}$ by $\mathbb{R}_{\ge 0}$.

10. Let W be a countably infinite well-ordered set (see §6.5), and let T be an infinite subset of W. (a) Show that the recursive construction of $f : \mathbb{Z}_{>0} \to T$ in the proof of Lemma 7.15 extends to this situation and produces a well-defined injective function f. (b) The proof that f is surjective does *not* extend to this setting. Explain the exact point in the proof that fails. (c) Consider the specific example $W = \{0, 1\} \times \mathbb{Z}_{>0}$, ordered lexicographically (see Exercise 12 in §6.5). Prove that the function f in part (a) is surjective iff $T \cap (\{1\} \times \mathbb{Z}_{>0}) = \emptyset$.

11. The proof that a union of countably many countable sets is countable requires the Axiom of Choice. Find the exact point in the proof where this axiom is used.

7.4   Uncountable Sets

One of the unintuitive aspects of cardinality theory is that there are many different "sizes" of infinite sets. We have already studied countably infinite sets, which are the "smallest" infinite sets. In this section, we produce some concrete examples of uncountably infinite sets, including: the set $\mathbb{R}$ of real numbers; any open interval (a, b) or closed interval $[a, b]$ with $a < b$; the set $\mathcal{P}(\mathbb{Z}_{>0})$ of all subsets of $\mathbb{Z}_{>0}$; and the set of infinite sequences of zeroes and ones. We also prove Cantor's Theorem, which says that $\mathcal{P}(X)$ always has strictly larger cardinality than X, and the Schröder–Bernstein Theorem, which asserts the antisymmetry of the relation $|X| \leq |Y|$.

Definition of Uncountable Sets

7.19. Definition: Uncountable Sets. For all sets X, X is uncountable iff X is not countable.

The following theorem is readily proved, using Theorem 7.16.

7.20. Theorem on Uncountable Sets. For all sets X and Y:

(a) If $|X| = |Y|$, then X is uncountable iff Y is uncountable.

(b) If $|X| \leq |Y|$ and X is uncountable, then Y is uncountable.

(c) If Y has an uncountable subset, then Y is uncountable.

Uncountable Sets of Sequences

To prove that an infinite set X is countable, it suffices to construct one particular bijection from $\mathbb{Z}_{>0}$ to X. On the other hand, proving the uncountability of X is harder, since we must show that every function from $\mathbb{Z}_{>0}$ to X fails to be bijective. To provide a specific example of an uncountable set, we need the notion of an infinite sequence.

7.21. Definition: Infinite Sequences. For any set X, an *infinite sequence with values in* X is a function $f : \mathbb{Z}_{>0} \to X$. When thinking of f as a sequence, we often write f_n instead of $f(n)$ and call f_n the *nth term of the sequence.*

7.22. Theorem on Uncountable Sets of Sequences. For any set X with more than one element, the set S of all infinite sequences with values in X is uncountable.

Proof. Fix a set X with more than one element, and fix two elements $a \neq b$ in X. To prove S is not countable, we must show S is not finite and S is not countably infinite. You can prove that S is not finite by exhibiting an injection from the infinite set $\mathbb{Z}_{>0}$ into S (see Exercise 5). To show that S is not countably infinite, we must prove there does not exist a bijection from $\mathbb{Z}_{>0}$ onto S. Our strategy is to prove that *every function* $g : \mathbb{Z}_{>0} \to S$ *is not surjective* (and hence not bijective).

Fix a function $g : \mathbb{Z}_{>0} \to S$. Surjectivity of g means $\forall f \in S, \exists m \in \mathbb{Z}_{>0}, g(m) = f$. Negating this, we must prove $\exists f \in S, \forall m \in \mathbb{Z}_{>0}, g(m) \neq f$. Note that for each $m > 0$, $g(m)$ is an infinite sequence, i.e., a function from $\mathbb{Z}_{>0}$ to X. The nth term of the sequence $g(m)$ is $g(m)(n)$, which we abbreviate as $g(m)_n$ or $g_{m,n}$. We can visualize g itself as an infinite table, where row m of the table contains the terms of the sequence $g(m)$, as shown here:

	$n = 1$	$n = 2$	$n = 3$	$n = 4$	$n = 5$	$\cdots$
$g(1)$:	$g_{1,1}$	$g_{1,2}$	$g_{1,3}$	$g_{1,4}$	$g_{1,5}$	$\cdots$
$g(2)$:	$g_{2,1}$	$g_{2,2}$	$g_{2,3}$	$g_{2,4}$	$g_{2,5}$	$\cdots$
$g(3)$:	$g_{3,1}$	$g_{3,2}$	$g_{3,3}$	$g_{3,4}$	$g_{3,5}$	$\cdots$
$g(4)$:	$g_{4,1}$	$g_{4,2}$	$g_{4,3}$	$g_{4,4}$	$g_{4,5}$	$\cdots$
$g(5)$:	$g_{5,1}$	$g_{5,2}$	$g_{5,3}$	$g_{5,4}$	$g_{5,5}$	$\cdots$
$\cdots$	$\cdots$	$\cdots$	$\cdots$	$\cdots$	$\cdots$	$\cdots$

Our goal is to build a sequence f that is unequal to every sequence in this table. For each $n \in \mathbb{Z}_{>0}$, define $f_n = b$ if $g_{n,n} = a$, and define $f_n = a$ if $g_{n,n} \neq a$. Fix an arbitrary $m \in \mathbb{Z}_{>0}$. We must prove $g(m) \neq f$. To prove that the sequence $g(m)$ is unequal to the sequence f, it is enough to find one position where these sequences disagree (by definition of equality of functions). By construction of f, we have $f_m \neq g_{m,m} = g(m)_m$, so that the sequence f and the sequence $g(m)$ disagree at position m. Thus $g(m) \neq f$, as needed. $\square$

Here is an example illustrating the preceding proof. Suppose $X = \{1, 2, 3\}$, and the table for g looks like this:

	$n = 1$	$n = 2$	$n = 3$	$n = 4$	$n = 5$	$\cdots$
$g(1)$:	[1]	1	1	1	1	$\cdots$
$g(2)$:	2	[2]	2	2	2	$\cdots$
$g(3)$:	3	3	[3]	3	3	$\cdots$
$g(4)$:	1	2	1	[2]	1	$\cdots$
$g(5)$:	3	1	2	2	[1]	$\cdots$
$\cdots$	$\cdots$	$\cdots$	$\cdots$	$\cdots$	$\cdots$	$\cdots$

To build a sequence f not appearing in this table, we read down the main diagonal of the table and change every entry. Taking $a = 1$ and $b = 2$, the sequence f for this g has first five terms $f_1 = 2$, $f_2 = 1$, $f_3 = 1$, $f_4 = 1$, and $f_5 = 2$. We see that $f \neq g(1)$ since these sequences disagree in position 1; $f \neq g(2)$ since these sequences disagree in position 2; and so on. This construction is called *Cantor's diagonal argument*.

Uncountable Sets of Real Numbers

Now that we have some uncountable sets available, we can use bijections and injections to show that certain subsets of $\mathbb{R}$ are uncountable.

7.23. Theorem on Uncountable Subsets of $\mathbb{R}$. The following sets are uncountable: the set $\mathbb{R}$ of real numbers; the open interval $(0, 1)$; the open interval (a, b) for all $a < b$ in $\mathbb{R}$; and the closed interval $[a, b]$ for all $a < b$ in $\mathbb{R}$.

Proof. First we show $(0, 1)$ is uncountable. We need the following fact about decimal representations of real numbers, which we state without proof: for every real number $r \in (0, 1)$, there exists exactly one infinite sequence f with values in $\{0, 1, 2, 3, 4, 5, 6, 7, 8, 9\}$ such that $r = \sum_{k=1}^{\infty} f(k)10^{-k}$, and $\sim \exists k, \forall m \geq k, f(m) = 9$. Informally, $f(k)$ is the kth digit after the decimal point in the decimal expansion of r. To ensure uniqueness, we forbid expansions that end in an infinite string of 9s. For example, taking $r = 1/3$, the sequence f has $f(k) = 3$ for all $k \in \mathbb{Z}_{>0}$. Taking $r = 1/2 = 0.5000\cdots = 0.4999\cdots$, the sequence f has $f(1) = 5$ and $f(k) = 0$ for all $k > 1$.

Now let S be the set of all infinite sequences with values in $X = \{4, 7\}$. We already know S is uncountable. Define $G : S \to (0, 1)$ by setting $G(f) = \sum_{k=1}^{\infty} f(k)10^{-k}$ for $f \in S$. Using calculus, you can check that this infinite series does converge to some number in $(0, 1)$. The fact cited above ensures that G is injective, so $|S| \leq |(0,1)|$. By part (b) of Theorem 7.20, $(0, 1)$ is uncountable.

Next, fix $a, b \in \mathbb{R}$ with $a < b$. The map $h : (0,1) \to (a,b)$, defined by $h(t) = a + (b-a)t$ for $0 < t < 1$, is readily seen to be a bijection. Thus $|(a,b)| = |(0,1)|$, so the open interval (a, b) is uncountable. It follows that $[a, b]$ and $\mathbb{R}$ itself are uncountable, since the uncountable set (a, b) is a subset of each of these sets. □

We now present another proof that closed intervals in $\mathbb{R}$ are uncountable. This proof does not rely on the existence and uniqueness of decimal expansions of real numbers. Instead, we need a much more elementary fact called the Nested Interval Property. This property says that for any sequence of nonempty closed intervals $[a_n, b_n]$ such that $[a_{n+1}, b_{n+1}] \subseteq [a_n, b_n]$ for all $n \in \mathbb{Z}_{>0}$, the intersection $\bigcap_{n=1}^{\infty} [a_n, b_n]$ is nonempty. See §8.5 for a proof of this property.

7.24. Theorem on Uncountability of Closed Intervals. For all real $a < b$, $[a, b]$ is uncountable.

Proof. Fix real $a < b$. We fix an arbitrary function (or sequence) $f : \mathbb{Z}_{>0} \to [a, b]$ and prove that f cannot be surjective. We recursively construct a sequence of nested intervals $[a_n, b_n]$ for $n \in \mathbb{Z}_{\geq 0}$ with the property that $[a_n, b_n] \subseteq [a, b]$, $a_n < b_n$, and $[a_n, b_n] \cap \{f(1), f(2), \ldots, f(n)\} = \emptyset$ for all $n \geq 0$. To begin, let $a_0 = a$ and $b_0 = b$. Now, fix $n \in \mathbb{Z}_{\geq 0}$ and assume $a_0, a_1, \ldots, a_n$, $b_0, b_1, \ldots, b_n$ have already been chosen in a way that satisfies the property above. Let $d = b_n - a_n > 0$ be the length of the interval $[a_n, b_n]$, and consider the three subintervals $[a_n, a_n + d/3]$, $[a_n + d/3, a_n + 2d/3]$, and $[a_n + 2d/3, b_n]$. These three intervals are subsets of $[a_n, b_n]$ and hence do not contain any of the real numbers $f(1), \ldots, f(n)$. Furthermore, at most two of the three subintervals can contain $f(n+1)$. [It is possible for $f(n+1)$ to belong to two of the intervals; this happens when $f(n+1) = a_n + d/3$ or $f(n+1) = a_n + 2d/3$.] Choose $[a_{n+1}, b_{n+1}]$ to be the first of the three intervals that does not contain $f(n+1)$. Then $[a_{n+1}, b_{n+1}] \subseteq [a_n, b_n]$, $a_{n+1} < b_{n+1}$, and $[a_{n+1}, b_{n+1}] \cap \{f(1), \ldots, f(n), f(n+1)\} = \emptyset$. By recursion, this defines $[a_n, b_n]$ for all $n \in \mathbb{Z}_{\geq 0}$. To finish, invoke the Nested Interval Property to see that there must exist a real number r in the intersection of all the intervals $[a_n, b_n]$. For fixed $n \in \mathbb{Z}_{>0}$, we know $r \in [a_n, b_n]$ and hence $r \neq f(n)$ by construction. Since n was arbitrary, we see that r is a number in $[a, b]$ that is outside the image of f. Thus, f is not surjective and so not bijective. So $[a, b]$ must be uncountable. □

Cantor's Theorem

We now prove the remarkable fact that for every set X, the power set $\mathcal{P}(X)$ (the set of all subsets of X) has strictly larger cardinality than X.

7.25. Cantor's Theorem. For any set X, $|X| < |\mathcal{P}(X)|$.

Proof. Fix a set X. We must show $|X| \leq |\mathcal{P}(X)|$ and $|X| \neq |\mathcal{P}(X)|$. On one hand, the map $k : X \to \mathcal{P}(X)$ given by $k(x) = \{x\}$ for $x \in X$ is readily seen to be an injection. So $|X| \leq |\mathcal{P}(X)|$. On the other hand, to show $|X| \neq |\mathcal{P}(X)|$, we must prove that every function $f : X \to \mathcal{P}(X)$ is non-bijective. Fix such a function f; it suffices to prove f is non-surjective. So, we must show $\exists S \in \mathcal{P}(X), \forall x \in X, S \neq f(x)$.

The rest of the proof is a variation of Cantor's diagonal argument for sequences. We choose $S = \{z \in X : z \notin f(z)\}$, which is a subset of X and hence a member of $\mathcal{P}(X)$. We prove $\forall x \in X, S \neq f(x)$ by contradiction. Assume, to get a contradiction, that $\exists x_0 \in X, S = f(x_0)$. Let us ask whether the object $z = x_0$ is in S. By definition of S and the fact that $S = f(x_0)$, $x_0 \in S$ iff $x_0 \notin f(x_0)$ iff $x_0 \notin S$. We now have reached the contradiction "$x_0 \in S \Leftrightarrow x_0 \notin S$." $\qquad\square$

Taking $X = \mathbb{Z}_{>0}$ in this theorem, we see that $\mathcal{P}(\mathbb{Z}_{>0})$ is uncountable. As another example, just as the uncountable set $\mathbb{R}$ is a larger infinite set than $\mathbb{Z}$ or $\mathbb{Q}$, we see from Cantor's Theorem that $\mathcal{P}(\mathbb{R})$ is an even larger set than $\mathbb{R}$. Iterating the theorem, we get a chain of uncountable sets, each strictly larger in cardinality than the previous one:

$$|\mathbb{R}| < |\mathcal{P}(\mathbb{R})| < |\mathcal{P}(\mathcal{P}(\mathbb{R}))| < |\mathcal{P}(\mathcal{P}(\mathcal{P}(\mathbb{R})))| < \cdots.$$

The mind-boggling tower of infinities never ends; for if we tentatively assumed there were a largest set X, Cantor's Theorem would provide an even larger set $\mathcal{P}(X)$.

Proof of the Schröder–Bernstein Theorem (Optional)

We conclude our introduction to cardinality by proving the Schröder–Bernstein Theorem. We begin with a special case of the main result.

7.26. Lemma. For all sets X and W, if $W \subseteq X$ and there exists an injective function $f : X \to W$, then there exists a bijective function $h : X \to W$.

Proof. Fix arbitrary sets X and W. Assume $W \subseteq X$ and $f : X \to W$ is a one-to-one function. When X is infinite, the function f may not be onto. The following ingenious construction uses f to build a bijection $h : X \to W$. Recursively define a sequence of sets

$$A_0 \supseteq A_1 \supseteq A_2 \supseteq \cdots \supseteq A_n \supseteq A_{n+1} \supseteq \cdots$$

by letting $A_0 = X$, $A_1 = W$ and $A_{n+2} = f[A_n]$ for all $n \in \mathbb{Z}_{\geq 0}$. We check by induction on n that $A_{n+1} \subseteq A_n$ for all $n \in \mathbb{Z}_{\geq 0}$. This is true when $n = 0$ because $W \subseteq X$ by assumption. This is true when $n = 1$ because $A_2 = f[X]$, $A_1 = W$, and we assumed f maps X into the codomain W. Now fix $n \in \mathbb{Z}_{\geq 2}$, and assume that $A_n \subseteq A_{n-1} \subseteq A_{n-2} \subseteq \cdots \subseteq A_0$ is already known. We must prove $A_{n+1} \subseteq A_n$. Since $A_{n-1} \subseteq A_{n-2}$, we conclude that $f[A_{n-1}] \subseteq f[A_{n-2}]$, so $A_{n+1} \subseteq A_n$ as needed.

The next step is to define $B_n = A_n - A_{n+1}$ for all $n \in \mathbb{Z}_{\geq 0}$ and $B = \bigcap_{n=0}^{\infty} A_n$. It is routine to check that the sets $B_0, B_1, \ldots, B_n, \ldots, B$ are pairwise disjoint and have union X, whereas the sets $B_1, B_2, \ldots, B_n, \ldots, B$ are pairwise disjoint and have union W. See Figure 7.8, where X is the entire box, and W is the part of the box below the thick line.

A crucial observation is that $f[B_n] = B_{n+2}$ for all $n \in \mathbb{Z}_{\geq 0}$. To verify this, note that injectivity of f implies $f[A - C] = f[A] - f[C]$ for all $A, C \subseteq X$ (see Exercise 14 in §5.8). So for each $n \in \mathbb{Z}_{\geq 0}$,

$$f[B_n] = f[A_n - A_{n+1}] = f[A_n] - f[A_{n+1}] = A_{n+2} - A_{n+3} = B_{n+2}.$$

Next, using injectivity of f, $f[B] = B$ follows from the computation

$$f[B] = f\left[\bigcap_{n=0}^{\infty} A_n\right] = \bigcap_{n=0}^{\infty} f[A_n] = \bigcap_{n=0}^{\infty} A_{n+2} = \bigcap_{j=2}^{\infty} A_j = B.$$

(The last equality uses the fact that $A_2 \subseteq A_1 \subseteq A_0$.)

Since f is injective, all the restrictions $f|_{B_n}$ and $f|_B$ are also injective. Since $f[B] = B$ and $f[B_n] = B_{n+2}$ for all $n \in \mathbb{Z}_{\geq 0}$, we can restrict the codomains of $f|_{B_n}$ and $f|_B$ to get

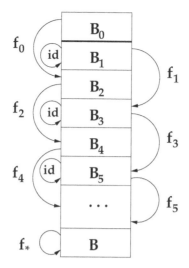

FIGURE 7.8
Visualizing the proof of the Schröder–Bernstein Theorem.

bijections $f_n : B_n \to B_{n+2}$ and $f_* : B \to B$. Now comes the final miraculous idea: define a function h by

$$h(x) = \begin{cases} f_n(x) & \text{if } x \in B_n \text{ where } n \in \mathbb{Z}_{\geq 0} \text{ is even;} \\ x & \text{if } x \in B_m \text{ where } m \in \mathbb{Z}_{>0} \text{ is odd;} \\ f_*(x) & \text{if } x \in B. \end{cases}$$

As shown on the left edge of the box in Figure 7.8, h is defined by gluing together the bijections $f_n : B_n \to B_{n+2}$ for n even, $\mathrm{Id}_{B_m} : B_m \to B_m$ for m odd, and $f_* : B \to B$. The domains of these bijections are the pairwise disjoint sets $B_0, B_1, B_2, B_3, B_4, \ldots, B$ whose union is X; and the codomains of these bijections are the pairwise disjoint sets $B_2, B_1, B_4, B_3, B_6, \ldots, B$ whose union is W. We see at once that h is a well-defined function mapping the domain X onto the codomain W, and h is one-to-one. So we have constructed the required bijection $h : X \to W$. $\qquad\qquad\square$

7.27. Schröder–Bernstein Theorem. For all sets X and Y, if $|X| \leq |Y|$ and $|Y| \leq |X|$ then $|X| = |Y|$.

Proof. Fix sets X and Y. Assume $|X| \leq |Y|$ and $|Y| \leq |X|$, which means there are injections $F : X \to Y$ and $G : Y \to X$. We must prove $|X| = |Y|$, which means there is a bijection $H : X \to Y$. Define $W = G[Y]$, which is a subset of X. Define $f : X \to W$ by $f(x) = G(F(x))$ for all $x \in X$. Note that f is injective, since it is obtained from the injective function $G \circ F : X \to X$ by shrinking the codomain to W. Similarly, if we shrink the codomain of the injective function G to $W = G[Y]$, we obtain a bijective function $G_1 : Y \to W$. Since $W \subseteq X$ and $f : X \to W$ is injective, Lemma 7.26 shows that there is a bijection $h : X \to W$. Then $H = G_1^{-1} \circ h : X \to Y$ is a bijection, since it is the composition of the bijections h and G_1^{-1}. $\qquad\qquad\square$

7.28. Corollary: No-Universe Theorem. There is no set of all sets.

Proof. Assume, to get a contradiction, that X is a set such that for all sets Y, $Y \in X$. Consider the power set $\mathcal{P}(X)$ consisting of all subsets of X. Every $Y \in \mathcal{P}(X)$ is a set, so every such Y is in X by hypothesis, so $\mathcal{P}(X) \subseteq X$. This implies $|\mathcal{P}(X)| \leq |X|$, and we saw earlier that $|X| \leq |\mathcal{P}(X)|$. Now $|X| = |\mathcal{P}(X)|$ follows from the Schröder–Bernstein Theorem; but this contradicts Cantor's Theorem. $\qquad\square$

Section Summary

1. *Uncountable Sets.* A set X is uncountable iff X is not countable. To prove that an infinite set is uncountable, it suffices to prove that every $f : \mathbb{Z}_{>0} \to X$ is not surjective. If $|X| = |Y|$, then X is uncountable iff Y is uncountable. If $|X| \leq |Y|$ and X is uncountable, then Y is uncountable. If Y has an uncountable subset, then Y is uncountable.

2. *Examples of Uncountable Sets.* $\mathbb{R}$ is uncountable. For all $a < b$ in $\mathbb{R}$, the intervals (a, b) and $[a, b]$ are uncountable. The set $\mathcal{P}(\mathbb{Z}_{>0})$ is uncountable. If X has more than one element, the set of infinite sequences with values in X is uncountable.

3. *Cantor's Theorem.* For every set X, $|X| < |\mathcal{P}(X)|$. The proof uses the diagonal method, starting with any given function $f : X \to \mathcal{P}(X)$ and constructing a subset of X not in the image of f. One consequence of Cantor's Theorem is that there is no set of all sets.

4. *Schröder–Bernstein Theorem.* For all sets X and Y, if $|X| \leq |Y|$ and $|Y| \leq |X|$, then $|X| = |Y|$. The proof combines injections $F : X \to Y$ and $G : Y \to X$ to build a bijection $H : X \to Y$. When $Y \subseteq X$ and $G : Y \to X$ is the inclusion map, the construction of H is visualized in Figure 7.8.

Exercises

1. For fixed $a < b$ in $\mathbb{R}$, show that $h : (0, 1) \to (a, b)$, given by $h(t) = a + (b - a)t$ for $t \in (0, 1)$, is a bijection.

2. Given that the interval $[0, 1]$ is uncountable, prove that every open interval (a, b) is uncountable without using Theorem 7.23.

3. Show that $f : (-1, 1) \to \mathbb{R}$, given by $f(x) = x/(1 - x^2)$ for $x \in (-1, 1)$, is a bijection.

4. Prove Theorem 7.20.

5. (a) Complete the proof of Theorem 7.22 by constructing an injection $g : \mathbb{Z}_{>0} \to S$. (b) The proof shows that the map g in part (a) cannot be surjective. Find the specific sequence f constructed by the proof that is outside the image of g.

6. Use calculus to show that the series $\sum_{k=1}^{\infty} f(k)10^{-k}$ in the proof of Theorem 7.23 converges to a real number in $(0, 1)$. [*Hint:* Compare to the geometric series $\sum_{k \geq 1}(4/10)^k$ and $\sum_{k \geq 1}(7/10)^k$.]

7. (a) What goes wrong with the proof of Theorem 7.23 if we use $X = \{0, 4\}$ instead of $X = \{4, 7\}$? (b) What goes wrong with the proof if we use $X = \{7, 9\}$ instead of $X = \{4, 7\}$?

8. Let X be a fixed countably infinite set, and let Y be a fixed uncountable set. Decide (with proof) whether each set of functions is countable or uncountable.
   (a) the set of all $f : \{1, 2, 3\} \to X$
   (b) the set of all $g : \{1, 2, 3\} \to Y$
   (c) the set of all $h : X \to \{1, 2, 3\}$
   (d) the set of all $k : Y \to \{1, 2, 3\}$

9. In the proof of Lemma 7.26, check that the sets $B_0, B_1, \ldots, B_n, \ldots, B$ are pairwise disjoint, the union of these sets is X, and the union of these sets excluding B_0 is W.

10. (a) Prove that for any injective function $f : X \to Y$ and all sets $A_n \subseteq X$, $f[\bigcap_{n=0}^{\infty} A_n] = \bigcap_{n=0}^{\infty} f[A_n]$. (b) In the proof of Lemma 7.26, explain in detail why $\bigcap_{j=2}^{\infty} A_j = B$.

11. Let $W = \{2k : k \in \mathbb{Z}\}$, and define an injection $f : \mathbb{Z} \to W$ by $f(n) = 4n$ for all $n \in \mathbb{Z}$. In the proof of Lemma 7.26, describe the sets A_n, B_n, B, and the bijection $h : \mathbb{Z} \to W$.

12. Define an injection $f : \mathbb{Z}_{\geq 0} \to \mathbb{Z}_{\geq 1}$ by $f(n) = n + 3$ for all $n \in \mathbb{Z}_{\geq 0}$. In the proof of Lemma 7.26, describe the sets A_n, B_n, B, and the bijection $h : \mathbb{Z}_{\geq 0} \to \mathbb{Z}_{\geq 1}$.

13. In Lemma 7.26, suppose $f : X \to W$ happens to be a bijection. Describe the sets A_n, B_n, B and the bijection $h : X \to W$ in this case.

14. In Lemma 7.26, suppose $W = X$. Describe the sets A_n, B_n, B, and the bijection $h : X \to W$ in this case.

15. Define $F, G : \mathbb{Z}_{>0} \to \mathbb{Z}_{>0}$ by $F(n) = 2n$ and $G(n) = 2n - 1$ for $n \in \mathbb{Z}_{>0}$. Trace through the proof of the Schröder–Bernstein Theorem to construct a bijection $H : \mathbb{Z}_{>0} \to \mathbb{Z}_{>0}$ from F and G.

16. For any sets X and Y, let Y^X be the set of all functions $f : X \to Y$. Prove: if $|X| = k$ and $|Y| = n$, then $|Y^X| = n^k$.

17. Use a bijection to prove: for all sets X, Y, Z, if $X \cap Y = \emptyset$ then $|Z^{X \cup Y}| = |Z^X \times Z^Y|$.

18. Use a bijection to prove: for all sets X, Y, Z, $|(Z^Y)^X| = |Z^{X \times Y}|$.

19. Given sets X and Y, a function $f : X \to Y$, and an injection $g : Y \to X$, show there exist sets A, B, C, D such that $\{A, B\}$ is a set partition of X, $\{C, D\}$ is a set partition of Y, $C = f[A]$, and $B = g[D]$. (*Hint:* Let $Z = X - g[Y]$ and $h = g \circ f$; then define A to be the intersection of all sets $U \subseteq X$ with $Z \cup h[U] \subseteq U$.)

20. Use the previous exercise to give another proof of the Schröder–Bernstein Theorem.

21. A set $S \subseteq \mathbb{Z}_{>0}$ is *closed under multiplication* iff $\forall a, b \in S, a \cdot b \in S$. Let X be the set of all subsets of $\mathbb{Z}_{>0}$ that are closed under multiplication. Is X countable or uncountable? Prove your answer.

22. A set $S \subseteq \mathbb{Z}_{>0}$ is *closed under addition* iff $\forall a, b \in S, a + b \in S$. Let Y be the set of all subsets of $\mathbb{Z}_{>0}$ that are closed under addition. Is Y countable or uncountable? Prove your answer.

Review of Functions, Relations, and Cardinality

Tables 7.1–7.6 on the following pages review the main definitions and theorems covered in the preceding chapters on functions, relations, and cardinality. All unquantified variables in the tables represent arbitrary objects, sets, functions, or relations (as required by context).

Functions and Relations

1. *Arrow Diagrams.* We can visualize a relation R from X to Y by drawing an arrow from a dot labeled x to a dot labeled y for each $(x, y) \in R$. The relation is the graph of a function $f : X \to Y$ iff each $x \in X$ has exactly one arrow leaving it. Such a function f is injective iff each $y \in Y$ has at most one arrow entering it; f is surjective iff each $y \in Y$ has at least one arrow entering it; and f is bijective iff each $y \in Y$ has exactly one arrow entering it. The relation R^{-1} is found by reversing all arrows. We get the arrow diagram of a composition $S \circ R$ by concatenating R-arrows followed by S-arrows.

2. *Cartesian Graphs.* We can visualize a relation R on $\mathbb{R}$ by the graph consisting of all points (x, y) in the plane with $(x, y) \in R$. The relation is the graph of a function $f : X \to Y$ iff the points drawn all lie within the rectangle $X \times Y$, and for each $x_0 \in X$ the vertical line $x = x_0$ intersects the graph in exactly one point. Such a function f is injective (resp. surjective, bijective) iff for all $y_0 \in Y$, the horizontal line $y = y_0$ intersects the graph in at most one (resp. at least one, exactly one) point. The relation R^{-1} is found by reflecting the graph in the line $y = x$.

3. *Digraphs.* We can visualize a relation R on X by drawing a single copy of X with arrows from x to y for each $(x, y) \in R$. R is reflexive on X iff there is a loop at each vertex $x \in X$; R is symmetric iff every arrow from x to y is accompanied by the reverse arrow from y to x; and R is transitive iff whenever we can go from x to y to z by following two arrows, the arrow directly from x to z is also present.

4. *Proving Functions are Well-Defined.* We often define functions $g : A \to B$ by giving a formula specifying $g(x)$ for each $x \in A$. To be sure we do have a well-defined function, we must check three things: $g(x)$ is in the codomain B for all $x \in A$; for each $x \in A$ there exists at least one associated output $g(x)$; and for all $x_1, x_2 \in A$, if $x_1 = x_2$, then $g(x_1) = g(x_2)$. The last condition can be restated: for all x, y, z, if (x, y) and (x, z) belong to the graph of g, then $y = z$. It is especially crucial to prove this *single-valuedness* condition when members of the domain A have multiple names (e.g., if A is a set of equivalence classes). If $f(x)$ is computed using one of the names of x, we must verify that changing the name of the input x does not change the value of the output $f(x)$ (although the name of the output could change).

5. *Proving Function Equality.* To prove that functions f and g are equal, first check that f and g share the same domain (say X) and the same codomain (say Y).

TABLE 7.1

Function Definitions. Here, f is a function with domain A and codomain B.

Concept	Definition
function [formal version]	ordered triple $f = (A, B, G)$ where A and B are sets, $G \subseteq A \times B$, and $\forall x \in A, \exists!\, y, (x, y) \in G$. A is the *domain* of f, B is the *codomain* of f, and G is the *graph* of f; $y = f(x)$ means $(x, y) \in G$.
$f : A \to B$	f is a function with domain A and codomain B, meaning: $\forall x, \forall y, y = f(x) \Rightarrow (x \in A \land y \in B)$, and $\forall x \in A, \exists!\, y, y = f(x)$.
graph of $f : A \to B$	$G = \{(x, f(x)) : x \in A\}$.
f is injective (one-to-one)	$\forall a \in A, \forall c \in A, f(a) = f(c) \Rightarrow a = c$.
f is surjective (onto)	$\forall y \in B, \exists x \in A, y = f(x)$.
f is bijective	f is one-to-one and f is onto; i.e., $\forall y \in B, \exists!\, x \in A, y = f(x)$.
direct image $f[C]$	$y \in f[C]$ iff $\exists x \in C, y = f(x)$.
preimage $f^{-1}[D]$	$x \in f^{-1}[D]$ iff $x \in A$ and $f(x) \in D$.
function equality	For $f : A \to B$ and $g : A' \to B'$, $f = g$ iff $A = A'$ and $B = B'$ and $\forall x \in A, f(x) = g(x)$.
identity function Id_A	$\text{Id}_A : A \to A$ where $\text{Id}_A(x) = x$ for all $x \in A$.
constant function c	$f : A \to B$ where $f(x) = c$ for all $x \in A$ ($c \in B$ is fixed).
inclusion map $j_{A,B}$	$j : A \to B$ where $j(x) = x$ for all $x \in A$.
pointwise sum $f + h$	$(f + h)(x) = f(x) + h(x)$ for $x \in A$ (codomain is $\mathbb{R}$).
pointwise product $f \cdot h$	$(f \cdot h)(x) = f(x)h(x)$ for $x \in A$ (codomain is $\mathbb{R}$).
composition $g \circ f$	For $f : A \to B$ and $g : B \to C$, $g \circ f : A \to C$ satisfies $(g \circ f)(x) = g(f(x))$ for $x \in A$.
inverse of $f : A \to B$	$f^{-1} : B \to A$ where $\forall x \in A, \forall y \in B, y = f(x) \Leftrightarrow x = f^{-1}(y)$.
right inverse of f	$g : B \to A$ with $f \circ g = \text{Id}_B$.
left inverse of f	$h : B \to A$ with $h \circ f = \text{Id}_A$.
invertible function	function that has a (two-sided) inverse.
restriction $f|_S$ (for $S \subseteq A$)	$f|_S : S \to B$ where $f|_S(x) = f(x)$ for $x \in S$.
characteristic function χ_S	$\chi_S(x) = 1$ for $x \in S$, $\chi_S(x) = 0$ for $x \notin S$.
$h = \text{glue}(h_i : i \in I)$	For $h_i : A_i \to B_i$ with graphs H_i agreeing on overlaps, $h : \bigcup_{i \in I} A_i \to \bigcup_{i \in I} B_i$ has graph $\bigcup_{i \in I} H_i$. So $h(x) = h_i(x)$ for all $x \in A_i$.
fiber of f over y	$f^{-1}[\{y\}] = \{x \in A : f(x) = y\}$.

TABLE 7.2
Relation Definitions.

Concept	Definition
R is a relation from X to Y	$R \subseteq X \times Y$.
R is a relation on X	$R \subseteq X \times X$.
infix notation aRb	aRb iff $(a, b) \in R$.
image $R[C]$	$y \in R[C]$ iff $\exists x \in C, (x, y) \in R$.
identity relation on X	$(a, b) \in I_X$ iff $a \in X$ and $a = b$.
inverse of relation R	$(y, x) \in R^{-1}$ iff $(x, y) \in R$.
composition of R and S	$(x, z) \in S \circ R$ iff $\exists y, (x, y) \in R \wedge (y, z) \in S$.
R is reflexive on X	$\forall a \in X, (a, a) \in R$.
R is symmetric	$\forall a, \forall b, (a, b) \in R \Rightarrow (b, a) \in R$.
R is transitive	$\forall a, \forall b, \forall c, ((a, b) \in R \wedge (b, c) \in R) \Rightarrow (a, c) \in R$.
R is antisymmetric	$\forall a, \forall b, ((a, b) \in R \wedge (b, a) \in R) \Rightarrow a = b$.
R is irreflexive on X	$\forall a \in X, (a, a) \notin R$.
equivalence relation on X	symmetric, transitive, reflexive relation on X.
equivalence class of a	$x \in [a]_R$ iff xRa (iff aRx).
quotient set X/R	$S \in X/R$ iff $\exists a \in X, S = [a]_R$.
equivalence relation $\sim_f$	for $f : X \to Y$ and $u, v \in X$, $u \sim_f v$ iff $f(u) = f(v)$.
congruence modulo n	For $a, b \in \mathbb{Z}$, $a \equiv_n b$ iff n divides $a - b$.
integers modulo n	$\mathbb{Z}/n = \{[a]_{\equiv_n} : a \in \mathbb{Z}\} = \{[0]_n, [1]_n, \ldots, [n-1]_n\}$, where $[r]_n = \{r + kn : k \in \mathbb{Z}\}$.
addition modulo n	$[a]_n \oplus_n [b]_n = [a + b]_n$ for $a, b \in \mathbb{Z}$.
multiplication modulo n	$[a]_n \otimes_n [b]_n = [ab]_n$ for $a, b \in \mathbb{Z}$.
set partition of X	a set P of nonempty subsets of X where $\forall x \in X, \exists S \in P, x \in S$ and $\forall S, T \in P, S \neq T \Rightarrow S \cap T = \emptyset$.
eqv. rel. $\sim_P$ of set ptn. P	For $a, b \in X$, $a \sim_P b$ iff $\exists S \in P, a \in S \wedge b \in S$.
partial order on X	antisymmetric, transitive, reflexive relation on X.
partially ordered set (poset)	pair $(X, \leq)$ where $\leq$ is a partial order on X.
total order on X	partial order where $\forall x, y \in X, x \leq y \vee y \leq x$.
z is greatest element of S	$z \in S \wedge \forall y \in S, y \leq z$.
z is least element of S	$z \in S \wedge \forall y \in S, z \leq y$.
well-ordering on X	partial order on X where each nonempty subset of X has a least element.

TABLE 7.3
Cardinality Definitions.

Concept	Definition
$|X| = |Y|$	There exists a bijection $f : X \to Y$.
$|X| \leq |Y|$	There exists an injection $g : X \to Y$.
X has size n $(n \in \mathbb{Z}_{\geq 0})$.	$|X| = n$ iff there exists a bijection $f : \{1, 2, \ldots, n\} \to X$.
X is finite.	$\exists n \in \mathbb{Z}_{\geq 0}, |X| = n$.
X is infinite.	X is not finite.
X is countably infinite.	There exists a bijection $f : \mathbb{Z}_{>0} \to X$.
X is countable.	X is finite or X is countably infinite.
X is uncountable.	X is not countable.

TABLE 7.4

Theorems on Relations.

Direct Images	**Setup:** R is a relation (or function) from X to Y.
Empty Image	$R[\emptyset] = \emptyset$.
Codomain Property	$R[A] \subseteq Y$.
Monotonicity	If $A \subseteq C$, then $R[A] \subseteq R[C]$.
Image of Union	$R[\bigcup_{i \in I} A_i] = \bigcup_{i \in I} R[A_i]$.
Image of Intersection	$R[\bigcap_{i \in I} A_i] \subseteq \bigcap_{i \in I} R[A_i]$.
Image of Difference	$R[A] - R[C] \subseteq R[A - C]$.
Inverses and Compositions	**Setup:** R is a relation from X to Y.
Inverse of Identity	$I_X^{-1} = I_X$.
Double Inverse	$(R^{-1})^{-1} = R$.
Inverse of Union	$(S \cup T)^{-1} = S^{-1} \cup T^{-1} = T^{-1} \cup S^{-1}$.
Inverse of Intersection	$(S \cap T)^{-1} = S^{-1} \cap T^{-1} = T^{-1} \cap S^{-1}$.
Composition with Identity	$R \circ I_X = R$ and $I_Y \circ R = R$.
Associativity	$T \circ (S \circ R) = (T \circ S) \circ R$.
Inverse of Composition	$(R \circ S)^{-1} = S^{-1} \circ R^{-1}$.
Monotonicity Properties	$X \subseteq Y \Rightarrow I_X \subseteq I_Y$; if $R \subseteq S$, then $R^{-1} \subseteq S^{-1}$, $T \circ R \subseteq T \circ S$, and $R \circ T \subseteq S \circ T$.
Empty Set Properties	$I_\emptyset = \emptyset$ and $\emptyset^{-1} = \emptyset$ and $R \circ \emptyset = \emptyset = \emptyset \circ R$.
Distributive Laws	$(S \cup T) \circ R = (S \circ R) \cup (T \circ R)$. $T \circ (S \cup R) = (T \circ S) \cup (T \circ R)$.
Partial Distributive Laws	$(S \cap T) \circ R \subseteq (S \circ R) \cap (T \circ R)$. $T \circ (S \cap R) \subseteq (T \circ S) \cap (T \circ R)$.

Then fix an arbitrary $x_0 \in X$, and use the definitions of f and g to prove $f(x_0) = g(x_0)$.

6. *Proving Invertibility.* We can prove that a function $f : X \to Y$ is invertible by proving f is one-to-one and onto. Alternatively, we can prove invertibility by defining a proposed inverse $g : Y \to X$, then proving $f \circ g = \mathrm{Id}_Y$ and $g \circ f = \mathrm{Id}_X$.

7. *Common Pitfalls.* Remember that composition of relations (or functions) is usually *not* commutative: $R \circ S \neq S \circ R$ for most choices of the relations (or functions) R and S. Do not confuse the preimage notation $f^{-1}[B]$ with the notation f^{-1} for an inverse function; preimages are defined for any function f, even when the inverse function f^{-1} does not exist. The two-sided inverse of f is unique when it exists, but left inverses and right inverses are not unique in general. Two functions with the same domain and the same values are not equal unless their codomains also agree.

Equivalence Relations, Set Partitions, and Partial Orders

1. *Equivalence Relations.* A relation R on a set X is an equivalence relation iff R is reflexive on X, symmetric, and transitive. This means that for all $a, b, c \in X$, aRa; if aRb then bRa; and if aRb and bRc then aRc. The equivalence class $[a]_R$ is the set of all elements of X related to a under R: $[a]_R = \{x \in X : xRa\}$.

2. *Theorem on Equivalence Classes.* Given an equivalence relation R on a set X and $a, b \in X$, $[a]_R = [b]_R$ iff aRb iff bRa iff $b \in [a]_R$ iff $a \in [b]_R$. Two equivalence classes of R are either disjoint or equal.

TABLE 7.5

Theorems on Functions.

Preimages	**Setup:** $f : X \to Y$.
Empty Preimage	$f^{-1}[\emptyset] = \emptyset$.
Domain Property	$f^{-1}[B] \subseteq X$.
Preimage of Codomain	$f^{-1}[Y] = X$.
Monotonicity	If $B \subseteq D$, then $f^{-1}[B] \subseteq f^{-1}[D]$.
Preimage of Union	$f^{-1}[\bigcup_{i \in I} B_i] = \bigcup_{i \in I} f^{-1}[B_i]$.
Preimage of Intersection	$f^{-1}[\bigcap_{i \in I} B_i] = \bigcap_{i \in I} f^{-1}[B_i]$.
Preimage of Difference	$f^{-1}[B - D] = f^{-1}[B] - f^{-1}[D]$.
Image of Preimage	$\forall B \subseteq Y$, $f[f^{-1}[B]] \subseteq B$; equality holds if f is surjective.
Preimage of Image	$\forall A \subseteq X$, $A \subseteq f^{-1}[f[A]]$; equality holds if f is injective.
Function Composition	**Setup:** $f : X \to Y$, $g : Y \to Z$, $h : Z \to W$.
Associativity	$(h \circ g) \circ f = h \circ (g \circ f)$.
Identity	$f \circ \mathrm{Id}_X = f = \mathrm{Id}_Y \circ f$.
Image under Composition	$(g \circ f)[A] = g[f[A]]$.
Preimage under Composition	$(g \circ f)^{-1}[C] = f^{-1}[g^{-1}[C]]$.
Injections/Surjections/Bijections	**Setup:** $f : X \to Y$ and $g : Y \to Z$.
Composing Injections	If f and g are injective, then $g \circ f$ is injective.
Composing Surjections	If f and g are surjective, then $g \circ f$ is surjective.
Composing Bijections	If f and g are bijective, then $g \circ f$ is bijective.
Backwards Rule for Injections	If $g \circ f$ is injective, then f is injective.
Backwards Rule for Surjections	If $g \circ f$ is surjective, then g is surjective.
Inverse Functions	**Setup:** $f : X \to Y$, $g : Y \to X$, $h : Y \to Z$.
Two-Sided Inverse	$g = f^{-1}$ iff $f \circ g = \mathrm{Id}_Y$ and $g \circ f = \mathrm{Id}_X$.
Bijectivity and Invertibility	f^{-1} exists iff f is bijective; in this case, f^{-1} is unique and bijective.
Inverse of Identity	Id_X is invertible, and $\mathrm{Id}_X^{-1} = \mathrm{Id}_X$.
Double Inverse	If f^{-1} exists, then $(f^{-1})^{-1} = f$.
Inverse of Composition	If f and h are invertible, then $(h \circ f)^{-1} = f^{-1} \circ h^{-1}$.
Left Inverse Criterion	For $X \neq \emptyset$, f has a left inverse iff f is injective.
Right Inverse Criterion	f has a right inverse iff f is surjective.
Pointwise Operations	**Setup:** $f, g, h : X \to \mathbb{R}$.
Commutativity	$f + g = g + f$ and $f \cdot g = g \cdot f$.
Associativity	$(f + g) + h = f + (g + h)$ and $(f \cdot g) \cdot h = f \cdot (g \cdot h)$.
Identity	$f + 0 = f = 0 + f$ and $f \cdot 1 = f = 1 \cdot f$.
Inverses	$f + (-f) = 0 = (-f) + f$ and (when $g(x)$ is never zero) $g \cdot (1/g) = 1 = (1/g) \cdot g$.
Distributive Law	$f \cdot (g + h) = (f \cdot g) + (f \cdot h)$.

TABLE 7.6

Theorems on Cardinality.

Cardinal Equality:	
Reflexivity	$\|A\| = \|A\|$.
Symmetry	$\|A\| = \|B\| \Rightarrow \|B\| = \|A\|$.
Transitivity	$(\|A\| = \|B\| \wedge \|B\| = \|C\|) \Rightarrow \|A\| = \|C\|$.
Disjoint Unions	If $\|A_i\| = \|A'_i\|$ for all $i \in I$ and $A_i \cap A_j = \emptyset = A'_i \cap A'_j$ for all $i \neq j$ in I, then $\left\|\bigcup_{i \in I} A_i\right\| = \left\|\bigcup_{i \in I} A'_i\right\|$.
Products	If $\|A_i\| = \|A'_i\|$ for $1 \leq i \leq n$, then $\|A_1 \times \cdots \times A_n\| = \|A'_1 \times \cdots \times A'_n\|$.
Power Sets	If $\|A\| = \|B\|$, then $\|\mathcal{P}(A)\| = \|\mathcal{P}(B)\|$.
Cardinal Ordering:	
Reflexivity	$\|A\| \leq \|A\|$.
Antisymmetry	$(\|A\| \leq \|B\| \wedge \|B\| \leq \|A\|) \Rightarrow \|A\| = \|B\|$ (Schröder–Bernstein Theorem).
Transitivity	$(\|A\| \leq \|B\| \wedge \|B\| \leq \|C\|) \Rightarrow \|A\| \leq \|C\|$.
Products	If $\|A_i\| \leq \|A'_i\|$ for $1 \leq i \leq n$, then $\|A_1 \times \cdots \times A_n\| \leq \|A'_1 \times \cdots \times A'_n\|$.
Finite Sets:	
Uniqueness of Size	For $m, n \in \mathbb{Z}_{\geq 0}$, $\|X\| = m$ and $\|X\| = n$ imply $m = n$.
Sum Rule	For pairwise disjoint finite $X_1, \ldots, X_s$, $\left\|\bigcup_{k=1}^{s} X_k\right\| = \sum_{k=1}^{s} \|X_k\|$.
Product Rule	For finite $X_1, \ldots, X_s$, $\|X_1 \times \cdots \times X_s\| = \prod_{k=1}^{s} \|X_k\|$.
Power Set Rule	For finite X, $\|\mathcal{P}(X)\| = 2^{\|X\|}$.
Preserving Finiteness	For finite $X, X_1, \ldots, X_s$, $\bigcup_{k=1}^{s} X_k$ and $\bigcap_{k=1}^{s} X_k$ and $X_1 \times \cdots \times X_k$ and $\mathcal{P}(X)$ and all subsets of X are finite.
Countable Sets:	
Countability Criteria	X is countably infinite iff $\|X\| = \|\mathbb{Z}_{>0}\|$. X is countable iff $\|X\| \leq \|\mathbb{Z}_{>0}\|$ iff $\exists$ bijection $f : X \to Y \subseteq \mathbb{Z}_{>0}$.
Countably Infinite Sets	$\mathbb{Z}_{\geq a}$, $\mathbb{Z}_{\leq a}$, $\mathbb{Z}$, $\mathbb{Q}$, $\mathbb{Z}^k$, $\mathbb{Q}^k$ (where $k \in \mathbb{Z}_{>0}$) are countably infinite.
Ordering Property	If $\|X\| \leq \|Y\|$ and Y is countable, then X is countable.
Subset Property	Any subset of a countable set is countable.
Product Property	A product of finitely many countable sets is countable.
Union Property	A union of countably many countable sets is countable.
Uncountable Sets:	
Ordering Property	If $\|X\| \leq \|Y\|$ and X is uncountable, then Y is uncountable.
Uncountable Sets	$\mathbb{R}$, (a, b), $[a, b]$, and $\mathcal{P}(\mathbb{Z}_{>0})$ are uncountable (for real $a < b$). For $\|X\| > 1$, the set of sequences $f : \mathbb{Z}_{>0} \to X$ is uncountable.
Cantor's Theorem	For all X, $\|X\| < \|\mathcal{P}(X)\|$.

3. *Set Partitions.* A set partition of X is a set of nonempty subsets of X such that every $a \in X$ belongs to exactly one set (block) in the set partition.

4. *Correspondence Between Equivalence Relations and Set Partitions.* Given an equivalence relation R on a set X, the quotient set $X/R = \{[a]_R : a \in X\}$ is a set partition of X. Given a set partition P of X, the relation $\sim_P$, defined for $a, b \in X$ by $a \sim_P b$ iff a and b belong to the same block of P, is an equivalence relation on X. The function sending R to X/R is a bijection from the set of equivalence relations on X to the set of set partitions of X. The inverse function sends a set partition P of X to $\sim_P$.

5. *Examples of Posets.* Number systems such as $\mathbb{R}$ are totally ordered by $\leq$. A collection of sets is partially ordered by $\subseteq$. A set of positive integers is partially ordered by divisibility. Other examples include subposets, product posets, and concatenation posets.

6. *Well-Ordering Property.* The poset $(\mathbb{Z}_{>0}, \leq)$ is well-ordered, meaning that every nonempty subset of positive integers has a least element. This property is equivalent to the Induction Axiom.

Review Problems

1. Complete the following definitions and proof templates:
   (a) "R is a relation from X to Z" means...
   (b) For a relation S and set A and object w, "$w \in S[A]$" means...
   (c) For relations R and S and objects x, y, "$(x, y) \in S \circ R$" means...
   (d) For a relation T and ordered pair (a, b), "$(a, b) \in T^{-1}$" means...
   (e) "F is a function from X to Y" means...
   (f) Given a function g with graph G and objects u, w, "$w = g(u)$" means...
   (g) Given a set Z and objects k, m, "$(k, m) \in I_Z$" means...
   (h) Given $f : X \to Y$ and $A \subseteq X$ and an object y, "$y \in f[A]$" means...
   (i) Given $f : X \to Y$ and $B \subseteq Y$ and an object $x \in X$, "$x \in f^{-1}[B]$" means...
   (j) "$h : A \to C$ is one-to-one" means...
   (k) "$p : D \to E$ is onto" means...
   (l) Given $F : Z \to W$ and $G : Z \to W$, outline a proof that $F = G$.
   (m) Given relations R and S from X to Y, describe two ways to prove that $R = S$.
   (n) Given $f, g, h : \mathbb{R} \to \mathbb{R}$ and $t \in \mathbb{R}$, define $(f \cdot g)(t + 1)$, $(f \circ g)(t + 1)$, and $(g + f)(t + 1)$.
   (o) "R is reflexive on X" means...
   (p) "S is symmetric" means...
   (q) "T is transitive" means...
   (r) "X has n elements" means...
   (s) "$|U| \leq |V|$" means...

2. Fix $a, b \in \mathbb{R}$ with $a \neq 0$. Define $f : \mathbb{R} \to \mathbb{R}$ by $f(x) = ax + b$ for all $x \in \mathbb{R}$.
   (a) Is f one-to-one? Is f onto? Prove your answers.
   (b) Repeat (a) replacing $\mathbb{R}$ by $\mathbb{Q}$ everywhere in the setup.
   (c) Repeat (a) replacing $\mathbb{R}$ by $\mathbb{Z}$ everywhere in the setup. Does the answer to (c) depend on the particular value of a?

3. For each set X and equivalence relation R on X, explicitly describe the requested equivalence classes.
   (a) $X = \{a, b, c, d\}$, $R = \{(a, a), (b, b), (c, c), (d, d), (b, c), (c, b), (a, d), (d, a)\}$; find $[a]_R$, $[b]_R$, $[c]_R$, and $[d]_R$.

(b) $X = \mathbb{Z}$, R is congruence mod 5; find $[7]_R$.
(c) $X = \{1, 2, 3, 4, 5\}$, $R = I_X$; find $[3]_R$.
(d) $X = \mathbb{R}_{\geq 0} \times \mathbb{R}_{\geq 0}$, $((x, y), (u, v)) \in R$ means $x + y = u + v$; draw $[(1, 2)]_R$.

4. Give a specific example of a function $f : \mathbb{Z}_{>0} \to \mathbb{Z}_{>0}$ with the stated properties. Prove that your examples work. (a) f is one-to-one and not onto. (b) f is onto and not one-to-one. (c) f is not one-to-one and not onto and not constant. (d) f is one-to-one and onto and $f \neq \mathrm{Id}_{\mathbb{Z}_{>0}}$.

5. Suppose S, T, U are finite sets with $|S| = a$ and $|T| = b$ and $U \subseteq S$. Which formulas are always true? Explain. (a) $|S \cup T| = a + b$. (b) $|S \times T| = ab$. (c) $|U| = c$ for some integer $c < a$. (d) $|\mathcal{P}(S)| = 2^a$. (e) $|S - T| = a - b$.

6. Which of the following sets are countably infinite? (a) $\emptyset$ (b) $\{1, 2, 3\}$ (c) $\mathbb{Z}_{\geq 0}$ (d) $\mathbb{Z}$ (e) $\mathbb{Q}$ (f) $\mathbb{Z} \times \mathbb{Z}$ (g) $\mathbb{R}$ (h) the set of all infinite sequences of 0s and 1s (i) $\mathcal{P}(\mathbb{Z})$.

7. (i) Draw an arrow diagram and a digraph for each relation. (ii) Is each relation reflexive on $X = \{1, 2, 3, 4\}$? symmetric? transitive? Explain. (iii) Compute $R[\{1, 3\}]$ and $R \circ R^{-1}$ for each R.
(a) $R = \{(1, 1), (2, 2), (3, 3), (4, 4), (1, 3), (3, 4), (1, 4), (2, 4)\}$.
(b) $R = \{(1, 2), (2, 1), (3, 4), (4, 3)\}$.
(c) $R = (X \times X) - I_X$.
(d) $R = \{(1, 1), (3, 3), (1, 3), (3, 1), (2, 2), (4, 4), (2, 4), (4, 2)\}$.

8. (a) Prove: for all $f, g, h : \mathbb{R} \to \mathbb{R}$, $(f \cdot g) \circ h = (f \circ h) \cdot (g \circ h)$.
(b) Disprove: for all $f, g, h : \mathbb{R} \to \mathbb{R}$, $f \circ (g \cdot h) = (f \circ g) \cdot (f \circ h)$.
(c) Give four examples of functions $f : \mathbb{R} \to \mathbb{R}$ such that for all $g, h : \mathbb{R} \to \mathbb{R}$, $f \circ (g \cdot h) = (f \circ g) \cdot (f \circ h)$.

9. Suppose R is an equivalence relation on $X = \{1, 2, 3, 4, 5, 6\}$ containing the ordered pairs $(1, 4)$, $(6, 3)$, and $(3, 2)$. (a) Draw the digraph of R. Give the answer with the fewest possible arrows. (b) Find all distinct equivalence classes of the relation R in (a). (c) What is the set partition X/R?

10. (a) Define $f : \mathbb{R}_{\neq 0} \to \mathbb{R}$ by $f(x) = \frac{3x-1}{x}$ for all real $x \neq 0$. Prove f is one-to-one but not onto. What is $f[\mathbb{R}_{\neq 0}]$? (b) Define $g : \mathbb{R} \times \mathbb{R} \to \mathbb{R} \times \mathbb{R}$ by $g((x, y)) = (x + y, 2 - y)$ for all $(x, y) \in \mathbb{R} \times \mathbb{R}$. Is g one-to-one? Is g onto? Prove your answers.

11. Suppose X and I are fixed sets, R_i is a fixed relation on X for $i \in I$, and $R = \bigcup_{i \in I} R_i$. Prove or disprove each statement. (a) R is a relation on X. (b) If every R_i is reflexive on X, then R must be reflexive on X. (c) If every R_i is symmetric, then R must be symmetric. (d) If every R_i is transitive, then R must be transitive.

12. Given functions $f : X \to U$ and $g : Y \to V$, define $h : X \times Y \to U \times V$ by $h((x, y)) = (f(x), g(y))$ for all $(x, y) \in X \times Y$.
(a) Prove: h is a function.
(b) Prove: if f is one-to-one and g is one-to-one, then h is one-to-one.
(c) Prove: if f is onto and g is onto, then h is onto.
(d) Prove: if f is a bijection and g is a bijection, then h is a bijection.

13. (a) Prove: for all relations R and all sets A and B, $R[A] - R[B] \subseteq R[A - B]$.
(b) Give an example to show that equality does not always hold in (a).
(c) Suppose R is a relation from X to Y and $A, B \subseteq X$. Which of the following conditions on R will force equality to hold in (a)? Select all that apply: R is the graph of a function from X to Y; R is the inverse of the graph of a function from X to Y; R is the graph of an injection from X to Y; R is the graph of a surjection from X to Y.

14. Prove: for all $f : X \to Y$ and $g : Y \to Z$ and all $C \subseteq Z$, $(g \circ f)^{-1}[C] = f^{-1}[g^{-1}[C]]$.

15. Prove: for all $f : X \to Y$, f is injective iff (for all $A, C \subseteq X$, $f[A \cap C] = f[A] \cap f[C]$).

16. For each set X and relation R, prove that R is an equivalence relation on X.
    (a) $X = \mathbb{Z}^3$, $(a, b, c)R(d, e, f)$ means $a + b + c = d + e + f$ for $a, b, c, d, e, f \in \mathbb{Z}$.
    (b) $X = \mathbb{R}$, $(x, y) \in R$ means $y - x \in \mathbb{Q}$ for $x, y \in \mathbb{R}$.
    (c) $X = \mathbb{R}_{>0} \times \mathbb{R}_{>0}$, $((x, y), (a, b)) \in R$ means there exists $r \in \mathbb{R}_{>0}$ with $a = rx$ and $b = ry$, for $x, y, a, b \in \mathbb{R}$.

17. Let $X = [0, 3] \times [0, 3]$. For each equivalence relation R below, draw a picture of X and the equivalence classes $[(1, 3)]_R$ and $[(2, 2)]_R$.
    (a) R is defined by: $((a, b), (c, d)) \in R$ iff $a = c$, for $(a, b), (c, d) \in X$.
    (b) R is defined by: $((a, b), (c, d)) \in R$ iff $b = d$, for $(a, b), (c, d) \in X$.
    (c) R is defined by: $((a, b), (c, d)) \in R$ iff $a + b = c + d$, for $(a, b), (c, d) \in X$.
    (d) R is defined by: $((a, b), (c, d)) \in R$ iff $ab = cd$, for $(a, b), (c, d) \in X$.

18. Let X and I be fixed, nonempty sets. For each $i \in I$, let R_i be a fixed equivalence relation on X. (a) Prove: $R = \bigcap_{i \in I} R_i$ is an equivalence relation on X. (b) Prove: for all $b \in X$, $[b]_R = \bigcap_{i \in I} [b]_{R_i}$.

19. Prove or disprove: for all sets X, Y, P, Q, if P is a set partition of X and Q is a set partition of Y, then $P \cup Q$ is a set partition of $X \cup Y$.

20. Let R be a relation from X to Y. For each identity below, find a condition on R that will guarantee the truth of the identity. (Possible conditions are: R is any relation whatsoever; R is the graph of a function; R is the graph of an injection; R is the graph of a surjection; R is the graph of a bijection.) Prove that your answer works, but give examples showing that more general relations do not work.
    (a) For all $B \subseteq Y$, $R[R^{-1}[B]] = B$.
    (b) For all $A \subseteq X$, $R^{-1}[R[A]] = A$.
    (c) For all $A, C \subseteq X$, $R[A \cup C] = R[A] \cup R[C]$.

21. (a) Define an explicit bijection $f : \mathbb{Z} \to \mathbb{Z}_{>0}$. (b) Prove: for all sets X and Y, if X and Y are disjoint countably infinite sets, then $X \cup Y$ is countably infinite. (c) Does (b) remain true if X and Y are not disjoint?

22. (a) Define $f : \mathbb{Z}_{\geq 0}^4 \to \mathbb{Z}_{>0}$ by $f(a, b, c) = 2^a 3^b 5^c 7^d$ for $a, b, c, d \in \mathbb{Z}_{\geq 0}$. Prove that f is injective but not surjective. (b) Use (a) and a theorem to prove that $\mathbb{Z}_{\geq 0}^4$ is countably infinite.

23. Let X be an infinite set, and let Z be the set of all $f : X \to X$ such that $f(x) \neq x$ for all $x \in X$. Prove that Z is uncountable.

24. Let Y be the set of all $f : \mathbb{Z}_{>0} \to \mathbb{Z}_{>0}$ such that $f(n) = n$ for all but finitely many $n \in \mathbb{Z}_{>0}$. Is Y countable or uncountable? Prove your answer.

8

Real Numbers (Optional)

8.1 Axioms for $\mathbb{R}$ and Properties of Addition

The goal of this optional chapter is to present an axiomatic development of the real number system. Such a development is needed to put the subject of calculus (also called *analysis*) on a firm logical footing. We assume no prior knowledge of real numbers or any other number system (even the integers), instead deducing all needed facts from the initial axioms. We do assume that facts about logic, proofs, sets, relations, and functions are already known. Thus we may invoke any results from previous sections excluding theorems about integers and rational numbers from Chapter 4. These theorems will become available later after we formally define $\mathbb{Z}$ and $\mathbb{Q}$.

Undefined Terms and Initial Definitions

Mathematical theories based on axioms were described in §2.1. Like all such theories, the axiomatic development of the real numbers begins with certain *undefined terms*. Our theory uses the following eight undefined terms:

(a) a set $\mathbb{R}$ called the *set of real numbers*

(b) a function $\mathrm{ADD} : \mathbb{R} \times \mathbb{R} \to \mathbb{R}$ called *addition*

(c) a real number $0 \in \mathbb{R}$ called *zero*

(d) a function $\mathrm{AINV} : \mathbb{R} \to \mathbb{R}$ called *additive inverse*

(e) a function $\mathrm{MULT} : \mathbb{R} \times \mathbb{R} \to \mathbb{R}$ called *multiplication*

(f) a real number $1 \in \mathbb{R}$ called *one*

(g) a function $\mathrm{MINV} : \mathbb{R}_{\neq 0} \to \mathbb{R}$ called *multiplicative inverse*

(h) a relation LEQ on $\mathbb{R}$ called the *ordering*.

The next definition introduces the standard arithmetical symbols for the functions and relations listed above.

8.1. Definition: Arithmetical Operators.

(a) For all $x, y \in \mathbb{R}$, define the *sum* of x and y to be $x + y = \mathrm{ADD}(x, y)$.

(b) For all $x \in \mathbb{R}$, define the *negative of x* to be $-x = \mathrm{AINV}(x)$.

(c) For all $x, y \in \mathbb{R}$, define the *difference* of x and y to be $x - y = x + (-y)$. The function sending (x, y) to $x - y$ is called *subtraction*.

(d) For all $x, y \in \mathbb{R}$, define the *product* of x and y to be $x \cdot y = \mathrm{MULT}(x, y)$. We often abbreviate $x \cdot y$ as xy.

(e) For all $x \neq 0$ in $\mathbb{R}$, define the *inverse of x* to be $x^{-1} = \mathrm{MINV}(x)$.

(f) For all $x, y \in \mathbb{R}$ with $y \neq 0$, define x *divided by y* to be $x/y = x \cdot (y^{-1})$. We also write $\frac{x}{y}$ for x/y. The function sending (x, y) to x/y is called *division*.

(g) For all $x, y \in \mathbb{R}$, define $x \leq y$ iff $(x, y) \in$ LEQ.

(h) For all $x, y \in \mathbb{R}$, define $x \geq y$ iff $y \leq x$.

(i) For all $x, y \in \mathbb{R}$, define $x < y$ iff ($x \leq y$ and $x \neq y$).

(j) For all $x, y \in \mathbb{R}$, define $x > y$ iff $y < x$.

(k) A chain of inequalities such as $x < y \leq z < w$ abbreviates "$x < y$ and $y \leq z$ and $z < w$."

Note that addition and the other arithmetical operations are assumed to be functions with codomain $\mathbb{R}$. This means, for example, that given any two real numbers x and y, there *exists* a *unique* real number $x + y$. The fact that the output $x + y$ must belong to the set $\mathbb{R}$ is called *closure* of $\mathbb{R}$ under addition. Thus, as part of the axiomatic setup, we are assuming that $\mathbb{R}$ is closed under addition, additive inverses, multiplication, and multiplicative inverses of nonzero real numbers. It follows that $\mathbb{R}$ is closed under subtraction and division by nonzero real numbers. Also, the special constants 0 and 1 belong to $\mathbb{R}$. Starting from 1, we can define some new real numbers as follows.

8.2. Definition of Some Specific Numbers. Define $2 = 1 + 1$, $3 = 2 + 1$, $4 = 3 + 1$, $5 = 4 + 1$, $6 = 5 + 1$, $7 = 6 + 1$, $8 = 7 + 1$, $9 = 8 + 1$, and $10 = 9 + 1$. By closure, the symbols 2 through 10 *are* real numbers, called *two, three, four, five, six, seven, eight, nine,* and *ten* (respectively).

Axioms

We list all the axioms for $\mathbb{R}$ in this subsection. We divide the axioms into four groups: axioms for addition, axioms for multiplication, basic ordering axioms, and the completeness axiom. The axioms in each group, and properties derived from them, are discussed in detail later. For emphasis, we list certain closure properties as axioms, even though these properties are already implicit in the initial descriptions of the undefined terms. We begin with the five axioms for addition.

8.3. Axioms for Addition.

A1. *Closure:* For all $x, y \in \mathbb{R}$, $x + y \in \mathbb{R}$ and $0 \in \mathbb{R}$ and $-x \in \mathbb{R}$.

A2. *Associativity:* For all $x, y, z \in \mathbb{R}$, $(x + y) + z = x + (y + z)$.

A3. *Additive Identity:* For all $x \in \mathbb{R}$, $x + 0 = x = 0 + x$.

A4. *Additive Inverses:* For all $x \in \mathbb{R}$, $x + (-x) = 0 = (-x) + x$.

A5. *Commutativity:* For all $x, y \in \mathbb{R}$, $x + y = y + x$.

There are seven axioms for multiplication. The first five are exact analogues of the five addition axioms, and the last two stipulate relationships between the operations of addition and multiplication.

8.4. Axioms for Multiplication.

M1. *Closure:* For all $x, y \in \mathbb{R}$, $x \cdot y \in \mathbb{R}$ and $1 \in \mathbb{R}$ and if $x \neq 0$, then $x^{-1} \in \mathbb{R}$.

M2. *Associativity:* For all $x, y, z \in \mathbb{R}$, $(x \cdot y) \cdot z = x \cdot (y \cdot z)$.

M3. *Multiplicative Identity:* For all $x \in \mathbb{R}$, $x \cdot 1 = x = 1 \cdot x$.

M4. *Multiplicative Inverses:* For all $x \in \mathbb{R}$, if $x \neq 0$, then $x \cdot (x^{-1}) = 1 = (x^{-1}) \cdot x$.

M5. *Commutativity:* For all $x, y \in \mathbb{R}$, $x \cdot y = y \cdot x$.

D. *Distributive Laws:* For all $x, y, z \in \mathbb{R}$, $x \cdot (y+z) = (x \cdot y) + (x \cdot z)$ and $(y+z) \cdot x = (y \cdot x) + (z \cdot x)$.

N. *Nontriviality:* $0 \neq 1$.

There are seven ordering axioms. The first four state that $(\mathbb{R}, \leq)$ is a totally ordered set; the next two relate this ordering to the algebraic operations of addition and multiplication.

8.5. Basic Ordering Axioms.

O1. *Reflexivity:* For all $x \in \mathbb{R}$, $x \leq x$.

O2. *Antisymmetry:* For all $x, y \in \mathbb{R}$, if $x \leq y$ and $y \leq x$, then $x = y$.

O3. *Transitivity:* For all $x, y, z \in \mathbb{R}$, if $x \leq y$ and $y \leq z$, then $x \leq z$.

O4. *Total Ordering:* For all $x, y \in \mathbb{R}$, $x \leq y$ or $y \leq x$.

O5. *Additive Property:* For all $x, y, z \in \mathbb{R}$, if $x \leq y$, then $x + z \leq y + z$.

O6. *Multiplicative Property:* For all $x, y, z \in \mathbb{R}$, if $x \leq y$ and $0 \leq z$, then $x \cdot z \leq y \cdot z$.

The final ordering axiom states a crucial technical property of $\mathbb{R}$ called *completeness*. To even formulate the axiom, we need the following concepts, which make sense for any partially ordered set.

8.6. Definition: Upper Bounds. For all $S \subseteq \mathbb{R}$ and $z_0 \in \mathbb{R}$, $\boxed{z_0 \text{ is an upper bound of } S}$ iff $\boxed{\forall y \in S, y \leq z_0.}$ For all $S \subseteq \mathbb{R}$, $\boxed{S \text{ is bounded above}}$ iff $\boxed{\exists z \in \mathbb{R}, \forall y \in S, y \leq z}$.

8.7. Definition: Least Upper Bounds. For all $S \subseteq \mathbb{R}$ and $z_0 \in \mathbb{R}$,

$\boxed{z_0 \text{ is a least upper bound of } S}$ (written $\boxed{z_0 = \operatorname{lub} S}$) iff

$$(\forall y \in S, y \leq z_0) \wedge (\forall x \in \mathbb{R}, \text{if } [\forall y \in S, y \leq x] \text{ then } z_0 \leq x).$$

8.8. Completeness Axiom O7. For all $S \subseteq \mathbb{R}$, if S is nonempty and S is bounded above then S has a least upper bound in $\mathbb{R}$.

Although not explicitly listed here, we are also assuming various *logical axioms* governing the logical symbols such as $\Rightarrow$, $\forall$, $\exists$, and $=$. Regarding equality, we are assuming that $=$ is reflexive, symmetric, and transitive, and satisfies various substitution properties such as: for all $x, y, z \in \mathbb{R}$, if $y = z$, then $x + y = x + z$. These properties will be used frequently without specifically invoking the underlying logical axioms.

8.9. Remark: Algebraic Structures. In abstract algebra, we study many algebraic systems that satisfy various subsets of the axioms listed above (with the set $\mathbb{R}$ replaced throughout by an appropriate set of objects). In particular, a system satisfying axioms A1 through A5 is called a *commutative group*; omitting A5 gives the axioms defining a *group*. A system satisfying all the additive and multiplicative axioms is called a *field*; omitting M4, N, and multiplicative inverses gives the axioms for a *commutative ring*; and further omitting axiom M5 gives the axioms for a *ring* (with identity). If we include all the axioms for $\mathbb{R}$ except completeness (O7), we get the definition of an *ordered field*; an *ordered commutative ring* is defined similarly. Here are a few examples using number systems not officially defined yet: $\mathbb{Z}$ (the set of integers) is an ordered commutative ring that is not a field; $\mathbb{Q}$ (the set of rational numbers) is an ordered field that does not satisfy the completeness axiom; $\mathbb{C}$ (the set of complex numbers) is a field that cannot be made into an ordered field; and the set of 2×2 matrices with real entries is a non-commutative ring.

Properties of Addition

We begin our development of the properties of $\mathbb{R}$ by deriving some consequences of the addition axioms A1 through A5. It is helpful to keep track of which axioms are needed to derive various results, since the same results also hold in other more general algebraic systems for which these particular axioms are true. In particular, we give proofs avoiding the commutativity axiom A5 when possible. Our first theorem is a basic but very useful uniqueness result.

8.10. Theorem: Left Cancellation Law for Addition.
For all $x, y, z \in \mathbb{R}$, if $x + y = x + z$, then $y = z$.

Proof. Fix arbitrary $x, y, z \in \mathbb{R}$. Assume $x + y = x + z$; prove $y = z$. By the substitution property of equality, we can add $-x$ to both sides of the assumption $x + y = x + z$, obtaining $(-x) + (x + y) = (-x) + (x + z)$. (This step also uses closure (A1), since we need to know $-x \in \mathbb{R}$, $x + y \in \mathbb{R}$, and $x + z \in \mathbb{R}$ in order for the sums $(-x) + (x + y)$ and $(-x)+(x+z)$ to be defined. We make constant use of substitution axioms and closure axioms when manipulating complicated expressions, but we seldom mention these axioms in the sequel.) By associativity (A2), the previous equation becomes $((-x)+x)+y = ((-x)+x)+z$. By the inverse axiom (A4), we get $0 + y = 0 + z$. Finally, the identity axiom (A3) turns this equation into $y = z$, which is the required conclusion. This proof used axioms A1, A2, A3, and A4. ∎

There is an analogous *right cancellation law*: for all $x, y, z \in \mathbb{R}$, if $y + x = z + x$, then $y = z$. This law follows from the left cancellation law via commutativity (A5), but it is better to adapt the proof given above (add $-x$ to both sides *on the right*) so that the proof only uses axioms A1 through A4. The next theorem lists some fundamental facts that can be deduced as special cases of the left cancellation law. Similar facts hold based on the right cancellation law, but we do not state these explicitly.

8.11. Theorem on Addition. For all $x, y, a, b \in \mathbb{R}$:

(a) *Uniqueness of Additive Identity:* If $x + y = x$, then $y = 0$.

(b) *Uniqueness of Additive Inverses:* If $x + y = 0$, then $y = -x$.

(c) *Negative of Zero:* $-0 = 0$.

(d) *Double Negative Rule:* For all $a \in \mathbb{R}$, $-(-a) = a$.

(e) *Negative of a Sum:* For all $a, b \in \mathbb{R}$, $-(a + b) = (-b) + (-a)$.

Proof. For (a), fix $x, y \in \mathbb{R}$; assume $x + y = x$; prove $y = 0$. Using the identity axiom (A3), our assumption becomes $x + y = x + 0$. Now use left cancellation, taking $z = 0$, to deduce $y = 0$. [This result says that zero is the *only* real number having the additive identity property stated in A3.]

For (b), fix $x, y \in \mathbb{R}$; assume $x + y = 0$; prove $y = -x$. Using the inverse axiom (A4), our assumption becomes $x + y = x + (-x)$. Now use left cancellation, taking $z = -x$, to deduce $y = -x$. [This result says that for each fixed real x, $-x$ is the *only* real number having the additive inverse property stated in A4.]

Now we prove (c). On one hand, taking $x = 0$ in the inverse axiom (A4) gives $0 + (-0) = 0$. On the other hand, taking $x = -0$ in the identity axiom (A3) gives $0 + (-0) = -0$. Thus, $-0 = 0 + (-0) = 0$.

For (d), fix $a \in \mathbb{R}$. By the inverse axiom (A4), $(-a) + a = 0$. Using part (b) with $x = -a$ and $y = a$, we deduce $y = -x$, i.e., $a = -(-a)$.

For (e), fix $a, b \in \mathbb{R}$. We compute

$$(a + b) + ([-b] + [-a]) = a + (b + ([-b] + [-a])) = a + ((b + [-b]) + [-a])$$
$$= a + (0 + [-a]) = a + [-a] = 0.$$

These equalities follow by associativity (A2), associativity again, the inverse axiom (A4), the identity axiom (A3), and the inverse axiom (A4). Now use part (b) with $x = a + b$ and $y = [-b] + [-a]$ to deduce $y = -x$, i.e., $[-b] + [-a] = -(a + b)$.

The above proofs use axioms A1 through A4 only.   □

8.12. Remark. If we allow ourselves to use commutativity (A5), we can rewrite part (e) of the last theorem as $-(a + b) = (-a) + (-b)$. We might attempt to prove this identity using the distributive axiom (D), as follows:

$$-(a + b) = (-1) \cdot (a + b) = (-1) \cdot a + (-1) \cdot b = (-a) + (-b).$$

However, this calculation requires the property $-a = (-1) \cdot a$, which has not yet been proved. Even worse, this proof refers to multiplication to prove a property involving addition alone. Thus the proof would not work in an algebraic structure that had an addition operation but no multiplication operation.

8.13. Theorem on Subtraction. For all $x, y, z \in \mathbb{R}$:
(a) $x - 0 = x$.
(b) $0 - x = -x$.
(c) $x - x = 0$.
(d) $-(x - y) = y - x = (-x) + y$.
(e) $(x - y) + (y - z) = x - z$.
(f) $(x + z) - (y + z) = x - y$.

Proof. We prove one identity and leave the others as exercises. Fix $x, y \in \mathbb{R}$. We compute

$$-(x - y) = -(x + [-y]) = -(-y) + (-x) = y + (-x) = y - x,$$

using the definition of subtraction, then 8.11(e), then 8.11(d), then the definition of subtraction. Using commutativity (A5), we can also write $y - x = y + (-x) = (-x) + y$.   □

The next result illustrates how specific arithmetical identities can be proved from the axioms.

8.14. Proposition. $2 + 4 = 6$ and $3 - 5 = -2$.

Proof. We gradually build up to the identity $2 + 4 = 6$ by proving simpler related facts. First, $2 + 2 = 2 + (1 + 1) = (2 + 1) + 1 = 3 + 1 = 4$ using the definition of 2, then associativity (A2), then the definition of 3, then the definition of 4. Second, $2 + 3 = 2 + (2 + 1) = (2 + 2) + 1 = 4 + 1 = 5$ using the definition of 3, then associativity (A2), then the first result, then the definition of 5. Third, $2 + 4 = 2 + (3 + 1) = (2 + 3) + 1 = 5 + 1 = 6$ using the definition of 4, then associativity (A2), then the second result, then the definition of 6.

Similarly, to prove $3 - 5 = -2$, observe that $2 + (3 - 5) = 2 + (3 + (-5)) = (2 + 3) + (-5) = 5 + (-5) = 0$ by definition of subtraction, then associativity (A2), then the second result above, then inverses (A4). Now use 8.11(b) with $x = 2$ and $y = 3 - 5$ to see that $y = -x$, i.e., $3 - 5 = -2$.   □

Any other arithmetical fact involving addition or subtraction of specific integers can be proved in a similar manner.

8.15. Remark: Generalized Associativity. Given three real numbers a, b, c, the expression $a + b + c$ is ambiguous: it could mean $(a + b) + c$ or $a + (b + c)$. However, associativity (A2) assures us that the latter two expressions must be equal, so it is safe to write $a + b + c$ with no parentheses. Similarly, given four real numbers a, b, c, d, the ambiguous expression $a + b + c + d$ could represent any of the following five quantities:

$$(a + (b + c)) + d, \ ((a + b) + c) + d, \ (a + b) + (c + d), \ a + (b + (c + d)), \ a + ((b + c) + d).$$

Each expression on this list can be seen to be equal to the next expression by an appropriate special case of associativity (A2). For example, letting $x = a + b$, $y = c$, and $z = d$ in A2 shows that the second expression equals the third. Because of this fact, we often write $a + b + c + d$ with no parentheses, and we can group adjacent terms together at will if needed in a particular calculation (as in the proof of Theorem 8.11(e)). A similar result holds for a sum of n terms: given $a_1, a_2, \ldots, a_n \in \mathbb{R}$, any possible parenthesization of $a_1 + a_2 + \cdots + a_n$ produces the same final output. This fact is called *generalized associativity*, and its proof is far from obvious, as the $n = 4$ case already suggests. Since we have not yet defined what a positive integer n is, we cannot even state this general fact formally, much less prove it. Luckily, in our initial development, we only need this fact for a few specific small values of n (like $n = 4$), where it can be justified by explicit repeated uses of A2.

There is a related fact called *generalized commutativity* based on A5. It states that we can rearrange the order of terms in $a_1 + a_2 + \cdots + a_n$ in any fashion without changing the final result. For example, $a + b + c + d = d + b + a + c = b + c + d + a$, and so on. Henceforth, we use generalized associativity and commutativity (for small values of n) without specific comment.

Section Summary

1. *Undefined Terms.* Our axiomatic development of the real number system uses these undefined concepts: the set $\mathbb{R}$, the operations $x + y$, $-x$, $x \cdot y$, and z^{-1}, the numbers 0 and 1, and the relation $x \leq y$. (Here, $x, y, z \in \mathbb{R}$ and $z \neq 0$.)

2. *Axioms.* Addition and multiplication satisfy the axioms of closure, associativity, identity, inverses, and commutativity. The distributive laws hold, and $0 \neq 1$. The relation $\leq$ is reflexive, antisymmetric, transitive, a total order, and is preserved by addition and multiplication by nonnegative numbers. Finally, the completeness axiom tells us that every nonempty subset of $\mathbb{R}$ that is bounded above has a least upper bound in $\mathbb{R}$.

3. *Basic Definitions.* We define $x - y = x + (-y)$, $x/z = x \cdot z^{-1}$, $x < y$ iff $x \leq y$ and $x \neq y$, $x \geq y$ iff $y \leq x$, and $x > y$ iff $y < x$ (for $x, y, z \in \mathbb{R}$ with $z \neq 0$). Define $2 = 1 + 1$, $3 = 2 + 1$, and so on.

4. *Properties of Addition.* The additive cancellation law holds: for all $x, y, z \in \mathbb{R}$, $x + y = x + z \Rightarrow y = z$. Zero is the unique additive identity; $-x$ is the unique additive inverse of $x \in \mathbb{R}$; $-0 = 0$; $-(-x) = x$; and $-(x + y) = (-y) + (-x)$ for all $x, y \in \mathbb{R}$. Addition satisfies generalized associativity and generalized commutativity.

Exercises

1. (a) Give a detailed proof of the right cancellation law without using axiom A5.
   (b) State and prove analogues of 8.11(a) and 8.11(b) based on the right cancellation law.

2. Complete the proof of Theorem 8.13.

3. Deduce 8.11(c) from 8.11(b).

4. Prove that $5 + 5 = 10$ and $2 - 6 = -4$ and $6 - 3 = 3$ (you may need to prove some other related identities first).

5. (a) Using only the addition axioms A1 through A5, prove: $\forall a, b, c, d \in \mathbb{R}, (a+c) + (b+d) = (a+b) + (c+d)$. (b) Using only axioms and theorems proved above, prove: $\forall a, b, c, d \in \mathbb{R}, (a+c) - (b+d) = (a-b) + (c-d)$.

6. Suppose we give meanings to the undefined symbols of our theory by letting $\mathbb{R} = \{a\}$ (where a is some fixed object), $a + a = a$, $-a = a$, $a \cdot a = a$, $0 = a$, $1 = a$, and LEQ $= \{(a, a)\}$. (Multiplicative inverses are not defined since nonzero elements of $\mathbb{R}$ do not exist.) Prove that this system satisfies all the axioms for the real numbers except the nontriviality axiom N.

7. Let $\mathbb{R}^3 = \mathbb{R} \times \mathbb{R} \times \mathbb{R}$. Define addition, zero, and negatives on this set as follows: for all $a, b, c, x, y, z \in \mathbb{R}$,

$$(a, b, c) + (x, y, z) = (a + x, b + y, c + z), \quad \mathbf{0} = (0, 0, 0), \quad -(a, b, c) = (-a, -b, -c).$$

Use known properties of $\mathbb{R}$ to show that $\mathbb{R}^3$ satisfies the additive axioms A1 through A5.

8. Suppose we redefine the multiplicative operations in $\mathbb{R}$ by setting $x \cdot y = 0$, $1 = 0$, and $z^{-1} = 0$ for all $x, y, z \in \mathbb{R}$ with $z \neq 0$. Which of axioms M1 through M5, D, and N are now true?

9. Suppose we redefine addition on $\mathbb{R}$ by setting $x + 0 = x = 0 + x$ for all $x \in \mathbb{R}$, and $y + z = 0$ for all nonzero $y, z \in \mathbb{R}$. (a) Show that axioms A3, A4, and A5 are true, but the associativity axiom A2 is false. (b) Show that additive inverses are not unique in this system.

10. Suppose we redefine addition on $\mathbb{R}$ by setting $x + y = x$ for all $x, y \in \mathbb{R}$. Which of the addition axioms A1 through A5 are true?

11. For $x, y \in \mathbb{R}$, define $x \star y = x + y + 2$. Show that the addition axioms A1 through A5 hold if we replace $+$ by $\star$, 0 by -2, and $-x$ by $-x - 4$.

12. Let X be a fixed set. Use theorems about bijections to prove that the set G consisting of all bijections $f : X \to X$ is a group if we replace addition (in axioms A1 through A4) by composition of functions, zero by Id_X, and additive inverses by inverse functions.

13. If possible, give an example of a field in which $0 = 3$.

14. Prove: for all $a, b, c \in \mathbb{R}$, $(a + b) + c = (b + a) + c = (c + b) + a = (b + c) + a = (a + c) + b = (c + a) + b$. Justify **every** step using axioms A2 and A5 only.

15. (a) Define real numbers 11, 12, 13, 14, 15, 16, 17, and 18. (b) Make an *addition table* with rows and columns labeled $0, 1, 2, \ldots, 9$, such that the entry in row a and column b is $a + b$. (c) Suppose you wrote a complete proof of your answer to (b). In such a proof, how many times would a definition, axiom, or previous result be invoked? Explain.

16. List all complete parenthesizations of the expression $a + b + c + d + e$, where $a, b, c, d, e \in \mathbb{R}$. Use the associativity axiom A2 to prove that all these expressions are equal.

8.2   Algebraic Properties of Real Numbers

We continue to develop properties of the arithmetical operations — addition, subtraction, multiplication, and division — based on the algebraic axioms A1 through A5, M1 through M5, D, and N. The theorems proved using these axioms are valid in any field (such as the rational numbers, the complex numbers, or the integers modulo a prime).

Multiplicative Properties of Zero

Before developing general properties of multiplication, we need to examine the relationship between zero (the *additive* identity) and the multiplication operation. Our first result is very familiar, but the proof is surprisingly tricky.

8.16. Theorem on Multiplication by Zero.
Z1. For all $a \in \mathbb{R}$, $0 \cdot a = 0$ and $a \cdot 0 = 0$.

Proof. Fix $a \in \mathbb{R}$. [How do we even begin? This is the hard part — the trick is to think of examining the quantity $0 \cdot a + 0 \cdot a$.] We compute $(0 \cdot a) + (0 \cdot a) = (0+0) \cdot a = 0 \cdot a$ using the distributive axiom (D) and the additive identity axiom (A3). Now recall Theorem 8.11(a): $\forall x, y \in \mathbb{R}, x + y = x \Rightarrow y = 0$. Taking $x = y = 0 \cdot a$ here, we have verified that $x + y = x$ and so we may conclude that $y = 0$, i.e., $0 \cdot a = 0$. We prove $a \cdot 0 = 0$ similarly, starting from the quantity $a \cdot 0 + a \cdot 0$. This proof used addition axioms A1 through A4, M1 (where?), and D. [Since D and N are the only axioms connecting multiplication to addition, we can retroactively motivate the introduction of $0 \cdot a + 0 \cdot a$ by noting that we needed some expression that would allow us to use the distributive law.] □

8.17. Theorem on Closure Properties of Nonzero Real Numbers.
Z2. For all $x, y \in \mathbb{R}$, if $x \neq 0$ and $y \neq 0$, then $x \cdot y \neq 0$.
Z3. *No Zero Divisors:* For all $x, y \in \mathbb{R}$, if $x \cdot y = 0$, then $x = 0$ or $y = 0$.
Z4. For all $x, y \in \mathbb{R}$, if $x \cdot y \neq 0$, then $x \neq 0$ and $y \neq 0$.
Z5. For all $x \in \mathbb{R}$, if $x \neq 0$, then $x^{-1} \neq 0$.
Z6. For all $x \in \mathbb{R}$, if $x \neq 0$, then $-x \neq 0$.

Proof. Z2. [We try proof by contradiction.] Assume, to get a contradiction, that there exist $x, y \in \mathbb{R}$ such that $x \neq 0$ and $y \neq 0$ and $x \cdot y = 0$. We know x^{-1} exists because $x \neq 0$; so we may multiply both sides of $x \cdot y = 0$ on the left by x^{-1}, obtaining $x^{-1} \cdot (x \cdot y) = x \cdot 0$. Using associativity (M2) and the previous result (Z1), this becomes $(x^{-1} \cdot x) \cdot y = 0$. By the inverse axiom (M4), we get $1 \cdot y = 0$. By the identity axiom (M3), we deduce $y = 0$. But this contradicts the assumption $y \neq 0$. So property Z2 holds.

Z3. The IF-statement here is the contrapositive of the IF-statement in Z2, so Z3 is logically equivalent to Z2.

Z4. Fix $x, y \in \mathbb{R}$. Using a contrapositive proof, we assume $x = 0$ or $y = 0$, and prove $x \cdot y = 0$. In the case where $x = 0$, $x \cdot y = 0 \cdot y = 0$ follows from Z1. Similarly, in the case where $y = 0$, $x \cdot y = x \cdot 0 = 0$ follows from Z1.

Z5. Assume, to get a contradiction, that there exists $x \in \mathbb{R}$ with $x \neq 0$ (so x^{-1} exists) and $x^{-1} = 0$. Multiply both sides of $x^{-1} = 0$ by x to get $x \cdot x^{-1} = x \cdot 0$. By the inverse axiom (M4) and Z1, we get $1 = 0$. This contradicts the nontriviality axiom N, which states $1 \neq 0$.

Z6. We use a contrapositive proof. Fix $x \in \mathbb{R}$; assume $-x = 0$; prove $x = 0$. By Theorem 8.11, $x = -(-x) = -0 = 0$. □

In abstract algebra, an *integral domain* is a commutative ring where $1 \neq 0$ and there are no zero divisors. In other words, an integral domain satisfies all the additive and multiplicative axioms omitting M4 and multiplicative inverses, and including Z3 as an additional axiom. Later, we define $\mathbb{Z}$ (the set of integers) and prove that $\mathbb{Z}$ is an integral domain. Our proof of Z3 from the field axioms shows that *every field is an integral domain*, but the converse is false ($\mathbb{Z}$ provides a counterexample).

Sign Rules; More Arithmetic

When first learning about negative numbers, many people are surprised to learn that the product of two negative numbers is positive. This fact follows from the *sign rules*, which relate the additive inverse operation to multiplication.

8.18. Theorem on Signs. For all $a, b, x, y, z \in \mathbb{R}$:
(a) *First Sign Rule:* $a \cdot (-b) = -(a \cdot b)$.
(b) *Second Sign Rule:* $(-a) \cdot b = -(a \cdot b)$.
(c) *Third Sign Rule:* $(-a) \cdot (-b) = a \cdot b$.
(d) *Inverses and Multiplication by -1:* $-x = (-1) \cdot x$.
(e) *Distributive Laws for Subtraction:* $x \cdot (y-z) = (x \cdot y) - (x \cdot z)$ and $(y-z) \cdot x = (y \cdot x) - (z \cdot x)$.

Proof. Fix $a, b \in \mathbb{R}$. For (a), we compute $(a \cdot b) + (a \cdot (-b)) = a \cdot (b + (-b)) = a \cdot 0 = 0$ using the distributive axiom (D), then additive inverses (A4), then property Z1. Using Theorem 8.11(b) with $x = a \cdot b$ and $y = a \cdot (-b)$, we conclude $y = -x$, i.e., $a \cdot (-b) = -(a \cdot b)$. The second sign rule is proved similarly. Using both of these sign rules and the double negative rule, we compute

$$(-a) \cdot (-b) = -[(-a) \cdot b] = -[-(a \cdot b)] = a \cdot b. \qquad \square$$

To prove (d), fix $x \in \mathbb{R}$. By the sign rule and the identity axiom M3, $(-1) \cdot x = -(1 \cdot x) = -x$. To prove (e), fix $x, y, z \in \mathbb{R}$. We compute

$$x \cdot (y - z) = x \cdot (y + (-z)) = (x \cdot y) + (x \cdot (-z)) = (x \cdot y) + -(x \cdot z) = (x \cdot y) - (x \cdot z),$$

by the definition of subtraction, the distributive law for addition, the sign rule, and the definition of subtraction. The other distributive law is proved similarly.

Using the distributive law, the sign rules, and theorems about addition, we can now prove more arithmetical facts about specific real numbers. For example, let us show that $2 \times 4 = 8$. (Here, $x \times y$ is defined to mean $x \cdot y$.) Assume that we have already proved $2 + 2 = 4$, $4 + 2 = 6$, and $6 + 2 = 8$, using the technique illustrated in Proposition 8.14. First,

$$2 \times 2 = 2 \times (1 + 1) = (2 \times 1) + (2 \times 1) = 2 + 2 = 4,$$

using the definition of 2, then the distributive axiom (D), then the identity axiom (M3), then the known result for addition. Second,

$$2 \times 3 = 2 \times (2 + 1) = (2 \times 2) + (2 \times 1) = 4 + 2 = 6,$$

using the definition of 3, then the distributive axiom (D), then previously proved results. Third,

$$2 \times 4 = 2 \times (3 + 1) = (2 \times 3) + (2 \times 1) = 6 + 2 = 8,$$

using the definition of 4, then the distributive axiom (D), then previously proved results. Now, the sign rules allow us to conclude

$$2 \times (-4) = -8 = (-2) \times 4; \qquad (-2) \times (-4) = 8.$$

Here is one more illustration of the power of the distributive law.

8.19. Theorem: FOIL Rule. For all $t, x, y, z \in \mathbb{R}$, $(t + x) \cdot (y + z) = ty + tz + xy + xz$.

Proof. Fix $t, x, y, z \in \mathbb{R}$. Set $a = y + z \in \mathbb{R}$. Using the distributive law three times, we find that $(t + x) \cdot (y + z) = (t + x) \cdot a = ta + xa = t(y + z) + x(y + z) = (ty + tz) + (xy + xz)$. $\square$

Using this rule, we can deduce algebraic identities such as $(x + y)^2 = x^2 + 2xy + y^2$ and $(x + y)(x - y) = x^2 - y^2$; here z^2 is defined to be $z \cdot z$, for any real z.

Properties of Multiplication and Inverses

We now derive properties of multiplication and multiplicative inverses that are based on the left and right cancellation laws. These properties and their proofs are analogous to the corresponding properties for addition, except we need to ensure that we never take the multiplicative inverse of zero.

8.20. Theorem on Multiplication. For all $a, b, c, x, y \in \mathbb{R}$:

(a) *Cancellation Laws for Multiplication:* If $a \neq 0$ and $a \cdot b = a \cdot c$, then $b = c$.
   If $a \neq 0$ and $b \cdot a = c \cdot a$, then $b = c$.

(b) *Uniqueness of Multiplicative Identity:* If $x \neq 0$ and $x \cdot y = x$, then $y = 1$.

(c) *Uniqueness of Multiplicative Inverses:* If $x \cdot y = 1$, then $x \neq 0$ and $y \neq 0$ and $y = x^{-1}$.

(d) *Inverse of Identity:* $1^{-1} = 1$.

(e) *Double Inverse Rule:* If $a \neq 0$, then $a^{-1} \neq 0$ and $(a^{-1})^{-1} = a$.

(f) *Inverse of a Product:* If $a \neq 0$ and $b \neq 0$, then $a \cdot b \neq 0$ and
   $(a \cdot b)^{-1} = (b^{-1}) \cdot (a^{-1})$.

(g) *Inverse of a Negative:* If $a \neq 0$, then $-a \neq 0$ and $(-a)^{-1} = -(a^{-1})$.

Proof. We prove (c), (d), and (g) as examples (assuming (a) has already been proved). For (c), fix $x, y \in \mathbb{R}$; assume $x \cdot y = 1$. Since $1 \neq 0$ by axiom N, we deduce $x \neq 0$ and $y \neq 0$ by Z4. Now we can use the inverse axiom (M4) to write $x \cdot y = x \cdot x^{-1}$. Using part (a) to cancel the nonzero real number x, we get $y = x^{-1}$. For (d), use axiom N to see that 1^{-1} exists; then use M3 and M4 to compute $1^{-1} = 1 \cdot 1^{-1} = 1$. For (g), fix $a \in \mathbb{R}$ with $a \neq 0$. We proved $-a \neq 0$ in Z6, so a^{-1} and $(-a)^{-1}$ both exist. Now $(-a) \cdot (-[a^{-1}]) = a \cdot a^{-1} = 1$ using the sign rules and M4. Invoking part (c) with $x = -a$ and $y = -[a^{-1}]$, we conclude $y = x^{-1}$, that is, $-[a^{-1}] = (-a)^{-1}$. $\square$

As in the case of addition, using commutativity (M5) we can write $(x \cdot y)^{-1} = x^{-1} \cdot y^{-1}$ when x, y are nonzero real numbers. We can also prove generalized associativity and generalized commutativity of multiplication by repeated use of M2 and M5. This justifies writing expressions such as $abcd$ without any parentheses. We can omit even more parentheses by agreeing that multiplication takes precedence over addition, so that $ab + cd$ means $(ab) + (cd)$ rather than $a(b + c)d$.

8.21. Theorem on Division. For all $x, y, z \in \mathbb{R}$, if $y \neq 0$ and $z \neq 0$, then:
(a) $x/1 = x$.
(b) $1/y = y^{-1}$.
(c) $y/y = 1$.
(d) $y/z \neq 0$.
(e) $(y/z)^{-1} = z/y = y^{-1} \cdot z$.
(f) $(x/y) \cdot (y/z) = x/z$.
(g) $(x \cdot z)/(y \cdot z) = x/y$.

Proof. We prove part (g) as an example. Fix $x, y, z \in \mathbb{R}$ with y and z nonzero. By Z2, $y \cdot z \neq 0$, so the division $(xz)/(yz)$ is defined. We compute

$$\frac{xz}{yz} = (xz) \cdot (yz)^{-1} = (xz)(z^{-1}y^{-1}) = x((zz^{-1})y^{-1}) = x(1y^{-1}) = xy^{-1} = \frac{x}{y},$$

using the definition of division, the rule for the inverse of a product, generalized associativity, the inverse axiom, the identity axiom, and the definition of division. $\square$

Fraction Rules

We now derive some algebraic rules for executing arithmetic operations on fractions.

8.22. Theorem on Fractions. For all $x, y \in \mathbb{R}$ and all nonzero $t, u, w \in \mathbb{R}$,

(a) *Addition Rules:* $\dfrac{x}{u} + \dfrac{y}{u} = \dfrac{x+y}{u}$, $\quad \dfrac{x}{u} + \dfrac{y}{w} = \dfrac{xw + uy}{uw}$.

(b) *Negation Rules:* $-\dfrac{x}{u} = \dfrac{-x}{u} = \dfrac{x}{-u}$, $\quad \dfrac{-x}{-u} = \dfrac{x}{u}$.

(c) *Subtraction Rules:* $\dfrac{x}{u} - \dfrac{y}{u} = \dfrac{x-y}{u}$, $\quad \dfrac{x}{u} - \dfrac{y}{w} = \dfrac{xw - uy}{uw}$.

(d) *Multiplication Rules:* $x \cdot \dfrac{y}{u} = \dfrac{x \cdot y}{u} = y \cdot \dfrac{x}{u}$, $\quad \dfrac{x}{u} \cdot \dfrac{y}{w} = \dfrac{x \cdot y}{u \cdot w}$.

(e) *Division Rules:* $\left(\dfrac{t}{w}\right)^{-1} = \dfrac{w}{t}$, $\quad \dfrac{x/u}{t/w} = \dfrac{x \cdot w}{u \cdot t}$.

Proof. We prove the addition rules, a multiplication rule, and a division rule as examples. Fix $x, y \in \mathbb{R}$ and nonzero $t, u, w \in \mathbb{R}$. First,

$$(x/u) + (y/u) = x \cdot u^{-1} + y \cdot u^{-1} = (x + y) \cdot u^{-1} = (x+y)/u,$$

using the definition of division, then the distributive axiom (D), then the definition of division. Second, recall the previous result $\frac{x}{u} = \frac{xw}{uw}$ and $\frac{y}{w} = \frac{uy}{uw}$. Combining this with what we just proved, we see that

$$\frac{x}{u} + \frac{y}{w} = \frac{xw}{uw} + \frac{uy}{uw} = \frac{xw + uy}{uw},$$

as needed. Third, compute

$$\frac{xy}{uw} = (xy)(uw)^{-1} = (xy)(w^{-1}u^{-1}) = (xu^{-1})(yw^{-1}) = \frac{x}{u} \cdot \frac{y}{w}$$

using the definition of division, the rule for the inverse of a product, generalized associativity and commutativity, and the definition of division. Finally, using the rule just proved, note that

$$\frac{t}{w} \cdot \frac{w}{t} = \frac{t \cdot w}{w \cdot t} = \frac{t \cdot w}{t \cdot w} = (tw)(tw)^{-1} = 1.$$

By Theorem 8.20(c), we deduce $(t/w)^{-1} = w/t$. $\square$

Section Summary

1. *Properties of Zero.* A product of real numbers is zero iff one of the factors is zero. The set of nonzero real numbers is closed under multiplication, inverses, and negatives.

2. *More Algebraic Properties.* The sign rules state $a(-b) = (-a)b = -(ab)$ and $(-a)(-b) = ab$ for all $a, b \in \mathbb{R}$. Nonzero factors can be cancelled: if $a \neq 0$ and $ab = ac$, then $b = c$. The number 1 is the unique multiplicative identity. For $a \neq 0$, a^{-1} is the unique multiplicative inverse of a. When the inverses are defined, we have $(a^{-1})^{-1} = a$, $(ab)^{-1} = b^{-1}a^{-1}$, and $(-a)^{-1} = -(a^{-1})$. The distributive law generalizes to give identities such as $(t + x)(y + z) = ty + tz + xy + xz$.

3. *Fraction Rules.* We can perform arithmetic on fractions via rules such as $(x/u) + (y/w) = (xw + uy)/(uw)$, $(x/u) \cdot (y/w) = (xy)/(uw)$, and $(t/w)^{-1} = w/t$ (assuming all denominators are nonzero).

Exercises

1. (a) Prove $\forall a \in \mathbb{R}, a \cdot 0 = 0$ by simplifying $(a \cdot 0) + (a \cdot a)$. (b) Prove the sign rule $\forall a, b \in \mathbb{R}, -(a \cdot b) = (-a) \cdot b$.

2. Finish the proof of Theorem 8.20.

3. Use Z3 (no zero divisors) to prove the cancellation property in Theorem 8.20(a) without using multiplicative inverses. (It follows that this cancellation property holds in all integral domains.)

4. Prove: (a) $3 \times 3 = 9$. (b) $5 \times (-2) = -10$. (c) $(-2) \times (-3) = 6$.

5. Prove: (a) $1/2 + 2/3 = 7/6$. (b) $1/3 - 3/5 = -4/15$. (c) $(3/8) \cdot (2/3) = 1/4$.

6. Finish the proof of Theorem 8.21.

7. Prove Theorem 8.22 parts (b) and (c).

8. Finish the proof of Theorem 8.22 parts (d) and (e).

9. Use the distributive law to prove the following identities for all $v, w, x, y, z \in \mathbb{R}$.
   (a) $(x + y)^2 = x^2 + 2xy + y^2$.
   (b) $(x - y)^2 = x^2 - 2xy + y^2$.
   (c) $(x + y)(x - y) = x^2 - y^2$.
   (d) $w(x + y + z) = wx + wy + wz$.
   (e) $(v + w)(x + y + z) = vx + wx + vy + wy + vz + wz$.
   (f) $(x + y)^3 = x^3 + 3x^2y + 3xy^2 + y^3$ (where u^3 means $u \cdot u \cdot u$, for $u \in \mathbb{R}$).

10. For $x, y \in \mathbb{R} - \{-1\}$, define $x \star y = x \cdot y + x + y$. (a) Show: $\forall x, y \in \mathbb{R} - \{-1\}$, $x \star y \in \mathbb{R} - \{-1\}$. (b) Prove that axioms M2, M3, and M5 hold with $\cdot$ replaced by $\star$ and 1 replaced by 0. (c) Define an operation x' (analogous to multiplicative inversion) and prove that for all $x \in \mathbb{R} - \{-1\}$, $x' \in \mathbb{R} - \{-1\}$ and $x \star x' = 0 = x' \star x$.

11. Define the set of matrices $M_2(\mathbb{R})$ to be the set $\mathbb{R}^4$, where we display $(a, b, c, d) \in \mathbb{R}^4$ as the 2×2 array $\begin{bmatrix} a & b \\ c & d \end{bmatrix}$. Define algebraic operations on matrices as follows: for $a, b, c, d, w, x, y, z \in \mathbb{R}$, let

$$\begin{bmatrix} a & b \\ c & d \end{bmatrix} + \begin{bmatrix} w & x \\ y & z \end{bmatrix} = \begin{bmatrix} a+w & b+x \\ c+y & d+z \end{bmatrix}, \quad -\begin{bmatrix} a & b \\ c & d \end{bmatrix} = \begin{bmatrix} -a & -b \\ -c & -d \end{bmatrix},$$

$$0_{M_2(\mathbb{R})} = \begin{bmatrix} 0 & 0 \\ 0 & 0 \end{bmatrix}, \quad 1_{M_2(\mathbb{R})} = \begin{bmatrix} 1 & 0 \\ 0 & 1 \end{bmatrix},$$

$$\begin{bmatrix} a & b \\ c & d \end{bmatrix} \cdot \begin{bmatrix} w & x \\ y & z \end{bmatrix} = \begin{bmatrix} aw + by & ax + bz \\ cw + dy & cx + dz \end{bmatrix},$$

$$\begin{bmatrix} a & b \\ c & d \end{bmatrix}^{-1} = \frac{1}{ad - bc} \begin{bmatrix} d & -b \\ -c & a \end{bmatrix} \text{ when } ad - bc \neq 0.$$

Use known algebraic properties of $\mathbb{R}$ to prove that $M_2(\mathbb{R})$ satisfies all the algebraic axioms except M5 and the requirement that *every* nonzero matrix have a multiplicative inverse.

12. Which of the properties Z1 through Z6 hold for matrices in $M_2(\mathbb{R})$? Explain. (Assume x^{-1} exists when considering Z5.)

8.3   Natural Numbers, Integers, and Rational Numbers

We are all informally acquainted with the natural numbers $\mathbb{N} = \mathbb{Z}_{>0} = \{1, 2, 3, \ldots\}$, the integers $\mathbb{Z} = \{\ldots, -2, -1, 0, 1, 2, \ldots\}$, and the rational numbers $\mathbb{Q}$ (ratios of integers). It is possible to give a rigorous construction of these number systems using set theory, defining first $\mathbb{N}$, then $\mathbb{Z}$, then $\mathbb{Q}$, and finally $\mathbb{R}$, and proving that each number system satisfies the appropriate algebraic and ordering properties. The details of this construction are very tedious and intricate, however, and we have chosen instead to begin with an axiom system for the final number system $\mathbb{R}$ on our list, which requires no prior knowledge of the more basic number systems.

However, we frequently need integers and rational numbers to prove theorems about real numbers and calculus. It turns out we can define $\mathbb{N}$, $\mathbb{Z}$, and $\mathbb{Q}$ as certain subsets of the set $\mathbb{R}$, and then derive the basic properties of these number systems fairly quickly from the axioms for $\mathbb{R}$. In particular, the Induction Axiom for $\mathbb{N}$ (discussed in §4.1) is *provable* from our definition of $\mathbb{N}$ given below. Aided by induction, we show that $\mathbb{N}$, $\mathbb{Z}$, and $\mathbb{Q}$ satisfy appropriate closure properties among those listed in axioms A1 and M1. The remaining algebraic and order properties of these number systems then follow automatically from the corresponding properties of $\mathbb{R}$.

Inductive Sets; Defining Natural Numbers

Intuitively, the set of natural numbers[1] is $\mathbb{N} = \{1, 2, 3, \ldots\}$. The trouble with this intuitive description is that it does not precisely define what is meant by the ... at the end. Given that real numbers and addition are already available to us, we can approach this problem via closure properties in $\mathbb{R}$. Restating the intuitive description, $\mathbb{N}$ should be the set of all real numbers that we can reach by starting with 1 and adding 1 repeatedly. This motivates the following definition.

8.23. Definition: Inductive Sets. Given $S \subseteq \mathbb{R}$,

$\boxed{S \text{ is inductive}}$ iff $\boxed{1 \in S \text{ and } \forall x \in \mathbb{R}, x \in S \Rightarrow (x+1) \in S}$.

Thus, an inductive set must be *closed under* 1 (meaning 1 is in the set) and *closed under adding* 1. For example, $\mathbb{R}$ itself is an inductive set due to the closure axioms A1 and M1. Later, when we discuss properties of order, we will see that $\mathbb{R}_{>0}$ and $\mathbb{R}_{\geq 1}$ are also inductive sets. Informally, $\mathbb{Z}$ and $\mathbb{Q}$ (not yet officially defined) are inductive sets. All of these inductive sets are too big to be $\mathbb{N}$ because they contain extra elements not reachable from 1 by adding 1. The following clever device lets us precisely define the idea that $\mathbb{N}$ should be the smallest possible inductive subset of $\mathbb{R}$.

8.24. Definition of the Natural Numbers $\mathbb{N}$. Let $\mathcal{I}$ be the set of all inductive subsets of $\mathbb{R}$. Define $\boxed{\mathbb{N} = \bigcap_{S \in \mathcal{I}} S}$.

Thus, $\mathbb{N}$ *is the intersection of all inductive subsets of* $\mathbb{R}$. Since $\mathbb{R}$ itself is inductive, $\mathcal{I}$ is nonempty, and the generalized intersection is defined. Expanding the definition, we see that for all objects x, $\boxed{x \in \mathbb{N}}$ iff $\boxed{\text{for every inductive set } S \subseteq \mathbb{R}, x \in S}$.

8.25. Theorem. $\mathbb{N}$ is an inductive set.

---

[1]In the next few sections, we use the notation $\mathbb{N}$, rather than $\mathbb{Z}_{>0}$, to avoid assuming in advance that *natural numbers* are the same as *positive integers*. This fact is proved as a theorem later.

Proof. For every inductive set S, $1 \in S$. So $1 \in \mathbb{N}$. Next we show $\mathbb{N}$ is closed under adding 1. Fix $x \in \mathbb{N}$ and prove $x + 1 \in \mathbb{N}$. For any inductive set S, we know $x \in S$, so $x + 1 \in S$ since S is inductive. Since this holds for all inductive sets S, we see that $x + 1 \in \mathbb{N}$. $\square$

We now prove two versions of the Induction Axiom for $\mathbb{N}$. The second version (involving open sentences) is the one discussed earlier in §4.1.

8.26. Theorem: Set-Theoretic Induction Principle. Suppose S is a set satisfying these conditions: (a) $S \subseteq \mathbb{N}$; (b) $1 \in S$; (c) for all $x \in S$, $x + 1 \in S$. Then $S = \mathbb{N}$.

Proof. Assume S satisfies the three conditions. By condition (a), it suffices to prove that $\mathbb{N} \subseteq S$. By conditions (b) and (c), S is an inductive set. Thus $\mathbb{N}$, which is the intersection of all inductive sets, must be a subset of S. $\square$

8.27. Theorem: Induction Principle. Suppose $P(n)$ is an open sentence such that $P(1)$ is true, and for all $n \in \mathbb{N}$, if $P(n)$ is true, then $P(n + 1)$ is true. Then for all $n \in \mathbb{N}$, $P(n)$ is true.

Proof. Given the open sentence $P(n)$, form the set $S = \{n \in \mathbb{N} : P(n) \text{ is true}\}$. By definition of S and the conditions on $P(n)$, S satisfies the three conditions in the Set-Theoretic Induction Principle. For example, to check condition (c), fix $x \in S$. We know $x \in \mathbb{N}$ and $P(x)$ is true, so we deduce $x + 1 \in \mathbb{N}$ and $P(x + 1)$ is true, hence $x + 1 \in S$. Now the Set-Theoretic Induction Principle tells us that $S = \mathbb{N}$, which means that $P(n)$ is true for all $n \in \mathbb{N}$. $\square$

Closure Properties of $\mathbb{N}$

Since $\mathbb{N}$ is an inductive set, we know that $\mathbb{N}$ contains 1 and is closed under adding 1. Using induction, we can now prove the stronger statements that $\mathbb{N}$ is closed under addition and multiplication.

8.28. Theorem on Closure of $\mathbb{N}$ under Addition. For all $m, n \in \mathbb{N}$, $m + n \in \mathbb{N}$.

Proof. Let m_0 be a fixed element of $\mathbb{N}$. For $n \in \mathbb{N}$, let $P(n)$ be the statement "$m_0 + n \in \mathbb{N}$." We use induction on n to prove $P(n)$ is true for every $n \in \mathbb{N}$. To see that $P(1)$ is true, note that $m_0 + 1 \in \mathbb{N}$ because $m_0 \in \mathbb{N}$ and $\mathbb{N}$ is closed under adding 1. Next, fix $n_0 \in \mathbb{N}$, assume $P(n_0)$, and prove $P(n_0 + 1)$. We have assumed $m_0 + n_0 \in \mathbb{N}$, and we must prove $m_0 + (n_0 + 1) \in \mathbb{N}$. Using associativity in $\mathbb{R}$ (A2), we know $m_0 + (n_0 + 1) = (m_0 + n_0) + 1$. Since $m_0 + n_0 \in \mathbb{N}$ and $\mathbb{N}$ is closed under adding 1, $(m_0 + n_0) + 1 \in \mathbb{N}$. Thus $m_0 + (n_0 + 1) \in \mathbb{N}$, as needed. $\square$

8.29. Theorem on Closure of $\mathbb{N}$ under Multiplication. For all $m, n \in \mathbb{N}$, $m \cdot n \in \mathbb{N}$.

Proof. Let m_0 be a fixed element of $\mathbb{N}$. We prove $\forall n \in \mathbb{N}, m_0 \cdot n \in \mathbb{N}$ by induction on n. When $n = 1$, the identity axiom (M3) gives $m_0 \cdot 1 = m_0 \in \mathbb{N}$. Now fix $n_0 \in \mathbb{N}$, assume $m_0 \cdot n_0 \in \mathbb{N}$, and prove $m_0 \cdot (n_0 + 1) \in \mathbb{N}$. By the distributive law in $\mathbb{R}$ and (M3), $m_0 \cdot (n_0 + 1) = m_0 \cdot n_0 + m_0 \cdot 1 = m_0 \cdot n_0 + m_0$. By induction hypothesis, $m_0 \cdot n_0 \in \mathbb{N}$. By closure of $\mathbb{N}$ under addition, $m_0 \cdot n_0 + m_0 \in \mathbb{N}$. So $m_0 \cdot (n_0 + 1) \in \mathbb{N}$, as needed. $\square$

On the other hand, $\mathbb{N}$ is *not* closed under additive inverses, subtraction, multiplicative inverses, or division. To prove these claims, we need order properties to be discussed later.

Defining Integers; Closure Properties of $\mathbb{Z}$

Informally, we obtain the set of integers from the set of natural numbers (positive integers) by adjoining zero and the negatives of all the positive integers. These negatives are known to exist in $\mathbb{R}$, by the closure axiom A1 for $\mathbb{R}$. Here is the official definition.

8.30. Definition of Integers. For all $x \in \mathbb{R}$, $\boxed{x \in \mathbb{Z}}$ iff $\boxed{x \in \mathbb{N} \text{ or } x = 0 \text{ or } -x \in \mathbb{N}.}$ Elements of $\mathbb{Z}$ are called *integers*.

Now we show that $\mathbb{Z}$ is closed under all the arithmetical operations except multiplicative inversion and division.

8.31. Theorem on Closure of $\mathbb{Z}$ under Negation. For all $y \in \mathbb{Z}$, $-y \in \mathbb{Z}$.

Proof. Fix $y \in \mathbb{Z}$; prove $-y \in \mathbb{Z}$. We know $y \in \mathbb{N}$ or $y = 0$ or $-y \in \mathbb{N}$, so consider three cases.
<u>Case 1.</u> Assume $y \in \mathbb{N}$. Then $-(-y) = y \in \mathbb{N}$ by the double negative rule, so $-y \in \mathbb{Z}$ because $-y$ satisfies the third alternative in the definition of $\mathbb{Z}$ (taking x there to be $-y$).
<u>Case 2.</u> Assume $y = 0$. Then $-y = -0 = 0$, so $-y \in \mathbb{Z}$ by the second alternative in the definition of $\mathbb{Z}$.
<u>Case 3.</u> Assume $-y \in \mathbb{N}$. Then $-y \in \mathbb{Z}$ by the first alternative in the definition of $\mathbb{Z}$ (taking x there to be $-y$). $\qquad\square$

8.32. Theorem on Closure of $\mathbb{Z}$ under Multiplication. For all $x, y \in \mathbb{Z}$, $x \cdot y \in \mathbb{Z}$.

Proof. Fix $x, y \in \mathbb{Z}$; prove $x \cdot y \in \mathbb{Z}$. We know $x \in \mathbb{N}$ or $x = 0$ or $-x \in \mathbb{N}$, and similarly for y, giving nine cases. Five of these cases can be handled by the observation that if $x = 0$ or $y = 0$, then $x \cdot y = 0 \in \mathbb{Z}$. Now we examine the remaining four cases.
<u>Case 1.</u> Assume $x \in \mathbb{N}$ and $y \in \mathbb{N}$. We already proved $x \cdot y \in \mathbb{N}$, so $x \cdot y \in \mathbb{Z}$.
<u>Case 2.</u> Assume $x \in \mathbb{N}$ and $-y \in \mathbb{N}$. By the sign rules and closure of $\mathbb{N}$ under multiplication, we deduce $-(x \cdot y) = x \cdot (-y) \in \mathbb{N}$. So $x \cdot y \in \mathbb{Z}$ by the third alternative in the definition of $\mathbb{Z}$.
<u>Case 3.</u> Assume $-x \in \mathbb{N}$ and $y \in \mathbb{N}$. As in Case 2, we get $-(x \cdot y) = (-x) \cdot y \in \mathbb{N}$, hence $x \cdot y \in \mathbb{Z}$.
<u>Case 4.</u> Assume $-x \in \mathbb{N}$ and $-y \in \mathbb{N}$. The sign rules and closure of $\mathbb{N}$ give $x \cdot y = (-x) \cdot (-y) \in \mathbb{N}$, so $x \cdot y \in \mathbb{Z}$. $\qquad\square$

Proving closure of $\mathbb{Z}$ under addition turns out to be trickier; we begin with a lemma.

8.33. Lemma. For all $x \in \mathbb{Z}$, $x - 1 \in \mathbb{Z}$.

Proof. Fix $x \in \mathbb{Z}$; prove $x - 1 = x + (-1) \in \mathbb{Z}$. We know $x \in \mathbb{N}$ or $x = 0$ or $-x \in \mathbb{N}$, so consider three cases.
<u>Case 1.</u> Assume $x \in \mathbb{N}$. We prove $x - 1 \in \mathbb{Z}$ by induction on x. If $x = 1$, then $1 - 1 = 0 \in \mathbb{Z}$. For the induction step, fix $x \in \mathbb{N}$, assume $x - 1 \in \mathbb{Z}$, and prove $(x + 1) - 1 \in \mathbb{Z}$. Since $(x + 1) - 1 = x + (1 - 1) = x + 0 = x \in \mathbb{N}$, we can conclude that $(x + 1) - 1 \in \mathbb{Z}$ without even using the assumption $x - 1 \in \mathbb{Z}$.
<u>Case 2.</u> Assume $x = 0$; prove $x - 1 \in \mathbb{Z}$. Here, $x - 1 = 0 - 1 = -1 \in \mathbb{Z}$ since $-(-1) = 1 \in \mathbb{N}$.
<u>Case 3.</u> Assume $-x \in \mathbb{N}$; prove $x - 1 \in \mathbb{Z}$. Write $-x = n \in \mathbb{N}$. Since $\mathbb{N}$ is closed under adding 1, we know $n + 1 \in \mathbb{N}$. Now $x - 1 = -(-x) + (-1) = (-n) + (-1) = -(n + 1)$, so $-(x - 1) = n + 1 \in \mathbb{N}$, so $x - 1 \in \mathbb{Z}$ by the third alternative in the definition of $\mathbb{Z}$. $\qquad\square$

8.34. Theorem on Closure of $\mathbb{Z}$ under Addition. For all $x, y \in \mathbb{Z}$, $x + y \in \mathbb{Z}$.

Proof. Fix $x, y \in \mathbb{Z}$. We know $x \in \mathbb{N}$ or $x = 0$ or $-x \in \mathbb{N}$, and similarly for y. If $x = 0$, then $x + y = 0 + y = y \in \mathbb{Z}$. If $y = 0$, then $x + y = x + 0 = x \in \mathbb{Z}$. So we can reduce to the following four cases.

<u>Case 1.</u> Assume $x, y \in \mathbb{N}$. We already proved $x + y \in \mathbb{N}$, so $x + y \in \mathbb{Z}$.

<u>Case 2.</u> Assume $-x, -y \in \mathbb{N}$. Write $n = -x$ and $m = -y$; since $n, m \in \mathbb{N}$, we know $m + n \in \mathbb{N}$. Then $-(x + y) = [-y] + [-x] = m + n \in \mathbb{N}$, so $x + y \in \mathbb{Z}$ by the third alternative in the definition of $\mathbb{Z}$.

<u>Case 3.</u> Assume $x \in \mathbb{N}$ and $-y \in \mathbb{N}$. We prove that $x + y \in \mathbb{Z}$ by induction on $n = -y$. Since $y = -n$, our goal is to show that $x - n \in \mathbb{Z}$ for each natural number n. If $n = 1$, then $x - 1 \in \mathbb{Z}$ by the lemma. For the induction step, fix $n \in \mathbb{N}$, assume $x - n \in \mathbb{Z}$, and prove $x - (n+1) \in \mathbb{Z}$. We know $x - (n+1) = x + (-(n+1)) = x + ((-n) + (-1)) = (x + (-n)) + (-1) = (x - n) - 1$. By hypothesis, $x - n \in \mathbb{Z}$. By the lemma, $(x - n) - 1 \in \mathbb{Z}$, completing the induction step for Case 3.

<u>Case 4.</u> Assume $-x \in \mathbb{N}$ and $y \in \mathbb{N}$. We can prove $(-x) + y \in \mathbb{Z}$ as in Case 3, or we can reduce this case to Case 3 by noting $(-x) + y = y + (-x)$. $\qquad \square$

It follows that $\mathbb{Z}$ is closed under subtraction, since $x - y = x + (-y)$. Since $\mathbb{Z} \subseteq \mathbb{R}$ by construction, we can also immediately conclude that the set $\mathbb{Z}$ satisfies all the axioms and theorems discussed so far that do not mention multiplicative inverses, division, or least upper bounds. For instance, since $x + y = y + x$ holds for all *real* numbers x and y by A5, this equation must also hold for all x and y in the subset $\mathbb{Z}$ of $\mathbb{R}$. As another example, properties Z2 and Z3 (absence of zero divisors) hold for $\mathbb{Z}$ even though their proofs used the concept of multiplicative inverses in $\mathbb{R}$. In summary, we have proved that $\boxed{\mathbb{Z} \text{ is an ordered integral domain.}}$ All the properties of integers proved in Chapter 4 (such as theorems on divisibility, primes, greatest common divisors, and unique prime factorizations) are now officially available for use in our formal development of $\mathbb{R}$.

Defining Rational Numbers

Now that we have formally constructed $\mathbb{Z}$ within $\mathbb{R}$, the definition of rational numbers given in §2.1 is valid. Recall that for all $x \in \mathbb{R}$, we defined

$$\boxed{x \in \mathbb{Q}} \quad \Leftrightarrow \quad \boxed{\exists m \in \mathbb{Z}, \exists n \in \mathbb{Z}, n \neq 0 \wedge x = m/n}.$$

Using the fraction rules, we can prove the following closure properties of $\mathbb{Q}$.

8.35. Theorem on Closure Properties of $\mathbb{Q}$. For all $x, y \in \mathbb{Q}$, $x + y \in \mathbb{Q}$, $-x \in \mathbb{Q}$, $x - y \in \mathbb{Q}$, and $x \cdot y \in \mathbb{Q}$; if $y \neq 0$, then $y^{-1} \in \mathbb{Q}$ and $x/y \in \mathbb{Q}$. Moreover, $\mathbb{Z} \subseteq \mathbb{Q}$, so $0 \in \mathbb{Q}$ and $1 \in \mathbb{Q}$.

Proof. We prove closure under subtraction as an example. Fix $x, y \in \mathbb{Q}$; write $x = a/b$ and $y = c/d$ for some $a, b, c, d \in \mathbb{Z}$ with $b, d \neq 0$. By the fraction rules, $x - y = (a/b) - (c/d) = (ad - bc)/(bd)$. By closure properties of $\mathbb{Z}$ proved earlier, we know ad, bc, $ad - bc$, and bd are in $\mathbb{Z}$ with $bd \neq 0$ (by Z2). So $x - y$ meets the criterion for membership in $\mathbb{Q}$, taking $m = ad - bc$ and $n = bd$ in the definition above.

To prove that $\mathbb{Z} \subseteq \mathbb{Q}$, note that $1 \in \mathbb{Z}$ (since $1 \in \mathbb{N}$) and $1 \neq 0$ by axiom N. So for any $m \in \mathbb{Z}$, writing $m = m/1$ shows that $m \in \mathbb{Q}$. $\qquad \square$

Having proved all the closure properties, we conclude that $\boxed{\mathbb{Q} \text{ is an ordered field}}$; in other words, the set $\mathbb{Q}$ with the restricted arithmetical operations and order relation satisfies all the axioms for $\mathbb{R}$ (and their consequences) except possibly completeness. In particular, the unique prime factorization theorem for $\mathbb{Q}$ (see §4.7) is now available. One consequence of that theorem was the fact that for any positive integer n that has an odd exponent in its prime factorization, the equation $x^2 = n$ has no solution x in $\mathbb{Q}$. It can be proved from the Completeness Axiom (O7) that this equation *does* have a solution x in $\mathbb{R}$; see Theorem 8.61 below. These facts combine to show that the number system $\mathbb{Q}$ does *not* satisfy the Completeness Axiom. This axiom thus reveals a crucial difference between the rational number system and the real number system. Indeed, many of the central theorems of calculus would not work if we used $\mathbb{Q}$ instead of $\mathbb{R}$.

Section Summary

1. *Natural Numbers.* An inductive set is a subset of $\mathbb{R}$ containing 1 and closed under adding 1. The set $\mathbb{N}$ is the intersection of all inductive sets. $\mathbb{N}$ itself is an inductive set closed under addition and multiplication. To prove $\forall n \in \mathbb{N}, P(n)$, we may prove that $P(1)$ is true and for all $n \in \mathbb{N}$, if $P(n)$ is true, then $P(n+1)$ is true.

2. *Integers.* The set $\mathbb{Z}$ consists of all real numbers x such that $x \in \mathbb{N}$ or $x = 0$ or $-x \in \mathbb{N}$. $\mathbb{Z}$ is closed under addition, negation, and multiplication. $\mathbb{Z}$ is an ordered integral domain.

3. *Rational Numbers.* The set $\mathbb{Q}$ consists of all real numbers x such that there exist $m, n \in \mathbb{Z}$ with $n \neq 0$ and $x = m/n$. $\mathbb{Q}$ contains $\mathbb{Z}$ and is closed under addition, negation, multiplication, and inverses of nonzero elements. $\mathbb{Q}$ is an ordered field that does not satisfy the Completeness Axiom.

Exercises

1. Finish the proof of Theorem 8.35.

2. Prove or disprove: the union of any nonempty collection of inductive subsets of $\mathbb{R}$ is an inductive set.

3. Decide (with proof) whether each set below is inductive. [You may assume that $1 > 0$ has already been proved.] (a) $\mathbb{Z}_{\geq 0}$ (b) $\mathbb{Z}_{>1}$ (c) $\mathbb{Z}_{\geq 1}$ (d) $\emptyset$ (e) the set of odd integers (f) $\{k/3 : k \in \mathbb{Z}\}$ (g) $\{1, 2, 3, 4, 5, 6, 7, 8, 9\}$ (your proof should use $1 > 0$ somewhere).

4. (a) Show that $\mathbb{R} - \mathbb{Z}$ is closed under adding 1. (b) Is $\mathbb{R} - \mathbb{Z}$ an inductive set? (c) Is $\mathbb{R} - \mathbb{Z}$ closed under addition? negation? multiplication? multiplicative inverses? When answering (c), assume we have proved that $2^{-1} \in \mathbb{R} - \mathbb{Z}$.

5. Repeat the previous exercise, replacing $\mathbb{R} - \mathbb{Z}$ by $\mathbb{R} - \mathbb{Q}$ and replacing 2^{-1} by $\sqrt{2}$.

6. Show that $\mathbb{Z}$ is the intersection of all subsets of $\mathbb{R}$ containing 0 and 1 and closed under addition and negation.

7. Show that $\mathbb{Z}$ is the intersection of all subsets of $\mathbb{R}$ containing 1 and closed under subtraction.

8. Show that $\mathbb{Z}$ is the intersection of all inductive subsets closed under subtracting 1. Use this to formulate an induction principle for proving statements of the form $\forall n \in \mathbb{Z}, P(n)$.

9. Show that $\mathbb{Q}$ is the intersection of all subsets of $\mathbb{R}$ containing 0 and 1 and closed under addition, subtraction, multiplication, and division by nonzero elements.

10. *Double Induction.* Suppose S is a set such that: (a) $S \subseteq \mathbb{N} \times \mathbb{N}$; (b) $(1, 1) \in S$; (c) $\forall (x, y) \in S, (x + 1, y) \in S \wedge (x, y + 1) \in S$. Prove that $S = \mathbb{N} \times \mathbb{N}$.

11. State an induction principle for proving statements of the form $\forall m \in \mathbb{N}, \forall n \in \mathbb{N}, P(m,n)$. Prove this principle using the previous exercise.

12. We know from the Division Theorem that there is a function $\text{MOD} : \mathbb{Z} \times \mathbb{Z}_{>0} \to \mathbb{Z}$ given by $\text{MOD}(a,b) = a \bmod b =$ the unique remainder when a is divided by b. For $n \in \mathbb{Z}_{>1}$, define the set of *integers modulo* n to be

$$\mathbb{Z}_n = \{0, 1, 2, \ldots, n-1\} = \{x \in \mathbb{Z} : 0 \le x < n\}.$$

Define algebraic operations and constants on $\mathbb{Z}_n$ as follows: for $x, y \in \mathbb{Z}_n$, let $x \oplus_n y = (x+y) \bmod n$, $\ominus_n x = (-x) \bmod n$, $x \otimes_n y = (x \cdot y) \bmod n$, $0_n = 0$, and $1_n = 1$. Prove that $\mathbb{Z}_n$ with these operations is a nonzero commutative ring (i.e., satisfies all algebraic axioms excluding M4). It can help to first prove this lemma: for all $a, b \in \mathbb{Z}_n$, $a = b$ iff n divides $a - b$. (An alternative construction of $\mathbb{Z}_n$, based on the equivalence relation $\equiv_n$, is given in §6.6.)

13. Continuing the previous exercise, suppose p is a prime. (a) Use gcds to prove: for all nonzero $k \in \mathbb{Z}_p$, there exists a unique $m \in \mathbb{Z}_p$ with $k \otimes_p m = 1 = m \otimes_p k$. Define the *multiplicative inverse of* k *in* $\mathbb{Z}_p$ to be this unique m. (b) Show that $\mathbb{Z}_p$ is a field. (c) Show that if n is not prime, then $\mathbb{Z}_n$ does not satisfy Z2 and therefore cannot be a field.

8.4 Ordering, Absolute Value, and Distance

Now that we understand the basic algebraic properties of $\mathbb{R}$, we begin to study the consequences of the order axioms O1 through O6. We develop fundamental laws of inequalities as well as various facts about the absolute value function. This leads to a discussion of the concept of distance on the real line, which is defined in terms of absolute value.

Ordering Properties

The basic ordering axioms were stated in §8.1. Axioms O1 through O4 postulate that $\mathbb{R}$ is totally ordered by the relation $\leq$. In detail, for all $x, y, z \in \mathbb{R}$, $x \leq x$ (reflexivity, axiom O1); if $x \leq y$ and $y \leq x$, then $x = y$ (antisymmetry, axiom O2); if $x \leq y$ and $y \leq z$, then $x \leq z$ (transitivity, axiom O3); and $x \leq y$ or $y \leq x$ (total ordering, axiom O4). We also defined $x \geq y$ to mean $y \leq x$, $x < y$ to mean $x \leq y$ and $x \neq y$, and $x > y$ to mean $y < x$.

Our first result derives the properties of the strict order $<$ that correspond to the properties of $\leq$ listed above. This result applies to any totally ordered set $(X, \leq)$, not just $X = \mathbb{R}$.

8.36. Theorem on Order Properties of $<$, $>$, and $\geq$. For all $x, y, z \in \mathbb{R}$:

(a) *Irreflexivity:* $x < x$ is false.

(b) *Transitivity:* If $x < y$ and $y < z$, then $x < z$.

(c) $x \leq y$ iff $x < y$ or $x = y$; and $x \geq y$ iff $x > y$ or $x = y$.

(d) *Trichotomy:* Exactly one of these three statements is true: $x < y$; $x = y$; $y < x$.

(e) Parts (a), (b), and (d) hold with $>$ in place of $<$.

(f) Axioms O1 through O4 hold with $\geq$ in place of $\leq$.

(g) If $x < y$ and $y \leq z$, then $x < z$. If $x \leq y$ and $y < z$, then $x < z$.
    Similar statements hold for $>$ and $\geq$.

Proof. Fix $x, y, z \in \mathbb{R}$. To prove (a), note that $x < x$ is defined to mean "$x \leq x$ and $x \neq x$." But $x = x$ is true, so this AND-statement is false.

To prove (b), assume $x < y$ and $y < z$; prove $x < z$. We have assumed $x \leq y$ and $x \neq y$ and $y \leq z$ and $y \neq z$; we must prove $x \leq z$ and $x \neq z$. On one hand, using $x \leq y$ and $y \leq z$ and transitivity (O3), we get $x \leq z$. On the other hand, if we had $x = z$, then $x \leq y$ and $y \leq z = x$ and antisymmetry (O2) would give $x = y$, which contradicts our assumption. Thus, $x \neq z$.

To prove (d) assuming (c) has been proved, use total ordering (O4) to see that $x \leq y$ or $y \leq x$ is true; i.e., "$x < y$ or $x = y$ or $y < x$ or $y = x$" is true. Thus *at least one* of the three statements in (d) must hold. If $x < y$ and $x = y$ were both true, we would deduce $x < x$, contradicting (a). Similarly $x = y$ and $y < x$ cannot both be true. Finally, if $x < y$ and $y < x$ were both true, (b) would give $x < x$, contradicting (a). Thus *exactly one* of the given alternatives holds.

We leave the remaining assertions as exercises in manipulating the definitions. $\qquad\square$

The next result uses trichotomy to deduce rules for forming useful denials of inequalities.

8.37. Theorem on Denial Rules for Inequalities. For all $x, y \in \mathbb{R}$:

(a) $\sim(x \leq y) \Leftrightarrow y < x \Leftrightarrow x > y$.

(b) $\sim(x < y) \Leftrightarrow y \leq x \Leftrightarrow x \geq y$.

(c) $\sim(x \geq y) \Leftrightarrow y > x \Leftrightarrow x < y$.

(d) $\sim(x > y) \Leftrightarrow y \geq x \Leftrightarrow x \leq y$.

Proof. We prove (a) as an example. Fix $x, y \in \mathbb{R}$. First assume $x \leq y$ is false. By Theorem 8.36(c), "$x < y$ or $x = y$" is false, so trichotomy forces $y < x$ to be true. This is equivalent to $x > y$ by definition. Conversely, assume $y < x$ is true. By trichotomy, $x < y$ is false and $x = y$ is false, so "$x < y$ or $x = y$" is false. By Theorem 8.36(c), we see that $x \leq y$ is false. $\qquad \square$

Algebraic Properties of Order

Next we investigate the consequences of axioms O5 and O6, which relate the algebraic operations of addition and multiplication to the ordering relation on $\mathbb{R}$. We recall that the additive axiom (O5) states

$$\forall x, y, z \in \mathbb{R}, x \leq y \Rightarrow x + z \leq y + z,$$

and the multiplicative axiom (O6) states

$$\forall x, y, z \in \mathbb{R}, (x \leq y \wedge 0 \leq z) \Rightarrow x \cdot z \leq y \cdot z.$$

8.38. Theorem on Algebraic Properties of Order. For all $w, x, y, z \in \mathbb{R}$:

(a) *Order Reversal:* $x \leq y$ iff $-y \leq -x$.

(b) *Adding Inequalities:* If $x \leq y$ and $w \leq z$, then $x + w \leq y + z$.

(c) *Multiplying by a Negative Number:* If $x \leq y$ and $z \leq 0$, then $y \cdot z \leq x \cdot z$.

(d) *Squares are Nonnegative:* $0 \leq w^2$.

(e) Parts (a), (b), (c), axiom O5, and axiom O6 hold with $\leq$ replaced by $<$ throughout. In place of (d), we have: $\forall w \in \mathbb{R}, w \neq 0 \Rightarrow 0 < w^2$. Similar properties hold for $\geq$ and $>$.

Proof. Fix $w, x, y, z \in \mathbb{R}$. To prove (a), first assume $x \leq y$. Taking $z = -x + (-y)$ in (O5), the assumption yields $x + (-x + (-y)) \leq y + (-x + (-y))$. This becomes $-y \leq -x$ after simplification via known algebra rules. Conversely, assume $-y \leq -x$. Adding $x + y$ to both sides (using O5) leads to $x \leq y$. Alternatively, the converse can be deduced from the direction already proved: from $-y \leq -x$ we get $-(-x) \leq -(-y)$, so that $x \leq y$ by the double negative rule.

To prove (b), assume $x \leq y$ and $w \leq z$. Adding w to both sides of $x \leq y$, we get $x + w \leq y + w$. Adding y to both sides of $w \leq z$, we get $y + w \leq y + z$. Now transitivity gives $x + w \leq y + z$.

To prove (c), assume $x \leq y$ and $z \leq 0$. By (a), $0 = -0 \leq -z$. So we may apply (O6) with z there replaced by $-z$, obtaining $x \cdot (-z) \leq y \cdot (-z)$. This becomes $-(xz) \leq -(yz)$ by the sign rules, so $yz \leq xz$ follows from (a).

Next we prove the version of (O6) for $<$, namely: if $x < y$ and $0 < z$, then $xz < yz$. Assume $x < y$ and $0 < z$. Then $x \leq y$ and $0 \leq z$, so the original O6 gives $xz \leq yz$. To conclude $xz < yz$, we show that $xz = yz$ cannot occur. Note that $0 < z$ forces $z \neq 0$, by trichotomy. So if $xz = yz$ were true, we could cancel z to get $x = y$, but this contradicts the assumption $x < y$. So $xz < yz$ does hold.

Next we prove the version of (d) for $<$. Assuming $w \neq 0$, we must have $0 < w$ or $w < 0$ by trichotomy.

Case 1. Assume $0 < w$. The version of (O6) just proved (taking $x = 0$, $y = w$, and $z = w$) gives $0 \cdot w < w \cdot w$, so $0 < w^2$.

<u>Case 2.</u> Assume $w < 0$, so $0 = -0 < -w$ by (a). Now take $x = 0$, $y = -w$ and $z = -w$ in the version of (O6) for $<$ to conclude $0 \cdot (-w) < (-w) \cdot (-w)$, so $0 < w^2$ by the sign rules. Thus, $0 < w^2$ holds in all cases.

The remaining assertions are left as exercises.   □

Note that $1^2 = 1 \cdot 1 = 1$ by the identity axiom (M3). Also $1 \neq 0$ by axiom N. Thus, as a corollary to the strict version of (d) above, we get $\boxed{0 < 1}$. This fact has surprisingly deep consequences, as we will see shortly. As one application, we can repeatedly add 1 to the inequality $0 < 1$ and use Theorem 8.38(a) to obtain

$$-10 < \cdots < -4 < -3 < -2 < -1 < 0 < 1 < 2 < 3 < 4 < \cdots < 10.$$

Positive and Negative Numbers

Now we can define positive and negative real numbers and study how these concepts relate to inequalities.

8.39. Definition. For all $x \in \mathbb{R}$, $\boxed{x \text{ is (strictly) positive}}$ iff $\boxed{x > 0}$; $\boxed{x \text{ is (strictly) negative}}$ iff $\boxed{x < 0}$; $\boxed{x \text{ is weakly positive}}$ iff $\boxed{x \geq 0}$; $\boxed{x \text{ is weakly negative}}$ iff $\boxed{x \leq 0}$. We write $\mathbb{R}_{>0}$ for the set of positive real numbers; $\mathbb{R}_{<0}$, $\mathbb{R}_{\geq 0}$, $\mathbb{R}_{\leq 0}$, $\mathbb{Z}_{>0}$, etc., are defined similarly.

Using the denial rules for inequalities, one sees that x is weakly positive iff x is nonnegative (not negative), and x is weakly negative iff x is nonpositive. For us, the unqualified term "positive" means "strictly positive." We warn the reader that some texts adopt the opposite convention: for them, "positive" means "weakly positive."

Here are some basic properties of positive numbers. You can formulate and prove analogous properties of negative numbers, nonnegative numbers, and nonpositive numbers.

8.40. Theorem on Positive Numbers.

(a) For all $x, y \in \mathbb{R}_{>0}$, $x + y \in \mathbb{R}_{>0}$, $xy \in \mathbb{R}_{>0}$, and $x^{-1}, y^{-1} \in \mathbb{R}_{>0}$.
      Also, $x < y$ iff $y^{-1} < x^{-1}$.

(b) For all $a \in \mathbb{R}$, a is positive iff $-a$ is negative; and a is negative iff $-a$ is positive.

(c) For all $a, b \in \mathbb{R}$, $ab \in \mathbb{R}_{>0}$ iff a and b are both positive or both negative.

Proof. We prove (c) and part of (a) as examples. For (a), fix $x, y \in \mathbb{R}_{>0}$. By trichotomy applied to x, $x \neq 0$, so x^{-1} exists. By trichotomy applied to x^{-1}, we know $x^{-1} < 0$ or $x^{-1} = 0$ or $x^{-1} > 0$. In the case $x^{-1} < 0$, we can multiply both sides of this inequality by the positive quantity x, obtaining $x \cdot x^{-1} < x \cdot 0$, or $1 < 0$. This contradicts trichotomy applied to 1, since we proved $0 < 1$ above. So $x^{-1} < 0$ is impossible. Similarly, in the case $x^{-1} = 0$, multiplying both sides by x leads to $1 = 0$, which violates axiom N. Thus, the third case $x^{-1} > 0$ must hold. Similarly, $y^{-1} > 0$. Now assume $x < y$. Multiply both sides by the positive quantity x^{-1} to get $1 < x^{-1}y$. Then multiply both sides by the positive quantity y^{-1} to get $y^{-1} < x^{-1}$. Conversely, assume $y^{-1} < x^{-1}$. The result just proved and the double inverse rule lead to $x < y$.

We prove (c) using proof by cases, assuming (b) has been proved. Fix $a, b \in \mathbb{R}$. If a and b are positive, then so is ab by part (a). If a and b are negative, then $-a$ and $-b$ are positive by (b), so $ab = (-a)(-b)$ is positive by (a) and the sign rules. If a or b is zero, then ab is zero, so ab is not positive. If a is positive and b is negative, then $-b$ is positive by (b), so $-ab = a(-b)$ is positive by the sign rules, so ab is negative by (b), so ab is not positive by trichotomy. The same conclusion holds if a is negative and b is positive. By trichotomy, all possible cases have now been considered, so (c) holds.   □

We can now deduce that $\mathbb{N}$ (the set of natural numbers, as defined in §8.3) coincides with $\mathbb{Z}_{>0}$ (the set of positive integers). Another useful technical fact contained in the next lemma is that there can be no integer strictly between two consecutive integers k and $k+1$.

8.41. Lemma on $\mathbb{N}$ and $\mathbb{Z}_{>0}$.
(a) For all $n \in \mathbb{N}$, $1 \leq n$.
(b) For all $m \in \mathbb{Z} - \mathbb{N}$, $m \leq 0$.
(c) $\mathbb{N} = \mathbb{Z} \cap \mathbb{R}_{>0} = \mathbb{Z}_{>0}$.
(d) For all $n, k \in \mathbb{Z}$, $k < n < k+1$ is false.
(e) For all $m, n \in \mathbb{Z}$, $m \leq n+1$ iff $m \leq n$ or $m = n+1$.

Proof. We prove (a) by induction on n. For $n = 1$, $1 \leq 1$ is true by reflexivity (O1). Fix $n \in \mathbb{N}$, assume $1 \leq n$, and prove $1 \leq n+1$. We know $0 \leq 1$; adding this inequality to the assumption $1 \leq n$ gives $1 + 0 \leq n+1$, or $1 \leq n+1$.

To prove (b), fix $m \in \mathbb{Z} - \mathbb{N}$. By definition of $\mathbb{Z}$, we know $m \in \mathbb{N}$ or $m = 0$ or $-m \in \mathbb{N}$. The first case cannot occur since we assumed $m \notin \mathbb{N}$. In the second case, $m = 0$, and $0 \leq 0$ by reflexivity (O1). In the third case, $0 < 1 \leq -m$ by part (a), so $0 < -m$ and hence $m < 0$ by order reversal. Thus $m \leq 0$ in this case as well.

To prove (c), fix $n \in \mathbb{N}$. Then $n \in \mathbb{Z}$ by definition, and $n \geq 1 > 0$, so $n \in \mathbb{R}_{>0}$. Thus, $\mathbb{N} \subseteq \mathbb{Z} \cap \mathbb{R}_{>0}$. On the other hand, fix $k \in \mathbb{Z} \cap \mathbb{R}_{>0}$. If k were not in $\mathbb{N}$, we would have $k \leq 0$ by (b). But we assumed $k > 0$ also, so this would contradict trichotomy. Thus, $k \in \mathbb{N}$, so $\mathbb{Z} \cap \mathbb{R}_{>0} \subseteq \mathbb{N}$. The equality $\mathbb{Z} \cap \mathbb{R}_{>0} = \mathbb{Z}_{>0}$ follows from the definition $\mathbb{Z}_{>0} = \{x \in \mathbb{Z} : x > 0\}$.

We prove (d) by contradiction. Assume, to get a contradiction, that there exist $n, k \in \mathbb{Z}$ with $k < n < k+1$. Adding $-k$ to these inequalities, we get $0 < n-k < 1$, where $m = n-k$ is an integer by closure of $\mathbb{Z}$ under subtraction. Since $m > 0$, we must have $m \in \mathbb{N}$ by (c). But then $m \geq 1$ by (a), and now $m < 1 \leq m$ contradicts trichotomy.

To prove (e), fix $m, n \in \mathbb{Z}$. First assume $m \leq n+1$, and prove $m \leq n$ or $m = n+1$. To do so, further assume $m \leq n$ is false, and prove $m = n+1$. We have assumed $n < m \leq n+1$. Since $n < m < n+1$ is impossible by (d), we conclude that $m = n+1$. For the converse, assume $m \leq n$ or $m = n+1$, and prove $m \leq n+1$. In the case where $m \leq n$, note that $n < n+1$ (as we see by adding n to $0 < 1$), so $m \leq n+1$ by transitivity. In the case $m = n+1$, $m \leq n+1$ holds by reflexivity (O1). $\square$

Absolute Value

We define a function ABS : $\mathbb{R} \to \mathbb{R}$, denoted $\text{ABS}(x) = |x|$ for $x \in \mathbb{R}$, using definition by cases.

8.42. Definition: Absolute Value. For $x \in \mathbb{R}$, let $|x| = x$ if $0 \leq x$, and let $|x| = -x$ if $x \leq 0$.

By total ordering (axiom O4), every real x satisfies at least one of the two cases in the definition. Moreover, the two cases overlap consistently: if $0 \leq x$ and $x \leq 0$ both hold, then $x = 0$ by antisymmetry (O2), and $x = -x = 0$, so $|0| = 0$. Thus, we do have a well-defined (single-valued) absolute value function. The next theorem gives the basic properties of this function.

8.43. Theorem on Absolute Value. For all $x, y, r \in \mathbb{R}$:

(a) $0 \leq |x|$, and $|x| = 0$ iff $x = 0$.

(b) $|x| = |-x|$.

(c) $|x \cdot y| = |x| \cdot |y|$.

(d) $|x| \leq y$ iff $-y \leq x \leq y$ (similarly for $<$).

(e) $|x - y| \leq r$ iff $y - r \leq x \leq y + r$ (similarly for $<$).

(f) $-|x| \leq x \leq |x|$.

Proof. We prove (b), (d) and (f), leaving the other parts as exercises. For (b), fix $x \in \mathbb{R}$. We know $0 \leq x$ or $x \leq 0$, so consider two cases.

Case 1. Assume $0 \leq x$, so $-x \leq 0$. Using the definition of absolute value twice, we see that $|x| = x$ and $|-x| = -(-x) = x$, so $|x| = |-x|$.

Case 2. Assume $x \leq 0$, so $0 \leq -x$. This time, the definition of absolute value gives $|x| = -x$ and $|-x| = -x$, so $|x| = |-x|$.

For (d), fix $x, y \in \mathbb{R}$. First, assume $|x| \leq y$, and prove $-y \leq x$ and $x \leq y$. By (a), $0 \leq |x|$, so $0 \leq y$ by transitivity, and $-y \leq 0$. In the case where $0 \leq x$, we know $|x| = x$, so $x \leq y$ by assumption, and moreover $-y \leq 0 \leq x$ gives $-y \leq x$ by transitivity. In the case where $x \leq 0$, we know $|x| = -x$, so the assumption becomes $-x \leq y$, or $-y \leq x$. Also $x \leq 0 \leq y$ gives $x \leq y$ by transitivity. Thus, $-y \leq x \leq y$ in both cases.

Conversely, assume $-y \leq x \leq y$, and prove $|x| \leq y$. In the case where $0 \leq x$, we know $|x| = x$, so we must prove $x \leq y$. We have just assumed this. In the case where $x \leq 0$, we know $|x| = -x$, so we must prove $-x \leq y$. We get this from the assumption $-y \leq x$ by order reversal and the double negative rule.

Finally, (f) follows from (d) by taking $y = |x|$ in (d). Since $|x| \leq |x|$ is known by reflexivity (O1), the equivalence in (d) lets us deduce that $-|x| \leq x \leq |x|$. $\qquad \square$

Part (a) of the next theorem contains a key property of absolute value called the *Triangle Inequality*. The other parts of this theorem give variations of this inequality that are frequently used in analysis.

8.44. Theorem on Triangle Inequality and Variations. For all $a, b, x, y, z \in \mathbb{R}$:

(a) $|a + b| \leq |a| + |b|$.

(b) $|x - y| \leq |x| + |y|$.

(c) $|x + y + z| \leq |x| + |y| + |z|$.

(d) $|x - y| \geq |x| - |y|$.

(e) $|x + y| \geq |x| - |y|$.

(f) $|x - z| \leq |x - y| + |y - z|$.

(g) $||x| - |y|| \leq |x - y|$.

Proof. We prove (a), (e), and (f), leaving the other parts as exercises. For (a), fix $a, b \in \mathbb{R}$. Part (d) of the previous theorem (with $x = a + b$ and $y = |a| + |b|$) tells us that the Triangle Inequality $|a + b| \leq |a| + |b|$ is logically equivalent to $-(|a| + |b|) \leq a + b \leq |a| + |b|$. Using part (f) of the previous theorem with $x = a$ and then $x = b$, we know that $-|a| \leq a \leq |a|$ and $-|b| \leq b \leq |b|$. Adding these inequalities and using algebra, we get $-(|a| + |b|) \leq a + b \leq |a| + |b|$, as needed.

To prove (e), fix $x, y \in \mathbb{R}$. The trick is to notice that $x = (x + y) + (-y)$, so that part (a) (with $a = x + y$ and $b = -y$) gives $|x| \leq |x + y| + |-y| = |x + y| + |y|$. Adding $-|y|$ to both sides and rearranging gives $|x + y| \geq |x| - |y|$, as needed.

To prove (f), fix $x, y, z \in \mathbb{R}$. Take $a = x - y$ and $b = y - z$ in part (a), noting that $a + b = x - z$, to see that $|x - z| \leq |x - y| + |y - z|$. $\qquad \square$

Distance

We can use absolute value to define the distance between any two real numbers, as follows.

8.45. Definition: Distance in $\mathbb{R}$. For all $x, y \in \mathbb{R}$, define the *distance from x to y* to be $d(x, y) = |x - y|$.

The distance function satisfies the following properties.

8.46. Theorem on Distance. For all $x, y, z \in \mathbb{R}$,

(a) *Positivity:* $d(x, y) \geq 0$, and $d(x, y) = 0$ iff $x = y$.

(b) *Symmetry:* $d(x, y) = d(y, x)$.

(c) *Triangle Inequality:* $d(x, z) \leq d(x, y) + d(y, z)$.

Proof. These properties all follow readily from corresponding properties of absolute value. For example, given $x, y \in \mathbb{R}$, $d(x, y) = |x - y| = |-(x - y)| = |y - x| = d(y, x)$. $\square$

In analysis, a set X together with a function $d : X \times X \to \mathbb{R}$ satisfying the three conditions in the last theorem (for all $x, y, z \in X$) is called a *metric space*. The real number system $\mathbb{R}$ provides one of the most basic examples of a metric space. More generally, one can show that $\mathbb{R}^n$ (with the *Euclidean distance function* $d(\vec{x}, \vec{y}) = \sqrt{\sum_{i=1}^{n}(x_i - y_i)^2}$) is a metric space, although the proof of the Triangle Inequality is harder.

Section Summary

1. *Ordering Properties.* The relation $<$ on $\mathbb{R}$ satisfies irreflexivity, transitivity, and trichotomy; for $x, y \in \mathbb{R}$, $x < y$ iff $x \leq y$ and $x \neq y$; $x \leq y$ iff $x < y$ or $x = y$; $\sim(x \leq y)$ iff $x > y$; x is positive iff $-x$ is negative; $\mathbb{R}_{>0}$ is closed under addition, multiplication, and multiplicative inversion; $0 < x < y$ iff $0 < y^{-1} < x^{-1}$; and so on.

2. *Properties Relating Order and Algebra.* Inequalities are preserved when we add a constant or multiply by a positive constant; they are reversed when we multiply by a negative constant. Nonzero squares are strictly positive, so $0 < 1$.

3. *Ordering of Integers.* Every $n \in \mathbb{N}$ satisfies $n \geq 1$, whereas every $m \in \mathbb{Z} - \mathbb{N}$ satisfies $m \leq 0$. So $\mathbb{N} = \mathbb{Z}_{>0}$; there are no integers k, n satisfying $k < n < k + 1$; and for all $m, n \in \mathbb{Z}$, $m \leq n + 1$ iff $m \leq n$ or $m = n + 1$.

4. *Absolute Value.* For all real x, if $x \geq 0$, then $|x| = x$; if $x \leq 0$, then $|x| = -x$. For $x, y, r \in \mathbb{R}$, $0 \leq |x|$, $|x| = |-x|$, $|xy| = |x| \cdot |y|$, and $|x - y| \leq r$ iff $y - r \leq x \leq y + r$. The Triangle Inequality says that $|a + b| \leq |a| + |b|$ for all $a, b \in \mathbb{R}$.

5. *Distance.* For $x, y \in \mathbb{R}$, the distance from x to y is $d(x, y) = |x - y|$. Distance satisfies positivity, symmetry, and the Triangle Inequality.

Exercises

1. Finish the proof of Theorem 8.36.

2. Finish the proof of Corollary 8.37.

3. Finish the proof of Theorem 8.38.

4. (a) Finish the proof of Theorem 8.40. (b) State and prove analogous properties for negative numbers, nonnegative numbers, and nonpositive numbers.

5. Finish the proof of Theorem 8.43.

6. Finish the proofs of Theorem 8.44 and Theorem 8.46.

7. Disprove: for all $a, b \in \mathbb{R}$, $|a - b| \leq |a| - |b|$.

8. For all $m, n \in \mathbb{Z}$, prove: (a) $m < n + 1$ iff $m \leq n$; (b) $m \geq n - 1$ iff $m \geq n$ or $m = n - 1$; (c) $m > n - 1$ iff $m \geq n$. (d) Prove these results do not always hold when $m, n \in \mathbb{R}$.

9. (a) Assume $<$ is a relation on $\mathbb{R}$ satisfying transitivity and trichotomy (properties (b) and (d) in Theorem 8.36). For $x, y \in \mathbb{R}$, define $x \leq y$ to mean $x < y$ or $x = y$. Prove that $\leq$ satisfies axioms O1, O2, O3, and O4.
   (b) Assume $<$ satisfies axioms O5 and O6. Prove that $\leq$ as defined in part (a) satisfies axioms O5 and O6.

10. In some developments of $\mathbb{R}$, the set $\mathbb{R}_{>0}$ of *positive real numbers* is taken as an undefined concept. Axioms O1 through O6 are replaced by these axioms: (P1) For all $x, y \in \mathbb{R}_{>0}$, $x + y \in \mathbb{R}_{>0}$. (P2) For all $x, y \in \mathbb{R}_{>0}$, $x \cdot y \in \mathbb{R}_{>0}$. (P3) For all $x \in \mathbb{R}$, exactly one of these statements is true: $x \in \mathbb{R}_{>0}$; $x = 0$; $-x \in \mathbb{R}_{>0}$. For $x, y \in \mathbb{R}$, $x < y$ is defined to mean $y - x \in \mathbb{R}_{>0}$, and $x \leq y$ is defined to mean $x < y$ or $x = y$. Show that axioms O1 through O6 follow from these definitions and axioms P1, P2, P3. [Use the previous problem.]

11. Define the set of complex numbers $\mathbb{C}$ to be the product set $\mathbb{R} \times \mathbb{R}$. Define algebraic operations on complex numbers as follows: for all $x, y, s, t \in \mathbb{R}$, let

$$(x, y) + (s, t) = (x + s, y + t), \qquad -(x, y) = (-x, -y), \qquad 0_{\mathbb{C}} = (0, 0),$$

$$1_{\mathbb{C}} = (1, 0), \qquad (x, y) \cdot (s, t) = (xs - yt, xt + ys),$$

$$(x, y)^{-1} = \left(\frac{x}{x^2 + y^2}, \frac{-y}{x^2 + y^2} \right) \text{ for } (x, y) \neq 0_{\mathbb{C}}.$$

Use known algebraic properties of $\mathbb{R}$ to show that $\mathbb{C}$ satisfies the field axioms (A1 through A5, M1 through M5, D, and N). It follows that all algebraic rules proved for $\mathbb{R}$ from these axioms also hold in $\mathbb{C}$.

12. (a) Show that the equation $z^2 = -1$ has exactly two solutions $z \in \mathbb{C}$. (b) Hence show that there does not exist an ordering relation $\leq$ on the field $\mathbb{C}$ satisfying axioms O1 through O6.

13. Show that for p prime, there does not exist on ordering relation $\leq$ on the field $\mathbb{Z}_p$ satisfying axioms O1 through O6. (See the exercises of §8.3 for the definition of $\mathbb{Z}_p$ and the algebraic operations on this set.)

14. Prove that $\mathbb{Z}$ is not a field.

8.5 Greatest Elements, Least Upper Bounds, and Completeness

Every nonempty finite subset of $\mathbb{R}$ has a greatest element and a least element; this can be proved by induction from the total ordering axiom (O4). However, an infinite subset of $\mathbb{R}$ may or may not have maximum and minimum elements. This leads to the concept of a least upper bound for a set of real numbers, which generalizes the idea of the maximum element of a set. The Completeness Axiom (O7) guarantees that least upper bounds exist whenever possible. This axiom has many striking consequences about the structure of the real number system, and it is absolutely fundamental for proving the major theorems of calculus.

After discussing the basic definitions, we use completeness to prove some familiar but technically subtle facts about how the number systems $\mathbb{Z}$ and $\mathbb{Q}$ appear within $\mathbb{R}$. In particular, we show that given any real number there is a larger integer (the Archimedean Property); any half-open interval $[x, x+1)$ of length 1 contains exactly one integer; and every open interval (a, b) contains a rational number. We also prove some strong versions of completeness that apply to sets of integers, such as the Well-Ordering Principle and the fact that if a nonempty set of integers is bounded above, then that set has a greatest element (not just a least upper bound). Finally, we use completeness to prove the Nested Interval Property of real numbers and the fact that every positive real number has a real square root.

Maximum and Minimum Elements

We recall the definitions of the maximum element and minimum element of a set; these definitions make sense in any partially ordered set.

8.47. Definition: Maximum and Minimum Elements. Fix $S \subseteq \mathbb{R}$ and $z_0 \in \mathbb{R}$.

(a) $\boxed{z_0 = \max S}$ iff $\boxed{z_0 \in S \text{ and } \forall x \in S, x \le z_0.}$

(b) $\boxed{S \text{ has a greatest element}}$ iff $\boxed{\exists z \in S, \forall x \in S, x \le z.}$

(c) $\boxed{z_0 = \min S}$ iff $\boxed{z_0 \in S \text{ and } \forall x \in S, z_0 \le x.}$

(d) $\boxed{S \text{ has a least element}}$ iff $\boxed{\exists z \in S, \forall x \in S, z \le x.}$

We stress that $\max S$ and $\min S$ need not exist; see the examples below. But if S has a greatest element, that element is unique. For suppose $y, z \in \mathbb{R}$ are both greatest elements of S. Since y is a greatest element and $z \in S$, we deduce $z \le y$. Since z is a greatest element and $y \in S$, we deduce $y \le z$. By antisymmetry (axiom O2), $y = z$. Similarly, $\min S$ is unique when it exists.

8.48. Example: Intervals. Fix $a < b$ in $\mathbb{R}$. The closed interval $S = [a, b] = \{x \in \mathbb{R} : a \le x \le b\}$ has greatest element $b = \max S$ and least element $a = \min S$. On the other hand, we claim the open interval $T = (a, b) = \{x \in \mathbb{R} : a < x < b\}$ has no greatest element and no least element. To prove that T has no maximum, we negate the definition and prove $\forall z \in T, \exists x \in T, z < x$. Fix $z \in T$, so $a < z < b$. Choose $x = (z+b)/2$. You can check that $a < z < (z+b)/2 < b$, so $z < x$ and $a < x < b$. Thus, x is a member of T larger than z. A similar proof shows that T has no least element. Observe, in particular, that a cannot be the least element of T since a does not belong to T.

8.49. Example. Every nonempty *finite* set $\{x_1, \ldots, x_n\} \subseteq \mathbb{R}$ has a maximum and minimum element, as can be proved by induction on n. However, the empty set $\emptyset$ has no maximum or minimum element, since the empty set has no elements in it at all.

8.50. Example. The entire set $\mathbb{R}$ has no least element and no greatest element. For, given any $x \in \mathbb{R}$, adding x to the inequalities $-1 < 0 < 1$ shows that $x - 1 < x < x + 1$, so there exist elements of $\mathbb{R}$ greater than x and less than x. The same proof shows that $\mathbb{Z}$ has no least element and no greatest element. On the other hand, $\mathbb{R}_{\geq 0}$ has least element 0, and $\mathbb{N} = \mathbb{Z}_{>0}$ has least element 1, by Lemma 8.41(a).

We just saw that the set $\mathbb{N}$ of natural numbers has a least element. In fact, $\mathbb{N}$ satisfies a much stronger condition called the *well-ordering* property: *every* nonempty subset of $\mathbb{N}$ has a least element. We use the following terminology in the proof: for each $n \in \mathbb{N}$, define $[n] = \{m \in \mathbb{N} : m \leq n\}$. We proved earlier that every natural number m satisfies $1 \leq m$, so we can also write $[n] = \{m \in \mathbb{N} : 1 \leq m \leq n\}$. It follows from Lemma 8.41(e) that $[n + 1] = [n] \cup \{n + 1\}$ for all $n \in \mathbb{N}$.

8.51. Theorem on Well-Ordering of $\mathbb{N}$. For all $S \subseteq \mathbb{N}$, if $S \neq \emptyset$, then S has a least element.

Proof. First we prove the following statement by induction on n: for all $S \subseteq \mathbb{N}$, if $S \cap [n]$ is nonempty then S has a least element. For the base case, consider $n = 1$. Fix $S \subseteq \mathbb{N}$ and assume $S \cap [1]$ is nonempty. Now, $[1] = \{m \in \mathbb{N} : 1 \leq m \leq 1\} = \{1\}$ by antisymmetry. The assumption $S \cap [1] \neq \emptyset$ forces $1 \in S$. For any $k \in S$, we know $k \in \mathbb{N}$, so $1 \leq k$. Thus, 1 is the least element of S in this case.

For the induction step, fix $n \in \mathbb{N}$ and assume: for all $S \subseteq \mathbb{N}$, if $S \cap [n] \neq \emptyset$ then S has a least element. We must prove: for all $S \subseteq \mathbb{N}$, if $S \cap [n + 1] \neq \emptyset$, then S has a least element. Fix $S \subseteq \mathbb{N}$ satisfying $S \cap [n + 1] \neq \emptyset$. Since $[n + 1] = [n] \cup \{n + 1\}$, there are two cases.
Case 1. Assume $S \cap [n] \neq \emptyset$. By induction hypothesis, S has a least element.
Case 2. Assume $S \cap [n] = \emptyset$, which forces $n + 1 \in S$. We claim $n + 1$ is the least element of S. To see this, fix $k \in S$. By trichotomy, we know $n + 1 \leq k$ or $k < n + 1$. The second alternative $k < n + 1$ is impossible, since it would imply $k \in S \cap [n + 1] = \{n + 1\}$, leading to the contradiction $k = n + 1$. Thus, $n + 1 \leq k$ for each $k \in S$, as needed.

To prove the theorem itself, fix $S \subseteq \mathbb{N}$ and assume S is nonempty. Then there exists a natural number $n \in S$. It follows that $S \cap [n]$ is nonempty since $n \in [n]$. We can now conclude that S has a least element. $\qquad\square$

Least Upper Bounds and Completeness

The concept of a *least upper bound* is a kind of substitute for the greatest element when the latter does not exist. For example, the open interval (a, b) has least upper bound b. To make this precise, we need some additional definitions.

8.52. Definition: Upper and Lower Bounds. Fix $S \subseteq \mathbb{R}$ and $z_0 \in \mathbb{R}$.

(a) $\boxed{z_0 \text{ is an upper bound for } S}$ iff $\boxed{\forall x \in S, x \leq z_0.}$

(b) $\boxed{S \text{ is bounded above}}$ iff $\boxed{\exists z \in \mathbb{R}, \forall x \in S, x \leq z.}$

(c) $\boxed{z_0 \text{ is a lower bound for } S}$ iff $\boxed{\forall x \in S, z_0 \leq x.}$

(d) $\boxed{S \text{ is bounded below}}$ iff $\boxed{\exists z \in \mathbb{R}, \forall x \in S, z \leq x.}$

As an example, consider the interval $S = [2, 4)$. Some upper bounds for S are 5, 7, 21/2, and 4. In fact, the set of all upper bounds of S is the interval $[4, \infty)$, and this set of upper bounds has least element 4. We therefore say 4 is the least upper bound (lub) of S and write $4 = \text{lub } S$. Similarly, the set of all lower bounds of S is the interval $(-\infty, 2]$. This set of lower bounds has greatest element 2, so we say 2 is the greatest lower bound (glb) of S.

Here is a related example that motivates the introduction of the Completeness Axiom. Consider the set $S = \{x \in \mathbb{R}_{\geq 0} : x^2 < 2\}$. Intuitively, for positive real x, the condition $x^2 < 2$ is equivalent to $x < \sqrt{2}$, so the set of upper bounds for S ought to be the interval $[\sqrt{2}, \infty)$. (This is only an intuitive argument for now, since we have not yet defined or proved the existence of $\sqrt{2}$.) Thus $\operatorname{lub} S$ should exist and be equal to $\sqrt{2}$. On the other hand, suppose we restrict ourselves to using rational numbers only. Intuitively, the restricted set $S' = \{x \in \mathbb{Q}_{\geq 0} : x^2 < 2\}$ has set of upper bounds $[\sqrt{2}, \infty) \cap \mathbb{Q}$. This set of upper bounds has *no* least element since $\sqrt{2}$ is not rational, but there are rational numbers arbitrarily close to $\sqrt{2}$ and larger than $\sqrt{2}$. Thus $\operatorname{lub} S'$ does *not* exist in the ordered set $\mathbb{Q}$, even though S' is bounded above. This example reveals that the number system $\mathbb{Q}$ has a "hole" in it where $\sqrt{2}$ should be. The Completeness Axiom for $\mathbb{R}$ says, intuitively, that $\mathbb{R}$ "has no holes" in the sense that least upper bounds exist in every case where this is possible. We now officially define $\operatorname{lub} S$ and $\operatorname{glb} S$ and state the Completeness Axiom.

8.53. Definition: Least Upper Bounds. Fix $S \subseteq \mathbb{R}$ and $z_0 \in \mathbb{R}$. Define

$\boxed{z_0 = \operatorname{lub} S}$ iff z_0 is an upper bound of S and for every upper bound y of S, $z_0 \leq y$.

Some texts denote $\operatorname{lub} S$ by $\sup S$ (the *supremum* of S).

8.54. Definition: Greatest Lower Bounds. Fix $S \subseteq \mathbb{R}$ and $z_0 \in \mathbb{R}$. Define

$\boxed{z_0 = \operatorname{glb} S}$ iff z_0 is a lower bound of S and for every lower bound y of S, $y \leq z_0$.

Some texts denote $\operatorname{glb} S$ by $\inf S$ (the *infimum* of S).

Expanding the definition of a least upper bound, we see that

$$z_0 = \operatorname{lub} S \quad \text{iff} \quad [\forall x \in S, x \leq z_0 \text{ and } \forall y \in \mathbb{R}, (\forall x \in S, x \leq y) \Rightarrow z_0 \leq y].$$

Taking the contrapositive of the final implication and using trichotomy, we arrive at the following phrasing of the definition that is often more convenient in proofs:

$$\boxed{z_0 = \operatorname{lub} S} \quad \Leftrightarrow \quad \boxed{[\forall x \in S, x \leq z_0] \wedge [\forall y \in \mathbb{R}, y < z_0 \Rightarrow \exists x \in S, y < x]}.$$

In words, z_0 is the least upper bound of S iff z_0 is an upper bound of S and each $y < z_0$ is not an upper bound of S. Similarly,

$$\boxed{z_0 = \operatorname{glb} S} \quad \Leftrightarrow \quad \boxed{[\forall x \in S, z_0 \leq x] \wedge [\forall y \in \mathbb{R}, z_0 < y \Rightarrow \exists x \in S, x < y]}.$$

Note that $\operatorname{lub} S$ is the minimum element of the set of upper bounds of S, and $\operatorname{glb} S$ is the maximum element of the set of lower bounds of S. Thus, $\operatorname{lub} S$ and $\operatorname{glb} S$ are unique when they exist. The Completeness Axiom postulates a basic sufficient condition for the existence of $\operatorname{lub} S$.

8.55. Completeness Axiom (O7). Every nonempty subset of $\mathbb{R}$ that is bounded above has a least upper bound in $\mathbb{R}$.

As a first illustration of this axiom, we use it to derive the analogous sufficient condition for the existence of greatest lower bounds.

8.56. Theorem on Greatest Lower Bounds. For all $S \subseteq \mathbb{R}$, if S is nonempty and bounded below, then S has a greatest lower bound in $\mathbb{R}$.

Proof. Fix $S \subseteq \mathbb{R}$ that is nonempty and bounded below by x_0. [The main idea of the proof is to use negation to flip the ordering, turning lower bounds into upper bounds and least upper bounds into greatest lower bounds.] Define a new set $T = \{-y : y \in S\}$, which is nonempty since S is nonempty. For all $y \in S$, we know $x_0 \le y$, so $-y \le -x_0$. This means that $-x_0$ is an upper bound for the set T. By the Completeness Axiom (O7), $z_0 = \text{lub}\, T$ exists in $\mathbb{R}$. We show that $-z_0 = \text{glb}\, S$. On one hand, for fixed $y \in S$, we know $-y \in T$, so $-y \le z_0$ and hence $-z_0 \le y$; this proves that $-z_0$ is a lower bound for the set S. On the other hand, consider a fixed real $w_0 > -z_0$. Then $-w_0 < z_0$, so $-w_0$ is not an upper bound for T; so there exists $t \in T$ with $-w_0 < t$. Writing $t = -y$ for some $y \in S$, we have $-w_0 < -y$, so $y < w_0$, so w_0 is not a lower bound for S. This means that $-z_0$ is the greatest lower bound for S, as needed.   □

The Archimedean Ordering Property

We often visualize $\mathbb{R}$ as a number line extending infinitely far to the left and right. Each real number x corresponds to exactly one point on this line, and the order relation $x < y$ means that the point for x is located to the left of the point for y. Our intuition tells us that no matter how far to the right we go on this line, there will always be positive integers even farther to the right. The next theorem uses the Completeness Axiom to justify this pictorial intuition.

8.57. Theorem on Archimedean Property of $\mathbb{R}$.

(a) $\forall x \in \mathbb{R}, \exists n \in \mathbb{Z}_{>0}, x < n$.

(b) $\forall y \in \mathbb{R}_{>0}, \exists n \in \mathbb{Z}_{>0}, 1/n < y$.

Proof. We prove (a) by contradiction. Assume, to get a contradiction, that $\exists x \in \mathbb{R}, \forall n \in \mathbb{Z}_{>0}, n \le x$. This assumption is precisely the statement that the set $\mathbb{Z}_{>0}$ is bounded above. By the completeness axiom for $\mathbb{R}$, the set $\mathbb{Z}_{>0}$ has a *least* upper bound, say $x_0 = \text{lub}\, \mathbb{Z}_{>0}$. Now $x_0 - 1 < x_0$, so $x_0 - 1$ is not an upper bound for $\mathbb{Z}_{>0}$. Thus there must exist a positive integer n with $x_0 - 1 < n$. Adding 1 to both sides, we get $x_0 < n + 1$, where $n + 1$ is also a positive integer. Since x_0 is an upper bound for $\mathbb{Z}_{>0}$, we have $n + 1 \le x_0$. But now $x_0 < n + 1 \le x_0$ contradicts trichotomy. This proves (a).

To prove (b), fix $y \in \mathbb{R}_{>0}$. Then $1/y$ is a positive real number, and by (a) there exists a positive integer n with $1/y < n$. By Theorem 8.40(a), we conclude that $1/n < 1/(1/y) = y$, as needed.   □

Ordering Properties of $\mathbb{Z}$

The Completeness Axiom tells us that any subset $S \subseteq \mathbb{R}$ that is nonempty and bounded above has a least upper bound. If S happens to be a subset of $\mathbb{Z}$, we can show that a stronger result holds: S actually has a maximum element. Similarly, if $S \subseteq \mathbb{Z}$ is nonempty and bounded below, then S has a minimum element. To prove these statements, we reduce to the well-ordering property of $\mathbb{N}$ established earlier.

8.58. Theorem on Strong Completeness of $\mathbb{Z}$.
(a) For all $S \subseteq \mathbb{Z}$, if S is nonempty and S is bounded below (in $\mathbb{R}$) then S has a least element. (b) For all $S \subseteq \mathbb{Z}$, if S is nonempty and S is bounded above (in $\mathbb{R}$) then S has a greatest element.

Proof. For (a), fix $S \subseteq \mathbb{Z}$ and assume S is nonempty and bounded below. Let $x \in \mathbb{R}$ be a lower bound for S. By the Archimedean Property, there is an integer N with $-x < N$, so $-N < x$. [The main idea of this proof is to translate $S \subseteq \mathbb{Z}$ to the right N units to get a subset of $\mathbb{N}$, thus reducing the result in (a) to the Well-Ordering Principle.] Consider the set $T = \{m + N : m \in S\}$ obtained from S by adding N to each element. Given any element $k = m + N \in T$ with $m \in S$, note that $-N < x \le m$, so $0 < m + N = k$, so $k \in \mathbb{Z}_{>0} = \mathbb{N}$.

So T is a subset of $\mathbb{N}$, and T is nonempty since S is nonempty. Since $\mathbb{N}$ is well-ordered, T has a least element $k_0 = m_0 + N$, where $m_0 \in S$. For any $m \in S$, we have $m + N \in T$. The fact that $m_0 + N = k_0 \leq m + N$ implies $m_0 \leq m$. Thus, m_0 is the least element of S.

For (b), fix $S \subseteq \mathbb{Z}$ and assume S is nonempty and bounded above. Let $x \in \mathbb{R}$ be an upper bound for S. [Here the idea is to flip the order by negating everything, as in the proof of Theorem 8.56.] Define $T = \{-m : m \in S\}$. Since $m \leq x$ for all $m \in S$, we have $-x \leq -m$ for all $-m \in T$. Thus, $-x$ is a lower bound for T. Also T is nonempty since S is nonempty. By (a), T has a smallest element, say $-m_0$ with $m_0 \in S$. Given any $m \in S$, we know $-m_0 \leq -m$, and hence $m \leq m_0$. Thus, m_0 is the largest element of S. $\qquad \square$

When we draw a number line to represent $\mathbb{R}$, we often make tick marks, spaced one unit apart, to represent the integers. Based on such a picture, it is highly believable that for every real x, the half-open interval $[x, x+1)$ must contain exactly one integer. The next theorem justifies this pictorial intuition.

8.59. Theorem on Distribution of $\mathbb{Z}$ in $\mathbb{R}$. $\boxed{\forall x \in \mathbb{R}, \exists! n \in \mathbb{Z}, x \leq n < x+1.}$

Proof. Fix $x \in \mathbb{R}$. To prove existence of n, let $S = \{m \in \mathbb{Z} : x \leq m\}$. By the Archimedean Property, S is nonempty. Also x is a lower bound for S, so we know S has a least element n. To get a contradiction, assume that $x + 1 \leq n$. Then $x \leq n - 1$ where $n - 1$ is an integer (by closure of $\mathbb{Z}$ under subtraction). But this means $n - 1 \in S$; since $n - 1 < n$, this contradicts the choice of n as the least element of S. Hence, $n < x + 1$ must hold. We also know $x \leq n$ since $n \in S$.

To prove uniqueness of n, suppose (to get a contradiction) that n_1 and n_2 are distinct integers that both belong to the interval $[x, x+1)$. We can choose notation so that $n_1 < n_2$, and hence $0 < n_2 - n_1$. On the other hand, $-n_1 \leq -x$ and $n_2 < x + 1$ combine to give $n_2 - n_1 < (x + 1) - x = 1$. Now $n_2 - n_1$ is an integer in the open interval $(0, 1)$, which contradicts Lemma 8.41(d). So the n found above is unique. $\qquad \square$

Density of $\mathbb{Q}$ in $\mathbb{R}$; Real Square Roots

The next result says that $\mathbb{Q}$ is dense in $\mathbb{R}$, meaning that arbitrarily small open intervals contain rational numbers.

8.60. Theorem on Density of $\mathbb{Q}$ in $\mathbb{R}$. $\boxed{\forall a, b \in \mathbb{R}, a < b \Rightarrow \exists q \in \mathbb{Q}, a < q < b.}$

Proof. Fix $a, b \in \mathbb{R}$ with $a < b$. Since $0 < b - a$, we can use the Archimedean property to find a positive integer n with $1/n < b - a$. Adding $a - (1/n)$ to both sides, we find that $a < b - (1/n)$; and $b - (1/n) < b$ since $0 < 1/n$. [The key idea is to magnify the interval $[b-1/n, b)$ by a factor of n to get the interval $[nb - 1, nb)$ of length 1.] Applying Theorem 8.59 to the real number $x = nb - 1$, we obtain a (unique) integer m with $nb - 1 \leq m < nb$. Multiplying these inequalities by the positive quantity $1/n$, we get $b - (1/n) \leq m/n < b$. Since $a < b - (1/n)$, transitivity gives $a < m/n < b$. As m and n are integers, $q = m/n$ is a rational number in (a, b). $\qquad \square$

The set of irrational numbers is also dense in $\mathbb{R}$, but to prove this we must first know that irrational numbers actually exist! We now use the Completeness Axiom to show that all positive real numbers have real square roots. Combining this with the earlier theorem that 2 has no *rational* square root, we deduce the existence of real numbers that are not rational. We may also conclude that $\mathbb{Q}$ cannot satisfy the Completeness Axiom (otherwise, the proof given below would apply to the number system $\mathbb{Q}$ and prove that 2 had a square root in $\mathbb{Q}$, which is not true).

8.61. Theorem on Existence of Real Square Roots. $\boxed{\forall a \in \mathbb{R}_{>0}, \exists x \in \mathbb{R}_{>0}, x^2 = a.}$

Proof. The proof uses the following fact, which is left as an exercise: for all $y, z \in \mathbb{R}_{\geq 0}$, $y < z$ iff $y^2 < z^2$. We first prove the result for a fixed $a \in \mathbb{R}_{\geq 1}$. Since $1 \leq a$ and a is positive, we see that $a \leq a^2$. [To use the Completeness Axiom to solve $x^2 = a$, we need a *set* of real numbers with least upper bound x.] Define the set $S = \{z \in \mathbb{R}_{\geq 0} : z^2 < a\}$. On one hand, S is nonempty because $0^2 = 0 < a$, so that $0 \in S$. On the other hand, we claim that S is bounded above with upper bound a. For given an arbitrary $z \in S$, we know $z^2 < a \leq a^2$, so $z < a$ by the initial fact.

We have now verified the hypotheses of the Completeness Axiom (O7), so we can conclude that $x = \text{lub}\, S$ exists in $\mathbb{R}$. Since $0 \in S$, $0 \leq x$. By trichotomy, $x^2 < a$ or $a < x^2$ or $x^2 = a$. The third case is the required result; we finish the proof by showing that the first two cases lead to contradictions.

First assume that $x^2 < a$. On one hand, this means $x \in S$, so $x < a$ as shown above. On the other hand, for any $n \in \mathbb{N}$, we know $1 \leq n$, so $n \leq n^2$ and $1/n^2 \leq 1/n$. Now calculate $(x + 1/n)^2 = x^2 + (2/n)x + (1/n^2) < x^2 + (2/n)a + (1/n) \leq x^2 + (2a + 1)/n$. By the Archimedean Property applied to the positive number $(a - x^2)/(2a + 1)$, we can choose a positive integer n with $1/n < (a - x^2)/(2a + 1)$. Then $x < x + 1/n$ and $(x + 1/n)^2 \leq x^2 + (2a + 1)/n < x^2 + (a - x^2) = a$. So $x + 1/n$ belongs to S and is larger than x; this contradicts the fact that x is an upper bound for the set S.

Now assume that $a < x^2$. This time, we calculate $(x - 1/n)^2 = x^2 - (2/n)x + (1/n^2) \geq x^2 - (2x/n)$. By the Archimedean Property, there exists a positive integer n with $1/n < x$ and $1/n < (x^2 - a)/(2x)$, so $x - (1/n) > 0$ and $(x - 1/n)^2 \geq x^2 - (2x/n) > a$. Given any $y \in S$, note that $y^2 < a < (x - 1/n)^2$, so $y < x - 1/n$ by the initial fact. But this means $x - 1/n$ is an upper bound for S, contradicting the choice of x as the *least* upper bound for the set S.

Finally, we prove the result for fixed a between 0 and 1. Since $1/a > 1$, we have just shown that there exists $y \in \mathbb{R}_{\geq 0}$ with $y^2 = 1/a$. Then $x = 1/y$ satisfies $x^2 = 1/y^2 = a$, as needed. $\qquad\square$

Nested Interval Property

As a final illustration of the Completeness Axiom, we prove a theorem stating that the intersection of any sequence of nested closed intervals is always nonempty. This is another way of formulating the intuitive notion that the real number system $\mathbb{R}$ has no holes.

8.62. Nested Interval Theorem. Suppose $\{[a_n, b_n] : n \in \mathbb{Z}_{>0}\}$ is a collection of nonempty closed intervals such that $[a_{n+1}, b_{n+1}] \subseteq [a_n, b_n]$ for all $n \in \mathbb{Z}_{>0}$. Then $\bigcap_{n=1}^{\infty}[a_n, b_n] \neq \emptyset$.

Proof. For each $n \in \mathbb{Z}_{>0}$, we have $a_n \leq b_n$ since $[a_n, b_n]$ is nonempty, $a_n \leq a_{n+1}$ since $a_{n+1} \in [a_{n+1}, b_{n+1}] \subseteq [a_n, b_n]$, and $b_{n+1} \leq b_n$ since $b_{n+1} \in [a_{n+1}, b_{n+1}] \subseteq [a_n, b_n]$. By induction, we readily deduce the following consequences of these inequalities: for all positive integers $i < j$, $a_i \leq a_j$ and $b_j \leq b_i$; and for all positive integers k, n, $a_k \leq b_n$.

Now consider the set $S = \{a_n : n \in \mathbb{Z}_{>0}\}$, which is a nonempty subset of $\mathbb{R}$. This set is bounded above by b_1, since any $a_n \in S$ satisfies $a_n \leq b_1$. By the Completeness Axiom (O7), there exists $r \in \mathbb{R}$ with $r = \text{lub}\, S$. We claim $r \in \bigcap_{n=1}^{\infty}[a_n, b_n]$, which proves that this intersection is nonempty. Fix $n \in \mathbb{Z}_{>0}$; we must show $r \in [a_n, b_n]$, which means $a_n \leq r$ and $r \leq b_n$. Since r is an upper bound for S, $a_n \leq r$ is true. Assume, to get a contradiction, that $r \leq b_n$ is false. Then $b_n < r = \text{lub}\, S$, so b_n is not an upper bound for S. So there exists $a_k \in S$ with $b_n < a_k$, contradicting the previous paragraph. Thus, $r \leq b_n$ is true. $\qquad\square$

One consequence of the Nested Interval Theorem is that for any sequence $(z_n : n \in \mathbb{Z}_{>0})$

and any interval $[a, b]$, there is a point in $[a, b]$ different from all z_n; in other words, $[a, b]$ is an uncountable set (see §7.4). The Nested Interval Theorem can also be used to prove the Intermediate Value Theorem for continuous real-valued functions. Given a continuous $f : [a, b] \to \mathbb{R}$ and $c \in \mathbb{R}$ between $f(a)$ and $f(b)$, the idea is to find $x \in \mathbb{R}$ solving $f(x) = c$ by repeatedly bisecting the interval $[a, b]$ and taking x in the intersection of the resulting sequence of nested closed intervals.

Section Summary

1. *Definitions.* For $S \subseteq \mathbb{R}$ and $z_0 \in \mathbb{R}$, z_0 is an upper bound of S iff $\forall x \in S, x \le z_0$; $z_0 = \max S$ iff $z_0 \in S$ and z_0 is an upper bound of S; $z_0 = \operatorname{lub} S$ (the least upper bound of S) iff z_0 is an upper bound of S and every $y < z_0$ is not an upper bound of S. Lower bounds, $\min S$, and the greatest lower bound glb S are defined similarly.

2. *Completeness and Related Properties.* The Completeness Axiom states that every nonempty subset of $\mathbb{R}$ that is bounded above has a least upper bound. Consequently, every nonempty subset of $\mathbb{R}$ that is bounded below has a greatest lower bound. A nonempty subset of $\mathbb{Z}$ that is bounded above (resp. below) has a greatest (resp. least) element. Every nonempty finite subset of $\mathbb{R}$ has a greatest and a least element. For a general subset S of $\mathbb{R}$, $\max S$, $\min S$, $\operatorname{lub} S$, and glb S need not exist; but each of these numbers is unique when it exists.

3. *Archimedean Property.* For each $x \in \mathbb{R}$, there is a positive integer n with $x < n$. For each $y \in \mathbb{R}_{>0}$, there is a positive integer n with $1/n < y$.

4. *Distribution of $\mathbb{Z}$ and $\mathbb{Q}$ in $\mathbb{R}$.* For each $x \in \mathbb{R}$, there is exactly one integer n such that $x \le n < x + 1$. For each $a < b$ in $\mathbb{R}$, there is a rational number q with $a < q < b$.

5. *Existence of Real Square Roots.* For each $a \in \mathbb{R}_{>0}$, there exists $x \in \mathbb{R}_{>0}$ with $x^2 = a$. This statement does not hold for $\mathbb{Q}_{>0}$, so $\mathbb{Q}$ does not satisfy the Completeness Axiom.

6. *Nested Interval Property.* Any collection of nonempty closed intervals $[a_n, b_n]$ that are nested (i.e., $[a_{n+1}, b_{n+1}] \subseteq [a_n, b_n]$ for all $n \in \mathbb{Z}_{\ge 0}$) must have nonempty intersection in $\mathbb{R}$.

Exercises

1. (a) Prove: for all $y, z \in \mathbb{R}_{\ge 0}$, $y < z$ iff $y^2 < z^2$. (b) Does either implication in (a) hold for all $y, z \in \mathbb{R}$?

2. (a) Prove: for all $x, y \in \mathbb{R}$, if $x < y$, then $x < (x + y)/2 < y$. (b) More generally, prove $\forall x, y, t \in \mathbb{R}$, if $x < y$ and $0 < t < 1$, then $x < tx + (1 - t)y < y$. (c) Prove: for all $a, b \in \mathbb{R}$, if $a < b$, then the open interval (a, b) has no least element.

3. Prove that $\mathbb{R}_{<0}$ has no maximum and no minimum.

4. Let $S = \{3 - 1/n : n \in \mathbb{Z}_{>0}\}$. Find (with proof) lub S and glb S. Does $\max S$ exist? Does $\min S$ exist?

5. (a) Prove: For all $S \subseteq \mathbb{R}$, if S is bounded above and bounded below, then glb $S \le$ lub S. (b) Find all sets $S \subseteq \mathbb{R}$ such that glb S and lub S exist and are equal.

6. Recall that $[n] = \{1, 2, \ldots, n\} = \{m \in \mathbb{Z}_{>0} : m \le n\}$. Prove: for all $n \in \mathbb{Z}_{>0}$ and all functions $f : [n] \to \mathbb{R}$, the image of f has a greatest element and a least

element. Deduce that every nonempty finite subset of $\mathbb{R}$ has a maximum and a minimum element.

7. (a) If S is not bounded above, what is the set U of all upper bounds of S? Does U have a least element? Does $\operatorname{lub} S$ exist? (b) Repeat part (a) for $S = \emptyset$ (the empty set).

8. Prove: $\forall y \in \mathbb{R}_{>0}, \forall \epsilon \in \mathbb{R}_{>0}, \exists n \in \mathbb{Z}_{>0}, n\epsilon > y$.

9. (a) Where is the flaw in this proposed proof of Theorem 8.57(a)? "Fix x_0 in $\mathbb{R}$. Choose $n = x_0 + 1$. Since we know $0 < 1$, adding x_0 to both sides gives $x_0 < x_0 + 1 = n$, as needed." (b) Where is the flaw in this proposed proof of Theorem 8.60? "Fix $a, b \in \mathbb{R}$ with $a < b$. Choose $q = (a + b)/2$. Since $a < b$, an earlier exercise gives $a < q < b$, as needed."

10. Prove: for all $x \in \mathbb{R}$, there exists a unique $n \in \mathbb{Z}$ with $x < n \le x + 1$.

11. Give a careful proof that $\{[n, n + 1) : n \in \mathbb{Z}\}$ is a set partition of $\mathbb{R}$.

12. Prove: for all $a, b \in \mathbb{R}$, if $a < b$, then there exists $x \in \mathbb{R} - \mathbb{Q}$ with $a < x < b$. [*Hint:* We know $\sqrt{2} \in \mathbb{R} - \mathbb{Q}$, and we know there is a rational number in $(a + \sqrt{2}, b + \sqrt{2})$.]

13. Prove that every $a \in \mathbb{R}_{>0}$ has a *unique* positive square root.

14. Let $(X, \le)$ be any partially ordered set satisfying the Completeness Axiom (O7). Prove that for all $S \subseteq X$, if S is nonempty and bounded below, then S has a greatest lowest bound. (This gives another proof of Theorem 8.56 not relying on the algebraic structure of $\mathbb{R}$.)

15. For $x = (x_1, \ldots, x_n)$ and $y = (y_1, \ldots, y_n)$ in $\mathbb{R}^n$, define

$$d(x, y) = \max\{|x_1 - y_1|, \ldots, |x_n - y_n|\}.$$

Prove that this distance function satisfies the metric space axioms in Theorem 8.46.

16. For $x = (x_1, \ldots, x_n)$ and $y = (y_1, \ldots, y_n)$ in $\mathbb{R}^n$, define $d(x, y) = \sum_{k=1}^{n} |x_k - y_k|$. Prove that this distance function satisfies the metric space axioms in Theorem 8.46.

Suggestions for Further Reading

In this text, we have covered fundamental concepts of logic, proofs, set theory, integers, relations, functions, cardinality, and real numbers that are absolutely essential for further pursuit of advanced mathematics. Each of the subjects we have studied is a vast subdiscipline of mathematics that goes far beyond the introductory material discussed here. We now suggest some books that interested readers may consult to learn more about these fascinating topics.

Logic and Set Theory

Mathematical logic is a beautiful and amazing subject that applies the precise tools of mathematical reasoning to explore the notions of proof, truth, and computability.

- *Logic for Mathematicians* by A. G. Hamilton (Cambridge University Press, 2001) is a very readable introduction to this field.

- *Gödel, Escher, Bach: An Eternal Golden Braid* by Douglas Hofstadter (Basic Books, 1999) is a remarkable, Pulitzer Prize-winning book that weaves together logic, art, music, mathematics, molecular biology, and much more. The book is written for a popular audience but is loaded with technical content including multiple explanations and metaphors for Gödel's celebrated Incompleteness Theorem.

- *Mathematical Logic* by Joseph R. Shoenfield (Association for Symbolic Logic, 1967) is a more advanced and challenging logic textbook, which is extremely carefully written.

Axiomatic set theory rigorously develops properties of sets beginning with an explicit list of axioms. We sketched the beginnings of this subject in the optional Section 3.7.

- *Naive Set Theory* by Paul Halmos (Springer, 1974) is a famous 100-page book by one of the greatest mathematical writers. Despite the title, this book really is about axiomatic set theory, although Halmos focuses on those aspects of set theory most relevant to other parts of mathematics.

- *Introduction to Set Theory* (third edition) by Karel Hrbacek and Thomas Jech (CRC Press, 1999) is a solid exposition of set theory appropriate for advanced undergraduates. One attractive feature of this book is that each new axiom of set theory is carefully motivated before it is introduced, giving readers intuition for why these particular axioms were chosen.

- *Introduction to Set Theory* by J. Donald Monk (McGraw-Hill, 1969) is an advanced but very well-written development of a version of set theory in which proper classes (such as the class of all sets) are allowed. After summarizing basic material on sets, relations, and functions (similar to what we covered in this text), the book provides an outstanding treatment of ordinal numbers, the axiom of choice, and cardinal numbers.

Analysis

Analysis is the branch of mathematics that rigorously develops the calculus of real-valued functions and their generalizations. This subject begins with the ordered field axioms for real numbers, which are covered in Chapter 8 of this text.

- *Advanced Calculus* (second edition) by Patrick Fitzpatrick (American Mathematical Society, 2006) gives a very readable account of single-variable and multivariable real analysis.

- *The Real Numbers and Real Analysis* by Ethan Bloch (Springer, 2011) is another nice advanced calculus textbook that starts with axiomatic developments of various number systems (integers, rational numbers, and real numbers).

- *Analysis on Manifolds* by James Munkres (Westview Press, 1991) is this author's all-time favorite mathematics book. It gives a wonderful treatment of multivariable calculus topics culminating in chapters on manifolds in $\mathbb{R}^n$, differential forms, and the Generalized Stokes Theorem.

- *Real Analysis* by Haaser and Sullivan (Dover, 1991) is an advanced textbook on abstract analysis. This book lucidly describes many topics including the formal construction of $\mathbb{R}$ from $\mathbb{Q}$ via Cauchy sequences, the theory of metric spaces, a development of the Lebesgue integral via the Daniell approach, Banach spaces, and Hilbert spaces.

Number Theory and Algebra

Number theory studies properties of the natural numbers. Our discussion of divisibility, primes, unique prime factorizations of integers, and modular arithmetic barely scratch the surface of this ancient subject, which has many unexpected practical applications to modern cryptography.

- *The Higher Arithmetic* (eighth edition) by Harold Davenport (Cambridge University Press, 2008) is a wonderful introduction to number theory with short chapters on many of the key topics in this area.

- *A Guide to Elementary Number Theory* by Underwood Dudley (MAA Press, 2009) is a fantastically clear exposition of the basic theorems of number theory that is only 140 pages long.

- *Elementary Number Theory and its Applications* (sixth edition) by Kenneth Rosen (Pearson, 2010) is a thorough, comprehensive undergraduate textbook on number theory.

Abstract algebra starts with familiar properties of integers and real numbers (like associativity) and considers these properties in far more general settings. What results is an elegant abstract theory that applies to any structure satisfying the initial axioms.

- *Ideals, Varieties, and Algorithms* (fourth edition) by Cox, Little, and O'Shea (Springer, 2015) is an amazing introduction to computational algebra that focuses on the algebra of multivariable polynomials. The book perfectly blends theory and computation.

- *A First Course in Abstract Algebra* (third edition) by Joseph Rotman (Pearson, 2005) is a nice introduction to groups, rings, and fields by one of the best writers on this subject.

Index